Hans-A. Bachor and Timothy C. Ralph
A Guide to Experiments in Quantum Optics

Second, Revised and Enlarged Edition

Hans-A. Bachor and Timothy C. Ralph

A Guide to Experiments in Quantum Optics

Second, Revised and Enlarged Edition

**WILEY-
VCH**

WILEY-VCH Verlag GmbH & Co. KGaA

Authors

Hans-A. Bachor
The Australian National University, Canberra
e-mail: hans.bachor@anu.edu.au

Timothy C. Ralph
University of Queenslad, Australia
e-mail: ralph@mail.physics.uq.edu.au

Cover Picture

Real experimental data for squeezed states,
generated at the ANU by Jin Wei Wu;
U. L. Andersen, B. C. Buchler, P. K. Lam,
J. W. Wu, J. R. Gao, H.-A. Bachor
Eur. Phys. J. D. **27**, No. 2 (2003)

1st Edition 2004
  1st Reprint 2006
  2nd Reprint 2009

Library of Congress Card No.: applied for
British Library Cataloging-in-Publication Data:
A catalogue record for this book is available from
the British Library

Bibliographic information published by
Die Deutsche Bibliothek
Die Deutsche Bibliothek lists this publication in
the Deutsche Nationalbibliografie; detailed
bibliographic data is available in the
Internet at <http://dnb.ddb.de>.

Printed on acid-free paper

Composition: Uwe Krieg, Berlin

ISBN 978-3-527-40393-6

Contents

Preface

The idea behind this guide

Some of the most interesting and sometimes puzzling phenomena in optics are those where the quantum mechanical nature of light is apparent. Recent years have seen a rapid expansion of experimental optics into this area known as *Quantum Optics*. Beautiful demonstrations and applications of the quantum nature of light are now possible and optics has been shown to be one of the best areas of physics to actually make use of quantum mechanical ideas. This book is intended to guide you through the many experiments published, to present and interpret them in one common style. It also provides a practical background in opto-electronics.

Several excellent textbooks have already been written on this topic. However, in the same way as this field of research has been initiated by theoretical ideas, most of these books have been written from a theoretical point of view. While this results in a very solid and reliable description of the field, we found that it frequently leaves out some of the important simpler pictures and intuitive interpretations of the experiments. Here we are using a different and complementary approach: This guide focuses on the actual experiments and what we can learn from them. It explains the underlying physics in the most intuitive way we could find. It addresses questions such as: what are the limitations of the equipment; what can be measured and what remains a goal for the future. The answers prepare the reader to make independent predictions of the outcome of their own experiments.

We assume that most of the readers will already have a fairly good understanding of optical phenomena, a reliable idea of how light propagates and how it interacts with components such as lenses, mirrors, detectors etc. The reader can picture these processes in action in an optical instrument.. It is useful in the everyday work of a scientist to have such pictures of what is going on inside the experiments. They have been shaped over many years by many lectures and through ones own experimentation. In optics three different pictures are used simultaneously: Light as waves, light as photons and light as the solution to operator equations. To these is added the picture of noise propagating through an optical system, a description using noise transfer functions. All of these pictures are useful; all of them are correct within their limits. They are based on specific interpretations of the one formalism, they are mathematically equivalent.

In this guide all these four descriptions are introduced rigorously and are used to discuss a series of experiments. These start with simple demonstrations, gradually getting more complex and include most of the experiments on quantum noise detection, squeezed states and quantum non demolition measurements published to date. All four pictures – waves, photons, operators and noise spectra – are used simultaneously. In each case the most appropriate and intuitive interpretation is highlighted, the limitations of all three interpretations are stated. Through this approach we hope to maintain and present a lot of the fascination of quantum optics that has captured us and many of our colleagues.

Preface to the 2nd edition

The development of the field has been rapid in recent years and the interest in quantum optics is growing. Experiments such as the demonstration of teleportation created headlines. The quest for communicating and storing quantum information, for quantum logic and quantum computing by optical means is a major driving force behind this popularity. Optics is an area that makes the concept of entanglement accessible. In the last few years several important demonstration experiments have been published and real applications seem to be very close.

The ideas are drawn equally from the techniques of single photon detection and single event logic as well as the ideas of continuous variable logic and the use of continuous coherent and squeezed beams. In the experiments on entanglement so far more work has been done with single photons. On the other hand the continuous variable approach overlaps well with the existing technology in optical communication and optical sensing systems. We can expect both areas to grow and to complement each other.

Thus, we felt that we should show more clearly the link between the single photon and the photo-current case. We have expanded and reorganized the theoretical section to represent both cases in equal detail. We have included a technical chapter on techniques for detecting single photons and the evaluation of correlations through coincidence counting. In regard to applications we included an introductory chapter on quantum information communication and processing. All technical chapters were updated to take into account the recent scientific developments and progress in technology. The book follows the well established idea of providing independent building blocks in theory and experimental techniques which are used in later chapters to discuss complete experiments and applications. Working as a team has allowed us to provide more rigour and to cover a wider range of topics to make this an even more useful book.

Hans Bachor and *Timothy Ralph*

September 2003

Acknowledgments

This book would not have been possible without the teaching we received from many of the creative and motivating pioneers of the field of quantum optics. We have named them in their context throughout the book. HAB is particularly in debt to Marc Levenson who challenged him first with ideas about quantum noise. From him he learned many of the tricks of the trade and the art of checking one's sanity when the experiment produces results that seem to be impossible. Over the years we learned from and received help for this book from many colleagues and students, including W. Bowen, B.C. Buchler, J. Close, G. Dennis, C. Fabre, P. Fisk, D. Gordon, M. Gray, C.C. Harb, D. Hope, J. Hope, E. Huntington, M. Hsu, P.K. Lam, G. Leuchs, P. Manson, D.E. McClelland, G.J. Milburn, G. Moy, W. Munro, C. Savage, R. Schnabel, R. Scholten, D. Shaddock, A. Stevenson, T. Symul, M. Taubman, N. Treps, A.G. White, H.M. Wiseman, J.W. Wu, and from the many others who are always keen to discuss at length the mysteries of quantum optics. We like to thank M. Colla for his expert technical support and A. Dolinska for the creative graphics work. TCR wishes to thank Bron Vincent and Bruce Harper for all their support and love and Isaac for putting up with a distracted Daddy. Most of all HAB wants to thank his wife Cornelia for the patience she showed with an absentminded and many times frustrated writer.

Hans Bachor and *Timothy Ralph*

1 Introduction

1.1 Historical perspective

Ever since the quantum interpretation of the black body radiation by M. Planck [Pla00], the discovery of the photoelectric effect by H. Hellwarth [Hel1898] and P. Lennard [Len02] and its interpretation by A. Einstein [Ein05] has the idea of photons been used to describe the origin of light. For the description of the generation of light a quantum model is essential. The emission of light by atoms, and equally the absorption of light, requires the assumption that light of a certain wavelength λ, or frequency ν, is made up of discrete units of energy each with the same energy $hc/\lambda = h\nu$.

This concept is closely linked to the quantum description of the atoms themselves. Several models have been developed to describe atoms as quantum systems, their properties can be described by the wave functions of the atom. The energy eigenstates of atoms lead directly to the spectra of light which they emit or absorb. The quantum theory of atoms has been developed extensively, it led to many practical applications [Tho00] and has played a crucial role in the formulation of quantum mechanics itself. Spectroscopy relies almost completely on the quantum nature of atoms. However, in this guide we will not be concerned with the quantum theory of atoms, but concentrate on the properties of the light, its propagation and its applications in optical measurements.

The description of optical phenomena, or physical optics, has developed largely independently from quantum theory. Almost all physical optics experiments can be explained on the basis of classical electromagnetic theory [Bow59]. An interpretation of light as a classical wave is perfectly adequate for the understanding of effects such as diffraction, interference or image formation. Even nonlinear optics, such as frequency doubling or wave mixing, are well described by classical theory. Even many properties of the laser, the key component of most modern optical instruments, can be described by classical means. Most of the present photonics and optical communication technologies are essentially applications of classical optics.

However, some experiments with extremely low intensities and those which are based on detecting individual photons raised new questions. For instance, the famous interference experiments by G.I. Taylor [Tay09] where the energy flux corresponds to the transit of individual photons. He repeated T. Young's double slit experiment [You1807] with typically less than one photon in the experiment at any one time. In this case the classical explanation of interference based on electromagnetic waves and the quantum explanation based on the interference of probability amplitudes can be made to coincide by the simple expedient of making the probability of counting a photon proportional to the intensity of the classical field. The experiment cannot distinguish between the two explanations. Similarly, the modern versions

A Guide to Experiments in Quantum Optics, 2nd Edition. Hans-A. Bachor and Timothy C. Ralph
Copyright © 2004 Wiley-VCH Verlag GmbH & Co. KGaA
ISBN: 3-527-40393-0

of these experiments, using the technologies of low noise current detection or, alternatively, photon counting, show the identity of the two interpretations.

The search for uniquely quantum optical effects continued through experiments concerned with intensity, rather than amplitude interferometry. The focus shifted to the measurement of fluctuations and the statistical analysis of light. Correlations between the arrival of photons were considered. It started with the famous experiment by R. Hanbury-Brown and R. Twiss [Bro56] who studied the correlation between the fluctuations of the two photocurrents from two different detectors illuminated by the same light source. They observed with a thermal light source an enhancement in the two-time intensity correlation function for short time delays. This was a consequence of the large intensity fluctuations of the thermal light and was called photon bunching. This phenomenon can be adequately explained using a classical theory which includes a fluctuating electro-magnetic field. Once laser light was available it was found that a strong laser beam, well above threshold shows no photon bunching, instead the light has a *Poissonian counting statistic*. One consequence is that laser beams cannot be perfectly quiet, they show noise in the intensity which is called *shot noise*, a name that reminds us of the arrival of many particles of light. This result can be derived from both classical and quantum models.

Next it was shown by R.J. Glauber in 1963 [Gla63] that additional, unique predictions could be made from his quantum formulation of optical coherence. One such prediction is *photon anti-bunching*, where the initial slope of the two-time correlation function is positive. This corresponds to greater than average separations between the arrival times of photons, the photon counting statistics may be sub-Poissonian and the fluctuations of the resulting photocurrent would be smaller than the shot noise. It was shown that a classical theory based on fluctuating field amplitudes would require negative probabilities in order to predict anti-bunching and this is clearly not a classical concept but a key feature of a quantum model. It was not until 1976, when H.J. Carmichael and D.F. Walls [Car76] predicted that light generated by resonance fluorescence from a two level atom would exhibit anti-bunching, that a physically accessible system was identified which exhibits non classical behaviour. This phenomenon was observed in experiments by H.J. Kimble, M. Dagenais and L. Mandel [Kim77], opening the era of quantum optics. More recently, the ability to trap and study individual ions allows the observation of both photon anti-bunching and sub-Poissonian statistics.

Initially these single particle effects were a pure curiosity. They showed the difference between the quantum and the classical world and were used to test quantum mechanics. In particular, the critique by A. Einstein and his collaborators B. Podolsky and N. Rosen and their EPR paradox [Ein35] required further investigation. Through the work of J.S. Bell [Bel64] it became possible to carry out quantitative tests. They are all based on the quantum concept of *entanglement*, that means pairs of particles or wave functions which have a common origin and have linked properties, such that the detection of one allows the perfect prediction of the other. Starting with the experiments by A. Aspect, P. Grangier and G. Roger [Asp82] with atomic sources, and later with the development of sources of *twin photon pairs* and experiments by A. Zeilinger [Zei00] and many others the quantum predictions have been extremely well tested. More recently, the peculiarities of single photon quantum processes have found technical applications, mainly in the area of communication. It was found that information sent by single photons can be made secure against eavesdropping, the unique quantum properties do not allow the copying of the information. Any eavesdropping will introduce noise and can therefore be detected. Quantum cryptography [Mil96] is now a viable technology, completely based on the weirdness of the concept of photons.

There is speculation of other future technological applications of photons. The superposition of different states of photons can be used to send more complex information, allowing the transfer not only of classical bits of information, such as 0 and 1, but of q-bits based on the superposition of states 0 *and* 1. The coherent evolution of several such quantum systems could lead to the development of quantum logic connections, quantum gates and eventually whole quantum computers which promise to solve certain classes of mathematical problems much more efficiently and rapidly [Mil98]. Even more astounding is the concept of *teleportation*, the perfect communication of the complete quantum information from one place to another. This is presently being tested in experiments, see Chapter 13. Optics is now a testing ground for future quantum technologies, for data communication and processing.

1.2 Motivation: Practical effects of quantum noise

Apart from their fundamental importance, quantum effects now increasingly play a role in the design and operation of modern optical devices. The 1960s saw a rapid development of new laser light sources and improvements in light detection techniques. This allowed the distinction between incoherent (thermal) and coherent (laser) light on the basis of photon statistics. The groups of A. Arecchi [Arr66], L. Mandel and R.E. Pike all demonstrated in their experiments that the photon counting statistic goes from super-Poissonian at threshold to Poissonian far above threshold. The corresponding theoretical work by R. Glauber [Gla63] was based on the concept that both the atomic variables and the light are quantised and showed that light can be described by a *coherent state*, the quantum analogy counterpart to a classical field. The results are essentially equivalent to a classical treatment of an oscillator. However, it is an important consequence of the quantum model, that any measurement of the properties of this state, intensity, amplitude or phase of the light, will be limited by *quantum noise*.

Quantum noise can impose a limit to the performance of lasers, sensors and communication systems and near the quantum limit the performance can be quite different. To illustrate this consider the result of a very simple and practical experiment: Use a laser, such as a laser printer with a few mW of power, detect all of the light with a photo-diode and measure the fluctuations of the photo-current with an electronic spectrum analyser, as shown in Fig. 1.1(a) This produces an electronic signal which represents the intensity noise at one single detection frequency. The fluctuations vary with the frequency, a plot of the noise power is called an *intensity noise spectrum* of the laser light. Such spectra will be discussed in detail in this book.

Figure 1.1(b) shows two different spectra. Trace (i) is the photo-current noise spectrum with light on the detector. Trace (ii) is the noise measured without light on the detector, the so called dark noise. It provides a noise background which contains all the electronic noise of the detector, the electronic amplifier, the spectrum analyser and any stray electrical signals picked up by the apparatus. It shows a strong, clearly identifiable spectral component at 5.5 [MHz] due to electric pickup, for example of a radio station, and a uniform, flat background noise slowly increasing with the frequency which is a property of the electronics used. The strong

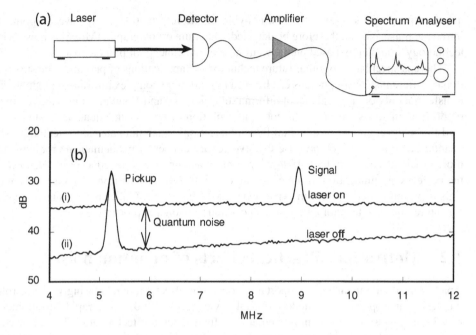

Figure 1.1: The intensity noise of a laser. (a) Schematical layout of the experiment. (b) The noise spectrum, for detection frequencies 4–12 [MHz]. The noise power is shown on a logarithmic scale. Two traces are shown: (i) laser on, (ii) laser off. Details of these measurements are discussed in Chapter 9.

individual components at 9 [MHz] are due to modulations of the light. This is the way information can be sent via a laser beam. The diagram shows that the quality of the information, given by the ratio between the size of the signal and the size of the noise, is limited.

The flat noise background in trace (i), which is frequency independent and thus classed as "white noise", is the feature of greatest interest in terms of quantum optics. This is *quantum noise* and represents a phenomenon which is not included in the classical electro-magnetic model of light. It is a direct consequence of the quantum theory of light and will be discussed in great detail in this book. This quantum noise represents some unavoidable fluctuations in the intensity, it forms the *quantum noise limit*, or *QNL*, for the intensity noise. It is an intrinsic property of the light and it appears for both laser and thermal light. The magnitude of the noise is directly linked to the intensity of the light detected, or the average of the photo-current, and is expressed by the shot noise formula.

Quantum noise can be interpreted in various ways. Firstly, it can be associated with the quantization of the light, can be regarded as a consequence of the Heisenberg uncertainty principle or of the properties of the operators used to describe the light field. Such a quantum model will be presented in detail and is used extensively in this book. This model is rigorous and will in all cases lead to the correct quantitative result. However, it is also very abstract and fits not directly to the concepts commonly used by many experimentalists. A related model is to view quantum noise as a consequence of the statistical property of a stream of photons.

These particles do not interact with each other and a certain degree of randomness, represented by a Poissonian distribution, is the natural state of this system. We will find that light tends to approach such a Poissonian distribution during the arrival times of the photons at the detector. As a consequence the photo-current, which is the quantity actually measured in the experiments, will display fluctuations of a magnitude dictated by the Poissonian distribution. Such a statistical model is very useful for the interpretation of intensity, but we will find that it has severe shortcomings whenever other properties of light, such as phase and interference, are investigated.

Alternatively, quantum noise can be regarded as a consequence of the photo-detection process, as a randomness of the stream of electrons produced in the photo-detector. This view prevailed for a long time, particularly in engineering text books. But this view is misleading as we find in the recent *squeezing* experiments, described in this book. It cannot account for situations where nonlinear optical processes modify the quantum noise, while the detector remains unaffected. In the squeezing experiments the quantum noise is changed optically and consequently we have to assume it is a property of the light, not of the detector.

Finally, it is possible to expand the concept of classical waves and to expand the concepts of beat notes and modulation, which we already used in the description of the origin of the discrete components contained in the noise spectrum, see Fig. 1.1. Consider any noise detected as the outcome of a beat experiment. A signal at frequency Ω corresponds to the beat, or product, of at least two waves with optical frequencies ν and $\nu \pm \Omega$. Since there is only one dominant frequency component in the spectral distribution of a laser beam, which is at the centre frequency ν_L of the laser, the noise at Ω can be regarded as the consequence of randomly fluctuating fields at the laser sidebands $\nu_L + \Omega$ and $\nu_L - \Omega$ beating with the centre component at ν_L. The randomness of the field in the sidebands is not included in the classical wave model, where the amplitude is strictly zero away from the centre component. However, the quantum effect can be incorporated by adding fluctuations of a fixed size to the otherwise noiseless classical sidebands. This idea is very successful in interpreting experiments concerned with quantum noise, for example the description of the properties of beamsplitters and interferometers. It uses all of the components familiar from physical optics and electronics (monochromatic waves, sidebands, phase difference, interference and beat signals) and adds only one extra component, namely a randomly varying field. This model, which is rarely used in most of the research literature, has a prominent position in this book due to its simplicity and practicality.

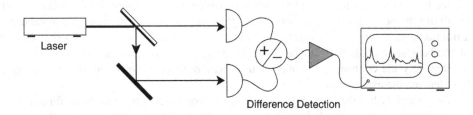

Difference Detection

Figure 1.2: First attempt to eliminate noise by dual detection.

It is worthwhile to explore the properties of quantum noise somewhat further. It was found that, in clear distinction to classical noise, no technical trick can eliminate quantum noise. To illustrate the differences consider the following schemes for the suppression of the noise. The first scheme, see Fig. 1.2, involves difference detection, as it is frequently employed in absorption experiments. A beamsplitter after the laser is used to generate a second beam, which is detected on a second detector. Any intensity modulation of the intensity of the laser beam will appear on both beams and results in changes of the two photo-currents. The modulations of the currents will be strongly correlated. Using a difference amplifier the common fluctuations will be subtracted. When the gain of the two amplifiers has been chosen appropriately the resulting difference current will contain no modulation. This idea works not only for modulations but also for technical noise, which can be regarded as a random version of the modulations. Such technical noise can be subtracted. However, this scheme fails for the suppression of quantum noise. Both light beams contain quantum noise, which leads to white noise in the two photo-currents. Experiments show that these fluctuations are not subtracted by the difference amplifier. The noise of the difference is actually larger than the noise of the individual beams, it is the quadrature sum of the noise from the two individual currents. This result remains unchanged if the difference is replaced by a sum or if the two currents are added with an arbitrary phase difference. The resulting noise is always the quadrature sum independent of the sign, or phase, of the summation. This is equivalent to the statement that the noise in the two photo-currents is not correlated.

At this stage it is not easy to identify the point in the experiment where this uncorrelated noise is generated. One interpretation assumes that the noise in the currents is generated in the photo-detectors. Another interpretation assumes that the beamsplitter is a random selector for photons and consequently the intensities of the two beams are random and thus uncorrelated. A distinction can only be made by further experiments with squeezed light, which are discussed later in this book.

An alternative scheme for noise suppression is the use of feedback control, as shown in Fig. 1.3. It achieves equivalent results to difference detection. The intensity of the light can be controlled with a modulator, such as an acousto-optic modulator or an electro-optic modulator. Using a feedback amplifier with appropriately chosen gain and phase lag the intensity noise can be reduced, all the technical noise can be eliminated [Rob86]. It is possible to get very close to the quantum noise limit, but the quantum noise itself cannot be suppressed [Mer93], [Tau95]. This phenomenon can be understood by considering the properties of photons. As mentioned before, the quantum noise measured by the two detectors is not correlated, thus the feedback control, when operating only on quantum noise on one detector, will not be able to control the noise in the beam that reaches the other detector. This can be explained using a full quantum theory. Alternatively, it can be interpreted as a consequence of the properties of the beamsplitter. It will randomly select the photons going to the control detector and consequently the control system has no information about the quantum fluctuations of the light leaving the experiment.

Only recently has the role of the photon generation process been explored further. It was found that the noise of the light may be below the standard quantum limit if the pumping process exhibits sub-Poissonian statistics. This is particularly easy to achieve for light sources driven directly by electric currents, namely light emitting diodes (LEDs) or semiconductor lasers. The currents driving these devices are classical, at the level of fluctuations we are

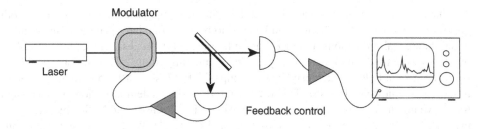

Figure 1.3: The second attempt to eliminate laser noise: An improved apparatus using a feed-back controller, or 'noise eater'.

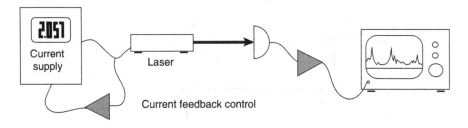

Figure 1.4: Feedback control of the current below the Poissonian limit. A laser with high quantum efficiency can convert the stabilised current into light with intensity noise suppressed below the quantum noise limit.

concerned with, and the fluctuations can be controlled with ease to levels well below the shot noise level. For sources with high quantum efficiencies the sub-Poissonian statistics of the drive current is transferred directly to the statistics of the light emitted. This is illustrated in the extension of our little experiment shown in Fig. 1.4. Such experiments were pioneered for the case of diode lasers by the group of Y. Yamamoto [Mac89]. They showed that intensity fluctuations can be suppressed in a high impedance semiconductor laser driven by a constant current and similar work was carried out with LEDs by several groups.

A two dimensional description of the light, with properties we will call *quadratures* will be necessary to explain this quantum effect. Noise can be characterized by the variance in both the amplitude and phase quadrature. This work shows one of the limits of the quantum models discussed above: they were restricted to one individual laser beam but should have included the entire apparatus. We will derive a simple formalism that allows us to predict the laser noise at any location within the instrument.

It took almost a decade after the observation of photon anti-bunching in atomic fluorescence to predict another quantum phenomenon of light – the suppression, or squeezing, of quantum fluctuations [Wal83]. For a coherent state the uncertainties in the quadratures are equal and minimise the product in Heisenberg's uncertainty principle. A consequence is that measurements of both the amplitude or the phase quadrature of the light show quantum noise. In a *squeezed state* the fluctuations in the quadratures are no longer identical. One quadrature may have reduced quantum fluctuations at the expense of increased fluctuations in the other quadrature. Such squeezed light could be used to beat the standard quantum limit. After the initial theoretical predictions, the race was on to find such a process. A number of nonlinear processes were tried simultaneously by several competing groups. The first observation of a squeezed state of light was achieved by the group of R.E. Slusher in 1985 in four-wave mixing in sodium atomic beam [Slu85]. This was soon followed by a demonstration of squeezing by four wave mixing in optical fibres by the group of M.D. Levenson and R. Shelby [She86] and in an optical parametric oscillator by the group of H.J. Kimble [Wu87]. In recent years a number of other nonlinear processes have been used to demonstrate the quantum noise suppression based on squeezing [Sp.Issues]. An generic layout is shown in Fig. 1.5. The experiments are now reliable and practical applications are feasible.

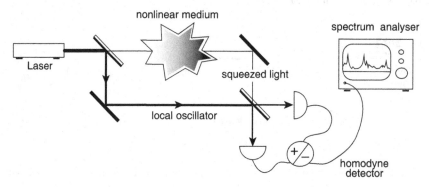

Figure 1.5: A typical squeezing experiment. The nonlinear medium generates the squeezed light which is detected by a homodyne detection scheme.

In the last few years the concept of *quantum information* has brought even more life to quantum optics. Plans for using the complexity of quantum states to code information, to transmit it without out losses, also known as teleportation, to use it for secure communication and cryptography, to store quantum information and possibly use it for complex logical processes and quantum computing have all been widely discussed. The concept of *entanglement* has emerged as one of the key qualities of quantum optics. As it will be shown in this guide, it is now possible to create entangled beams of light either from pairs of individual photons or from the combination of two squeezed beams and the demand and interest in non-classical states of light has sharply risen. We can see quantum optics playing a large role in future communication and computing technologies [Mil96, Mil98].

For this reason the guide covers both single photon and CW beam experiments parallel to each other. It provides a unified description and compares the achievements as well as tries to predict the future potential of these experiments.

1.3   How to use this guide

This guide leads through experiments in quantum optics, experiments which deal with light and which demonstrate, or use, the quantum nature of light. It shows the practicalities and challenges of these experiments and gives an interpretation of their results. One of the current difficulties in understanding the field of quantum optics is the diversity of the models used. On the one hand, the theory, and most of the publications in quantum optics, are based on a rigorous quantum model which is rather abstract. On the other hand, the teaching of physical optics and the experimental training in using devices such as modulators, detectors and spectrum analysers are based on classical wave ideas. This training is extremely useful, but frequently does not include the quantum processes. Actually, the language used by these two approaches can be very different and it is not always obvious how to relate a result from a theoretical model to a technical device designed and vice versa. As an example compare the schematical representation of a squeezing experiment, given in Fig. 1.6, both in terms of the theoretical treatments for photons and laser beams and in an experimental description. The purpose of this guide is to bridge the gap between theory and experiment. This is done by describing the different building blocks in separate chapters and combining them into complete experiments as described in recent literature.

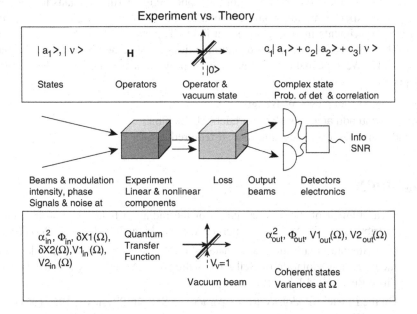

Figure 1.6: Comparison between an experiment (middle) and the theoretical description for few photon states (top) and laser beams (bottom)

We start with a classical model of light (Chapter 2). Experiments reveal that we require a concept of photons (Chapter 3), which is expanded into a quantum model of light (Chapter 4). The properties of optical components and devices (Chapter 5) and a detailed description of lasers (Chapter 6) are given. Next is a detailed discussion of photodetection for single photons and beams (Chapter 7). This is followed by a discussion of complete experiments. The technical details required for reliable experimentation with quantum noise, including techniques such as cavity locking and feedback controller, given in Chapter 8. The concept of squeezing is central to all attempts to improve optical devices beyond the standard quantum limit and is introduced and discussed in Chapter 9. It also describes the various squeezing experiments and their results are discussed, the different interpretations are compared. In a similar way Chapter 11 discusses quantum non-demolition experiments. Finally, the potential applications of squeezed light are described in Chapter 10. In Chapter 12 experiments which test the fundamental concepts of quantum mechanics are discussed. Finally, the concepts of quantum information and the present state of art of experiments using either single photon or CW beams are presented in Chapter 13.

This guide can be used in different ways. A reader who is primarily interested in learning about the ideas and concepts of quantum optics would best concentrate on Chapters 2, 3, 4, 9, 12, and 13 but may leave out many of the technical details. For these readers Chapter 5 would provide a useful exercise in applying the concepts introduced in Chapter 4. In contrast, a reader who wishes to find out the limitations of optical engineering or wants to learn about the intricacies of experimentation would concentrate more on Chapters 2, 6, 5, 8 and for an extension into experiments involving squeezed light Chapters 9, 10 can be added. A quick overview of the possibilities opened by quantum optics can be gained by reading Chapters 3, 4, 6, 9, 11, and 12. We hope that in this way our book provides a useful guide to the fascinating world of quantum optics.

This book is accompanied by a Web page
http://photonics.anu.edu.au/qoptics/people/bachor/book.html
which provides updates, exercises, links and, if required, errata.

Bibliography

[Arr66]   Measurement of the time evolution of the radiation field by joint photocurrent distributions, F.T. Arecchi, A. Berne, A. Sona, Phys. Rev. Lett. 17, 260 (1966)

[Asp82]   Experimental realisation of Einstein-Podolsky-Rosen-Bohm Gedankenexperiment: A new violation of Bell's inequalities, A. Aspect, P. Grangier, G. Roger, Phys. Rev. Lett. 49, 91 (1982)

[Bel64]   On the Einstein-Podolsky-Rosen paradox, J.S. Bell, Physics 1, 195 (1964)

[Ben93]   Teleporting an unknown quantum state via dual classical and Einstein-Podolsky-Rosen channels, C.H. Bennett, G. Brassard, C. Crepau, R. Jozsa, A. Peres, W.K. Wootters, Phys. Rev. Lett. 70, 1895 (1993)

[Bow59]   *The principles of Optics*, M. Born and E. Wolf (1959)

[Bro56]   Correlation between photons in two coherent beams of light, R. Hanbury-Brown, R.Q. Twiss, Nature 177, 27-29 (1956)

[Car76]   A quantum-mechanical master equation treatment of the dynamical Stark effect, H.J. Carmichael, D.F. Walls, J. Phys. B 9, 1199 (1976)

[Ein05]   Über einen die Erzeugung und Verwandlung des Lichtes betreffenden heuristischen Gesichtspunkt, A. Einstein, Annalen der Physik 17,132 (1905)

[Ein35]   Can a quantum-mechanical description of physical reality be considered complete?, A. Einstein, B. Podolsky, N. Rosen, Phys. Rev. A 47, 777 (1935)

[Gla63]   The quantum theory of optical coherence, R.J. Glauber, Phys. Rev. Lett. 130, 2529 (1963)

[Hel1898] Elektrometrische Untersuchungen, W. Hallwachs, Annalen der Physik und Chemie, neue Folge Band XXIX (1886)

[Kim77]   Photon antibunching in resonance fluorescence, H.J. Kimble, M. Dagenais, L. Mandel, Phys. Rev. Lett. 39, 691 (1977)

[Len02]   Ueber die lichtelektrische Wirkung, P. Lenard, Ann. Physik 8, 149 (1902)

[Mac89]   Observation of amplitude squeezing from semiconductor lasers by balanced direct detectors with a delay line, S. Machida, Y. Yamamoto, Opt. Lett. 14, 1045 (1989)

[Mer93]   Photon noise reduction by controlled deletion techniques, J. Mertz, A. Heidmann, J. Opt. Soc. Am. B 10, 745 (1993)

[Mil96]   *Quantum technology*, G. Milburn, Allen & Unwin (1996)

[Mil98]   *The Feynman processor*, G. Milburn, Allen & Unwin (1998)

[Pla00]   Zur Theorie des Gesetzes der Energieverteilung im Normalspektrum, M. Planck, Verhandlungen der Deutschen Physikalischen Gesellschaft 2, 237 (1900)

[Rob86]   Intensity stabilisation of an Argon laser using an electro-optic modulator: performance and limitation, N.A. Robertson, S Hoggan, J.B. Mangan, J. Hough, Appl. Phys. B 39, 149 (1986)

[She86]   Generation of squeezed states of light with a fiber-optics ring interferometer, R.M. Shelby, M.D. Levenson, D.F. Walls, A. Aspect, G.J. Milburn, Phys. Rev. A33, 4008 (1986)

[Slu85]   Observation of squeezed states generated by four-wave mixing in an optical cavity, R.E. Slusher, L.W. Hollberg, B. Yurke, J.C. Mertz, J.F. Valley, Phys. Rev. Lett. 55, 2409 (1985)

[Sp.Issues] *Special issues on squeezed states of light*, J. Opt. Soc. Am. B4 (1987) and J. Mod. Opt. 34 (1987)

[Tap91]   Sub-shot-noise measurement of modulated absorption using parametric downconversion, P.R. Tapster, S.F. Seward, J.G. Rarity, Phys. Rev. A44, 3266 (1991)

[Tau95]   Quantum effects of intensity feedback, M.S. Taubman, H. Wiseman, D.E. McClelland, H-A. Bachor, J. Opt. Soc. Am. B 12, 1792 (1995)

[Tay09]   Interference fringes with feeble light, G.I. Taylor, Proc. Cambridge Phil. Soc.15, 114 (1909)

[Tho00]   *Spectrophysics, principles and applications*, 3rd edition, A. Thorne, U. Litzen, S. Johansson, Springer Verlag (1999)

[Wal83]   Squeezed states of light, D. Walls, Nature 306, 141 (1983)

[Wu87]   Squeezed states of light from an optical parametric oscillator, Ling-An Wu, Min Xiao, H.J. Kimble, J. Opt. Soc. Am. B 4, 1465 (1987)

[You1807] *Course of lectures on natural philosophy and mechanical arts*, Th. Young, London (1807)

[Zei00]   Quantum cryptography with entangled photons, A. Zeilinger, Phys. Rev. Lett. 84, 4729 (2000)

2 Classical models of light

The development of optics was driven by the desire to understand and interpret the many optical phenomena which were observed. Over the centuries more and more refined instruments were invented to make observations, revealing ever more fascinating details. All the optical phenomena investigated during the 19th century were observed either directly or with photographic plates. The spatial distribution of light was the main attraction. The experiments measured long term averages of the intensity, no time resolved information was available. The formation of images and the interference patterns were the main topics of interest. In this context the classical optical models, namely *geometrical optics*, dating back to the Greeks and developed ever since, and *wave optics*, introduced through Huygens' ideas and Fresnel's mathematical models, were sufficient.

A revolution came with the invention of the laser. This changed optics in several ways: It allowed the use of coherent light, bringing the possibilities of wave optics to their full potential, for example through holography. Lasers allowed spectroscopic investigations with unprecedented resolution, testing the quantum mechanical description of the atom in even more detail. The fast light pulses generated by lasers allowed the study of the dynamics of chemical and biological processes. In addition, a range of unforeseen effects emerged: The high intensities resulted in nonlinear optical effects. One fundamental assumption of the wave model, namely that the frequency of the light remained unchanged, was found to be seriously violated in this new regime. The sum-, difference- and higher harmonic frequencies were generated. Finally, optics was found to be one of the best testing grounds of quantum mechanics. The combination of sensitive photoelectric detection with reliable, stable laser sources meant not only that interesting quantum mechanical states could be produced but also that quantum mechanical effects had to be accounted for in technical applications.

Light is now almost exclusively observed using photo-electric detectors which can register individual events or produce photocurrents. A description of these effects requires the field of *statistical optics*.

We need a description of the beam of light which emerges from a laser. A laser beam is a well confined, coherent electro-magnetic field oscillating at a frequency ν, or angular frequency $\omega = 2\pi\nu$. This frequency is constant for all times and locations and can only be changed by the interaction of light with nonlinear materials. Furthermore, the beam propagates and maintains its shape unperturbed by diffraction. In this chapter we will develop classical models suited to describing this type of light. In following chapters we will see how these must be modified to account for quantum mechanical effects.

A Guide to Experiments in Quantum Optics, 2nd Edition. Hans-A. Bachor and Timothy C. Ralph
Copyright © 2004 Wiley-VCH Verlag GmbH & Co. KGaA
ISBN: 3-527-40393-0

2.1 Classical waves

2.1.1 Mathematical description of waves

An electromagnetic wave in an isotropic, insulating medium is described by the wave equation

$$\nabla^2 \mathbf{E}(\mathbf{r}, t) - \frac{1}{c} \frac{\partial^2}{\partial t^2} \mathbf{E}(\mathbf{r}, t) = 0 \qquad (2.1.1)$$

with a solution for the electric field vector $\mathbf{E}(\mathbf{r}, t)$, given by

$$\mathbf{E}(\mathbf{r}, t) = E_0 \left[\alpha(\mathbf{r}, t) \exp(i2\pi\nu t) + \alpha^*(\mathbf{r}, t) \exp(-i2\pi\nu t) \right] \quad \mathbf{p}(\mathbf{r}, t) \qquad (2.1.2)$$

The wave oscillates at a frequency ν in [Hz]. For now we consider only a single frequency component but in general the solution can be an arbitrary superposition of such components. The electric field vector is always perpendicular to the local propagation direction. The orientation of the electric field on the plane tangential to the local wavefront is known as the polarisation and is described by the vector $\mathbf{p}(\mathbf{r}, t)$. The dimensionless complex amplitude $\alpha(\mathbf{r}, t)$ of the wave can be written as a magnitude $\alpha_0(\mathbf{r}, t)$ and a phase $\phi(\mathbf{r}, t)$.

$$\alpha(\mathbf{r}, t) = \alpha_0(\mathbf{r}, t) \exp\{i\phi(\mathbf{r}, t)\} \qquad (2.1.3)$$

The magnitude $\alpha_0(\mathbf{r}, t)$ will change in time when the light is modulated or pulsed. It will also vary in space, describing the extent of the wave. The wave will typically be constant in space for dimensions of the order of several wavelengths. The direction and shape of the wave front is determined by the phase term $\phi(\mathbf{r}, t)$ and the spatial distribution of the $\phi(\mathbf{r}, t)$ describes the curvature of the waves.

Simple examples of specific solutions are monochromatic plane waves and spherical waves. These represent the extreme cases in terms of angular and spatial confinement. The monochromatic plane wave is the ideal case in which the complex amplitude magnitude α_0 is a constant and the phase is given by

$$\phi(\mathbf{r}) = -\mathbf{k} \cdot \mathbf{r} \qquad (2.1.4)$$

The wave propagates in a direction and with a wavelength λ represented by the wave vector $\mathbf{k}$, where $\mathbf{k}/k$ is a unit vector that provides the direction of the local wave front and $k = 2\pi/\lambda$. The wavelength is related to the frequency via $\lambda = c/n\nu$, where n is the refractive index of the medium in which the wave propagates and c is the speed of light in the vacuum. Thus we can write

$$\alpha(z) = \alpha_0 \exp\{-ikz\} \qquad (2.1.5)$$

where we have assumed that the wave propagates in the z direction. The wave has a constant amplitude in any plane perpendicular to this optical axis. Substituting Equation (2.1.5) into Equation (2.1.2) and assuming α_0 as real we obtain the familiar solution

$$\mathbf{E}(z, t) = E_0 \cos(\nu t - kz) \quad \mathbf{p}(z, t) \qquad (2.1.6)$$

Linear polarisation means that $\mathbf{p}(\mathbf{r}, t)$ has a fixed direction perpendicular to $\mathbf{k}$. Circular polarisation means that in any one location $\mathbf{p}(\mathbf{r}, t)$ is rotating in the plane perpendicular to $\mathbf{k}$. Equation (2.1.6) represents a wave which has no angular spread. In contrast the spherical wave originates from one singular point. All its properties are described by one parameter r, the distance to its point of origin. The amplitude magnitude reduces with $1/r$, and the phase is constant on spheres around the origin,

$$\alpha(\mathbf{r}, t) = \alpha(r) = \frac{\alpha_0}{r} \exp\{ikr\} \tag{2.1.7}$$

2.1.2   The Gaussian beam

Both of these models are not physical, they are only crude approximations to a beam of light emitted by a real laser. Such a beam is confined to a well defined region and should be as close as possible to the ideal of a spatially localized but non divergent wave. It has to be immune to spreading by diffraction and therefore has to be a stable solution of the wave equation. An example of such solutions are the Gaussian modes. The simplest of these is circularly symmetric around the optical axis, thus depending only on the radial distance ρ from the beam axis

$$\rho^2 = x^2 + y^2$$

This special solution is the *Gaussian beam* or fundamental mode, with a complex amplitude distribution of

$$\alpha(\mathbf{r}, t) = \frac{\alpha_0}{q(z)} \exp\left(\frac{-ik\rho^2}{2q(z)}\right), \qquad q(z) = z + iz_0 \tag{2.1.8}$$

where z_0 is a constant. The Gaussian beam is a good spatial representation of a laser beam. Laser light is generated inside a cavity, constructed of mirrors, by multiple interference of waves and simultaneous amplification. These processes result in a well defined, stable field distribution inside the cavity, the so called cavity modes [Kog66, Sie86, Tei96]. The freely propagating beam is the extension of this internal field. It has the remarkable property that it does not change its shape during propagation. Both the internal and the external fields are described by Equation (2.1.8). By taking the square of the complex amplitude $\alpha(\mathbf{r}, t)$ the intensity is found, resulting in a Gaussian transverse intensity distribution.

$$I(\rho, z) = I_0 \left(\frac{W_0}{W(z)}\right)^2 \exp\left(\frac{-2\rho^2}{W^2(z)}\right)$$

$$\text{and} \quad W(z) = W_0\sqrt{1 + \left(\frac{z}{z_0}\right)^2} \tag{2.1.9}$$

The properties of this Gaussian beam are the following: The shape of the intensity distribution remains unchanged by diffraction. The size of the beam is given by $W(z)$. The beam has its narrowest width at $z = 0$, the so-called waist of the beam. It corresponds to the classical focus, but rather than point-like, the Gaussian beam has a size of W_0. This value gives the radial distance where the intensity has dropped from $I_0 = I(0, 0)$ at the centre to $I(W_0, 0) = I_0/e$.

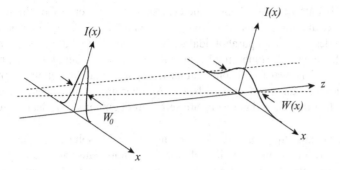

Figure 2.1: Properties of a Gaussian beam.

At the waist the phase front is a plane, $\phi(\rho)$ is a constant. Moving away from the waist the beam radius $W(z)$ expands, the intensity at the axis reduces and, at the same time, the wave front curves. Along the optical axis the Gaussian beam scales with the parameter z_0 which is known as the Rayleigh length

$$z_0 = \frac{\pi W_0^2}{\lambda}$$

It describes the point where the intensity on axis has dropped to half the value at the waist, $I(0, z_0) = \frac{1}{2} I_0$, and the beam radius has expanded by a factor of $\sqrt{2}$. This can be regarded as the spatial limit to the focus region, the value $2z_0$ is frequently described as the depth of focus or as the confocal parameter. Far away from the waist, for $z \gg z_0$, the Gaussian beam has essentially the wave front of a spherical wave originating at the waist. Obviously, the Gaussian beam is still centered on the beam axis and has only a finite extent $W(z)$. At these large distances the size of the beam $W(z)$ depends linearly on z, the beam has a constant divergence angle of

$$\Xi = \frac{2\lambda}{2\pi W_0} \tag{2.1.10}$$

In addition to the wave front curvature a gradual phase shift occurs on the beam axis between a uniform plane wave with identical phase at $z = 0$ and the actual Gaussian beam. The phase difference increases to $\pi/4$ at $z = z_0$ and asymptotically approaches $\pi/2$.

Table 2.1: Properties of a Gaussian beam

Location	$z = 0$	$z = z_0$	$z \gg z_0$
Beam Waist	$W = W_0$	$W = \sqrt{2}W_0$	$W(z) = z/z_0 W_0$
Intensity	I_0	$I_0/2$	$I_0(z_0/z)^2$
Wave front	plane wave	curved wave front	spherical wave
Phase retardation	0	$\pi/4$	$\pi/2$

As known from experiments with laser beams, the Gaussian beam is very robust in regard to propagation through optical components. Any component which is axially symmetric and the effect of which can be described by a paraboloidal wave front will keep the Gaussian characteristic of the beam. This means the emerging beam will be Gaussian again, however, now with a new position $z = z'$ of the waist. All spherical lenses, spherical and parabolic mirrors transform a Gaussian beam into another Gaussian beam. Another important consequence of this transformation is the fact that Gaussian beams are self reproducing under reflection with spherical mirrors.

However, the Gaussian beam, Equation (2.1.8), is not the only stable solution of the paraxial wave equation. There are many other spatial solutions with non-Gaussian intensity distributions which can be generated by a laser and propagate without change in shape. One particular set of solutions are the Hermite-Gaussian beams. These beams have the same underlying phase, except from an excess term $\zeta_{l,m}(z)$ which varies slowly with z. They have the same paraboloidal wave front curvature as the Gaussian beam, which matches the curvature of spherical optical components aligned on the axis. Their intensity distribution is fixed and scales with $W(z)$. The distribution is two dimensional, with symmetry in the x and y directions and usually not with a maximum on the beam axis. The intensity on the beam axis decreases with the same factor $W(z)$ as the Gaussian beam along the z axis. The various solutions can be represented by a product of two Hermite-Gaussian functions, one in the x the other in the y direction, which directly leads to the following notation

$$\alpha_{l,m}(x,y,z) = \alpha_0 A_{l,m} \left(\frac{W_0}{W(z)} \right) G_l \left(\frac{\sqrt{2}x}{W(z)} \right) G_m \left(\frac{\sqrt{2}y}{W(z)} \right)$$

$$\times \exp\left\{ -i \left(kz + \frac{kr^2}{2R(z)} - \zeta_{l,m}(z) \right) \right\} \quad \text{for} \quad l, m = 0, 1, 2, \ldots$$

$$(2.1.11)$$

Here $A_{l,m}$ is a constant and $G_l(u)$ and $G_m(u)$ are standard mathematical functions, the Hermite polynomials of order l, m respectively, which can be found in tables [Sie86, Abr72]. In particular, $G_0(u) = \exp\{-u^2/2\}$ and thus the solution $\alpha_{0,0}(x, y, z)$ is identical to the Gaussian beam described above. This set of solutions forms a complete basis for all functions with paraboloidal wave fronts. Any such beam can be described by a superposition of Hermite-Gaussian beams. All these components propagate together, maintaining the amplitude distribution. The only component with circular symmetry is the Gaussian beam $\alpha_{0,0}(x, y, z)$. However, it is possible to describe other circularly symmetric intensity distributions. Consider for example the superposition of two independent beams with $\alpha_{1,0}(x, y, z)$ and $\alpha_{0,1}(x, y, z)$ and equal intensity; if these beams have random phase their interference can be neglected and the sum of the intensities results in a donut shaped circularly symmetric function, occasionally observed in lasers with high internal gain.

2.1.3   Quadrature amplitudes

The description of a wave as a complex amplitude, Equation (2.1.3), with amplitude $\alpha_0(\mathbf{r}, t)$ and phase distribution $\phi(\mathbf{r}, t)$, is useful for the description of interference and diffraction.

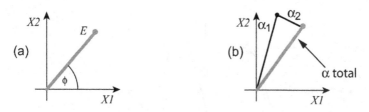

Figure 2.2: Phasor diagram (a) for a constant single wave, and (b) the representation of interference of several coherent waves.

As shown above, the phase distribution $\phi(\mathbf{r}, t)$ describes the shape of the wavefront as well as the absolute phase, in respect to a reference wave. However, it is useful to separate the two, to consider the absolute phase independently from the phase distribution, or wave front curvature. An equivalent form of Equation (2.1.2) can be given in terms of the quadrature amplitudes, $X1$ and $X2$, the amplitudes of the associated sine and cosine waves

$$E(\mathbf{r}, t) = E_0 \left[X1(\mathbf{r}, t) \cos(2\pi \nu t) + X2(\mathbf{r}, t) \sin(2\pi \nu t) \right] \quad \mathbf{p}(\mathbf{r}, t) \qquad (2.1.12)$$

The quadrature amplitudes are proportional to the real and imaginary parts of the complex amplitude.

$$
\begin{aligned}
X1(\mathbf{r}, t) &= \alpha(\mathbf{r}, t) + \alpha^*(\mathbf{r}, t) \\
X2(\mathbf{r}, t) &= i \left[\alpha(\mathbf{r}, t) - \alpha^*(\mathbf{r}, t) \right]
\end{aligned}
\qquad (2.1.13)
$$

In this notation the absolute phase ϕ_0 of the wave is associated with the distribution of the amplitude between the quadratures $X1$ and $X2$.

$$\phi_0 = \tan^{-1}\left(\frac{X2}{X1} \right) \qquad (2.1.14)$$

For $X2 = 0$ the wave is simply a cosine wave, Equation (2.1.6). We can regard this wave arbitrarily as the phase reference. A common description of classical waves whenever interference or diffraction is described is *phasor diagram*, a two dimensional diagram of the values $X1$ and $X2$ as shown in Fig. 2.2(a). In such a diagram, which represents the field at one point in space and time, each wave corresponds to one particular vector of length α from the origin to the point $(X1, X2)$, and this vector could be measured as the complex amplitude averaged over a few optical cycles of length $1/\nu$. The magnitude of the wave is given by the distance α of the point from the origin, the relative phase, ϕ_0 is given by the angle with the $X1$ axis, since the absolute phase is arbitrary. It is useful to choose α as real ($\phi = 0$), that means the vector parallel to the $X1$ axis.

If several waves are present at one point at the same time, they will interfere. Each individual wave is represented by one vector, the total field is given by the sum of the vectors, as shown in Fig. 2.2(b). These types of diagrams are a useful visualisation of the interference effects. After the total amplitude α_{total} has been determined we can again select a phase for this total wave and choose α_{total} as being real. All forms of interference can be treated this way and we obtain a value for the average, or DC, value of the optical amplitude.

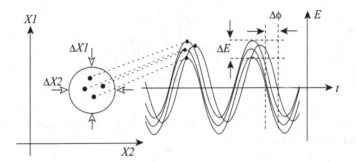

Figure 2.3: Fluctuations shown in phasespace: the field changes in time and moves from point to point in the phasor diagram. This corresponds to waves with different amplitude and phase.

In addition, the wave can have fluctuations or modulations in both amplitude and phase. Such variations correspond to fluctuations in the quadrature values. We can describe the time dependent wave with the complex amplitude

$$\alpha(t) = \alpha + \delta X1(t) + i\,\delta X2(t) \qquad (2.1.15)$$

In the limit $\alpha \gg \delta X1(t)$, that means for waves that change or fluctuate only very little around a fixed value α, we see that $\delta X1(t), \delta X2(t)$ describe the changes in the amplitude and phase respectively. This is the regime which we use to describe beams of light detected by detectors which produce a photocurrent. We will call them the *amplitude and phase quadratures*. The direct link between variations in the quadrature values and those of the field magnitude and phase is shown diagrammatically in Fig. 2.3.

2.1.4   Field energy, intensity, power

The energy passing through a small surface area element dx, dy around the point x, y in a short time interval dt around the time t is obtained by integrating the square of Equation (2.1.2). Note that because the field is propagating in the z-direction this is equivalent to the energy in the volume element $dx, dy, c\,dt$. We get

$$\mathcal{E}(x,y,t) = E_0^2\,2\,\alpha^*\alpha\,dt\,dx\,dy \qquad (2.1.16)$$

where we have assumed that $dx\,\,dy\,\,dt$ is sufficiently small that α does not vary over the interval, but that $dt \gg 1/\Omega$ is satisfied and the time average will be over a large number of optical cycles. As a result only the phase independent terms survive. The intensity of the beam is

$$I(x,y) = \int\limits_{1\,\text{second}} E_0^2\,2\,\alpha^*\alpha\,dt \qquad (2.1.17)$$

What is normally measured by a photo-detector is the power P of the light, the total flux of energy reaching the detector in a second, integrated over the area of the detector

$$P = \int I(x,y)\,dx\,dy \qquad (2.1.18)$$

The energy in a small volume element, Equation (2.1.16), can also be expressed in terms of the quadratures as

$$\mathcal{E}(x, y, t) = E_0^2 \left(X1^2 + X2^2 \right) dt \, dx \, dy \tag{2.1.19}$$

where we have used Equation (2.1.13). The parameter E_0^2 is proportional to the frequency of the light, ν, specifically:

$$E_0^2 = \frac{\hbar 2\pi\nu}{4} \tag{2.1.20}$$

where $\hbar$ is Planck's constant. It is the power in Equation (2.1.18) which is measured in most experiments and from this we infer the amplitude, the phase and the fluctuations of the beam.

2.1.5   A classical mode of light

The idea of a phasor diagram can now be combined with the concept of spatial modes as described in Equation (2.1.3). Let us concentrate on the simplest type of beam, a Gaussian beam. Equation (2.1.8) relates the relative amplitude and phase at any one point back to a point at $z = 0$ at the centre of the waist. Once the size W_0 is given the entire spatial distribution is well defined. We can assign one overall optical power, or peak intensity I_0, one overall phase to the entire spatial distribution. Thus the entire laser beam can be described by only six parameters.

Table 2.2: The parameters of a mode of light.

Parameters required to describe one mode of light	
Intensity of the mode	I_0
Frequency of optical field	ν
Absolute phase of this mode	ϕ_0
Direction of propagation	z
Location of the waist ($z = 0$)	$\mathbf{r_0}$
Waist diameter of the beam	W_0
Direction of polarisation	$\mathbf{P}$

Such a mode of light is strictly speaking a constant in time. It corresponds to a laser which has a stable, time independent output, it produces an intensity distribution which is not changing at all in time. More realistically the light mode will not be perfectly constant. Experiments show that there are always fluctuations. Ultimately quantum mechanics limits how small the fluctuations can be.

In addition, we wish to communicate some sort of information with the light and that means we must modulate it either in time or in space. Any such changes in the intensity have to be considered separately by introducing time variable parameters such as $I(t)$, $\nu(t)$ etc. . We will encounter two different types of experiments. In the case of continuous wave (CW) experiments the laser intensity is considered to be constant in the long term. Any time

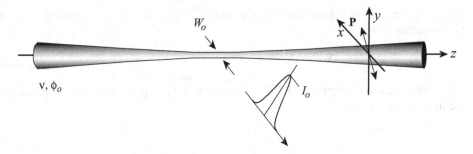

Figure 2.4: Graphical representation of the parameters of a mode of light.

dependence of the parameters is slow compared to the time constant of the detection process. That means we can actually resolve and process the fluctuations and modulations of the light.

Pulses of light are the other extreme case. Laser pulses are frequently shorter than the response time of the detector. Consequently only the integrated energy of many pulses is measured and the details of the time development of an individual pulse are lost. However, it is still possible to determine the fluctuations of the beam and this becomes a measurement of the statistics of the energy of the pulses. This special case will be discussed in Section 9.9.

2.1.6   Classical modulations

Information carried by a beam of light can be described in the form of modulation sidebands. Note that the equations used here are identical to those describing radio waves where AM and FM (or PM) modulation are the familiar, standard techniques. Many of the technical expressions, such as *carrier* for the fundamental frequency components and *sidebands*, are taken directly from this much older field of technology.

A beam of light which is amplitude modulated by a fraction M at one frequency Ω_{mod} can be described by

$$
\begin{aligned}
\alpha(t) &= \alpha_0 \left(1 - \frac{M}{2}\left(1 - \cos(2\pi\Omega_{\mathrm{mod}}t)\right) \right) \exp(i2\pi\nu_L t) \\
&= \alpha_0 \left(1 - \frac{M}{2} \right) \exp(i2\pi\nu_L t) \\
&\quad + \alpha_0 \frac{M}{4} \left[\exp\left(i2\pi(\nu_L + \Omega_{\mathrm{mod}})t\right) + \exp\left(i2\pi(\nu_L - \Omega_{\mathrm{mod}})t\right) \right]
\end{aligned}
\tag{2.1.21}
$$

We see that the effect of the modulation is to create new frequency components at $\nu_L + \Omega_{\mathrm{mod}}$ and $\nu_L - \Omega_{\mathrm{mod}}$. These are known as the upper and lower sidebands respectively whilst the component remaining at frequency ν_L is known as the carrier Fig. 2.5(a). For completely modulated light, that means $M = 1$, the sidebands have exactly $\frac{1}{2}$ the amplitude of the fundamental component. The sidebands are perfectly correlated, they have the same amplitude and phase, that means the two beat signals, generated by the sidebands at $\nu_L + \Omega_{\mathrm{mod}}$ and $\nu_L - \Omega_{\mathrm{mod}}$ beating with the carrier, are in phase. A second representation is in the time domain through oscillating vectors $\delta\alpha$ in the phasor diagram Fig. 2.5(b). The two vectors $\delta\alpha_{+\Omega}$

and $\delta\alpha_{-\Omega}$ rotate in opposite directions with angular frequency $2\pi\Omega$. A third representation is in frequency space in the phasor diagram Fig. 2.5(c). Notice that for a given modulation depth M the sideband amplitudes scale with the amplitude of the carrier, α_0. Thus even very small modulation depths, M, can result in significant sideband amplitude if the carrier amplitude is large. The detected intensity is given by

$$I(t) = I_0 \ |\alpha(t)|^2 \ = \ I_0 \left(1 - \frac{M}{2}(1 - \cos(2\pi\Omega_{\mathrm{mod}}t))\right)^2 \tag{2.1.22}$$

and oscillates at the frequency Ω_{mod} around the long term average I_0. The resulting photocurrent has two components, a long term average i_{DC} and a constant component at frequency Ω_{mod}, labelled $i(\Omega_{\mathrm{mod}})$. A common (and useful) situation is when the modulation depth is very small, $M \ll 1$. Then the resulting intensity modulation is linear in the modulation depth

$$I(t) \approx I_0 \left[1 - M(1 - \cos(2\pi\Omega_{\mathrm{mod}}t))\right] \tag{2.1.23}$$

A light beam with modulated phase can be described in a similar way. The amplitude can be represented by

$$\alpha(t) = \alpha_0 \exp\left[iM \cos(2\pi\Omega_{\mathrm{mod}}t)\right] \ \exp(i2\pi\nu_L t) \tag{2.1.24}$$

Clearly the intensity will not be modulated as $|\alpha(t)|^2 = |\alpha_0|^2$. Expanding α_t we obtain

$$\alpha(t) = \alpha_0 \left(1 + iM \cos(2\pi\Omega_{\mathrm{mod}}t) - \frac{M^2}{2} \cos^2(2\pi\Omega_{\mathrm{mod}}t) + \cdots\right) \exp(i2\pi\nu_L t)$$

$$= \alpha_0 \left[\left(1 - \frac{M^2}{4} + \cdots\right) \exp(i2\pi\nu_L t)\right.$$

$$+ i\left(\frac{M}{2} + \cdots\right) \left(\exp(i2\pi(\nu_L + \Omega_{\mathrm{mod}})t) + \exp(i2\pi(\nu_L - \Omega_{\mathrm{mod}})t)\right) \tag{2.1.25}$$

$$- \left(\frac{M^2}{8} + \cdots\right) \left(\exp(i2\pi(\nu_L + 2\Omega_{\mathrm{mod}})t) + \exp(i2\pi(\nu_L - 2\Omega_{\mathrm{mod}})t)\right)$$

$$\left. + \cdots\right]$$

Suppose we measure the quadrature amplitudes ($X1$ and $X2$) of this field relative to a cosine reference wave (i.e. we take α to be real). The $X2$ quadrature will have pairs of sidebands separated from the carrier by $\pm\Omega_{\mathrm{mod}}, \pm3\Omega_{\mathrm{mod}}, \ldots$. The $X1$ quadrature on the other hand will have sidebands separated from the carrier by $\pm2\Omega_{\mathrm{mod}}, \pm 4\Omega_{\mathrm{mod}}, \ldots$. Again a simple, and useful, situation is for the modulation depth to be very small, $M \ll 1$. The amplitude is approximately

$$\alpha(t) \approx \alpha_0 \left[\exp(i2\pi\nu_L t) + i\frac{M}{2}\left[\exp(i2\pi(\nu_L + \Omega_{\mathrm{mod}})t) + \exp(i2\pi(\nu_L - \Omega_{\mathrm{mod}})t)\right]\right] \tag{2.1.26}$$

and the sidebands will only appear on measurements of the $X2$ quadrature. Exact expressions for larger M involve n'th order Bessel functions [Lou73]. The phasor diagram corresponding to this double sideband approximation is shown in Fig. 2.5(b) and Fig. 2.5(d). The sidebands have the same magnitude. The relative phase of the amplitudes is such that the two resulting beat signals have the opposite sign and the beat signals cancel. Thus, as we have seen, phase modulation cannot be detected with a simple intensity detector. Another way of saying this is that the sidebands for phase modulation are perfectly anti-correlated. The treatment of frequency modulation proceeds in a similar way. Any information carried by the light can be expressed as a combination of amplitude and phase modulation. Any modulation, in turn, can be expressed in terms of the corresponding sidebands.

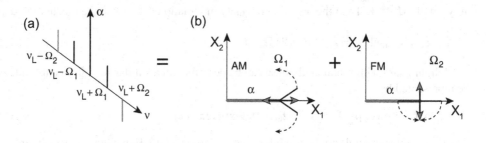

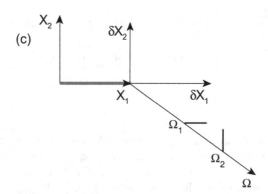

Figure 2.5: Three representations of amplitude (AM) and frequency (FM) modulation with the frequencies sidebands Ω_1 and Ω_2 respectively. (a) shows the sidebands $\nu_L \pm \Omega_1$ and $\nu_L \pm \Omega_2$, (b) shows the same modulations in the time domain and in a phasor diagram, and (c) shows the phasor diagram in frequency space.

2.2 Statistical properties of classical light

2.2.1 The origin of fluctuations

The classical model so far assumes that light consists of a continuous train of electro-magnetic waves. This is unrealistic: atoms emitting light have a finite lifetime, they are emitting bursts of light. Since these lifetimes are generally short compared to the detection interval t_d, these bursts will not be detected individually. The result will be a well defined average intensity, which will fluctuate on time scales $> t_d$. Similarly in most classical, or thermal, light sources there will be a large number of atoms, each emitting light independently. In addition, the atoms will move, which leads to frequency shifts, and they can collide, which disrupts the emission process and leads to phase shifts. All these effects have two major consequences:

1. Any light source has a spectral line shape. The frequency of the real light is not perfectly constant, it is changing rapidly in time. The description of any light requires a spectral distribution which is an extended spectrum, not a delta-function.

2. Any light source will have some intensity noise. The intensity will not be constant. The amplitude of the light will change in time. There will be changes in the amplitude due to the fluctuations of the number of emitting atoms but also due to the jumps, or discontinuities, in the phase of the light emitted by the individual atoms.

The first point leads to the study of spectral line shapes, an area of study which can reveal in great detail the condition of the atoms. The study of spectra is a well established and well documented area of research. However, in the context of this guide we will not expand on this subject. The relevant results are cited, where required. For the analysis of our quantum optics experiments we are not concerned as much with the spectral distribution of the light. We are more interested in the noise, and thus the statistical properties, of intensity, amplitude and phase. Only the intensity fluctuations can be directly measured in the experiments. However, with the homodyne detector, which will be explained in Section 8.1.4, we have a tool to investigate both the phase and amplitude of the light.

The measurement of the statistical properties of light can be carried out in two alternative ways. Either through a measurement of the fluctuations themselves in the form of temporal noise traces or noise spectra. Or, alternatively, through the measurement of coherence or correlation functions. While requiring entirely different equipment, the two concepts are related and in this chapter predictions for both types of measurement will be derived. The emphasis here is on the results and the underlying assumptions and not on the details of the derivation. Many of the details can be found in the excellent monographs by Loudon, Louisell and Gardiner [Lou73, Loi73, Gar91]. In the experimental Chapter 8 it will be shown how these statistical properties can actually be measured and how they have been used to distinguish between classical and quantum properties of the light.

2.2.2 Coherence

Light emitted by a point-like source has properties closely resembling one single wave. The amplitudes at two closely spaced points are linked with each other, as part of the same wave. In contrast, the amplitudes at two very different points or at two very different times will not

be identical. The question how far the detectors can be separated in space, or in time, before
the amplitudes will differ depends on an additional property of the light: its coherence. Any
source of light has a coherence length and a coherence time. The amplitude at two points
within the coherence length and coherence time is in phase and has the same fluctuations.
Interference can be observed. Light at two points separated by more than the coherence length,
or the coherence time, has independent amplitudes which will not interfere.

One clear example of this concept is the technique of measuring the size of a star, which
has a small angular size but is not a point source, using a stellar interferometer. The size of a
distant star can usually not be resolved with a normal optical telescope. The starlight forms
an image in the focal plane of the telescope which is given by the diffraction of the light.
The image depends on the geometrical shape and the size of the aperture of the telescope and
has little to do with the actual shape and size of the object, the star. Diffraction limits the
resolution. Only a bigger aperture of the telescope, larger than technically feasible, would
increase the resolution. To overcome this limitation, A.H. Fizeau proposed the idea of a
stellar interferometer which was first used by A.A. Michelson and his colleagues [Mic21].
The concept of the interferometer is shown in Fig. 2.6.

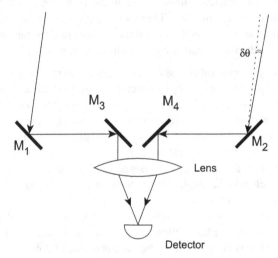

Figure 2.6: Stellar interferometer as proposed and developed by A. Michelson [Mic21]

In this apparatus the starlight is collected by two mirrors, M_1 and M_2, which direct the
light via the mirrors M_3 and M_4 to a telescope. There the two rays of light are combined
in the focal plane. In addition we have optical filters in the instrument which select only a
limited range of optical wavelengths. The mirrors M_1 and M_2 are separated by a distance d
and the apparatus is carefully aligned such that the distances from the focal plane to M_1 and
to M_2 are identical. As a consequence we have interference of the two partial waves in the
focal plane. An interference system will appear, due to the fact that the mirrors M_1 and M_2
are not exactly at 45 degrees to the axis of the telescope. This arrangement is equivalent to
a Michelson interferometer with an optical wedge in one arm and it measures the first order
coherence of the light.

For an extended source we get light from different parts of the source reaching the telescope under slightly different directions, let us say with a very small angular difference of $\delta\theta$. Each wave will produce an interference system, but at a slightly different location in the focal plane. This is due to the additional optical path $\Delta l = d\sin(\delta\theta) \approx d\,\delta\theta$ inside the interferometer. We assume that the light from the different parts of the star is independent. Consequently, it will interfere with light from the same part of the star, but not with light from other parts and the intensity of the interference patterns from the different directions are added. The interference pattern in the focal plane will be visible as long as the path difference is small compared to an optical wavelength. If the separation d between M_1 and M_2 is increased the total interference pattern will eventually disappear, since the individual patterns are now shifted in respect to each other.

An estimate for the separation at which fringes are no longer visible is $d\,\delta\theta = \lambda$. In this situation the two individual patterns from the edges of the star are shifted by one fringe period and all other patterns lie in between, washing out the total pattern. A more detailed analysis predicts the change in the visibility of the interference patterns as a function of the separation d. The interferometer measures the coherence of the light. In this case the coherence is limited due to the independence of the different parts of the star and the coherence length is the path difference for the light travelling from different edges of the star to the earth. The corresponding coherence time for the entire starlight is the difference in travel time. We find that interferometry can be used to measure the coherence length and coherence time.

Michelson built an instrument based on this idea and was able to measure star diameters down to about 0.02 arc seconds [Mic21, Pea31]. There are improved versions of this instrument in operation, in particular an instrument called SUSI (Sydney University Stellar Interferometer) which have achieved resolutions as good as 0.004 arc seconds. However, there are technical limitations. It is very easy to lose the visibility of the fringes, already the wavelength is restricted to an interval of about 5 nm and even then the path difference inside the interferometer has to be kept to less than 0.01 mm. A wider spectrum, containing more light, would require an even better tolerance. It is very difficult to maintain such a small path difference for the entire time of the observation, while the mirrors track the star and the separation between the mirrors is changed by up to several meters.

To avoid these complications, R. Hanbury Brown & R.Q. Twiss proposed an entirely different scheme to measure coherence. Theirs is a very elegant approach which uses intensity fluctuations rather than interference, an idea that had shortly before been used in radio astronomy. In this way they founded a new discipline based on the statistical properties of light. They set out to measure the correlation between the intensities measured at different locations rather than the amplitude and thereby avoided all the complications associated with building an optical interferometer. In their experiment they replaced the two mirrors M_1 and M_2 (Fig. 2.6) with curved mirrors which collected the light onto two different photo-multipliers, as shown in Fig. 2.7 [Bro56]. The photo-currents from these two detectors are amplified independently and then multiplied with each other and the time average of the product is recorded. This signal reflects the correlation between the intensities. This apparatus does not require the mechanical stability of an interferometer. However, some questions arise: In which way is this experiment related to an interferometer? How can it provide similar information if we have disbanded the crucial phase measurement?

To answer these questions, we remind ourselves that stellar interferometers measure the spatial coherence of light, transverse to the direction of observation. At any one point the

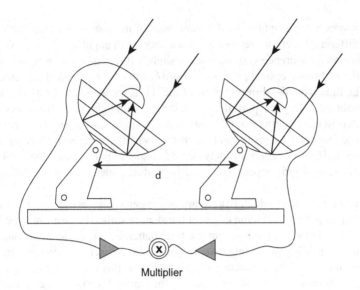

Figure 2.7: Intensity interferometer proposed and developed by Hanbury Brown & Twiss [Bro56]

phase of the light fluctuates wildly and so does the amplitude of light from a thermal source. Generally the amplitude and the phase are not directly linked and both fluctuate independently. However, for two points within a coherence length the phase, and also the amplitude, are linked to each other. The light will interfere and the intensity fluctuations are correlated to each other. For two points separated by more than a coherence length the interference will disappear and so will the intensity correlation.

Fortunately for Hanbury Brown and Twiss, and for us, it is a characteristic property of thermal light that the coherence length over which the phase is constant is approximately the same length in which the amplitude changes very little. Thus a measurement of the path length over which the intensity correlation disappears is in most cases identical to a measurement of the phase coherence length. Hanbury Brown and Twiss measured this distance and thereby were able to determine the size of the source. Already in 1956 they measured the diameter of the star Sirius. An improved apparatus was installed in Narrabri, Australia, with an extended baseline. The detectors are moved on a circular track with a diameter of 396 m [Bro64] and the star diameters can be determined down to 0.0005 arc seconds.

How does a simple multiplier reveal the correlation between the signal? We can describe the intensity at one detector as the sum of a constant average value, $\langle I \rangle$, and a zero mean fluctuating term, $\delta I_1(t)$ by $I_1(t) = \langle I \rangle + \delta I_1(t)$. Similarly the intensity at the other detector can be written $I_2(t) = \langle I \rangle + \delta I_2(t)$. While both $\langle \delta I_1(t) \rangle$ and $\langle \delta I_2(t) \rangle$ are equal to zero, for correlated waves we get $\langle \delta I_1(t) \delta I_2(t) \rangle \neq 0$. If the intensities are perfectly correlated then $\delta I_1(t) = \delta I_2(t)$ and the time average of the product will be $\langle I_1(t) I_2(t) \rangle = \langle I^2 + \delta I_1(t)^2 \rangle$. If the separation of the sources exceeds the coherence length then the intensities are not correlated and the product $\delta I_1(t) \delta I_2(t)$ randomly changes sign and the average $\langle \delta I_1(t) \delta I_2(t) \rangle$ vanishes. That means $\langle I_1(t) I_2(t) \rangle = \langle I \rangle^2$, which is distinctly different to the correlated case.

A reduction in the correlation signal indicates that the coherence length has been reached. A study of the correlation signal for different receiver separations allows a measurement of the second order coherence of the light.

2.2.3 Correlation functions

After this introduction and motivation we will now examine first and second order coherence, and the correlation functions that describe them, quantitatively. First order coherence is important when two light beams are physically superimposed. As a generic example let us consider the case of two beams being combined on a 50:50 beamsplitter, eg. a semi transparent mirror, followed by direct detection of the intensity of one output beam. We allow a time delay, τ to be imposed on one of the beams before the beamsplitter. We can describe the input fields to the beamsplitter by their complex amplitudes $\alpha_1(t)$ and $\alpha_2(t + \tau)$. We assume the beams are in single, identical spatial modes which are completely detected, thus we can ignore the spatial dependence of the amplitudes. For our purposes here we can assume that the action of the beamsplitter is simply to produce the output $\alpha_{out}(t) = (1/\sqrt{2})(\alpha_1(t) + \alpha_2(t + \tau))$, an equal, linear superposition of the inputs. The detector produces a photocurrent proportional to the intensity of the output field, which in turn is proportional to the absolute square of the amplitude, $|\alpha_{out}|^2$. Our measurement result will then be proportional to $|\alpha_{out}|^2$ averaged over some measurement interval, say t_d, such that

$$
\begin{aligned}
\langle |\alpha_{out}|^2 \rangle &= \frac{1}{t_d} \int |\alpha_{out}|^2 \, dt \\
&= \frac{1}{2} \langle |\alpha_1(t)|^2 + |\alpha_2(t + \tau)|^2 + \alpha_1(t)\alpha_2^*(t + \tau) + \alpha_1^*(t)\alpha_2(t + \tau) \rangle
\end{aligned}
\tag{2.2.1}
$$

where we have used $\langle \ldots \rangle$ to indicate the time average. Suppose that α_1 and α_2 both derive from the same continuous source oscillating at frequency ν such that $\alpha_1(t) = \alpha_2(t) = \alpha(t)\exp(i2\pi\nu t)$. Then

$$
\langle |\alpha_{out}|^2 \rangle = \langle |\alpha|^2 \rangle + \cos(2\pi\nu\tau + \phi) \; \langle \alpha^*(t)\alpha(t + \tau) \rangle
\tag{2.2.2}
$$

There is interference between the waves resulting in a modulation of the photocurrent as a function of the path difference. This basic effect is the source of fringes in all first order interferometry including double slit type experiments. The temporal coherence of the light is quantified by the first order correlation function, $g^{(1)}(\tau)$, defined via

$$
g^{(1)}(\tau) = \frac{\langle \alpha^*(t) \, \alpha(t + \tau) \rangle}{\langle |\alpha|^2 \rangle}
\tag{2.2.3}
$$

To see that $g^{(1)}(\tau)$ measures the coherence of the light consider the depth of the modulation in the above example. This is given by the visibility $\mathcal{V}$, defined by

$$
\mathcal{V} = \frac{I_{max} - I_{min}}{I_{max} + I_{min}}
\tag{2.2.4}
$$

Maximum modulation results in $\mathcal{V} = 1$ whilst no modulation results in $\mathcal{V} = 0$. For the output of Equation (2.2.2) we find

$$
\begin{aligned}
\mathcal{V}(\tau) &= \frac{|\langle \alpha^*(t)\,\alpha(t + \tau)\rangle|}{\langle |\alpha|^2\rangle} \\
&= |g^{(1)}(\tau)|
\end{aligned}
\tag{2.2.5}
$$

Thus for this simple example the first order correlation is equal to the visibility. If the source was a perfect sinusoid then α would be a constant in time and we would have $g^{(1)}(\tau) = 1$ regardless of the time delay. This corresponds to an infinite coherence length. More realistically the source will have some bandwidth. Suppose that the source has a Gaussian spread of frequencies $\Delta\nu$ around ν. This can be represented in Fourier (or frequency) space by

$$
\tilde{\alpha}(\Omega) = \alpha_c \exp\left(-2\pi(\nu - \Omega)^2/\Delta\nu^2\right)
\tag{2.2.6}
$$

where $\tilde{\alpha}(\Omega)$ is the Fourier transform of $\alpha(t)$ and α_c is a constant. Taking the inverse transform gives

$$
\alpha(t) = \alpha_c \frac{\Delta\nu}{\sqrt{2}} \exp(-\pi\Delta\nu^2 t^2)
\tag{2.2.7}
$$

for our amplitude in time space. Thus our first order correlation function becomes

$$
\begin{aligned}
g^{(1)}(\tau) &= \frac{\dfrac{1}{t_d}|\alpha_c|^2 \exp\left(-\pi\Delta\nu^2\tau^2\right)\dfrac{\Delta\nu^2}{2}\displaystyle\int \exp\left(-\pi\Delta\nu^2 t^2\right)dt}{\dfrac{1}{t_d}|\alpha_c|^2 \dfrac{\Delta\nu^2}{2}\displaystyle\int \exp\left(-\pi\Delta\nu^2 t^2\right)dt} \\
&= \exp\left(-\frac{\pi}{2}\Delta\nu^2\tau^2\right),
\end{aligned}
\tag{2.2.8}
$$

a Gaussian function of τ with a width inversely proportional to the frequency spread. For a narrow frequency spread, $\Delta\nu$ small, we have a large coherence length whilst for a broad spectrum, $\Delta\nu$ large, the coherence length becomes very short. It is this first order coherence length which is measured by the Stellar interferometer.

Now let us consider intensity correlation measurements. As a generic example we consider direct detection of the intensity of two beams that have not been optically mixed, but one of which has a variable time delay imposed on it before detection. The two photocurrents from the detectors are then sent into a correlator which multiplies the photocurrents together. Suppose firstly that, as in our previous discussion, the two beams originate from the same source. The output of the correlator will then be proportional to the second order correlation function, $g^{(2)}$, which is defined by

$$
g^{(2)}(\tau) = \frac{\langle I(t)I(t + \tau)\rangle}{\langle I(t)\rangle^2} = \frac{\langle \alpha^*(t)\alpha(t)\alpha^*(t + \tau)\alpha(t + \tau)\rangle}{\langle \alpha^*(t)\alpha(t)\rangle^2}
\tag{2.2.9}
$$

where the I are the detected intensities.

From the symmetry of this definition follows the property $g^{(2)}(-\tau) = g^{(2)}(\tau)$ and measurements need only to be made for positive values of τ. Whilst the degree of first order coherence $g^{(1)}(\tau)$ takes on values between 0 and 1, and is unity for $\tau = 0$, the second order correlation is not bounded above. Remembering that the variance of I is given by $\Delta I^2 = <I^2> - <I>^2$, we can write the second order correlation at $\tau = 0$ as

$$g^{(2)}(0) = \frac{\Delta I^2}{\langle I \rangle^2} + 1 \qquad (2.2.10)$$

Given that the variance obeys $\Delta I^2 \geq 0$ we find that the value of the second order coherence at zero time delay is

$$g^{(2)}(0) \geq 1 \qquad (2.2.11)$$

and it is not possible to derive an upper limit. This proof is only valid for $\tau = 0$ and cannot be applied to other delay times. The only restriction stems from the positive nature of the intensity, resulting in

$$\infty \geq g^{(2)}(\tau) \geq 0 \quad \text{for all} \quad \tau \neq 0. \qquad (2.2.12)$$

It is possible to relate the degree of second order coherence at later times to the degree of coherence at zero delay time by using the Schwarz inequality in the form

$$\langle I(t_1)I(t_1 + \tau)\rangle^2 \leq \langle I(t_1)^2\rangle\langle I(t_1 + \tau)^2\rangle \qquad (2.2.13)$$

For ergodic or stationary systems, two-time averages depend only on the time difference not the absolute time. For such systems we then have $\langle I(t_1)^2\rangle = \langle I(t_1 + \tau)^2\rangle$. Consequently we have

$$\langle I(t_1)I(t_1 + \tau)\rangle \leq \langle I(t_1)^2\rangle$$

yielding

$$g^{(2)}(\tau) \leq g^{(2)}(0) \qquad (2.2.14)$$

For a perfectly stable wave (i.e. zero variance) the equal sign applies and $g^{(2)}(\tau) = 1$ for all delay times. Equation (2.2.14) means that the degree of coherence for any classical wave should always be less than $g^{(2)}(0)$. This is a result which has been contradicted for quantum states of light and has been used to demonstrate experimentally the quantum nature of light as shown in detail in Chapter 3.

2.2.4 Noise spectra

A very direct way of measuring the fluctuations, or the noise, of the light is to record the time history of the intensity $I(t)$ and to evaluate the variance $\Delta I^2(t)$ for a certain detection time. The variance of a classical value is defined as

$$\Delta I^2(t) = \text{Var}(\Delta I(t)) = \langle (I(t) - \langle I \rangle)^2\rangle - \langle I(t) - \langle I \rangle\rangle^2 = \text{Var}(I(t)) \qquad (2.2.15)$$

where $\langle I \rangle$ is the average intensity and the brackets $\langle \ldots \rangle$ denote averaging over a detection interval of length t_d. Similarly, we define the root mean square value (RMS) as:

$$RMS(I(t)) = \sqrt{\text{Var}(I(t))} = \sqrt{\Delta I^2(t)} \qquad (2.2.16)$$

As we have seen in Section 2.1.6 information can be carried on a light beam as small amplitude or phase modulations. The variance gives limited information about the level of noise at a particular modulation frequency and thus how successfully information can be carried.

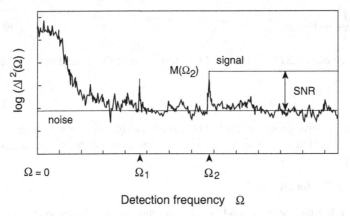

Figure 2.8: Example of a noise spectrum showing noise at many frequencies and two modulations at Ω_1 and Ω_2. This plot has a logarithmic scale and the signal to noise ratio can be read off it directly if $M(\Omega) \gg \text{Var}(I(\Omega))$.

More useful is the spectral variance or noise spectrum, which gives the noise power as a function of frequency from the carrier. The spectral variance of a variable y is given by the Fourier transform of the autocorrelation function for that variable,

$$V(\Omega) = \frac{1}{2\pi} \int_{-\infty}^{\infty} e^{i2\pi\Omega\tau} \langle y(t)y(t+\tau)\rangle d\tau, \tag{2.2.17}$$

and can be evaluated experimentally using a spectrum analyser (see Chapter 7). Most commonly we will be evaluating the spectral variances of the intensity, $y = \alpha^*\alpha$, and the quadrature amplitudes, $y = \alpha + \alpha^*$ and, $y = i(\alpha - \alpha^*)$. In many practical situations we can write our variable in the form of

$$y(t) = y_0 + \delta y(t) \tag{2.2.18}$$

where y_0 is the average value of the variable and all the signals and noise are fluctuations around the average, carried by the term $\delta y(t)$ which is assumed to have zero mean (i.e. $\langle \delta y(t)\rangle = 0$). Under these conditions the spectrum is simply given by

$$V(\Omega) = <|\delta\tilde{y}(\Omega)|^2> \tag{2.2.19}$$

where the tilde indicates a Fourier transform.

A simple example is illustrative at this point. Suppose the fluctuations in our variable are made up of two parts: a deterministic signal, $\delta y_s = g\cos(2\pi\Omega t)$ and randomly varying noise which we represent by the stochastic function $\delta y_n = f\zeta(t)$ (f and g are positive constants). The signal and noise are uncorrelated so we can treat them separately. For the signal we use Equation (2.2.19). The Fourier transform of a cosine is a pair of delta functions at $\pm\Omega$. Squaring and averaging over a small range of frequencies (say $\delta\Omega$) around $\pm\Omega$ we end up with two signals of size g^2. For the noise it is easier to work from Equation (2.2.17). We can take the autocorrelation of the noise to be $\langle \zeta(t)\zeta(t+\tau)\rangle = \delta(\tau)$. That is the noise is random on all time scales, there is no correlation between how it behaves from instant to instant. The Fourier transform of a delta function is a constant. Thus after integration over the small bandwidth the noise contributes $\delta\Omega f^2$ to the spectrum at all frequencies. The signal to noise ratio is defined by

$$SNR + 1 = \frac{g^2}{\delta\Omega f^2} \tag{2.2.20}$$

In Fig. 2.8 an actual intensity spectrum with AM modulation is plotted. In this case the coefficient g corresponds to the modulation $M\alpha_0$. The relative intensity noise RIN is the ratio of the variance to the average intensity. When stated as a function of the detection frequency Ω, the integration time t_d, or bandwidth has to be taken into account. For a bandwidth of 1 [Hz] it is given by

$$RIN(\Omega) = Var(I(\Omega))/\langle I\rangle^2 \tag{2.2.21}$$

In classical theory it is in principle possible to make any noise arbitrarily small in the amplitude and phase such that the amount of information that can be carried by a beam of light is limited only by technical considerations. We shall find that this is not true in the quantum theory of light. Strict bounds apply to the amount of information that can be encoded on light of a particular power.

2.2.5 An idealized classical case: Light from a chaotic source.

In most practical cases classical light will be noisy. Only a perfect oscillator would emit an electro-magnetic wave with perfectly constant amplitude and phase. There are some relatively simple models of realistic light sources. Chaotic light is the idealized approximation for the light generated by independent sources emitting resonance radiation. A practical example is a spectral lamp. In contrast, thermal light is an approximation of the light emitted by many interacting atoms that are thermally excited and together emit a broad, non resonant spectrum of light. A practical example is a hot, glowing filament. Thermal light will be discussed in the next chapter.

The source of chaotic light is a large ensemble of atoms which are independent and thermally excited. They all emit light at the same resonance frequency. For simplicity, we assume they are sufficiently slow so that the effect of Doppler shifts can be neglected. The atoms will collide with each other, thus the waves of radiation emitted by each atom will be perturbed by collisions. The noise properties are best derived by analysing the total complex amplitude $\alpha(t)$ of the wave which is the sum of the amplitudes emitted by the individual atoms. We can evaluate the coherence function of the total amplitude and from there we can determine the intensity noise spectrum. For a large number M of atoms the total wave amplitude is

$$\alpha(t) = \sum_{j=0}^{M} \alpha_j(t) = \alpha_0 \exp(i2\pi\nu_0 t)\sum_{j=0}^{M}\left(\exp(i\phi_j)\right) = |\alpha(t)|\exp\left(i\phi(t)\right) \tag{2.2.22}$$

where ϕ_j represents the phase of the individual contributions. For simplicity we assume that all atoms emit into the same polarisation. For a large number of atoms $|\alpha(t)|$ and $\phi(t)$ are effectively randomly varying functions which describe the magnitude and the phase of the total amplitude at any time t.

If the average time between collisions is τ_0, one can derive [Lou73] a probability $P(\tau)$ that an atom is still in free flight after a time τ since its last collision which is given by

$$P(\tau)d\tau = \exp\left(-\frac{\tau}{\tau_0}\right) \tag{2.2.23}$$

The time interval τ_0 is usually very short. If we assume that a single measurement takes a time larger than τ_0, the coherence function can be evaluated as:

$$
\begin{aligned}
\langle \alpha^*(t)\alpha(t+\tau)\rangle &= \alpha_0^2 \exp(i2\pi\nu_0\tau)\\
&\quad \langle\, [\exp(-i\phi_1(t)) + \cdots + \exp(-i\phi_M(t))]\\
&\qquad [\exp(i\phi_1(t+\tau)) + \cdots + \exp(i\phi_M(t+\tau))]\, \rangle
\end{aligned}
$$

In multiplying out the large brackets many cross terms appear which correspond to waves emitted by different atoms. Since the phase between the light from different atoms is random they cancel and give no contribution to the total average. Thus

$$
\begin{aligned}
\langle \alpha^*(t)\alpha(t+\tau)\rangle &= |\alpha_0|^2 \exp(i2\pi\nu_0\tau) \sum_{j=1}^{M} \exp[\, i(\phi_j(t+\tau) - \phi_j(t))\,]\\
&= M\langle \alpha_j^*(t)\alpha_j(t+\tau)\rangle
\end{aligned}
$$

where $\alpha_j(t)$ is the amplitude emitted by a single atom and thus the entire correlation is completely described by the single atom correlation function. A single atom emits a steady train of waves until a random phase jump occurs due to a collision. If one neglects the small amount of light that is emitted during the collision (impact approximation) one gets the correlation terms

$$
\begin{aligned}
\langle \alpha_j^*(t)\alpha_j(t+\tau)\rangle &= |\alpha_0|^2 \exp(-i2\pi\nu_0\tau)P(\tau)\\
&= |\alpha_0|^2 \exp\left(-i2\pi\nu_0\tau - \frac{\tau}{\tau_0}\right)
\end{aligned}
$$

and the first order correlation function

$$g^{(1)}(\tau) = \exp\left(-i2\pi\nu_0\tau - \frac{\tau}{\tau_0}\right) \tag{2.2.24}$$

For all types of chaotic light it is possible to derive the following relationship between the first and second order correlation coefficient [Lou73]:

$$g^{(2)}(\tau) = 1 + \left|g^{(1)}(\tau)\right|^2 \tag{2.2.25}$$

The exact form of both $g^{(2)}(\tau)$ and $g^{(1)}(\tau)$ depends on the type and the statistical distribution of collisions.

For illustration a numerical example is given in Fig. 2.9 showing the various properties and descriptions of a fluctuating light beam. Here the chaotic light is simulated by calculating explicitly the emission from 20 independent sources, all with the same amplitude. The initial phase of these sources is selected at random. The initial conditions are given in Fig. 2.9(a), all twenty partial waves $\alpha_i(t = 0)$ and the total amplitude $\alpha(t = 0)$ are shown. The phase evolution of the emitters is simulated by adding to each emitter a change in phase which can randomly vary in the range -0.13π to 0.13π for each time step. Thus each emitter has a random development of its phase, occasionally with a phase jump. For each time interval the total amplitude can be described as one point in a polar diagram, this leads to the distribution of points shown in Fig. 2.9(b). The vector describes a convoluted path, it moves from one point to another seemingly at random. The density of points can be interpreted as a probability of finding a certain amplitude. The points are scattered around the origin. While the amplitudes would average out, the most probable intensity is not zero.

The full time development of the magnitude of the total amplitude $|\alpha(t)|$, the total phase $\phi(t)$ and the scaled intensity $I(t) = I_0 \alpha^* \alpha$ are all shown in Figs. 2.9(c), 2.9(d) and 2.9(e). The magnitude $|\alpha(t)|$ and the intensity $I(t)$ show dramatic fluctuations, occasionally dropping close to zero. The intensity can be described by the long term average $\langle I \rangle$ and the RMS value, both included in Fig. 2.9(d). For this numerical example the correlation functions can be calculated directly. Equations (2.2.3) and (2.2.9) have been evaluated, with the results shown in Figs. 2.9(f) and 2.9(g). The results are very close to the predictions for a chaotic source. Over the time interval shown, the first order correlation functions change smoothly from $g^{(1)}(0) = 1$ down to 0.15. In the same time interval $g^{(2)}(\tau)$ declines from $g^{(2)}(0) = 1.8$ to 1.0. There are some oscillations in $g^{(2)}(\tau)$, which is typical for such a short statistical sample. The analysis of the longer time sample would approach the theoretically predicted smooth decline of the values of $g^{(2)}(\tau)$. The link between $g^{(2)}$ and $g^{(1)}$ given in Equation (2.2.25) is clearly reproduced in this example.

In order to predict the intensity noise of the chaotic source of M atoms we consider the instantaneous intensity $I(t)$ and the averaged intensity $\langle I \rangle$. We find

$$\langle I \rangle = \langle I(t) \rangle = \frac{1}{2} E_0^2 \epsilon_0 c |\alpha_0|^2 \left\langle \left| \sum_{j=1}^{M} \exp(i\phi_j(t)) \right|^2 \right\rangle = \frac{1}{2} E_0^2 \epsilon_0 c |\alpha_0|^2 M \quad (2.2.26)$$

since all the cross terms average out due to the randomness of the phase functions. The total average intensity is simply the sum of the average intensity emitted by a single atom. In order to evaluate the variance we require a value for

$$\langle I(t)^2 \rangle = \left(\frac{1}{2} E_0^2 \epsilon_0 c |\alpha_0|^2 \right)^2 \left\langle \left| \sum_{j=1}^{M} \exp(i\phi_j(t)) \right|^4 \right\rangle \quad (2.2.27)$$

The only terms which will contribute to this value are those where each factor is multiplied by

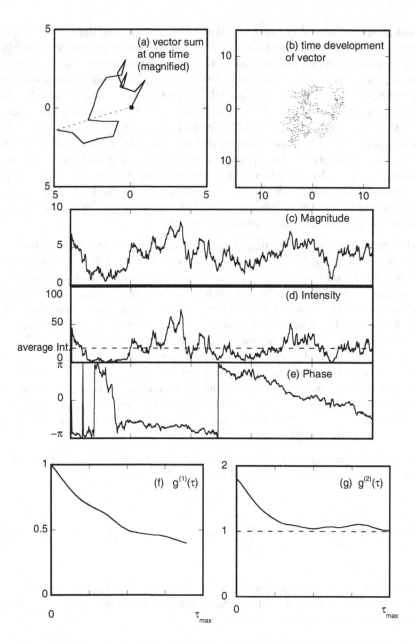

Figure 2.9: An example of the statistical properties of chaotic light. The light is simulated as the total field emitted by 20 independent emitters, each with constant amplitude but randomly evolving phase. Part (a) shows the individual vectors at $t = 0$ and (b) the resulting total phase vector for various times. The time history is shown for the magnitude (c), the intensity (d) and the phase of the field (e). This leads to the first order (f) and second order (g) correlation function.

its own complex conjugate. These terms result in

$$
\begin{aligned}
\langle I(t)^2 \rangle &= \left(\frac{1}{2} E_0^2 \epsilon_0 c |\alpha_0|^2 \right)^2 \left\langle \left(\sum_j^M e^{i\phi_j} \right)^2 \left(\sum_j^M e^{-i\phi_j} \right)^2 \right\rangle \\
&= \left(\frac{1}{2} E_0^2 \epsilon_0 c |\alpha_0|^2 \right)^2 \left\langle \sum_j^M e^{i(2\phi_j - 2\phi_j)} + 2 \sum_{j \neq k}^M e^{i(\phi_j + \phi_k - \phi_j - \phi_k)} + \cdots \right\rangle \\
&= \left(\frac{1}{2} E_0^2 \epsilon_0 c |\alpha_0|^2 \right)^2 \langle M + 2M(M-1) + \cdots \rangle \\
&= \left(\frac{1}{2} E_0^2 \epsilon_0 c |\alpha_0|^2 \right)^2 (M + 2M(M-1))
\end{aligned}
$$

where $\cdots$ represents terms which average to zero. With Equation (2.2.26) we obtain

$$
\langle I(t)^2 \rangle = \left(\frac{\langle I \rangle}{M} \right)^2 (2M^2 - M) = \langle I \rangle^2 \left(2 - \frac{1}{M} \right)
$$

and using $\langle I(t)^2 \rangle = 2\langle I \rangle^2$ for large M, the variance is

$$
\Delta I(t)^2 = (2\langle I \rangle^2 - \langle I \rangle^2) = \langle I \rangle^2 \tag{2.2.28}
$$

This is a simple and important result: for a chaotic source the variance of the time resolved intensity is equal to the square of the average intensity. Calculations for more practical situations, which include Doppler broadening and realistic models of the line broadening, show basically the same result. The intensity noise spectrum will be confined to the spectral linewidth. Outside the linewidth this classical description predicts an arbitrarily low intensity noise spectrum.

Summary

We have now a set of tools to describe the properties is of a beam of light: it is a mode with a set of DC values and we can describe the power which can be measured both in space and in time. In addition, we can communicate information with the beam through modulations, using the AC sidebands. Next to making and sending images, communication of data is now the most widespread use of light, it has led to a whole industry of *photonics*.

Finally, we can describe the fluctuations of the light, as a spectrum or through correlation functions. This noise is intrinsic to the way the light is generated and we distinguish between incoherent light, such as sunlight, coherent light from a laser, and all stages in between. Correlation measurements have opened new ways in determining the properties of the light, with many applications ranging from astronomy to microscopy. The field of classical optics is wide, versatile and beautiful.

Further reading

The principles of optics, M. Born, E. Wulf (1959)
Fundamentals of Photonics, M.C. Teich, B.E.A. Saleh, Wiley (1996)

Quantum theory of light, R. Loudon, Oxford University Press, Oxford (1973)
Optical Coherence and Quantum Optics, L. Mandel, E. Wolf, Cambridge (1995)

Bibliography

[Abr72]   Handbook of mathematical functions, M. Abramovitz, I. Stegun, Dover, New York (1972)

[Bro56]   Correlation between photons in two coherent beams of light, R. Hanbury Brown, R.Q. Twiss Nature 177, 27 (1956)

[Bro64]   The stellar interferometer at Narrabri observatory R. Hanbury Brown, Sky and Telescope 28, 64 (1964)

[Cah69]   Density operators and quasi-probability distributions K.E. Cahill, R.J. Glauber, Phys. Rev. 177, 1882 (1969)

[Cav80]   C.M. Caves in Coherence, Cooperation and Fluctuations, eds. F. Haake, L.M. Narducci, D.F. Walls, Cambridge University Press (1980)

[For55]   Photoelectric mixing of incoherent light, A.T. Forrester, R.A. Gudmundsen, P.O. Johnson, Phys. Rev. 99, 1691 (1955)

[Gar91]   Quantum noise, C.W. Gardiner, Springer, Berlin (1991)

[Kog66]   Laser beams and resonators, H. Kogelnik, T. Li, Appl. Opt. 5, 1550 (1966)

[Law27]   The instantaneity of the photo-electric effect, E.O. Lawrence, J.W. Beams, Phys. Rev. 29, 903 (1927)

[Len02]   Ueber die lichtelektrische Wirkung, P. Lenard, Ann. Physik 8, 149 (1902)

[Loi73]   Quantum statistical properties of radiation, W.H. Louisell, Wiley, New York (1973)

[Lor10]   Alte and neue Fragen der Physik, H.A. Lorentz, Phys. Zeitschr. 11, 1234 (1910)

[Lou73]   Quantum theory of light, R. Loudon, Oxford University Press, Oxford (1973)

[Mic21]   Measurement of the diameter of α orionis with the the interferometer, A.A. Michelson, F.G. Pease, Astrophys. J. 53, 249 (1921)

[Pau30]   General principles of quantum mechanics, W. Pauli, Springer (1930)

[Pau83]   Light absorption by a dipole, H. Paul, R. Fischer, Usp. Fiz. Nauk 141, 375 (1983)

[Pau95]   Photonen, eine Einführung in die Quantenoptik, H. Paul, Teubner, Stuttgart (1995)

[Pav99]   The Sydney University stellar interferometer - the instrument, Davis, W.J. Tango, Mon. Not. R. Astrom. Soc. 303, 773-782 (1999)

[Pea31]   Interferometer methods in astronomy, F.G. Pease, Ergebnisse der exakten Naturwissenschaften 10, 84 (1931)

[Pla00]   Zur Theorie des Gesetzes der Energieverteilung im Normalspektrum, M. Planck, Verh. Dt. Phys. Ges. 2, 237 (1900)

[Pow64]   Introductory Quantum Electrodynamics, E.A. Power, Longmans (1964)

[Sie86]   Lasers, A.E. Siegman, University Science Books (1986)

[Tan80]   Michelson stellar interferometry, W.J. Tango, R.Q. Twiss, Progress in Optics XVII, 239 (1980)

[Tei96]   Introduction to Photonics, M.C. Teich, B.E.A. Saleh, Wiley (1996)

3 Photons – the motivation to go beyond classical optics

Although classical optics is very successful in describing many aspects of light, it is not all in-clusive. Many fascinating effects are due to the quantum nature of the light, they are the foun-dation of *quantum optics*. This chapter introduces the concept of the photon as the smallest detectable quantity of light and develops some of the basics required for modelling quantum optics experiments.

3.1 Detecting light

So far we have considered light to be just one form of electro-magnetic waves, much like radio waves. The detection of radio waves and of light is fundamentally different. Radio waves are detected with an antenna, which is a macroscopic structure made out of many atoms. The waves induce an electric field in the antenna, a movement of many electrons. This leads to an electric signal which can be amplified and processed. The signal is proportional to the magnitude of the electric field. In contrast, the detection of light relies on the interaction with individual atoms. The microscopic, or atomic, properties of the material of the detector determine the detection process. A photo-detector does not measure the field, it cannot follow the frequency ν of the oscillations, typically of the order of 10^{15} [Hz]. It can, however, determine a transfer of energy from the radiation field to the atoms in the detector. And from this process we can conclude properties of the light intensity.

What are the experimental details of the process of photodetection? One fundamental fact is the resonance like property of the interaction. Atoms interact only with radiation that matches their eigen-frequencies. The absorption and emission of light results in very specific spectra. For free atoms, as those in a gas, these spectra have a resonant structure; they consist of several narrow well defined lines. Thus, free atoms are not suitable for a general purpose detector. Consequently, we use a material with many interacting atoms with a broad absorp-tion spectrum. But the fact remains, individual atoms absorb the radiation that matches their individual eigen-frequencies.

One well established detector is the photographic plate. The atoms in the photo-sensitive emulsion can absorb light. One individual absorption process can lead to the chemical change of an entire grain in the emulsion and thereby create one dark dot on the plate. This technique has one major technical problem: long exposure times. Not every photon incident leads to the change of a grain. It requires a lot of intensity to expose a plate. It is also far more convenient to deal with electric signals. Thus we use the photoelectric effect. Very early experiments with this effect have already shown that radiation interacts with such a detector by the absorption of multiples of the energy $h\nu$ [Len02]. The probability $\mathcal{P}_{\mathrm{det}}$ per unit time of detecting one

A Guide to Experiments in Quantum Optics, 2nd Edition. Hans-A. Bachor and Timothy C. Ralph
ISBN: 3-527-40393-0

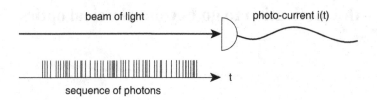

Figure 3.1: Photoelectric detection.

individual photoelectron averaged over the detection time interval t_1 scales with the integral of the intensity

$$\mathcal{P}_{\text{det}} = \frac{D}{h\nu t_1} \int_0^{t_1} I(t')dt' \tag{3.1.1}$$

This equation applies to any form of the photoelectric effect. The response time is very similar. The conversion efficiency D depends on the frequency of the detected light and on the material used.

Each individual detection process results in one individual electron. This is not sufficient to produce a signal which can be processed. Several techniques have been invented to turn one electron into a measurable pulse of many electrons. This has led to the development of a range of modern photo-detectors, such as photo-multipliers, photo-diodes and charge coupled devices (CCDs), [Kin95]. For technical detail see Section 7.4. While different in their technology they all produce a photo-current which is proportional to the light intensity and the detector circuitry averages over a time t_1, which is much longer than a cycle of the oscillation. The link between the photo-current $i(t)$ and intensity $I(t)$ is given by

$$i(t) = \frac{D'}{t_1} \int_0^{t_1} I(t')dt' \tag{3.1.2}$$

where D' describes the sensitivity and the gain of the detector. Detection times as short as 10^{-10} [sec] have been achieved. Such a detector can only record changes of the intensity which occur on time scales longer than t_1. The electric signal will contain a direct current (DC) term, proportional to the long term average intensity and alternating current (AC) components with frequencies up to $1/t_1$.

Please note that in our description so far there has been no need for the quantization of the radiation. We need a quantised detection process, but Equation (3.1.1) could also be used for a classical electro-magnetic field. And indeed, at high intensities such a classical interpretation is valid. However, at low intensities we find experimental evidence that the light field has to be quantised, that we need the concept of particles of light, of photons. For example, consider a situation where a pulse of light contains energy less than $h\nu$. In this case the classical Equation (3.1.1) would predict a probability which is larger than zero, while the experimental evidence shows that there is no detection at all. This can be taken into account by choosing a quantised theory of the detection process [Gla62] which replaces Equation (3.1.1) with

$$\mathcal{P}_{\text{det}} = \frac{D'}{h\nu \, t_1} \int_0^{t_1} \langle \tilde{\mathbf{A}}^\dagger \tilde{\mathbf{A}} \rangle dt' \tag{3.1.3}$$

where $\tilde{\mathbf{A}}$ and $\tilde{\mathbf{A}}^\dagger$ are operators which describe the complex amplitude of the electric field and its conjugate. We will find that there are some peculiar features of these operators compared to the classical functions. For example, the sequence of the operators in the integral is important. The sequence in Equation (3.1.3) corresponds to the normal ordering which has been chosen such that absorption of photons is the detection process, a different ordering would lead to completely unrealistic effects. This quantum model is described in Chapter 4.

One motivation for introducing quantised radiation stems from the question: How quickly, or with what response time, can low intensity light pulses be detected? In a classical theory there would have to be a delay in order to accumulate sufficient energy in one atom to release an electron. The response time t_{min} for light with a photon flux density of N [photons/(m^{2}s)] and for an atom with cross-section A [m^2] can be estimated as $t_{min} = 1/NA$. This is the time required for the energy equivalent to one photon to be transmitted through one atom, it would be a minimum for the response time. Is this realistic?

Let us consider sunlight as an example. Green light, with a wavelength of 500 [nm], is selected with an optical filter that transmits about 1% of the entire spectrum. This green light has a photon flux density of about 10^{19} [photons/(m^{2}s)] as can be calculated from Planck's radiation law using the well known data for sun (surface temperature $= 6000$ [K], diameter 32 [arc minutes]). Assuming an atomic cross-section of 10^{-18} [m^2] the detection time is about 0.1 [s]. For less intensive beams the time would increase to fantastic values – while we know that our eye can be fairly fast even at intensities much lower than full sunlight.

Actually, the sensitivity of the human eye for green light is about 5×10^{-17} [W] which corresponds to a photon flux of about 120 [photons/s] into the eye, or a photon flux density of about 10^{10} [photons/(m^2 s)]. This low level is sufficient to release photoelectrons and the detection process certainly does not take longer than one second. If the simple classical model was correct we could not see low intensity images, such as the impressive display of star patterns in a dark night. Furthermore, the vision process would be entirely different to our experience. The main difference between bright and dim objects would be the time it takes to see them, the detected intensity would always appear to be similar. Thus we conclude that our personal experience of vision requires the concept of photons.

It was pointed out that this simplified model overestimates the response time. A faster response can be estimated by assuming that atoms can draw energy from a wider area. This would be the case when the atoms are separated, or when only some of the densely packed atoms contribute to the detection process. Estimates for such a situation have been made for classical oscillators by Lorentz [Lor10] and can be combined with the assumptions of modern atomic theories [Pau83, Pau95]. These calculations show the ability of atoms to draw energy from a wide area. One physical interpretation of this effect is that the incoming field induces a dipole moment in the atom which in turn emits a wave that destructively interferes with the incoming wave. In this way the energy is concentrated in the atom. For the case of sunlight a response time of about 10^{-7} [s] and for the sensitivity limit of the eye a response of 10^{-1} [s] can be calculated.

While these values are short enough to be comparable with the response time of the human eye they are still orders of magnitudes too long compared to experiments. Fast response times have been measured even before the advent of modern short pulse laser technology. In 1927 Lawrence & Beams [Law27] used an electric spark to generate a light pulse which was detected with a photo cathode which could be gated using an additional internal grid. The timing of the grid could be delayed by propagating the electric gating pulses along various lengths of cable and it was observed that the photo electrons from the detector appeared simultaneously with the light pulse, within an accuracy of 10^{-9} [s]. Forrester et al. [For55] used the detection of beat signals between two monochromatic thermal sources and concluded that the detectors can follow the electro-magnetic field with accumulation times less than 10^{-10} [s]. Nowadays even shorter pulses, femtoseconds, and higher modulation frequencies are routinely used in optical instruments. All these observations show the need for a concept of light as a quantised system – the need for *photons*.

3.2   The concept of photons

We really need more than one concept of photons. So far we have taken the view of an experimentalist who would talk about *detecting photons* from a propagating beam of light. Photons have properties quite different from those of classical particles. They have energy $h\nu$, zero rest mass, momentum $\hbar k$, spin $\hbar$, they are non-interacting particles following the Bose statistics. A laser beam contains a large number of these particles (see Table 3.2) and the measurement of the beam intensity corresponds to a measurement of the flux of photons. The photons, which were created by the emission process, that is the de-excitation of the atoms via stimulated or spontaneous emission, are destroyed in the detection process. This in turn creates an electron-hole pair in the detector material. For the experimentalist photons are particles that can be localized, for example at the detector. The experimentalist takes the consequence of the detection, the electron created or the grain in a photographic plate that is darkened, as the evidence of the photon.

It is tempting from this point of view to envisage a beam of light as a stream of photons travelling at the speed of light. But this is incorrect and as we shall see in the following sections cannot successfully explain many experimental results. A different concept for photons, perhaps more familiar to a theoretician, is needed to describe the propagation of light. Here we would talk about the *photon number of a mode*, where "modes" refers to the radiation field modes described in the last chapter (Section 2.1). Each mode has energy states like a harmonic oscillator and the state of excitation of a mode is said to be the number of photons in the mode. In this concept photons are distributed, they do not exist at any specific location, they are a property of the entire mode. The only possible processes are the annihilation and creation of a photon in which the entire mode is affected and to describe this we need the formal rules of quantum mechanics.

The wonder of quantum mechanics is that these apparently contradictory points of view – localized, particle-like emission and detection but delocalized field-like propagation (often referred to as wave-particle duality) – can coexist. A laser beam can be described by a single mode propagating from the laser, with physical properties much like the Gaussian mode in Equation (2.1.8). The photon number of the laser mode is linked to the flux of photons in the external laser beam, which can be detected. Laser modes typically contain a large number of photons, the flux is high. For thermal light, on the other hand, the number of possible modes can exceed the photon number and on average there will be less than one photon in any mode, see Table 3.2. A more detailed review of the concept of photons can be found in books by H. Paul [Pau95] and R. Loudon [Lou73].

Table 3.1: Examples of the photon flux and the photon density of typical optical fields. For simplicity it is assumed that all fields have the same wavelength of $\lambda = 500$ [nm] and thus all photons have the same energy $h\nu = 2.510^{-19}$ [J]. In the case of the pulsed laser the photon density inside the pulse has been evaluated. The table shows that we can consider a mode to contain a large number of photons only for laser beams.

Type of light	Intensity I $[W/m^2]$	Elec. Field E $[V/m]$	photon density $[m^{-3}]$	photons/mode
White light ($T = 6000$ K)	10^3	10^3	10^{13}	10^{-4}
Spectral lamp	10^4	3×10^3	10^{14}	10^{-2}
CW laser	10^5	10^4	10^{15}	10^{10}
Pulsed laser	10^{13}	10^8	10^{23}	10^{18}

3.3 Light from a thermal source

We will now look at the example of light emitted by a broadband thermal source. We will assume that the emission properties of the atoms are quantised, as first proposed by Planck [Pla00], but will treat the field in a classical way. The term photon will be used in this section to describe the fact that the light is emitted in "lumps" by the atoms.

The light emitted by a thermal source can be analysed by considering a finite volume of space with zero field boundary conditions. The region will support a discrete number of standing waves of the electro-magnetic field. We first derive the number and the spectral density of these modes and from there we determine the distribution of photons in the light emitted by the source. If the region has the physical dimension L it can support a variety of modes with the wave vectors $k_x = \pi \nu x/L, k_y = \pi \nu y/L, k_z = \pi \nu z/L$ resulting in a mode density

$$\rho_\nu d\nu = \frac{8\pi\nu^2 d\nu}{c^3} \qquad (3.3.1)$$

These modes are thermally excited according to a Boltzmann distribution

$$\mathcal{P}(E_n) = \frac{\exp(-E_n/k_bT)}{\sum \exp(-E_n/k_bT)} \qquad (3.3.2)$$

Assuming the condition $E_n = (n+\frac{1}{2})h\nu$ for the energy of a mode we can derive a probability distribution for n photons in one mode.

$$\mathcal{P}_n(\nu, T) = \frac{1 - \exp(-h\nu/k_bT)}{\exp(n \, h\nu/k_bT)} \qquad (3.3.3)$$

We can now evaluate the average number $\bar{n}(\nu, T)$ of photons in a mode

$$\bar{n}(\nu, T) = \sum_{n=0}^{\infty} n\mathcal{P}_n(\nu, T) = \frac{1}{\exp(h\nu/k_bT) - 1} \qquad (3.3.4)$$

which is the well known black body formula. It explains the existence of the black body spectrum of a solid at temperature T. The fluctuations in a single mode of this radiation can be rather large. For normal thermal conditions and for radiation in the visible or near infrared the average excitation is very low. For example, at $T = 300$ [K] and for $\nu = 10^{15}$ [Hz] the average excitation is about 7×10^{-4} photons in each mode. The probability of finding a fixed number of photons, for example $n = 0, 1, 2, \ldots$ in one mode follows a Bose-Einstein distribution which actually has a maximum at $n = 0$.

However, this calculation is only valid if we count photons for less than the coherence time and for only one mode. This result is not applicable in the situation where we detect an intensity sufficient to create a measurable photo-current. In order to describe a beam of light with measurable power we have to consider many modes at different frequencies. They all contribute to the measured intensity. We have to evaluate the probability $\mathcal{P}_m$ that m photons are emitted in the detection interval t_d. The interval is chosen so large that $m \gg 1$. Further we assume that the average intensity of the light is not time dependent. The light source sends out a stream of photons, and at all times we measure the same average intensity, the same average number of photons. The result from classical statistics for many independent sources is

$$\mathcal{P}_m(t_d) = \frac{\langle m \rangle^m}{m!} \exp(-\langle m \rangle) \quad \text{with} \quad \langle m \rangle = \frac{t_d \langle I \rangle}{h\nu} \tag{3.3.5}$$

which is known as the Poissonian distribution. One important property of this distribution is $\Delta m^2 = \langle m \rangle$, which states that the width of the distribution, measured by the variance, is equal to the mean value of the distribution. The probability distribution in Equation (3.3.5) describes the fluctuations of the intensity emitted by a thermal source.

The nature of these fluctuations is such that they are independent of the time intervals from which the photon number data is collected. These time intervals can be close and regularly spaced or widely separated and intermittent, the results will be the same provided a sufficiently large sample is taken. Noise which behaves in this way is known as stationary noise. As a consequence $g^{(1)}(\tau)$ and $g^{(2)}(\tau)$ are independent of τ. In particular, the noise spectrum $I(\Omega)$ (averaged over a small frequency range $\delta\Omega$ about Ω) is independent of the detection frequency Ω, i.e. all frequency components carry the same intensity, ($I(\Omega)$ is flat). This type of noise is referred to as *white noise*. The value of the variance is a fixed for a given average intensity and depends only on the length of the detection interval,

$$\text{Var}(I(\Omega)) = \frac{\langle I \rangle h\nu}{t_d} \tag{3.3.6}$$

This statistical result can be applied to any other situation where events generated by independent processes are counted and the average number of such events is not dependent on time. Examples are the measurement of particles from radioactive decay or the number of raindrops, particles of rain, in a steady downpour. The spectrum of the noise for all these cases is equivalent to the sound of a heavy rain shower on a metal roof. It was this analogy, namely the dropping of lead shot on a metal surface, that was the origin for the name of this statistical phenomenon: *shot noise*.

The classical description provides the tools for describing the statistical properties of light. Chapter 4 gives a complete quantum mechanical description. The classical description can serve as a reference and the interesting question will be: How do the classical predictions compare with the more complete results from quantum theory? We will find that the predictions for thermal light, that is light emitted by many independent atoms, are identical in both the classical and the quantum models. The lesson here is that it is sometimes sufficient to quantize the emitting (and absorbing) atoms, while still retaining a classical description of the light (this is often referred to as a semi-classical description). In contrast, the results from a quantum mechanical description of the laser will be quite different from the results for a classical oscillator. A detailed comparison between the two models will show up several features which are uniquely due to the quantum nature of the light.

3.4 Interference experiments

The measurement of interference, made accessible through the use of instruments such as a Michelson interferometer, clearly demonstrates the wave nature and the coherence properties of light. However, the same type of interference experiments carried out at very low intensities also document some of the quantum properties. This was done in very early experiments by Taylor [Tay09]. These experiments used photographic plates and extremely long exposure times – up to three months. The experiments showed that the visibility of the fringes remained constant, at a given total exposure, independent of how long it took and thus how many photons could have been in the instrument at any one time. It should be noted that all these interference experiments are most impressive achievements. The interferometers are very sensitive devices and nothing would have been easier than wiping out the interference fringes through a small perturbation of the apparatus.

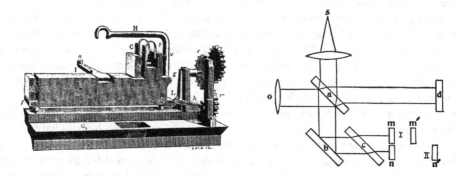

Figure 3.2: Historical example: Interferometer at the turn of the century (from the notes of A. Michelson [Mic27]).

More recent experiments, [Jan58] [Har93], using photoelectric detection, confirmed these results. In addition, some of these experiments used interferometers with total optical paths far exceeding the coherence length of the light source. For example in the experiment by Janossy each arm of the Michelson interferometer was 14.5 m long, requiring a very quiet and temperature stable ($\Delta T < 0.001$ [K]) environment which was achieved in an underground tunnel. Again, the visibility remained unchanged – independent of the actual intensity. Complementary results were obtained with the synchrotron radiation emitted by a storage ring. Here the number of electrons generating the radiation can be determined separately from the intensity of the radiation – and again the visibility was shown to be independent of the photon number.

All these experiments support the statement by P. Dirac:

> *"The new theory, which connects the wave function with probabilities for one photon, gets over the difficulty by making each photon go partly into each of the two components. Each photon then interferes only with itself. Interference between two different photons never occurs"*. [Dir58]

The beamsplitter in the interferometer does not divide the light into smaller parcels – any detection of the light shows quanta of the same amount of energy. This result cannot be derived if we think of photons as small local parcels of energy – miniature billiard balls – with the properties of classical particles. It reveals one of the consequences of quantum mechanics, namely that a conclusion about the specific outcome of an experiment cannot be drawn before the experiment is carried out. This consequence of quantum mechanics becomes very clear by considering the various possible outcomes of an interference experiment where detectors and counters are used in an attempt to measure the presence of the photon in the one or the other of the interferometer arms. Any such attempt would fail, as it has been so elegantly expressed by Feynman in his lecture notes [Fey63]. The duality of the wave-particle model is a clear conclusion from these experiments: light is neither exclusively a wave nor exclusively a stream of particles. The type and condition of the experiment determines which one of these models will dominate, which one will give the simpler interpretation. All the time light is something more inclusive, more complicated than either of these models would suggest.

The statement that *"a photon will interfere with itself"* should not be taken too literally in the sense that it could be tested through the observation of individual events. In the case of the photographic recording each detection event produces one clear dark dot on the photographic plate. This dot could be part of an interference pattern or of a random distribution of dots. It is the distribution of many dots – a frequent repeat of the same measurement – that reveals the result of the experiment. Theoretically, one single dot measured exactly at the location on the plate where there should be absolutely no intensity according to the calculation would have an enormous significance. Such an event would prove the calculation wrong. But we have to note that in reality there will always be some other imperfection, such as an imbalance of the beamsplitter, finite coherence length or background noise, which could have resulted in this particular, single event. It is the distribution of the dots, or of the counts in the electronic instrument, that is crucial for the test of a model. As is common for all tests of quantum mechanics, a large number of individual experiments is required to allow a valid conclusion.

The nature of the light seems to depend on the type of experiment: A single beamsplitter can often be well described by the photon picture, a complete interferometer requires the wave picture. An interesting extension of the demonstration described above are the so called "delayed choice experiments". Here we are trying to trick the light into the wrong behaviour by delaying the decision of which type of experiment is being performed until it has already passed the first beamsplitter. Will the photon be able to predict the type of experiment that is lying ahead and react in the appropriate way? While this view is certainly naive, it poses a valid question: Does it matter whether we have a delay built into the interferometer? One example for such an experiment [Hel87], shown in Fig. 3.3, uses pulses of light of short

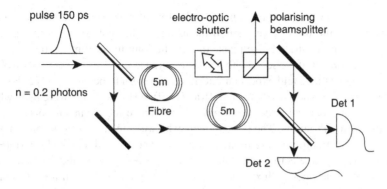

Figure 3.3: Schematics of a modern delayed choice experiment,after Hellmuth et al. [Hel87].

duration (150 [ps]). The intensity was chosen sufficiently low that on average only every fifth pulse contained one photon.

Both arms of the interferometer contain 5 m of single mode fibre. One arm contained an electro-optical switch which allowed this arm to be blocked and the interference to be turned off through an electronic signal. In order to put the photons in the above mentioned dilemma, the interferometer was operated such that the switch was operated after the pulse had passed through the first beamsplitter. However, the delay had no effect on the visibility of the interference pattern. The outcome of the experiment depended only on the configuration of the experiment which was present at the time of detection, independent of the earlier state of the apparatus. A more thorough discussion of these types of experiments is given in Chapter 12.

This raises the question as to whether photons from two completely independent sources can interfere with each other. It is well known that light from different sources, or from different parts of the same source, does not interfere when observed for a long time. This is a consequence of the fact that the phase of the electro-magnetic wave emitted by the two sources has two independent phase terms. In a normal interferometer, which has wavefront splitting mirrors, this does not matter as long as the path length difference is less than the coherence length. Similarly, for independent sources the fringes should also appear if the measurement takes less time than the coherence time t_{coh}. During such a short time interval the phase will not vary enough to reduce the visibility. This has been shown in an experiment with two independent lasers [Mag63], in this early experiment two ruby lasers. Spatial interference patterns from single pulses of the lasers were recorded, clearly showing the interference between photons from different sources.

An alternative is to measure the beat between two lasers, this was first demonstrated for two independent He-Ne lasers [Jav62]. The experiment operated continuously and the narrow line widths of the gas lasers made the observations straight forward. The major problem was the drifts in the laser frequency due to mechanical and temperature variations. Once this was overcome this experiment demonstrated clearly that interference exists between beams from independent sources provided that the detection time t_1 is less than the coherence time t_{coh}. With modern technology, such as tunable diode lasers and radiowave spectrum analysers, this type of experiment is almost trivial. The effect is common knowledge for a laser physicist.

However, the interesting question remains: Should this interference effect between independent lasers persist in the limit of very small intensities? The first answer might be *NO*, since we cannot be certain of the relative phase during the long measurement time required for the weak beams. However, if we select only data measured at times when the relative phase is within a set small range one could expect an interference pattern to become visible. Exactly this experiment was carried out by Radloff et al. [Rad71]. Two He-Ne lasers were used which had been built in one mechanical structure in order to minimise the technical noise and the phase fluctuations due to vibrations. The beat frequency between the beams was continuously monitored. A very small fraction of the light was selected and formed a spatial interference pattern. This was gated by an electro optic shutter and only opened when the frequency was within a preset range. The photon flux was as low as 10^5 [photons/s] with a typical shutter time of 10^{-5} [s]. After exposures of about 30 minutes clearly recognizable interference patterns became visible. These patterns were present even at arbitrarily low intensities. The particle picture is not at all suitable for this situation. It is extremely unlikely that one photon from each laser will pass together through the shutter. The wave model on the other hand provides a very simple explanation. The only problem arises when we want to understand the quantization of the energy together with interference.

These results for the interference of photons from independent sources are in some contradiction to the complete statement by Dirac who said *"Each photon interferes with itself. Interference between two different photons does not occur."* The context shows that Dirac made this statement while considering only interference experiments known at his time. It should not be applied to the experiments involving independent sources. Nevertheless, attempts have been made to rescue Dirac's statement by invoking additional processes that guarantee that photons are emitted simultaneously by both lasers. But these attempts are really not all that helpful in understanding the quantum nature of light.

In summary, we can say that the only conditions under which interference can be observed are those where it is impossible to detect which path the photon has taken. Any attempt to measure the path, or even to reconstruct the path after the event, would wipe out the interference pattern. A discussion of these types of experiments has led to the postulate of Heisenberg's uncertainty principle. For this particular case the principle states that $\Delta x \Delta p \geq h/4\pi$. It enforces that we can either measure the interference pattern, by determining the position x of the photon, or alternatively the path of the photon, as measured by its momentum p. But we cannot measure both simultaneously [Pau95]. Discussion about these results have played a prominent role in the development of quantum mechanics as a whole.

3.5   Modelling single photon experiments

We now turn to a more idealized description of single photon experiments in order to see how they may be modelled. We will introduce a number of basic optical elements as "black boxes"; describing and contrasting their operation for classical light waves and single photons. Descriptions of how these optical elements can be constructed will be found in later chapters. Similarly, the simple quantum mechanical models introduced here will also later be expanded.

The average number of photons per second in a beam of light is proportional to its intensity. However, the quantization of the field introduces a degree of freedom not present in the classical theory, the photon statistics. Suppose we had a "photon gun", that is, a light source which emits just one photon when we pull the trigger. Such a photon gun can be produced through pulsed excitation of a single molecule, though presently not conveniently. Such a source would share some characteristics with the highly attenuated sources described in the previous section. However, whilst an attenuated laser may produce one photon on average in some time interval, there will be fluctuations: sometimes there will be two or more photons, sometimes there will be none. On the other hand, we will always detect just one photon in the measurement interval coming from our photon gun. There is no way to describe this difference in the photon statistics with a classical description of light. We must introduce a new concept, that of the *quantum state* of the light. The quantum state describes all that is worth knowing about a light beam. The photon gun produces light in a single photon *number state* which we will write as $|1\rangle$. Studying the behaviour of such a light state in simple situations illustrates some of the basics of the quantum description of light.

3.5.1 Polarization of a single photon

Let us first consider how a single polarized photon behaves. We will describe this behaviour using single photon polarization states. We will see how it is possible to build a consistent description of light phenomena in spite of wave-particle duality. To do so, however, we will need to introduce some quite different concepts to those found in classical theory.

Recall that in our classical description the polarization of the field was described by the vector $\mathbf{p}(t)$ which gave the direction of the electric field tangential to the wavefront. For linearly polarized light $\mathbf{p}$ will be fixed. For circularly polarized light $\mathbf{p}$ will rotate at the optical frequency. It is possible to design optics which will decompose a light field into two orthogonal modes. The amplitudes of the two modes will be equal to the projection of $\mathbf{p}$ onto the chosen axes. For example, a horizontal/vertical polarizing beamsplitter will project the field onto a horizontally polarized mode and a vertically polarized mode. These two modes will exit through different ports of the beamsplitter. Similarly, one can use a diagonal/anti-diagonal polarizing beamsplitter to decompose the field into linearly polarized fields at 45 degrees and -45 degrees away from horizontal. With slightly more effort it also possible to create a left/right circular polarizing beamsplitter which decomposes the field into clockwise and anti-clockwise rotating circular polarizations.

Suppose we fire a string of single photons at a horizontal/vertical polarizing beam-splitter. As we have noted, classically such a device will direct horizontally polarized light in one direction and vertically polarized light in the other. Thus we will define photons that exit through the horizontal port of this beamsplitter as being in the horizontal state,

$$|H\rangle \tag{3.5.1}$$

Similarly we define photons that exit through the vertical port as being in the vertical state,

$$|V\rangle \tag{3.5.2}$$

By "exit through the horizontal port" we mean that a photon detector placed at the output of the horizontal port will detect a photon, or equally a photon detector placed at the vertical port does not detect a photon. These situations are illustrated in Fig. (3.4). The symbol for

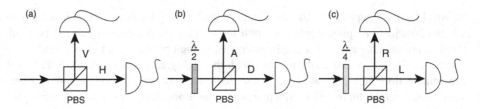

Figure 3.4: Operational definition of different polarisation states. PBS is a polarizing beamsplitter, $\lambda/2$ is a half-wave plate, $\lambda/4$ is a quarter-wave plate. (a) horizontal-vertical polarization, (b) diagonal polarization, and (c) left-right circular polarization.

the state, $|-\rangle$, is referred to as a *ket*. These definitions are sensible because a photon which exits through the horizontal port of the beamsplitter and is passed through another polarizing beamsplitter will certainly exit through the horizontal port. A similar definition applies to vertical photons. The diagonal single photon state, $|D\rangle$, and the anti-diagonal state, $|A\rangle$ can be defined in a similar way by analysing the beams with a diagonal/anti-diagonal polarizing beamsplitter. So far this is straightforward. The physics of classical polarized beams and single photons looks the same. Things become more interesting when we start to mix up the polarizations.

If we send a diagonally polarized classical beam of light into a horizontal/vertical polarizing beamsplitter then half the beam will exit through the horizontal port and half will exit through the vertical port. What will happen if we send single photons in the state $|D\rangle$ through this beamsplitter? As we have discussed in the previous sections the photons cannot be divided, in the sense that a detector placed at one of the output ports will either detect a whole photon or no photon. Instead they must go one way or the other. To be consistent with the classical result for many photons it must be that they go one way 50% of the time and the other way the other 50% of the time. But can we tell in which direction an individual photon will go? To answer this first remember how $|D\rangle$ photons behave at a diagonal/anti-diagonal polarizing beamsplitter: they all exit through the diagonal port. That is they all behave perfectly predictably and identically. Yet when these identically behaving photons are sent into a horizontal/vertical beamsplitter we have argued some must take one path and others take the other. Thus the path an individual photon in the state $|D\rangle$ takes through the horizontal/vertical beamsplitter is not specified. All that is specified is that on average half the photons will go one way and half will go the other.

This example illustrates the fact that in quantum mechanics it is probabilities of outcomes, not particular outcomes, that are predicted. Even though we can say with certainty that a photon in the state $|D\rangle$ will emerge from the diagonal port of a diagonal/anti-diagonal beamsplitter, its reaction to a horizontal/vertical beamsplitter is as random as a flip of a coin. It is tempting to think that maybe there are other variables (perhaps hidden to us) that do determine in a precise way the polarization behaviour of individual photons under all conditions. However such a possibility can be ruled out experimentally, see Chapter 12. We are forced to accept the intrinsic indeterminacy of the quantum world.

Continuing our discussion of polarization states, we can also introduce the right circular single photon polarization state, $|R\rangle$, and the left circular state, $|L\rangle$, in an analogous way to the other states. A photon in state $|R\rangle$ will randomly take one port or the other when sent into either a horizontal/vertical or a diagonal/anti-diagonal beamsplitter. A photon in state $|L\rangle$ will behave in the same way. Similarly a photon in state $|H\rangle$ will give a random result for both diagonal/anti-diagonal and right/left circular polarizing beamsplitters and so on for the other states. We will now outline the mathematical formalism for treating these situations.

3.5.2 Some mathematics

The state kets we have introduced are vectors. They live in a vector space called *Hilbert space*. Hilbert space is a complex vector space: when we add vectors we are allowed to do so using complex coefficients. The dimensions of the vector space are equal to the number of degrees of freedom of the property being described. In the polarization example our vector space is discrete and has dimension 2. That is in all the experiments we have described there are only two possible outcomes; the photon comes out one port or the other. It can only take two discrete values. This is in stark contrast to the classical case in which the amount of the field which comes out of one or other port can vary continuously between all and nothing. In general, discrete systems can have any number of dimensions and continuous systems can also be described. A continuous variable can give any outcome in some range and their vector space will have infinite dimensions. We will look at such examples in optics in the next chapter.

As for other vector spaces we can reach any vector in our space through linear superpositions of n orthogonal basis kets, where n is the dimension of the system. To define orthogonality we introduce *bra* vectors which are written $\langle x|$. Bras and kets are related by the hermitian conjugate, an operation similar to complex conjugation and indicated by the "dagger" symbol, †. Thus we can write

$$z^* \langle x| = (z|x\rangle)^\dagger. \tag{3.5.3}$$

where z is just a complex number. Two vectors $|x\rangle$ and $|y\rangle$ are orthogonal if

$$\langle x| \times |y\rangle \equiv \langle x|y\rangle = 0. \tag{3.5.4}$$

In general the object $\langle x|y\rangle$ is called a *bracket* and is similar to the dot product in real vector spaces. A ket is said to be normalized if $\langle x|x\rangle = 1$. We will generally assume this property.

Simple physical questions can be asked in the following way. If we prepare our system in the state $|\psi\rangle$ and then use an analyser to ask the question: Is my system in the state $|\sigma\rangle$? The probability, P, that we get the answer "yes" is given by

$$P = |\langle \psi|\sigma\rangle|^2. \tag{3.5.5}$$

The bracket $\langle \psi|\sigma\rangle$ is referred to as the *probability amplitude* and will in general be a complex number. The orthogonality condition then says that if we prepare our system in $|\sigma\rangle$ our analyser will never find it in $|\psi\rangle$ if theses two states are orthogonal. Normalization reflects the fact that if we prepare our state in $|\psi\rangle$ we will certainly ($P = 1$) find it there with our analyser.

We will try in this book to minimize the mathematics by introducing concepts on a need to use basis. However, a deep understanding of the workings of quantum optics requires a good understanding of the mathematical structure of quantum mechanics. Readers unfamiliar with the basics of quantum mechanics are encouraged to study other texts such as references [Mey89] and [Sak85].

3.5.3 Polarization states

The concepts introduced above allow us to identify our polarization states $|H\rangle$ and $|V\rangle$ as orthogonal states. That is $\langle H|V\rangle = \langle V|H\rangle = 0$ which means that a photon in state $|H\rangle$ never comes out the vertical port and vice versa. We will assume also that our states are normalized such that $\langle H|H\rangle = \langle V|V\rangle = 1$. This means that we should be able to express all our polarisation states as linear superpositions of the horizontal and vertical states. Consider first the diagonal states. From our discussion in Section 3.5.1 we know that they must have the following properties:

$$
\begin{aligned}
\langle D|A\rangle &= \langle A|D\rangle = 0 \\
|\langle D|H\rangle|^2 &= |\langle D|V\rangle|^2 = .5 \\
|\langle A|H\rangle|^2 &= |\langle A|V\rangle|^2 = .5
\end{aligned}
\tag{3.5.6}
$$

or the diagonal states are orthogonal and a diagonal state placed through a horizontal/vertical polarizing beamsplitter has a 50% probability of exiting through either port. It is straightforward to check that

$$
\begin{aligned}
|D\rangle &= \frac{1}{\sqrt{2}}(|H\rangle + |V\rangle) \\
|A\rangle &= \frac{1}{\sqrt{2}}(|H\rangle - |V\rangle)
\end{aligned}
\tag{3.5.7}
$$

satisfy the conditions of Equations (3.5.6). Now consider the circularly polarized states. We need to satisfy

$$
\langle R|L\rangle = \langle L|R\rangle = 0
$$

along with

$$
\begin{aligned}
|\langle R|H\rangle|^2 &= |\langle R|V\rangle|^2 = .5 \\
|\langle L|H\rangle|^2 &= |\langle L|V\rangle|^2 = .5 \\
&\quad\quad and \\
|\langle R|D\rangle|^2 &= |\langle R|A\rangle|^2 = .5 \\
|\langle L|D\rangle|^2 &= |\langle L|A\rangle|^2 = .5
\end{aligned}
\tag{3.5.8}
$$

This looks impossible until we remember that Hilbert space is a *complex* vector space. Thus we are allowed to form the complex coefficient superpositions

$$
\begin{aligned}
|R\rangle &= \frac{1}{\sqrt{2}}(|H\rangle + i|V\rangle) \\
|L\rangle &= \frac{1}{\sqrt{2}}(|H\rangle - i|V\rangle)
\end{aligned}
\tag{3.5.9}
$$

which satisfy the above requirements. It follows that an arbitrary polarization state can be represented by

$$|\omega\rangle = x|H\rangle + e^{i\phi}y|V\rangle \qquad (3.5.10)$$

where $x^2 + y^2 = 1$ and ϕ, x and y are real numbers (absolute phase factors have no physical significance). Notice that such a representation is not unique. We could equally well write all our states as superpositions of diagonal and anti-diagonal states or circular states. Any two orthogonal states can span the Hilbert space.

We have seen that in order to describe the behaviour of a quantized light field we have had to introduce two new concepts: (i) the state of the light, and (ii) that predictions are made of probabilities of events, not actual events. The fact that these concepts lead us to represent, for example, a diagonal photon as an equal superposition of horizontal and vertical photon states may seem relatively benign at this point. That such a superposition has no analogue for classical particles will be demonstrated more graphically in the next section. In Chapter 12 and 13 we will discuss in detail photon experiments which use polarization as their degree of freedom. In particular, we will investigate the unique communications and information processing abilities these quantum states can exhibit.

3.5.4 The single photon interferometer

We will now describe a simple single photon interference experiment using a Mach-Zehnder interferometer, similar in concept to the experiments described in Section 3.4. A more general description of this device can be found in Section 5.2. Here we will illustrate the basic physics of the arrangement in the extreme situation of only a single photon present in the interferometer at a time, and how the mathematics introduced in the previous sections can describe it.

A Mach-Zehnder interferometer consists of a 50/50 beamsplitter, which divides the input light into two equal parts, followed by a series of mirrors which direct the two beams back together again on a second 50/50 beamsplitter which recombines them. The path length difference between the two arms of the interferometer can be adjusted. This then corresponds to the situation discussed in Section 5.2 and thus classically we expect the amount of light exiting one port to change as a function of the path length difference. In particular if our path length difference remains well within the coherence length we will be able to pick a path length such that all the light exits from a single port. This set-up is depicted schematically in Fig. 3.5.

Let us now describe the behaviour of this interferometer when single photons pass through it. We will assume that the path length difference is held in the situation described above such that for classical beams all the light enters and exits through a single port. We will also assume our beamsplitters and mirrors are polarization independent so that the polarization of our single photons is irrelevant and we can simply label a single photon in a beam by the number state ket $|1\rangle$. There are however two ways in which the photon can enter the interferometer: via the beam path labelled a in Fig. (3.5) or by the beam path labelled b. Suppose we in fact send our photon in by beam path a. We can describe this initial situation by the state ket

$$|\sigma\rangle = |1\rangle_a|0\rangle_b. \qquad (3.5.11)$$

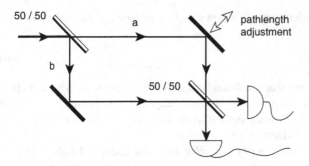

Figure 3.5: Schematic of single photon interferometer

meaning there is one photon in beam a and no photon in beam b. There is an even probability
of the photon exiting from either port of the first beamsplitter. The output state from the
beamsplitter can thus be written

$$|\sigma'\rangle = \frac{1}{\sqrt{2}}(|1\rangle_a|0\rangle_b + |0\rangle_a|1\rangle_b) \tag{3.5.12}$$

The path length condition means that, apart from an overall phase, this state remains the same
as it propagates through the interferometer. The time symmetric nature of light propagation
through a beamsplitter means that when the beams reach the second beamsplitter they must
return to their original state of a photon in a and no photon in b, see Eq. (3.5.11). Thus we just
get the photon equivalent of the classical result, the photons always exit through a single port.

Now let us consider the physical significance of our description. In particular, what does
Equation (3.5.12) mean? It says that the photon is in a superposition of being in each path.
It is tempting to think that this just means the photon is in one or the other path with a 50%
probability. Certainly the probability of finding the photon in path a

$$P_a = |\langle 1|\sigma'\rangle_a|^2 = .5$$

is 50%. But is the interference effect, i.e. the fact that the photons always exit from the a
port (see Fig. (3.5)), consistent with the photons taking one path or the other? Well, the
second beamsplitter is identical to the first one so our first puzzle is, why should the photon
only exit through one port? We might imagine that it picks up some kind of "label" at the
first beamsplitter that tells it what to do at the second. So suppose we find our photon in
beam a in the middle of the interferometer, but then we "release" it and let it finish its journey
to the second beamsplitter. Experimentally we find that now there is no interference, the
photon is just as likely to come out from either of the output ports. Quantum mechanically,
because we now know that the photon is in beam a, the state inside the interferometer becomes
$|\sigma'\rangle_m = |1\rangle_a|0\rangle_b$ and so the photon just behaves as it did at the first beamsplitter and has a
50/50 chance of emerging from either port.

We might argue that when we "caught" the photon it forgot its label and so just behaved
randomly at the second beamsplitter. So consider another possibility. Suppose we look for the
photon in beam b but we *do not* find it. Now quantum mechanically the situation is the same

as above; we know that the photon must be in beam a so the state again becomes $|\sigma'\rangle_m = |1\rangle_a|0\rangle_b$ and the interference should disappear. But any common sense description would say this cannot be. We have not interacted with the photon, so any label it has must be unaffected. Thus we would expect the interference to still appear. Experiments, however, confirm the quantum mechanical result, see Chapter 12. We are forced to conclude that the superposition state of Equation (3.5.12) represents in some objective sense the presence of the photon in both paths of the interferometer. Any experiment which causes the photon to manifest itself in one or the other path, whether an actual detection occurs or not, destroys this superposition and in turn wipes out the interference. This is wave particle duality and we will encounter it in many guises throughout this book.

3.6   Intensity correlation, bunching, anti-bunching

The quantum properties of light can be investigated using the techniques of photon counting. We will find that the statistics of photons are considerably different from the statistics of independent particles. Again, we have to combine the particle picture with the wave picture. For these statistical experiments it is necessary to measure quantitatively the correlation between different beams of light. This approach has already been discussed in Section 2.2.2. In particular the discussion of the Hanbury Brown & Twiss experiment for thermal light, in their case starlight, showed a way for determining the correlation between two fluctuating currents $i_a(t)$ and $i_b(t)$ by recording and analysing the product $i_a(t)i_b(t)$ of these currents. This technique can be used for any type of light, including laser light. In principle this technique also applies to any intensity of light.

At very low intensities the photo-detector becomes a photon counter. For a photon counter the probability of a count is proportional to the intensity of the light. Now the number $n(t, t_{\text{dec}})$ of counts recorded at time t within a time interval t_{dec} represents the intensity of the light. The detection time interval t_{dec} should be shorter than the coherence time t_{coh} of the light source, otherwise the fluctuations are simply averaged out. As before we determine the intensity correlation by using two detectors Det_a and Det_b, measuring the photon counts $n_a(t, t_{\text{dec}})$ and $n_b(t, t_{\text{dec}})$ independently and then forming the product $n_a(t, t_{\text{dec}})n_b(t, t_{\text{dec}})$.

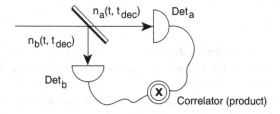

Figure 3.6: Schematic layout of an optical intensity correlator.

Actually medium intensities, ranging from a few to several tens of counts per detection interval, are very challenging and no particularly satisfactory technology for this range has yet emerged. On the other hand, at very low intensities, where the probability of more than a single count per detection interval can be neglected, good and reliable technology exists, see Chapter 12. A moment's thought shows that the probability of coincidences, that means events where both detectors register a photon within a short time interval, is proportional to the value of $n_a(t, t_{dec})n_b(t, t_{dec})$. It is technically very attractive to build a circuit that only responds to such coincidences. This technique shows an increase of the coincidence rate, under the same conditions where big correlations are measured. Going back to the Hanbury Brown & Twiss experiment, the coincidence rate is high if the separation between the detectors is sufficiently low. This is an excess of coincidences – compared to the numbers one would expect from completely random counting events. In other words, whenever a photon has reached detector Det_a the probability for another photon to be at Det_b is enhanced. This could not be understood by regarding photons naively as independent particles, thereby ignoring the wave properties of the light. The photons emitted from the different part of the source would not know of each other. Their arrival time would not be correlated.

In contrast, the wave picture predicts that the instantaneous intensity on the detector is influenced by all atoms in the source. Each part of the source is making a small contribution. Thus both detectors see related contributions where light is originated from one atom in the source. Again, this means we cannot consider individual photons, each with one well defined birthplace. Atoms cannot be localized in that way.

These observations are concerned with spatial effects. Intensity information can also be gained from temporal effects within a single beam. To measure the time correlation of a beam of light, we can use a trick which was used by Hanbury-Brown & Twiss as a precursor to the stellar observations. We can use a beamsplitter to generate two light beams with each one being detected individually, as shown in Fig. 3.6. The coincidence rate can now be directly measured using detectors which are gated. Only events which happen within the gate time at both detectors are recorded. By moving one of the detectors either further away or closer to the beamsplitter we introduce a delay time τ and can determine the coincidence rate as a function of the delay time. Alternatively, we can leave the detectors stationary and use electronic delays. For thermal light the coincidence rate for $\tau < t_{coh}$ should be higher than for $\tau > t_{coh}$. This is a direct analogy to the properties of the second order correlation coefficient. It corresponds to

$$g^{(2)}(\tau < t_{coh}) \quad > \quad g^{(2)}(\tau > t_{coh}) \tag{3.6.1}$$

as shown in Equation (2.2.9). Since the coincidence rate for $\tau \gg t_{coh}$ can only be those events which are completely random, we simply have to look for excess coincidences. The long delay time measurements serve as a normalization.

One practical difficulty is that the detection time t_{det} has to be less than the coherence time. For normal thermal light the linewidth is very large, the coherence time is extremely short, and the response of the detectors is too slow. A direct measurement of the coincidences for thermal light is not possible. In contrast, laser light has a small linewidth, long coherence time, and the experiment is feasible. One clever experimental trick is to simulate thermal light through scattering of laser light on rapidly and randomly moving objects. This was first done by Arrecchi and collaborators [Arr66]. The experiments used light from a gas laser with long coherence time, which was scattered by polystyrene particles of various sizes suspended in water and detected by two detectors. The coincidence rate, or second order correlation coefficient, was measured. For comparison the laser light itself was investigated and the results show a clear excess of coincidences. The shape of $g^{(2)}(\tau)$ fits properly to the prediction of a thermal light source with a Gaussian linewidth and the difference between thermal light and coherent laser light is clearly demonstrated.

An alternative interpretation of these results is to say that the photons from the thermal light arrive in bunches, the probability for photons to arrive together, in coincidence, is large for short delay time intervals. This phenomenon is known as *photon bunching*. It is typical for thermal light which has naturally large fluctuations. In contrast, for coherent light there is no preferred time interval between photons. Coherent light is not bunched. The coincidences show random, Poissonian statistics on all time scales. Can there be anti-bunched light, where the photons keep a distance, in time, to each other? This would be light that clearly violates the classical conditions. The second order correlation function would be less than one, which is not possible for classical statistics, as already shown in Section 2.2. A measurement of anti-bunched light would be a clear demonstration of quantum behaviour, of non-classical light. Well in fact the single photon quantum state we discussed in the last section would be just such a state. Recall from Eq. (3.5.12) that such a beam will send a photon in one or other direction at a beamsplitter. Thus at zero time delay it is impossible to record a coincidence from such a state in the set-up of Fig. 3.6. Hence $g^{(2)}(0) = 0$. A pioneering experiment by Kimble, Dagenais & Mandel succeeded in 1977 in demonstrating the generation of non-classical light [Kim77]. They used the fluorescence of single atoms to generate light with the statistical properties of anti-bunching. The experiment was carried out with Na atoms which were first prepared into a two level system and then excited by two resonant CW laser beams, as shown in the schematic layout in Fig. 3.7(a). The fluorescence light from a very small volume, containing on average not more than a single atom, was imaged through a 50/50 beamsplitter on to two photo-detectors, D_1 and D_2. Upon arrival of a photon the detectors generated pulses which were processed, via counting logic, and produced the probability function $\mathcal{P}(t, t + \tau)$ for the arrival of photons at different detectors with a time delay τ. This information was converted into the second order correlation function $g^{(2)}(0)$, see Section 2.2.

The excitation of individual atoms is determined by the quantum mechanical evolution of the atomic wavefunction. After the single atom has been excited by a laser beam it will emit one photon, let's say at time $t = 0$. The atom cannot immediately emit another photon, it requires time for re-excitation. Consequently the probability for the generation of pairs of photons with small time separation is very small. The fluorescence light has a value of $g^{(2)}(0)$ close to zero, as shown in Fig. 3.7(b), which is not possible for classical light. This light clearly shows quantum properties – it is *anti-bunched*. Since this first demonstration several experiments, for example reported by J. Cresser et al. [Cre82], J.G. Walker & E. Jakeman [Wal85] and P. Grangier et al. [Gra86], have been able to reveal the properties of similar quantum systems in more and more detail.

By placing the atoms inside a resonator further interesting properties of the atom-light system can be studied. This idea led to a series of beautiful experiments with the micromaser built at Garching in the group around H. Walther. In this system so called Rydberg atoms, that is atoms in excited states with very high quantum numbers (n > 40) and which have very

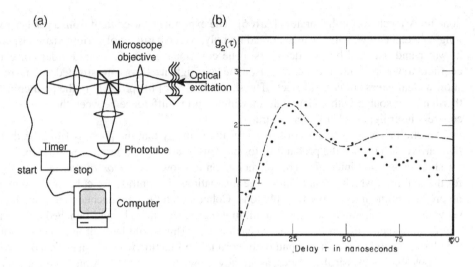

Figure 3.7: The experiment by Kimble, Dagenais and Mandel [Kim77]. (a) Schematic outline, (b) Measured values for $g^{(2)}(\tau)$ derived from the data. The broken curve shows the theoretically expected form for a single atom, arbitrarily normalized to the same peak.

simple, long lived sates, are placed inside a microwave cavity of extremely high quality. The atoms couple to the microwave and the coupled systems evolve in an elegant way determined by quantum mechanics. By observing the atoms that leave from the resonator conclusions can be drawn about the state of the light-atom field inside. These systems can show features such as the collapse and revival of the wavefunction of the atom, as reviewed for example by P. Meystre & M. Sargent III [Mey89]. We will come back to these experiments when we discuss in Chapter 12 different ways of generating strong nonclassical states of light.

3.7   Single photon Rabi frequencies

Similar ideas have recently led to a different, very direct demonstration of the quantum nature of light carried out by groups in Paris [Bru96] and Müenchen [Rem87]. These authors point out that the most direct evidence of the quantization of the radiation field, namely the discrete nature of the photo-current, could be explained by a classical description of the radiation field provided that the detector is linear and a quantum system. They set out to find a more direct proof for the discreteness of the field and analysed the light interacting with atoms contained inside a cavity. The atom can be approximated as a system with just two energy levels, a two level system. The coupling of this system with a single mode of the radiation field leads to a dynamical evolution of the atomic excitation in the form of nutations at the Rabi frequency. This simple model for the interaction is referred to as the Jaynes-Cummings model [Jay63], is well described in the literature [Coh77] and extensively used in coherent optics with light or microwaves (NMR). In a strong coherent field, that means a beam with many photons, the coupling leads to a nutation at a frequency associated with the mean photon number n_α, which is a continuously variable number. However, for a weak field ($n_\alpha \approx 1$) the Rabi frequencies should be discrete, linked directly to the photon numbers in the field.

Early experiments with Rydberg atoms by G. Rempe et al. [Rem87] in a micromaser built in Garching have revealed an oscillation of the atomic population in a thermal field with ($1.5 < n_\alpha < 3.8$) and in the micromaser field. Similarly, the series of experiments in Paris; led by S. Haroche [Bru96] has achieved extreme conditions. Here the Rydberg atoms traverse a cavity where both atom-cavity interaction time and cavity damping time are very long. This requires a cryogenic cavity ($T = 0.8$ [K]) of very high quality ($Q = 710^7$) and a very stable source of slow atoms. The average delay between successive atoms was set to 2.5 [ms], much longer than the cavity decay time. In the cavity the radiation field and the atoms are coupled together, resulting in a time evolution that is dominated by oscillations at the Rabi frequency and this determines the state of the atom at the end of its travel through the field in the resonator. After leaving the cavity the atoms were detected by state selective field ionization. From these data, accumulated for many atoms, the dynamics of the atom-cavity evolution was reconstructed, which showed a mixture of oscillations in time. Taking the Fourier transform of the traces at low average photon numbers the experiments clearly showed discrete frequency components which correspond to the discrete Rabi oscillations associated with the photon numbers $n = 1, 2, 3, 4$, etc.. The discreteness of the radiation field is directly evident.

Summary

There is compelling experimental evidence that photons exist and that the classical wave picture has to be extended. Some of these observations are linked directly to the detection process, while others show in detail the quantum nature of light, as we formally introduce it in the next Chapter 4. Optical experiments are now used to test and to create new applications based on quantum ideas.

Further reading

Photonen, eine Einführung in die Quantenoptik, H. Paul, Teubner, Stuttgart (1995)
Optical coherence and quantum optics, L. Mandel, E. Wolf, Cambridge (1995)
Elements of quantum optics, P. Meystre, M. SargentIII, Springer (1989)
Quantum interferometry, F. De Martini, G. Denardo, Y. Shih, VCH Verlag (1996)

Bibliography

[Arr66]   Measurement of time evolution of the radiation field by joint photo-current distributions, F.T. Arrecchi, A. Berne, A. Sona, Phys. Rev. Lett. 17, 260 (1966)

[Asp82]   Experimental test of Bell's inequalities using time-varying analysers, A. Aspect, J. Dalibard, G. Roger, Phys. Rev. Lett. 49, 1804 (1982)

[Bru96]   Quantum Rabi oscillation: a direct test of field quantization in a cavity, M. Brune, F. Schmidt-Kaler, A. Maali, J. Dreyer, E. Hagley, J.M. Raimond, S. Haroche, Phys. Rev. Lett. 76, 1800 (1996)

[Coh77]   *Quantum Mechanics*, C. Cohen-Tannoudj, Bernard Diu, Franck Laloë, Wiley (1980)

[Cre82]   J.D. Cresser, J. Haeger, G. Leuchs, M. Rateike, H. Walther in *Dissipative systems in quantum optics*, ed. R. Bonifacio, 21, Springer (1982)

[Dir58]   *The Principles of Quantum Mechanics*, 4th edition, P.A.M. Dirac, Oxford University Press (1958)

[Ein35]   Can a quantum-mechanical description of physical reality be considered complete?, A. Einstein, B. Podolsky, N. Rosen, Phys. Rev. A 47, 777 (1935)

[Fey63]   *The Feynman Lectures on Physics*, vol. 3, R. Feynman, R.B. Leighton, M. Sands, Addison and Wesley (1963)

[For55]   Photoelectric mixing of incoherent light, A.T. Forrester, R.A. Gudmundsen, P.O. Johnson, Phys. Rev. 99, 1691 (1955)

[Fry96]   A loophole free test of Bell inequalities, E.S. Fry, T. Walther in *Quantum interferometry*, Eds. F. De Martini, G. Denardo, Y. Shih, VCH p. 107 (1996)

[Gla62]   Coherent and incoherent states of the radiation field, R.J. Glauber, Phys. Rev. 131, 2766 (1962)

[Gra86]   Observation of photon antibunching in phase-matched multiatom resonance, P. Grangier, G. Roger, A. Aspect, A. Heidmann, S. Reynaud, Phys. Rev. Lett. 57, 687 (1986)

[Gra02]   Single Photon Quantum Cryptography, A. Beveratos, R. Brouri, T. Gacoin, A. Villing, J-P. Poizat, and P. Grangier, Phys. Rev. Lett. 89, 187901 (2002)

[Hal1886] Elektrometrische Untersuchungen, W. Hallwachs, Annalen der Physik und Chemie, neue Folge Band XXIX (1886)

[Har93]   Interference of independent laser beams at the single-photon level, P. Hariharan, N. Brown, B.C. Sanders, J. Mod. Opt. 40, 113 (1993)

[Hel87]   Delayed-choice experiments in quantum interference, T. Hellmuth, H. Walther, A. Zajonc, W. Schleich, Phys. Rev. A 35, 2532 (1987)

[Jan58]   Investigation into interference phenomena at extremely low light intensities by means of a large Michelson interferometer, L. Janossy, Z. Naray, Nuovo Cimento, Suppl. 9, 588 (1958)

[Jav62]   Frequency characteristics of a continuous wave He-Ne optical maser, A. Javan, E.A. Ballik, W.L. Bond, J. Opt. Soc. Am. 52, 96 (1962)

[Jay63]   Comparison of quantum and semiclassical radiation theories with application to the beam maser, E.T. Jaynes, F.W. Cummings, Proc. IEEE 51, 89 (1963)

[Kie93]   Einstein-Poldosky-Rosen-Bohm experiment using pairs of light quanta produced by Type-II parametric down conversion, T.E. Kiess, Y.H. Shih, A.V. Sergienko, C.O. Alley, Phys. Rev. Lett. 71, 3893 (1993)

[Kim77]   Photon antibunching in resonance fluorescence, H.J. Kimble, M. Dagenais, L. Mandel, Phys. Rev. Lett. 39, 691 (1977)

[Kin95]   *Optical sources, detectors and systems*, R.H. Kingston, Academic Press (1995)

[Kwi95]   New high-intensity source of polarization entangled photon pairs, P.G. Kwiat, K. Mattle, H. Weinfurter, A. Zeilinger, A.V. Sergienko, Y. Shih, Phys. Rev. Lett. 75, 4337 (1995)

[Law27]   The instantaneity of the photoelectric effect, E.O. Lawrence, J.W. Beams, Phys. Rev. 29, 903 (1927)

[Len02]   Ueber die lichtelektrische Wirkung, P. Lenard, Ann. Physik 8, 149 (1902)

[Lor10]   Alte und neue Fragen der Physik, H.A. Lorentz, Phys. Ztschr. 11, 1234 (1910)

[Lou73]   *Quantum theory of light*, R. Loudon, Oxford University Press, Oxford (1973)

[Mag63]   Interference fringes produced by the superposition of two independent maser light beams, G. Magyar, L. Mandel, Nature 198, 255 (1963)

[Man82]   Squeezed states and sub-Poissonian photon statistics, L. Mandel, Phys. Rev. Lett. 49, 136 (1982)

[Man95]   *Optical coherence and quantum optics*, L. Mandel, E. Wolf, Cambridge (1995)

[Mer93]   Photon noise reduction by controlled deletion techniques, J. Mertz, A. Heidmann J. Opt. Soc. Am. B10, 745 (1993)

[Mey89]   *Elements of quantum optics*, P. Meystre, M. SargentIII, p. 448, Springer (1989)

[Mic27]   *Studies in Optics*, A.A. Michelson, University of Chicago Press (1927)

[Ou88]   Violation of Bell's inequality and classical probability in a two-photon correlation experiment, Z.Y. Ou, L. Mandel, Phys. Rev. Lett. 61, 50 (1988)

[Pau83]   Light absorption by a dipole, H. Paul, R. Fischer, Usp. Fiz. Nauk 141, 375 (1983)

[Pau95]   *Photonen, eine Einführung in die Quantenoptik*, H. Paul, Teubner, Stuttgart (1995)

[Pla00]   Zur Theorie des Gesetzes der Energieverteilung im Normalspektrum, M. Planck, Verhandlungen Dt. Phys. Ges. 2, 237 (1900)

[Rad71]   Zur Interferenz unabhängiger Lichtstrahlen geringer Intensität, W. Radloff, Ann. Physik 26, 178 (1971)

[Rem87]   Observation of quantum collapse and revival in a one-atom maser, G. Rempe, H. Walther, N. Klein, Phys. Rev. Lett. 58, 353 (1987)

[Sak85]   *Modern Quantum Mechanics*, J.J. Sakurai, Addison-Wesley (1985)

[Tay09]   Interference fringes with feeble light, G.I. Taylor, Proc. Cambridge Phil. Soc. 15, 114 (1909)

[Wal85]   Photon-antibunching by use of a photoelectron event triggered optical shutter, J.G. Walker, E. Jakeman, Opt. Acta 32, 1303 (1985)

[Wei97]   Towards a long distance Bell experiment with independent observers, G. Weihs, H. Weinfurter, A. Zeilinger in *Experimental Metaphysics*, ed. R.S. Cohen, 239 (1997)

4 Quantum models of light

4.1 Quantization of light

4.1.1 Some general comments on quantum mechanics

In this chapter we will develop quantum mechanical models for light which will help us understand the various experimental results and techniques we will encounter in this book. Let us begin by reviewing the types of predictions one can make about experimental outcomes of quantum mechanical systems and how these are obtained from the mathematics.

In classical mechanics, an observable quantity is represented by a variable such as x, the position of a particle. This is just a number which may vary as a function of time and/or other parameters. The outcome of an experiment which measures position will just be the value of x.

In general, for quantum mechanical systems, the result of an experiment will not be unique. Instead, a range of possible outcomes will occur with some probabilities. Consequently, a single number is insufficient to describe all the possible outcomes of a particular experiment. Thus, in quantum mechanics, variables are replaced by operators such as $\mathbf{x}$, the position operator, describing the type of measurement to be made, and states, represented by kets, $|\sigma\rangle$. As we saw in the last chapter, these contain the information about the possible outcomes. Each operator representing an observable quantity will have a set of eigenstates associated with it whose eigenvalues represent the spectrum of possible results that a measurement of this observable can produce. Typically a system will be in a state that is a superposition of various eigenstates of some observable. A special case is when the system is actually in an eigenstate of the observable, in which case a definite result is predicted. Most generally the system will be in a statistical mixture of various quantum states.

Quantum theory makes predictions about the results of large ensembles of measurements. The average value of an observable is given by the expectation value of the observable's operator:

$$\langle \hat{\mathbf{x}} \rangle = \langle \sigma | \hat{\mathbf{x}} | \sigma \rangle \tag{4.1.1}$$

Equation (4.1.1) predicts the average value of measurements performed on a large ensemble of systems all prepared in the same state, $|\sigma\rangle$, of the observable represented by the operator $\mathbf{x}$. We can also consider expectation values of higher moments such as $\langle \mathbf{x}^2 \rangle$ which naturally leads to us to define the variance of an observable as

$$\langle \Delta \hat{\mathbf{x}}^2 \rangle = \langle \hat{\mathbf{x}}^2 \rangle - \langle \hat{\mathbf{x}} \rangle^2 \tag{4.1.2}$$

A Guide to Experiments in Quantum Optics, 2nd Edition. Hans-A. Bachor and Timothy C. Ralph
Copyright © 2004 Wiley-VCH Verlag GmbH & Co. KGaA
ISBN: 3-527-40393-0

and similarly for higher order properties. In classical mechanics a non-zero variance, or noise, is associated with imperfect preparation of the system with the desired property. In quantum mechanics even ideally prepared systems can exhibit non-zero variance due to the probabilistic nature of the theory. Such non-zero fluctuations are thus often referred to as quantum noise. The characteristics of a system with respect to a particular observable can be completely characterized by the expectation values of all the moments of the observables operator. However, in many cases of interest in this book the systems will exhibit Gaussian statistics. In such cases, the system is completely characterized by just the first and second order moments.

A powerful link between classical variables and quantum operators exists in the procedure of canonical quantization, first pioneered by Dirac. If a system is described classically by the canonically conjugate variables x and p (in the sense of Hamilton's Equations [Gol80]) then the correct quantum mechanical description can be obtained by simply replacing x and p with the operators $\mathbf{x}$ and $\mathbf{p}$ in the Hamiltonian and replacing the classical Poisson bracket with the commutation relation $[\mathbf{x}, \mathbf{p}] = \mathbf{x}\mathbf{p} - \mathbf{p}\mathbf{x} = i\hbar$. In the following section we will use this procedure to convert the classical description of light introduced in Chapter 2 into a quantum mechanical description.

4.1.2 Quantization of cavity modes

The classical description of light in Section 2.1 assumed as a model for a laser beam a single frequency wave. Although an idealization, for classical light it is not an unreasonable one ad it would neglect modulations and fluctuations of the light. Anticipating this we start our discussion by considering the discrete optical modes in a ring cavity. A sufficiently high finesse cavity will only support fields with wavelengths λ according to

$$\lambda = L/n, \qquad n = 0, 1, 2, 3, \ldots \tag{4.1.3}$$

where L is the cavity round trip length. This discrete frequency approximation is valid in both the classical and quantum domains.

Recall first from Section 2.1.4, that the energy in a classical light field is proportional to the sum of the squares of the quadrature amplitudes $X1$ and $X2$. Specifically for polarized modes in a high finesse cavity the energy is given classically by

$$H = \Sigma_k \frac{\hbar 2\pi\nu_k}{4}(X1_k^2 + X2_k^2) \tag{4.1.4}$$

where k labels the different frequency modes. Moreover we can identify suitably scaled versions of the quadratures as the canonical variables of the classical light wave. Specifically q_k and p_k are canonical variables, where

$$q_k = \sqrt{\frac{\hbar 2\pi\nu_k}{2}} X1_k$$

$$p_k = -\sqrt{\frac{\hbar}{4\pi\nu_k}} X2_k \tag{4.1.5}$$

We can now apply canonical quantization by introducing the operators $\tilde{\mathbf{q}}_k$ and $\tilde{\mathbf{p}}_k$ with $[\tilde{\mathbf{q}}_k, \tilde{\mathbf{p}}_l] = i\hbar\delta_{kl}$. This immediately implies that the operator equivalents of the quadrature

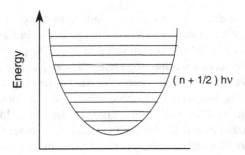

Figure 4.1: Energy levels of a harmonic oscillator.

amplitudes $\tilde{\mathbf{X}}\mathbf{1}_k$ and $\tilde{\mathbf{X}}\mathbf{2}_k$ obey the commutator

$$[\tilde{\mathbf{X}}\mathbf{1}_k, \tilde{\mathbf{X}}\mathbf{2}_l] = 2i\delta_{kl} \tag{4.1.6}$$

The energy operator or Hamiltonian for the system is just

$$\hat{\mathbf{H}} = \sum_k \frac{h\nu_k}{4}\left(\tilde{\mathbf{X}}\mathbf{1}_k^2 + \tilde{\mathbf{X}}\mathbf{2}_k^2\right) \tag{4.1.7}$$

4.1.3   Quantized energy

The Hamiltonian is comprised of a sum of components with a whole range of frequencies. These different waves are independent in the sense that different frequency components commute with each other, see Eq. (4.1.6) – much like we are used to considering classical waves of different frequencies to be independent. Like in a classical cavity the frequency is a discrete variable, only frequencies ν associated with the spatial modes are possible. In fact the Hamiltonian is mathematically equivalent to a sum of uncoupled Harmonic oscillators, one for each frequency ν. Each term describes one mode. For each mode we can solve the operator equation. Before carrying out the calculations in detail let us consider the overall properties of the quantised harmonic oscillator. One important consequence of the quantization of the oscillator is the discreteness of the possible energy states of this system. The system has a series of possible energies which are equally spaced

$$\mathcal{E}_n = \left(n + \frac{1}{2}\right)h\nu \tag{4.1.8}$$

Similarly, the quantised field has discrete energies, $\mathcal{E}_n$. The energy of the field can only be increased by multiples of the *energy quantum* of $h\nu$. This quantization suggests that the field is made up of particles – *photons*. However, one has to be careful to use only those properties of particles which can be directly derived from the quantised field equations. Care has to be taken with this picture to avoid an over-interpretation and to regard photons too much as particles behaving like small classical billiard balls.

It is worthwhile to contemplate one peculiar case, the lowest possible energy $\mathcal{E}_0 = \frac{1}{2}h\nu$, known as the zero point energy. The value of $\mathcal{E}_0$ is not zero. There is remaining energy.

However, this does not upset our measurements since it is not possible to extract this energy from the system. It is the lowest energy which can be achieved with this particular mode of light. All measurements of the energy of the mode refer back to it. The fact that we can only measure changes in the energy, not the absolute energy, means that the ground state is not directly accessible. Further discussions on this may be found in E.A. Power [Pow64]. It is noteworthy that the value of $\mathcal{E}_0$ is unique to the optical frequency. A field which consists of many modes has a total zero point energy of $\frac{1}{2}h\nu$ per mode. Since in theory there is no upper bound on the frequencies, the energy of the ground state is infinite – this creates a conceptual problem. Mathematically this can be overcome by re-normalization [Vog94] and in practice any real experiment will have a finite detection bandwidth and the infinity is avoided. It is possible to measure the effect of the zero point energy directly. One can create an experimental situation where the number of allowed modes in a multi mode system can be changed – and as a consequence the total zero point energy changes. This increase in energy results in a force which tries to change the system towards the lowest total zero point energy. Such effects were investigated by H.B.G. Casimir [Cas48]. Consider a cavity with length d. The cavity supports modes for all wavelengths $\lambda = 2d/n = L/n$. The longest supported wavelength is L. The size of the cavity determines the energy of the allowed low frequency modes and thus the total zero point energy. That means that for a smaller cavity, shorter L, the total zero point energy is smaller than for a larger cavity. The change in total zero point energy is proportional to the change in number of modes. Casimir [Cas48] worked this out for a system of two conducting plates. In this case the total zero point energy varies with L^{-3} and the resulting force F_{cas} is given by

$$F_{cas} = \frac{A\pi hc}{4.8 10^{-20} d^4} \ [\text{N}] \tag{4.1.9}$$

where A is the area of the plates and $d = L/2$ is the separation. This force does not depend on the electron charge and it does not require any electric field between the plates. It is a quantum mechanical effect and does apply to two plates in vacuum. The forces can be significant for very small plate separations. As an example values of $A = 10^{-4}$ [m^2] and $d = 0.01$ [mm] result in $F_{cas} = 1.3 \times 10^{-11}$ [N]. This pure quantum force was demonstrated in careful measurements, where all other forces were excluded [Kit57]. Without elaborating, it is worth noting that similar forces are part of the explanation of the well known van de Waals forces [Pow64]. These forces are essential for the explanation of a range of intermolecular effects and include such practical effects as the stability of colloids in paint suspensions. The zero point energy can play an important role.

4.1.4   The quantum mechanical harmonic oscillator

Just as, given certain assumptions about the spatial modes, we could describe the essential characteristics of a classical light field via its complex amplitude, α_k (see Chapter 2), so its quantised equivalent, $\tilde{a}_k$, plays a similar role in a quantum description of light. The operator $\tilde{a}$ and its adjoint, $\tilde{a}^\dagger$, defined in frequency space, raise and lower the excitation of a specific mode by one increment respectively. It follows from Equation (4.1.6) and the relationships, $\tilde{X}1 = \tilde{a} + \tilde{a}^\dagger$ and $\tilde{X}2 = i(\tilde{a}^\dagger - \tilde{a})$ that they obey the Boson commutation rules

$$[\tilde{a}_\nu, \tilde{a}_{\nu'}^\dagger] = (\tilde{a}_\nu \tilde{a}_{\nu'}^\dagger - \tilde{a}_\nu^\dagger \tilde{a}_{\nu'}) = \delta_{\nu\nu'} \tag{4.1.10}$$

and

$$[\tilde{a}_\nu, \tilde{a}_{\nu'}] = [\tilde{a}_\nu^\dagger, \tilde{a}_{\nu'}^\dagger] = 0 \tag{4.1.11}$$

We can rewrite the Hamiltonian operator of the radiation field of a single mode in terms of these operators as

$$
\begin{aligned}
\tilde{H} &= \frac{1}{2}h\nu\left(\tilde{a}^\dagger\tilde{a} + \tilde{a}\tilde{a}^\dagger\right) \\
&= h\nu\left(\tilde{a}^\dagger\tilde{a} + \frac{1}{2}\right) = h\nu\left(\tilde{n} + \frac{1}{2}\right)
\end{aligned}
\tag{4.1.12}
$$

where

$$\tilde{n} = \tilde{a}^\dagger\tilde{a}$$

is the photon number operator. The Hamiltonian above represents the number of photons in one mode multiplied by the energy of one photon in that mode. For a multimode field the total Hamiltonian would be the sum of several terms such as given in Equation (4.1.12), one for each frequency component. Note the specific ordering of the operators $\tilde{a}^\dagger$ and $\tilde{a}$. The ordering is important since the operators do not commute. The choice of ordering the operators is fixed, namely the creation operator appears left of the annihilation operator, see Equation (3.1.3).

Finally, the link between the position and momentum operators $\tilde{q}$, $\tilde{p}$ and the creation and annihilation operators $\tilde{a}^\dagger$, $\tilde{a}$ is given by

$$
\begin{aligned}
\tilde{q} &= \sqrt{\frac{\hbar 2\pi\nu_k}{2}}\ (\tilde{a} + \tilde{a}^\dagger) \\
\tilde{p} &= -i\sqrt{\frac{\hbar}{4\pi\nu_k}}\ (\tilde{a}^\dagger - \tilde{a})
\end{aligned}
\tag{4.1.13}
$$

4.2 Quantum states of light

4.2.1 Number or Fock states

The Hamiltonian for a single mode of the radiation field (Equation (4.1.12)) has the eigenvalues $h\nu(n + \frac{1}{2})$. The corresponding eigenstates can be represented by the notation $|n\rangle$. These are known as the Number or Fock states and are eigenstates of the number operator

$$\tilde{a}^\dagger\tilde{a}|n\rangle = \tilde{n}|n\rangle = n|n\rangle \tag{4.2.1}$$

The ground state, also known as the vacuum state, is defined by

$$\tilde{a}|0\rangle = 0 \tag{4.2.2}$$

It cannot be lowered any further. The energy of the ground state has already been discussed in Section 4.1.3 and is given by

$$\langle 0|\hat{H}|0\rangle = \frac{1}{2}\sum_k h\nu_k \tag{4.2.3}$$

where we sum over all frequencies. The operators $\tilde{a}$ and $\tilde{a}^\dagger$ are raising and lowering operators on a series of equally spaced eigenstates.

$$\tilde{a}|n\rangle = \sqrt{n}|n-1\rangle \quad \text{and} \quad \tilde{a}^\dagger|n\rangle = \sqrt{(n+1)}|n+1\rangle \tag{4.2.4}$$

They can be used to generate all possible number states starting with the ground state $|0\rangle$ by a successive application of the creation operator

$$|n\rangle = \frac{(\tilde{a}^\dagger)^n}{\sqrt{n!}}|0\rangle \tag{4.2.5}$$

They are orthogonal and complete. The number of photons n is exactly defined. It corresponds to the eigenvalue of the corresponding photon number operator. The variance of the photon number is zero for a number state

$$V_n = \langle \Delta n^2 \rangle = \langle n|\tilde{\Delta n}^2|n\rangle = \langle n|\tilde{n}\tilde{n}|n\rangle - \langle n|\tilde{n}|n\rangle^2 = 0 \tag{4.2.6}$$

We encountered the single photon number state in Section 3.5 where we found that a consideration of their properties in an interferometer led us to a radically different view of light from the classical one. We can illustrate this further by contrasting the average value of the quadrature amplitudes for classical light and light in a number state. Recall that in a classical description of light an arbitrary quadrature amplitude, X^θ, can be described as a function of the phase angle, θ, according to

$$X^\theta = \cos[\theta]X1 + \sin[\theta]X2 \tag{4.2.7}$$

We can convert this equation to the correct quantum mechanical form by simply replacing the quadrature variables by quadrature operators. The average value of the quadrature amplitude for light in a particular quantum state is then found by taking the expectation value over the state of interest. However what we find for a number state is that

$$\langle \tilde{X}^\theta \rangle_n = \langle n|\tilde{X}^\theta|n\rangle = 0 \tag{4.2.8}$$

where the orthogonality of the number states ensure that all terms like $\langle n|\tilde{a}|n\rangle$ and $\langle n|\tilde{a}^\dagger|n\rangle$ are zero, see Eq. (4.2.4). The results in Equations 4.2.7 and 4.2.8 could hardly be more different. In the classical case the quadrature amplitude oscillates as a function of the quadrature angle, θ, whilst for the number state the average value is always zero, independent of the quadrature angle. It is clear that number states represent quite exotic states of light, quite dissimilar to what is produced directly by a laser. We will return to consider them again later but for now we look for light states with more familiar behaviour, which in turn turn out to be those most naturally produced. For now we use the number states as a basis set – and find more realistic superpositions.

However, one particular number state plays a major role in all quantum optical experiments: the *vacuum state* $|0\rangle$.

4.2.2 Coherent states

A disadvantage of the number states is that they do not represent a well defined phase. This makes them impractical for the description of real laser beams. What we require is a state which has a more precisely defined phase. This is the *coherent state* $|\alpha\rangle$ which is the closest quantum approximation to the field generated by a laser.

Consider, in an idealized way, how a light wave is produced in the classical picture. Initially the energy in the electric and magnetic fields is zero. Then, by some means, the electric field is displaced to some non-zero value, say $|\alpha|$. This in turn displaces the magnetic field and so on producing a propagating light field. Suppose we follow the same idealized recipe in quantum mechanics. This suggests that we should displace the quantum mechanical vacuum state to some non-zero value. We might hope that this will produce an oscillating field. However, because the quantum mechanical vacuum has non-zero energy we also expect uniquely quantum features.

The *displacement operator* $\tilde{\mathbf{D}}$ acting on the vacuum ket has the effect of displacing it by an amount α, where α is a complex number which can be directly associated with the complex amplitude in classical optics, see Chapter 2. That is

$$|\alpha\rangle = \tilde{\mathbf{D}}(\alpha)|0\rangle$$

The displacement operator is given by

$$\tilde{\mathbf{D}}(\alpha) = \exp(\alpha\tilde{\mathbf{a}}^\dagger - \alpha^*\tilde{\mathbf{a}}). \tag{4.2.9}$$

It is a unitary operator, i.e.

$$\tilde{\mathbf{D}}^\dagger(\alpha) = \tilde{\mathbf{D}}^{-1}(\alpha) = \tilde{\mathbf{D}}(-\alpha)$$

and has the properties

$$\begin{aligned}
\tilde{\mathbf{D}}^\dagger(\alpha)\tilde{\mathbf{a}}\tilde{\mathbf{D}}(\alpha) &= \tilde{\mathbf{a}} + \alpha \\
\tilde{\mathbf{D}}^\dagger(\alpha)\tilde{\mathbf{a}}^\dagger\tilde{\mathbf{D}}(\alpha) &= \tilde{\mathbf{a}}^\dagger + \alpha^*
\end{aligned} \tag{4.2.10}$$

The family of states $|\alpha\rangle$ are called coherent states and, as we will now show, they behave in many ways like classical laser light. Due to the unitarity of the displacement operator, they are normalized to unity

$$\langle\alpha|\alpha\rangle = 1 \tag{4.2.11}$$

Also the coherent states are eigenstates of the annihilation operator $\tilde{\mathbf{a}}$, as can be verified from

$$\tilde{\mathbf{D}}^\dagger(\alpha)\tilde{\mathbf{a}}|\alpha\rangle = \tilde{\mathbf{D}}^\dagger(\alpha)\tilde{\mathbf{a}}\tilde{\mathbf{D}}(\alpha)|0\rangle = (\tilde{\mathbf{a}} + \alpha)|0\rangle = \alpha|0\rangle$$

Multiplying both sides with $\tilde{\mathbf{D}}(a)$ results in the eigenvalue condition

$$\tilde{\mathbf{a}}|\alpha\rangle = \alpha|\alpha\rangle \tag{4.2.12}$$

Using Eq. (4.2.12) we now see that the expectation value for the quadrature amplitude operator in a coherent state, $\langle\alpha|\tilde{\mathbf{X}}^\theta|\alpha\rangle = \langle\tilde{\mathbf{X}}^\theta\rangle_\alpha$ are identical to the classical variables provided we identify the displacement α with the classical complex amplitude. That is

$$\langle\tilde{\mathbf{X}}^\theta\rangle_\alpha = \cos[\theta]X1 + \sin[\theta]X2 \tag{4.2.13}$$

Hence we see that the average value of the field oscillates as a function of quadrature angle. Similarly the expectation value for the photon number in a coherent state is

$$\langle \tilde{a}^\dagger \tilde{a} \rangle_\alpha = \alpha^* \alpha \tag{4.2.14}$$

which again is the same as the classical value. Thus we see that the average values of the observables for a coherent state are identical to those for a classical field. However, unlike a number state or a classical field which contain a definite amount of energy, a coherent state does not contain a definite number of photons. This follows from the fact that a coherent state $|\alpha\rangle$ is made up of a superposition of number states

$$|\alpha\rangle = \sum_{n=0}^{\infty} |n\rangle\langle n|\alpha\rangle \tag{4.2.15}$$

Using Equation (4.2.5) and the completeness of the number states one obtains

$$\langle n|\alpha\rangle = \frac{\alpha^n}{\sqrt{n!}} \exp\left(-\frac{1}{2}|\alpha|^2\right) \tag{4.2.16}$$

which is the contribution of the various number states.

Now consider the relationship between two different coherent states. Their scalar product is

$$
\begin{aligned}
|\langle \alpha_1|\alpha_2\rangle|^2 &= |\langle 0| \; \tilde{\mathbf{D}}^\dagger(\alpha_1)\tilde{\mathbf{D}}(\alpha_2) \; |0\rangle|^2 \\
&= |\exp(-\frac{1}{2}(|\alpha_1|^2 + |\alpha_2|^2) + \alpha_1^*\alpha_2)|^2 \\
&= \exp(-|\alpha_1 - \alpha_2|^2)
\end{aligned}
\tag{4.2.17}
$$

which is not zero. That means two different coherent states are not orthogonal to each other. However, for two states with vastly different photon numbers, that means $|\alpha_1 - \alpha_2| \gg 1$, the product is very small. Such states are approximately orthogonal. One consequence is that the coherent states form a basis which is over complete. But despite this, the coherent states can be used as a basis on which any state $|\Psi\rangle$ can be expanded [Wal94].

In summary, the coherent states have properties similar to those of classical coherent light and in fact represent its closest quantum mechanical approximation. Just as under certain conditions laser light can be approximated as idealized classical coherent light, as in situations where a quantum mechanical description is needed, laser light can often be well approximated by a coherent state. We will find, in Section 5.1 that coherent states are stable in regard to dissipation, that means one coherent state is transformed into another coherent state when attenuated. Consequently, the coherence properties of the light remain similar after attenuation. Coherent states will be used extensively in our modelling of laser experiments.

Being a complex number, the range of possible values of α, and thus the range of possible coherent states, requires a two dimensional representation. This is only natural, since we can distinguish different states of light experimentally by two parameters, intensity and phase. But we note that only one of them, the intensity, has a simple quantum mechanical equivalent (i.e. the number operator). A similarly simple quantum mechanical operator for the phase is not available. Phase operators can only be approximated in a rather complex way through a limit of a series of operators [Wal94, Peg88, Fre95] and their definition have stimulated a lively debate amongst theoreticians. However, the real and imaginary parts of the annihilation operator $\tilde{a}$, i.e. the quadrature amplitude operators $\tilde{X1}$ and $\tilde{X2}$ are both well behaved operators. This leads to a pair of Hermitian operators that describe the field.

4.2.3 Mixed states

So far we have only considered pure field states. More generally we can have fields that are in an incoherent mixture of different pure states. This is referred to as a mixed state. A mixed state is normally represented by a density operator, $\tilde{\rho}$, such that

$$\tilde{\rho} = w_1|\sigma_1\rangle\langle\sigma_1| + w_2|\sigma_2\rangle\langle\sigma_2| + \ldots$$

represents a mixture of the states $|\sigma_1\rangle$, $|\sigma_2\rangle$,... with fractional weightings w_1, w_2, ... respectively. It is important not to confuse mixtures of states and superpositions of states, they are very different. Consider the equal superposition state $|\phi\rangle = 1/\sqrt{2}(|a\rangle + |b\rangle)$. Written as a density operator this state is

$$\begin{aligned}\tilde{\rho}_{\mathbf{p}} &= |\phi\rangle\langle\phi| \\ &= \frac{1}{2}|a\rangle\langle a| + \frac{1}{2}|a\rangle\langle b| + \frac{1}{2}|b\rangle\langle a| + \frac{1}{2}|b\rangle\langle b|\end{aligned} \tag{4.2.18}$$

In contrast an equal mixture of the states a and b is just

$$\tilde{\rho}_{\mathbf{m}} = \frac{1}{2}|a\rangle\langle a| + \frac{1}{2}|b\rangle\langle b| \tag{4.2.19}$$

The coherences between the states which are present for the superposition are absent in the case of the mixture. Expectation values for mixed states are given by

$$\langle\tilde{x}\rangle = \text{Tr}\{\tilde{x}\tilde{\rho}\}$$

where the the trace, Tr, is taken over a complete set of eigenstates. An important example of a mixed state in optics is the thermal state which can be represented as a mixture of number states by:

$$\tilde{\rho}_{\text{th}} = \frac{1}{G}\left(|0\rangle\langle 0| + \frac{G-1}{G}|1\rangle\langle 1| + \left(\frac{G-1}{G}\right)^2|2\rangle\langle 2| + \ldots\right) \tag{4.2.20}$$

where G is a parameter which ranges from one for a zero temperature thermal field (the vacuum) up to a high temperature thermal field for $G \rightarrow \infty$.

4.3 Quantum optical representations

4.3.1 Quadrature amplitude operators

Let us now look in some detail at the quadrature amplitude operators defined by,

$$\tilde{X1} = \tilde{a} + \tilde{a}^\dagger \quad \text{and} \quad \tilde{X2} = i(\tilde{a}^\dagger - \tilde{a}) \tag{4.3.1}$$

We will sometimes refer to them as the *amplitude quadrature* and the *phase quadrature* respectively, for reasons that will become apparent when we discuss bright fields (Section 4.5). Please note that over the years a shift in notation has occurred in recent publications $\tilde{X}^+ = \tilde{X}1$ and $\tilde{X}^- = \tilde{X}2$ is very common.

We can define eigenstates of these operators via $\tilde{X}1|x1\rangle = x1|x1\rangle$ and similarly $\tilde{X}2|x2\rangle = x2|x2\rangle$. The eigenvalues xi form an unbounded, continuous set. Because of the non-zero commutation relation between $\tilde{X}1$ and $\tilde{X}2$ it is not possible to be in a simultaneous eigenstate of $\tilde{X}1$ and $\tilde{X}2$. The resulting uncertainty principle constrains the variances of the operators in any state to

$$\langle \Delta \tilde{X}1^2 \rangle \langle \Delta \tilde{X}2^2 \rangle \geq 1 \tag{4.3.2}$$

It follows that exact eigenstates of $\tilde{X}1$ or $\tilde{X}2$ are unphysical as infinite uncertainty in an unbounded spectrum implies infinite energy. This is analogous to the situation for position and momentum of a free particle.

We saw in the previous section that the expectation values of the quadrature operators in a coherent state were equal to their classical values. It is straightforward to show that the uncertainty product for the quadratures in a coherent state is given by

$$\langle \Delta \tilde{X}1^2 \rangle_\alpha \langle \Delta \tilde{X}2^2 \rangle_\alpha = 1 \tag{4.3.3}$$

which means a coherent state is a minimum uncertainty state. Equally important is the fact that the uncertainty of the coherent state is split equally between the two quadratures. The uncertainty relationship is symmetric

$$\langle \Delta \tilde{X}1^2 \rangle = \langle \Delta \tilde{X}2^2 \rangle = 1 \tag{4.3.4}$$

Hence the coherent state can be thought of as the best allowed compromise for a simultaneous eigenstate of the two quadratures.

The quadrature operators allow us to represent a beam of light graphically. These so called *phasor diagrams* are very popular in quantum optics – ever since their introduction by C. Caves [Cav80]. Any state of light can be represented on a phasor diagram of the operators, that is a plot of $\tilde{X}2$ versus $\tilde{X}1$. In such a diagram any particular coherent state $|\alpha\rangle$ is equivalent to an area centered around the point given by the value $(\langle \tilde{X}1 \rangle, \langle \tilde{X}2 \rangle)$ for this state. The size of the area is given by the variances $\langle \Delta \tilde{X}1^2 \rangle$ and $\langle \Delta \tilde{X}2^2 \rangle$ and is usually represented just by a circle with diameter $\langle \Delta \tilde{X}1^2 \rangle$, the area of the circle describes the extent of the uncertainty distribution. This phasor diagram is a qualitative description which shows several important results:

1. The uncertainty area is symmetric. For coherent light any measurement of the fluctuations will yield the same result, independent of the projection angle θ chosen.

2. The uncertainty area is of constant size, independent of the intensity α^2 of the state.

3. Two states are only distinguishable if their average coordinates in the $\tilde{X}1$, $\tilde{X}2$ plane are separated by more than $\langle \Delta \tilde{X}1^2 \rangle$.

The last result is significant. It gives us a practical rule when two states are measurably different. The uncertainty area sets a minimum change in amplitude or phase which we have to make before we can distinguish two states. In a classical system such an uncertainty region would be associated with noise, due to technical imperfections. However, here the uncertainty area is a consequence of the quantum rules – *not* technical imperfections. Consequently, the uncertainty area represents the effect of *quantum noise* on a measurement, whether it is amplitude, phase or any other property of the light. Experimental measurements cannot be more precise than the quantum noise.

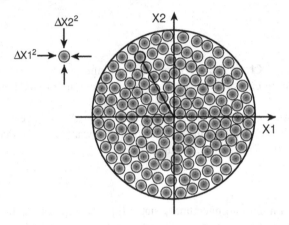

Figure 4.2: Uncertainty areas can cover the entire parameter space like tiles.

The uncertainty area is 2-dimensional. One can imagine the entire parameter space being covered, or tiled, with circular uncertainty areas of uniform size, see Fig. 4.2. The number of tiles, which are not all overlapping, approximately determines the number of states which can be distinguished. As a practical example, consider the light in an optical communication system. The information is transmitted by changing either the intensity or phase of the light at one modulation frequency. We assume the intensity is limited. How many distinguishable states of light, that means different pieces of information could we send? The light is represented by a 2-dimensional plane with a circle that indicates the largest possible intensity I_{max}. Inside this circle we can place our tiles. The number of tiles is an upper limit to the number of states we can use, that is the number of different pieces of information which a beam of light modulated at one frequency can contain. As described in Section 10.1 present day systems are far away from this quantum resolution – but ultimately technology will reach this limit.

4.3.2 Probability and quasi-probability distributions

The uncertainty area is a simple, useful tool which can, under certain circumstances, be used to completely characterize the state. To understand under what conditions this is possible we need to look at the probability distributions associated with measuring $\tilde{X}1$ and $\tilde{X}2$. Such measurements can be carried out using the technique of balanced homodyne detection which will be described in detail in Chapter 11. The probability of obtaining a result in a small interval, δx, around a particular value x_1 from a measurement of $\tilde{\mathbf{X}}\mathbf{1}$ performed on a system in the state $|\Psi\rangle$ is given by

$$P_\Psi(x_1) = |\langle x_1|\Psi\rangle|^2 \delta x \tag{4.3.5}$$

and similarly for a measurement of $\tilde{\mathbf{X}}\mathbf{2}$ the probability of obtaining results in a small interval around x_2 is

$$P_\Psi(x_2) = |\langle x_2|\Psi\rangle|^2 \delta x \tag{4.3.6}$$

In other words $|\langle x_i|\Psi\rangle|^2$ is the probability density at the point x_i. These distributions can be evaluated for various states using the relationships

$$\langle x_1|\tilde{\mathbf{X}}\mathbf{2}|\Psi\rangle = -i\frac{d}{dx_1}\langle x_1|\Psi\rangle, \quad \langle x_2|\tilde{\mathbf{X}}\mathbf{1}|\Psi\rangle = i\frac{d}{dx_2}\langle x_2|\Psi\rangle \tag{4.3.7}$$

For example for a coherent state we can show

$$
\begin{aligned}
\langle x_1|\tilde{\mathbf{a}}|\alpha\rangle &= \alpha\langle x_1|\alpha\rangle \\
&= \frac{1}{2}\langle x_1|\tilde{\mathbf{X}}\mathbf{1} + i\tilde{\mathbf{X}}\mathbf{2}|\alpha\rangle \\
&= \frac{1}{2}\left(x_1 + \frac{d}{dx_1}\right)\langle x_1|\alpha\rangle
\end{aligned} \tag{4.3.8}
$$

Solving the relevant differential equation then gives

$$|\langle x_1|\alpha\rangle|^2 = \frac{1}{\sqrt{\pi}}e^{-(\bar{x}_1 - x_1)^2} \tag{4.3.9}$$

where $\bar{x}_1 = (\alpha + \alpha^*)/2$ is the real part of α. We see that our x_1 results form a Gaussian distribution. The width, or variance of the distribution is 1 as expected. Similarly we find

$$|\langle x_2|\alpha\rangle|^2 = \frac{1}{\sqrt{\pi}}e^{-(\bar{x}_2 - x_2)^2} \tag{4.3.10}$$

for the probability distribution of x_2 results. Here $\bar{x}_2 = i(\alpha - \alpha^*)/2$ is the imaginary part of α. Again, we find a Gaussian distribution of width 1. Because the coherent states are Gaussian they can be completely characterized by the first and second order moments of $\tilde{\mathbf{X}}\mathbf{1}$ and $\tilde{\mathbf{X}}\mathbf{2}$ and thus the uncertainty area of the previous section.

It turns out that most experimentally accessible optical quantum states have Gaussian quadrature distributions and therefore are quite easily characterized. This is not in general true however. For example the quadrature distributions for a single photon number state are given by

$$|\langle x_1|1\rangle|^2 = \frac{1}{\sqrt{\pi}}2x_1^2 e^{-x_1^2} \tag{4.3.11}$$

and similarly

$$|\langle x_2|1\rangle|^2 = \frac{1}{\sqrt{\pi}}2x_2^2 e^{-x_2^2} \tag{4.3.12}$$

These distributions are clearly non-Gaussian.

A two dimensional probability distribution for a state $|\Psi\rangle$ is given by the Q-function defined as

$$Q_\Psi(\alpha) = \frac{|\langle\alpha|\Psi\rangle|^2}{\pi} \tag{4.3.13}$$

which is positive and bounded by

$$Q_\Psi \leq \frac{1}{\pi}$$

Physically the Q-function represents the probability density of obtaining the result $\alpha = (x_1 + ix_2)/2$ from an optimal simultaneous measurement of $\tilde{X}1$ and $\tilde{X}2$. Because $\tilde{X}1$ and $\tilde{X}2$ do not commute such a measurement is necessarily non-ideal. By optimal we mean that the resolution of the measurement is as good as the uncertainty principle allows. Such an optimal measurement can be achieved using the techniques of heterodyne or dual homodyne detection which will be discussed in Chapter 8.

For a coherent state $|\beta\rangle$ the Q-function is simply

$$Q_\beta = \frac{|\langle\alpha|\beta\rangle|^2}{\pi} = \frac{\exp(-|\alpha - \beta|^2)}{\pi} \tag{4.3.14}$$

It is a Gaussian function centered around the point $(\mathcal{R}e(\beta), \mathcal{I}m(\beta))$. The width of the Gaussian is two, that is twice that of what we found when the quadratures were measured individually. The additional noise is a result of the non-ideal measurement in which one extra unit of quantum noise is added. Note that this function has a width that is independent of the average photon number of the state. A strong field has the same sized Q-function as a weak field.

The Q-function for a number state is quite different to that of a coherent state and is given by

$$Q_n = \frac{|\langle\alpha|n\rangle|^2}{\pi} = \frac{|\alpha|^{2n}}{\pi n!} \exp(-|\alpha|^2) \tag{4.3.15}$$

This distribution has the shape of a ring, see Fig. 4.3. The maximum probability is a circle around the origin, the distribution in the radial direction is Gaussian centered at the average photon number with a width which decreases very rapidly, with $n!$. That means for large photon numbers the radial distribution is approaching a delta-function. A number state contains no phase information at all. The ring shape does not favour any particular phase.

A two dimensional distribution more closely related to our homodyne distributions is the Wigner function which can be written

$$W_\Psi(\alpha) = \frac{2\langle\Psi|\tilde{D}(\alpha)\,\tilde{\Pi}\,\tilde{D}(-\alpha)|\Psi\rangle}{\pi} \tag{4.3.16}$$

where $\tilde{\Pi}$ is the parity operator. The parity operator measures the symmetry of the probability amplitude distribution of a state. The number states are parity eigenstates with eigenvalues $+1$ for n even and -1 for n odd. So the action of the parity operator on an arbitrary superposition of number states is

$$\tilde{\Pi}(c_0|0\rangle + c_1|1\rangle + c_2|2\rangle + \cdots) = (c_0|0\rangle - c_1|1\rangle + c_2|2\rangle - \cdots)$$

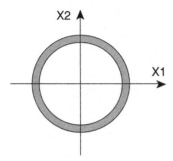

Figure 4.3: Uncertainty area of a number state.

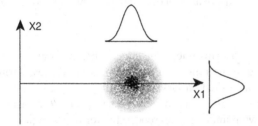

Figure 4.4: The randomness of the quadrature values represented by a weighted random distribution of dots. Each dot represents one particular instantaneous value. The histogram of all dots, both in the $\tilde{X}1$ and $\tilde{X}2$ direction correspond to the projection of the Wigner function.

This is effectively a π rotation on the phasor diagram. Thus the Wigner function is the overlap between the state, displaced by $-\alpha$, and itself, also displaced by $-\alpha$ but then also rotated by 180 degrees. For a coherent state $|\beta\rangle$ we will obtain

$$W_\beta(\alpha) = \frac{2\langle \beta - \alpha | \alpha - \beta \rangle}{\pi} = \frac{2e^{-2|\alpha-\beta|^2}}{\pi} \tag{4.3.17}$$

This is a symmetric Gaussian with width one, centered at the point $(\mathcal{R}e(\beta), \mathcal{I}m(\beta))$. Its projections along the $X1$ and $X2$ directions correspond with the homodyne probability distributions of Equations (4.3.9) and (4.3.10). In this way the noise introduced by the quantum fluctuations can be interpreted as producing "classical" noise distributions.

However, the Wigner function is a quasi-probability function. Although for Gaussian states it is always positive and generally well behaved it can become negative for non-Gaussian states, such as the number state, and thus cannot be interpreted as a true probability distribution. For such states the quantum fluctuations do not have a "classical" noise interpretation. Producing and analysing exotic states with negative Wigner functions is technically challenging. Such attempts are described in Section 9.13. For the rest of this chapter we will focus on Gaussian states.

The uncertainty areas which are shown in the phasor diagrams are simplified representations of the Wigner function. They are contours drawn at the half maximum value of the full

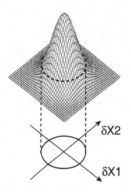

$\delta X2$

$\delta X1$

Figure 4.5: A 2-dimensional Gaussian distribution and its contour. This is the origin of the circular uncertainty area.

2-dimensional Wigner function, see Fig. 4.5. This line is the edge of the uncertainty area. It is described by only two values, the size of the minimum and the maximum axes. From an experimental point of view, we use the measured variance, for a certain angle θ, as a value for the uncertainty of this particular quadrature. That means we select one θ, measure the variance of this quadrature and obtain two points. We can repeat this for many angles, and in a real experiment we would actually do this to check the theory and reduce the experimental errors. All these points together describe the edge of the uncertainty area.

While the Wigner function of a coherent state is circularly symmetric this is not the only option. The width of the distribution in orthogonal directions could be different. The only requirement is that the product of the two widths, that means the product of the variances, satisfies the uncertainty relationship. We shall see in Chapter 9 that such states can be generated in optical experiments, they are known as *squeezed states*. It follows from the special properties of Gaussian functions that for such a state the contour at the half maximum value is always an ellipse. The diameter along the minor and major axes corresponds to the minimum and maximum quantum noise. The various distributions we have looked at can be generalized to mixed states such that

$$P_\Psi(x_i) = \langle x_i | \tilde{\rho_\Psi} | x_i \rangle$$

$$Q_\Psi(\alpha) = \frac{\langle \alpha | \tilde{\rho_\Psi} | \alpha \rangle}{\pi}$$

and

$$W_\Psi(\alpha) = \frac{2}{\pi} \mathrm{Tr}\{\tilde{\mathbf{D}}(-\alpha)\, \tilde{\rho_\Psi}\, \tilde{\mathbf{D}}(\alpha)\, \tilde{\mathbf{\Pi}}\} \qquad (4.3.18)$$

where in each case $\tilde{\rho_\Psi}$ is the density operator of the state to be characterized. For example the Wigner function for a thermal state, see Eq. (4.2.20), is given by

$$W_{th}(\alpha) = \frac{2e^{-2|\alpha|^2/\sigma}}{\pi} \qquad (4.3.19)$$

which is again a symmetric Gaussian with width $\sigma = 2G - 1$.

Finally, let's make a comment on other possible representations. The most obvious representation would generate a delta function for any particular coherent state. A representation with this property is the P-representation. A plot of the P-function would immediately show which coherent state is present. However, there are a number of mathematical problems which arise with the P-function once squeezed states are considered. Since for a squeezed state the variance of one of the quadratures is smaller than for a coherent state, and since a delta function cannot be made any narrower, the P-function will get negative values and even singularities. A modified P - function, the positive P-representation, avoids some of these mathematical problems and is found in many theoretical papers analysing quantum optical systems. It can actually be shown that all three representations (Q, P and W) are only limiting cases of a whole continuum of possible representations.

4.3.3 Photon number distributions, Fano factor

In the previous section we looked at results of measurements of the quadrature amplitudes. We now consider photon number measurements of a single optical mode. These are modelled as measurements of the observable $\tilde{n} = \tilde{a}^\dagger \tilde{a}$. For a photon number state there is only one photon number, it will be exactly the same in every measurement. However, for a coherent state one expects to find a spread in the results of many measurements of the photon number. We can use the expansion in Equation (4.2.16) to determine the probability $\mathcal{P}_\alpha(n)$ of finding n photons in the coherent state $|\alpha\rangle$.

$$\mathcal{P}_\alpha(n) = |\langle n|\alpha\rangle|^2 = \frac{|\alpha|^{2n}}{n!} \exp(-|\alpha|^2) \tag{4.3.20}$$

This is a Poissonian distribution with a mean value of $|\alpha|^2$, which represents the average number n_α of photons in this mode. The intensity fluctuates according to this distribution. The noise is not a result of technical imperfections of the experiment. This is a quantum property of the light. Mathematically speaking it is a direct consequence of the quantum mechanical commutation rules.

One important property of any Poissonian distribution is that the variance of the distribution is equal to the mean value. Consequently, the width of the distribution Δn is equal to $\Delta n = \sqrt{n_\alpha}$. Examples of three such distributions, for $n_\alpha = 4$, $n_\alpha = 36$ and $n_\alpha = 1000$ are shown in Fig. 4.6. Evaluating the relative uncertainty of the mean photon number of a coherent state results in $\Delta n/n = \sqrt{n_\alpha}/n_\alpha = 1/\sqrt{n_\alpha}$ which decreases with n_α. A state with a high photon number is better defined than that of a state with a small number of photons. We can interpret this result in the photon picture: A photo-detector will not see a regular stream of photons, it will see a somewhat randomized stream such that the measurement of the absorption events – for a perfect detector equal to the number of photons incident – has a distribution given by $\mathcal{P}_\alpha(n)$. Each event creates a fixed number of electrons, the stream of electrons leaving the detector forms the photo-current. The distribution $\mathcal{P}_\alpha(n)$ is directly represented by the variation of the photo-current. The intensity noise properties of the light could be expressed in terms of the distribution function $\mathcal{P}_\alpha(n)$. Thinking in the time domain this means that the photons arrive in unequal time intervals as shown in Fig. 4.7(d).

For large intensities $\mathcal{P}_\alpha(n)$ tends to a normal Gaussian function. In contrast, for low intensities the Poissonian distribution differs markedly from a Gaussian function.

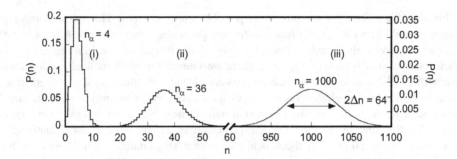

Figure 4.6: Poissonian distribution of photons for two coherent states (i) $n_\alpha = 4$, (ii) $n_\alpha = 36$ and (iii) $n_\alpha = 1000$.

A convenient way to classify the photon number distribution is via its width relative to Poissonian. A coherent state is not the only type of light that can be produced. In many practical situations light has more fluctuations than a coherent state, its photon distribution is wider than a Poissonian function. In this case we would call it *Super-Poissonian*. On the other hand, we will find that nonlinear processes can be used to make the distribution narrower. This we will call *Sub-Poissonian* light. In order to quantify the noise one defines the *Fano factor*, which is given by the ratio of the actual width of the distribution to the width of a Poissonian distribution with the same mean photon number. For some arbitrary state α_β with mean photon number n_β and distribution $V n_\beta$ we define the Fano factor as

$$f_\beta = \frac{V n_\beta}{n_\beta} \tag{4.3.21}$$

which classifies the states in the following way:

$f = 0$ number state
$f < 1$ Sub-Poissonian state
$f = 1$ coherent state
$f > 1$ Super-Poissonian state.

Note that the Fano factor is determined only by the variance. In principle many different photon distributions could have the same Fano factor. However, in many practical situations, the Fano factor is a complete description of the intensity properties of a state, even a squeezed state. Finally, we note that the Fano factor is identical to the normalized variance of the photon flux which again for many practical situations, is equivalent to the variance $V1$ of the quadrature amplitude, as we shall see in Section 4.5. In much of this guide we will use $V1$ interchangeably with the Fano factor.

Summary of different representations of quantum noise

Quantum noise is the most easily observable consequence of the quantization of light. It is a direct result of the use of operators for describing the amplitude of the electromagnetic field and of the commutation rules applied to these operators. In other words, the minimum uncertainty principle applied to light leads to the quantum noise that affects every optical measurement. A useful quantum state, which also has the minimum uncertainty, is the coherent state. It is an idealized state of light representing the output of a perfect light source. In mathematical terms it is the eigenstate of the annihilation operator. A physical consequence of this is that attenuation changes one coherent state into another coherent state, just with lower energy.

The characteristic properties of the coherent state are preserved. The coherent states can be used as a basis set, all types of light can be described as a superposition of these basic states. This expansion can be represented by a probability distribution for measuring particular α, the Q-function, which includes the intrinsic measurement penalty, or by a quasi probability distribution such as the Wigner function which includes only the intrinsic noise of the state.

The characteristic properties of a coherent state of light can be summarized in Table 4.1. The corresponding graphical representations of quantum noise are shown in Fig. 4.7.

4.4 Propagation and detection of quantum optical fields

Up till now we have been implicitly considering a stationary single mode in a cavity and the behaviour of observables of that mode without specifying how they would be measured. Now we wish to address the experimentally relevant situation of propagating fields and photodetection. We begin with a quite general discussion of the problem, whilst still limiting our discussion to that of single frequency mode fields. In free space there is an infinite continuum of frequency modes which must all be quantised. The following approach can be a satisfactory approximation in situations where, over the detection bandwidth, the frequency spectrum of the source and the frequency response of the optical elements is flat. That is, although there are many modes present, they are all "doing the same thing". In the following section we will discuss a more specific approach in which the effects of non-uniform spectra and spectral responses are included.

4.4.1 Propagation in quantum optics

The propagation of quantum optical modes through optical elements is modelled via unitary transformation operators, $\tilde{\mathbf{U}}$, where $\tilde{\mathbf{U}}^{-1} = \tilde{\mathbf{U}}^\dagger$. Two distinct pictures can be employed: (i) the Schrödinger picture, in which the evolution of the state of the mode $|\Psi\rangle \rightarrow |\Psi'\rangle$ is calculated via

$$|\Psi'\rangle = \tilde{\mathbf{U}}|\Psi\rangle$$

and the system observables remain stationary, or; (ii) the Heisenberg picture in which the evolution of a mode observable $\tilde{\mathbf{A}} \rightarrow \tilde{\mathbf{A}}'$ is calculated via

$$\tilde{\mathbf{A}}' = \tilde{\mathbf{U}}^\dagger \tilde{\mathbf{A}} \tilde{\mathbf{U}} \tag{4.4.1}$$

Table 4.1: Properties of quantum noise.

Properties of quantum noise	

Average complex amplitude: $\qquad$ $\alpha = |\alpha|e^{i\phi}$

Average photon number: $\qquad$ $n_\alpha = |\alpha|^2$

Quadrature amplitude operators: $\qquad$ Operators $\tilde{X}1$, $\tilde{X}2$ corresponding to a complex amplitude α

Wigner function: $\qquad$ Quasi probability distribution function is given by a two dimensional Gaussian function of unity width. Contour of the Wigner function is a circle.

Fluctuations of the quad. amplitude: $\qquad$ Variance of the operator is given by $V_\alpha(\theta) = V1_\alpha = 1$, independent of quadrature and detection frequency. Standard quantum limit (SQL).

Intensity noise: $\qquad$ $Vn_\alpha = n_\alpha$, which is equivalent to the shot noise formula. The intensity noise spectrum is flat, independent of detection frequency.

Photon number distribution: $\qquad$ Poissonian : $\mathcal{P}_\alpha(n) = |\langle n|\alpha\rangle|^2 = \frac{|\alpha|^{2n}}{n!}\exp(-|\alpha|^2)$

Photon statistics: $\qquad$ Time sequence of photon arrival is irregular as shown in Fig. 4.7(c). The statistics of the number of photons arriving in a time interval times follows a Poissonian distribution.

Fano factor: $\qquad$ $f_\alpha = 1$ for all detection frequencies.

Classical field: $\qquad$ Carrier component has an amplitude $|\alpha|$, with optical phase $\arg(\alpha)$. It has completely uncorrelated noise sidebands.

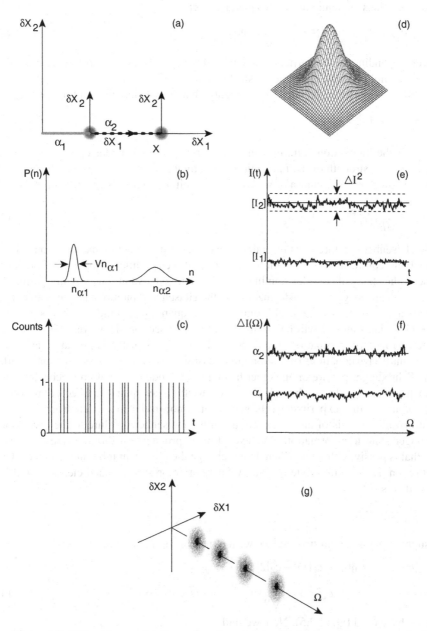

Figure 4.7: Different representations of a coherent state of light, shown for two states $|\alpha_1\rangle, |\alpha_2\rangle$ with different intensities. (a) phase space diagram, (b) photon number distribution, (c) arrival times of photons, (d) Wigner function, (e) time evolution of the intensities, (f) variance of the intensities, and (g) the noise at different frequencies.

and the state of the system is stationary. The two pictures are equivalent in the sense that all expectation values are equal in the two pictures because

$$\langle\Psi'|\tilde{\mathbf{A}}|\Psi'\rangle = \langle\Psi|\tilde{\mathbf{U}}^\dagger\tilde{\mathbf{A}}\tilde{\mathbf{U}}|\Psi\rangle = \langle\Psi|\tilde{\mathbf{A}}'|\Psi\rangle \tag{4.4.2}$$

However, depending on the particular situation, the two pictures offer different levels of computational complexity and different physical insights.

For propagation through stationary elements (that do not change with time) we can write

$$\tilde{\mathbf{U}} = e^{i\,\tilde{\mathbf{H}}\,\tau/\hbar} \tag{4.4.3}$$

where $\tilde{\mathbf{H}}$ is the interaction Hamiltonian characterizing the particular optical element and τ characterizes the strength of the interaction or the time for which it acts. For the latter case it is easy to show that the equation of motion for an arbitrary Heisenberg operator is then

$$\dot{\hat{A}} = \frac{1}{i\hbar}[\hat{\mathbf{A}}, \hat{\mathbf{H}}] \tag{4.4.4}$$

The Heisenberg picture most resembles the classical treatment thanks to canonical quantization which requires that the equations of motion of the annihilation operator $\tilde{a}$ are identical to those of the classical complex amplitude α. Thus in many situations it is straightforward to write down the propagation transformations in the Heisenberg picture and one is able to retain a considerable amount of classical intuition in the quantum treatment. This is particularly true for Gaussian states as will be discussed in the next section. However, this is not always the case and in certain situations the higher order moments of the annihilation and creation operators may have equations of motion quite dissimilar from their classical counterparts.

The Schrödinger picture on the other hand gives a uniquely quantum mechanical view of the transformations and as such tends to be most useful when quantum effects dominate, such as in photon counting experiments. In general it is not trivial to convert from a potentially straightforward analysis of an optical circuit in the Heisenberg picture to the corresponding Schrödinger state transformation. An exception is propagation through linear optical elements, that is passive elements which do not change the photon number in the modes. Clearly the action on the vacuum mode of unitary operators representing such elements will be the identity, that is

$$\tilde{\mathbf{U}}_{\text{lin}}|0\rangle = |0\rangle \tag{4.4.5}$$

Now suppose we write out the state we wish to transform in the number basis

$$\begin{aligned}
|\psi\rangle &= c_0|0\rangle + c_1|1\rangle + c_2|2\rangle + \cdots \\
&= (c_0 + c_1\tilde{a}^\dagger + c_2\frac{1}{\sqrt{2}}(\tilde{a}^\dagger)^2 + \cdots)|0\rangle
\end{aligned} \tag{4.4.6}$$

where we have used Eq. (4.2.5). Now we find

$$\begin{aligned}
\tilde{\mathbf{U}}_{\text{lin}}|\psi\rangle &= \tilde{\mathbf{U}}_{\text{lin}}(c_0 + c_1\tilde{a}^\dagger + c_2\frac{1}{\sqrt{2}}(\tilde{a}^\dagger)^2 + \cdots)\tilde{\mathbf{U}}^\dagger_{\text{lin}}\tilde{\mathbf{U}}_{\text{lin}}|0\rangle \\
&= (c_0 + c_1\tilde{a}''^\dagger + c_2\frac{1}{\sqrt{2}}(\tilde{a}''^\dagger)^2 + \cdots)|0\rangle
\end{aligned} \tag{4.4.7}$$

where $\tilde{a}'' = \tilde{U}\tilde{a}\tilde{U}^\dagger$ is the inverse Heisenberg picture annihilation operator, obtained by inverting the standard Heisenberg equations of motion. This technique is easily generalized to optical elements which mix more than one mode and, where applicable, is a convenient tool for moving between the two pictures.

4.4.2 Detection in quantum optics

The basic detection tool in quantum optics is photo-detection. One way or another almost all measurements of optical observables rely on photo-detection. For bright fields photo-detection measures the intensity of the field, for weak fields it measures the photon number of the field. Although quite distinct technologies must be employed in the bright and weak field limits the process can be modelled in both cases as the measurement of the photon number observable $\tilde{a}^\dagger\tilde{a}$.

As for propagation, detection in quantum optics is modelled differently in the Heisenberg and Schrödinger pictures. In the Heisenberg picture photo-detection can be viewed as the transformation of the quantum mode $\tilde{a}'$ into the classical mode $\tilde{N}_a = \tilde{a}'^\dagger\tilde{a}'$. The mode $\tilde{N}_a$ represents the photo-current or, more generally, the information generated by the detector and is classical in the sense that $[\tilde{N}_a, \tilde{N}_a^\dagger] = 0$. Nevertheless $\tilde{N}_a$ is an operator and must be so to correctly predict all behaviour. The statistics of $\tilde{N}_a$ can be investigated by calculating expectation values of its various moments

$$\langle \tilde{N}_a^j \rangle = \langle \Psi | \tilde{N}_a^j | \Psi \rangle \tag{4.4.8}$$

where $|\Psi\rangle$ is the initial state of the mode $\tilde{a}$. Using the technique of homodyne detection, see Section 8.1.4, photo-detection can also be used to measure the quadrature observable $\tilde{a}'e^{i\theta} + \tilde{a}'^\dagger e^{-i\theta}$. This then can be modelled as a transformation of the quantum mode $\tilde{a}'$ into the classical mode $\tilde{X}_a^\theta = \tilde{a}'e^{i\theta} + \tilde{a}'^\dagger e^{-i\theta}$. Once again although $\tilde{X}_a^\theta$ represents a classical mode it must be modelled as an operator. As for $\tilde{N}_a$ the statistics of the quadrature amplitude can be calculated by taking expectation values over the initial state of mode $\tilde{a}$. The operators representing the photo-currents can be manipulated as they might be in an experiment. For example correlation measurements can be modelled by adding, subtracting or multiplying photo-current operators representing the measurement of different modes.

In the Schrödinger picture photo-detection can be viewed as projection onto the eigenstates of $\tilde{a}^\dagger\tilde{a}$, i.e. the number states $|n\rangle$. Thus we can predict the probability that a particular result n will occur via

$$P_n = |\langle n | \Psi' \rangle|^2 \tag{4.4.9}$$

Average values, variances etc. can be calculated from the probability distribution thus generated. Similarly if the experiment is configured such that a quadrature observable is measured then the projection is onto the eigenstates of $\tilde{a}e^{i\theta} + \tilde{a}^\dagger e^{-i\theta}$, $|x_\theta\rangle$, and the probabilities are given, as was used in Section 4.3.2, by

$$P_\Psi(x_\theta) = |\langle x_\theta | \Psi \rangle|^2 \delta x \tag{4.4.10}$$

It is important to note that the predictions about the photo-current generated in the Schrödinger picture are just numbers, unlike the operators generated in the Heisenberg picture.

A major conceptual difference arises between the two approaches when multi-mode systems are considered. Suppose a system initially describable by a pure state in a single mode interacts with a number of other modes, also initially in pure states, during propagation. In the Heisenberg picture there will now be a number of output modes which are all a function of the input modes. Now suppose we have "lost" all but one of these modes, $\tilde{a}'$. Nevertheless we proceed as before by analysing the properties of $\tilde{N}_a = \tilde{a}'^\dagger \tilde{a}'$, just as we would with classical modes. This scenario looks quite different in the Schrödinger picture however. After propagation we are left with a state which describes the global properties of all the modes. If we have lost the other modes we must describe that explicitly by averaging over all the possible detection results the other modes may have given. This is achieved by tracing over all the lost modes such that:

$$\tilde{\rho}_\mathbf{a} = \mathrm{Tr}_{c \neq a}\{\tilde{\rho}'\} \tag{4.4.11}$$

where ρ' is the global state of the system, whilst ρ_a is the reduced density operator that describes the state of only mode $\tilde{a}$. The subscript $c \neq a$ indicates that the trace is over all modes except $\tilde{a}$. Under certain conditions the trace will not change the state of mode $\tilde{a}$. The state is then said to be separable as it indicates that the portion describing mode $\tilde{a}$ can be factored out. More generally though the state will be inseparable or *entangled*, meaning no factorization is possible. Then inevitably ρ_a will be a mixed state and

$$P_n = \langle n|\tilde{\rho}_\mathbf{a}|n\rangle \tag{4.4.12}$$

will give the probability of a particular detection result n.

If the results of measurements on the other modes *are* known then we must insert them explicitly so

$$|\Psi_a\rangle = \{\langle n_b|\langle n_c|\langle n_d|\ldots\}|\Psi'\rangle \tag{4.4.13}$$

where now $|\Psi_a\rangle$ is the conditional state of the system given that the results $n_b, n_c, n_d, \ldots$ were obtained for the other modes.

4.4.3 An example: The beamsplitter

To illustrate some of the ideas introduced in this section we consider a simple example: that of a single photon number state passing through a beamsplitter. A more general discussion of beamsplitters will be given in Chapter 5. The beamsplitter is assumed to have transmittance η. The other mode entering the beamsplitter is unoccupied. We label the mode containing the single photon state $\tilde{a}$ and the unoccupied mode $\tilde{b}$. Thus the initial state can be written: $|1\rangle_a|0\rangle_b$. Let us consider the Heisenberg approach first. As described in Section 5.1, the classical equations for mixing two modes with amplitudes α and β on a beamsplitter can be written:

$$\begin{aligned}
\alpha_{\mathrm{out}} &= \sqrt{\eta}\,\alpha + \sqrt{1-\eta}\,\beta \\
\beta_{\mathrm{out}} &= \sqrt{1-\eta}\,\alpha - \sqrt{\eta}\,\beta
\end{aligned} \tag{4.4.14}$$

Canonical quantization then allows us to immediately write

$$\tilde{a}' = \sqrt{\eta}\,\tilde{a} + \sqrt{1-\eta}\,\tilde{b}$$
$$\tilde{b}' = \sqrt{1-\eta}\,\tilde{a} - \sqrt{\eta}\,\tilde{b} \tag{4.4.15}$$

Hence if we photo-detect $\tilde{a}'$ we obtain

$$\tilde{N}_a = \tilde{a}'^\dagger \tilde{a}'$$
$$= \eta \tilde{a}^\dagger \tilde{a} + (1-\eta)\tilde{b}^\dagger \tilde{b} + \sqrt{\eta(1-\eta)}(\tilde{a}^\dagger \tilde{b} + \tilde{b}^\dagger \tilde{a}) \tag{4.4.16}$$

The average photon number detected is then

$$\bar{n} = \langle \tilde{N}_a \rangle = \langle 1|_a \langle 0|_b \tilde{a}'^\dagger \tilde{a}'|0\rangle_b|1\rangle_a$$
$$= \eta\langle 1|_a \tilde{a}^\dagger \tilde{a}|1\rangle_a + (1-\eta)\langle 0|_b \tilde{b}^\dagger \tilde{b}|0\rangle_b +$$
$$\sqrt{\eta(1-\eta)}(\langle 1|_a \tilde{a}^\dagger|1\rangle_a \langle 0|_b \tilde{b}|0\rangle_b + \langle 0|_b \tilde{b}^\dagger|0\rangle_b \langle 1|_a \tilde{a}|1\rangle_a)$$
$$= \eta \tag{4.4.17}$$

where we have used the relationships of Eq. (4.2.4). In a similar way the variance is

$$Vn = \langle \tilde{N}_a^2 \rangle - \langle \tilde{N}_a \rangle$$
$$= \eta(1-\eta)\langle 1|_a \tilde{a}^\dagger \tilde{a}|1\rangle_a \langle 0|\tilde{b}\tilde{b}^\dagger|0\rangle$$
$$= \eta(1-\eta) \tag{4.4.18}$$

Whilst the result for the average photon number can be understood from a classical field point of view: the intensity is reduced according to the transmission of the beamsplitter; the non-zero variance after some loss at the beamsplitter reflects the particle aspect of light. More on this is said in the next chapter.

Now let us consider the same problem in the Schrödinger picture. Using the technique of Equation (4.4.7) we find the state evolution through the beamsplitter is given by

$$|\Psi'\rangle = \tilde{a}''|0\rangle_{a'}|0\rangle_{b'}$$
$$= (\sqrt{\eta}\,\tilde{a}' + \sqrt{1-\eta}\,\tilde{b}')|0\rangle_{a'}|0\rangle_{b'}$$
$$= \sqrt{\eta}|1\rangle_{a'}|0\rangle_{b'} + \sqrt{1-\eta}|0\rangle_{a'}|1\rangle_{b'} \tag{4.4.19}$$

In order to consider only detection of the $\tilde{a}$ mode we need to trace over the $\tilde{b}$ mode as per Eq. (4.4.11). Writing $\tilde{\rho}' = |\Psi'\rangle\langle\Psi'|$ we find

$$\tilde{\rho}_{a'} = \langle 0|_{b'} \tilde{\rho}'|0\rangle_{b'} + \langle 1|_{b'} \tilde{\rho}'|1\rangle_{b'}$$
$$= \eta|1\rangle_{a'}\langle 1|_{a'} + (1-\eta)|0\rangle_{a'}\langle 0|_{a'} \tag{4.4.20}$$

Using Eq. (4.4.12) we find detection of this state gives the probabilities

$$P_1 = \langle 1|\tilde{\rho}_{a'}|1\rangle = \eta$$
$$P_0 = \langle 0|\tilde{\rho}_{a'}|0\rangle = 1 - \eta \tag{4.4.21}$$

which leads to the same average value and variance as predicted from the Heisenberg result.

Now, instead of "losing" the $\tilde{b}$ mode let us suppose that we have photo-detected it. If the photodetection of the $\tilde{b}$ mode gave the result "0" then the conditional state of the $\tilde{a}$ mode will be

$$|\Psi\rangle_{a'} = \{\langle 0|_{b'}\}|\Psi'\rangle = \frac{1}{\sqrt{2}}|1\rangle \tag{4.4.22}$$

indicating that a photon will be found in mode $\tilde{a}$ with certainty. The factor of $1/\sqrt{2}$ indicates that the probability of this state being prepared (i.e. the probability of the measurement of the $\tilde{b}$ mode giving the "0" result is $(1/\sqrt{2})^2 = 1/2$. Similarly if the measurement of the $\tilde{b}$ mode gave the result "1" then the conditional state of the $\tilde{a}$ mode would be

$$|\Psi\rangle_{a'} = \{\langle 1|_{b'}\}|\Psi'\rangle = \frac{1}{\sqrt{2}}|0\rangle$$

indicating that it is certain that no photon will be found in the $\tilde{a}$ mode. How could we predict this latter result in the Heisenberg picture? Well, we found that whenever a photon appeared in mode $\tilde{b}$ none was not found in $\tilde{a}$ and vice versa. Another way of saying this is that the two modes are anti-correlated. To test for anti-correlation in an experiment we might add the photo-currents obtained from the photo-detection of the two modes and look for reduced noise (variance). Adding the photo-currents we would obtain (using Equation (4.4.15))

$$\begin{aligned}\tilde{N}_a + \tilde{N}_b &= \tilde{a}'^{\dagger}\tilde{a}' + \tilde{b}'^{\dagger}\tilde{b}' \\ &= \tilde{a}^{\dagger}\tilde{a} + \tilde{b}^{\dagger}\tilde{b}\end{aligned} \tag{4.4.23}$$

which has zero variance indicating perfect anti-correlation as expected. Although the results match in the two pictures, as they must, the calculational techniques and the physical pictures suggested are quite different.

4.5 Quantum transfer functions

As we saw in Section 4.3, in general a complete description of a state of light, including its quantum properties, can be carried out through quasi probability distributions, such as the Wigner function, Q-function, or P-representation [Wal94]. For the description inside an optical cavity a single function will suffice. In contrast, for a freely propagating beam the situation is more involved [Fab92]. We will find that we need a whole continuum of distribution functions. These functions can be derived in the Schrödinger picture as the solutions of complex stochastic differential equations, in many cases master equations [Wal94, Scu97, Gar91]. While complete, this approach is often more general than required in practice. Here we present an alternative route based in the Heisenberg picture. We restrict our attention to systems that are linear, or can be linearized. Liberalization can be done if the modulations and the fluctuations we are concerned with are small compared to the amplitude of the laser beam. This is correct for situations with practical detectable optical powers (larger than micro watts) and where we are using CW or pulsed lasers and electronic spectrum analysers to measure the fluctuations. Working in the Heisenberg picture means the operator equations are very similar to the classical equations encountered in Section 2.2.4, and similar interpretations can be made. This approach is not directly applicable to experiments that resolve single photons using photon counting technology. However, in Chapter 12 we will adapt this approach to treat such experiments.

4.5.1 A linearized quantum noise description

Consider a propagating laser beam in a single spatial mode. We will represent the beam by the operator $\hat{A}$, where we use capitals to remind us that this is a propagating beam. We can break this operator up into two contributions: (i) the mean value of the amplitude, given by one complex number α and representing the single frequency part of the mode, and (ii) the fluctuations and modulations, described by the operator $\delta\hat{\mathbf{A}}(t)$, representing the continuum of modes surrounding the single frequency *carrier*. We thus write

$$\hat{\mathbf{A}} = \alpha + \delta\hat{\mathbf{A}}(t) \tag{4.5.1}$$

Note the similarity here with the classical treatment of small fluctuations, Eq. (2.1.15), the difference being that now the fluctuating term is represented as an operator. If the beam is detected by a photodetector then the photo-current $\hat{\mathbf{N}}$ is given by

$$\begin{aligned}
\hat{\mathbf{N}} &= \hat{\mathbf{A}}^\dagger(t)\hat{\mathbf{A}}(t) \\
&= (\alpha + \delta\hat{\mathbf{A}}^\dagger(t))(\alpha + \delta\hat{\mathbf{A}}(t)) \\
&= \alpha^2 + \alpha\,\delta\hat{\mathbf{A}}^\dagger(t) + \alpha\,\delta\hat{\mathbf{A}}(t) + \delta\hat{\mathbf{A}}^\dagger(t)\,\delta\hat{\mathbf{A}}(t) \\
&= \alpha^2 + \alpha\,\delta\hat{\mathbf{X}}\mathbf{1}(t)
\end{aligned} \tag{4.5.2}$$

where we have taken the arbitrary phase of α real and linearized by assuming the fluctuations are small and hence neglecting higher than first order terms in the fluctuations. By "small" we mean that ultimately the contribution to the calculated expectation values due to these terms will be negligible. We see that from the quantum point of view we are effectively measuring the in-phase quadrature operator $\delta\hat{\mathbf{X}}\mathbf{1} = \delta\hat{\mathbf{A}}^\dagger(t) + \delta\hat{\mathbf{A}}(t)$. For this reason $\delta\hat{\mathbf{X}}\mathbf{1}$ is often referred to as the amplitude quadrature operator. By introducing another bright beam as a reference it is also possible to measure the fluctuations in the phase of the beam, see Chapter 8. Given the linear approximation these turn out to be proportional to the out-of phase quadrature, $\delta\hat{\mathbf{X}}\mathbf{2} = i(\delta\hat{\mathbf{A}}^\dagger(t) - \delta\hat{\mathbf{A}}(t))$, which is thus often referred to as the phase quadrature. This link is illustrated in Fig. 4.7(a). Recall that $\delta\hat{\mathbf{X}}\mathbf{1}$ and $\delta\hat{\mathbf{X}}\mathbf{2}$ have continuous eigenvalue spectra. This implies that under such measurement conditions discrete, single photon events cannot be observed, regardless of the resolution of the detectors.

The discussion is more naturally continued in frequency space. The Fourier transform of the operator in Eq. (4.5.1) is

$$\tilde{\mathbf{A}}(\Omega) = \alpha\,\delta(\Omega) + \delta\tilde{\mathbf{A}}(\Omega) \tag{4.5.3}$$

where $\delta(\Omega)$ is the Dirac delta function. Recall that α is defined in a frame rotating at the optical frequency ν, see Section 2.1. Thus Ω is the frequency difference from the mean optical frequency. The mean coherent amplitude of the field appears as a delta function spike, $\alpha\,\delta(\Omega)$, which is only non-zero for $\Omega = 0$, i.e. at the mean optical frequency. We can consider the Fourier transformed fluctuation operator, in some small frequency interval around radio frequency Ω, as approximately describing a single quantum mode at this frequency. Such modes obey the commutation relations

$$[\tilde{\mathbf{A}}(\Omega), \tilde{\mathbf{A}}^\dagger(\Omega')] = \delta(\Omega - \Omega'), \quad [\tilde{\mathbf{A}}(\Omega), \tilde{\mathbf{A}}(\Omega')] = 0 \tag{4.5.4}$$

in agreement with Equation (4.1.10).

This is a useful and practical picture because experimentally we can perform Fourier analysis on the photo-current produced by photo-detection using a Spectrum analyser, discussed in Chapter 8. We can also place information on the laser beams in the form of modulation at various different frequencies Ω_{mod}, which is measured in [Hz] and each frequency can be thought of as an independent information channel. From Equation (4.5.2) we find the operator describing the photo-current in Fourier space is

$$\tilde{\mathbf{N}}_{out} = \alpha^2 \delta(\Omega) + \alpha \, \delta\tilde{\mathbf{X1}}_{out}(\Omega) \tag{4.5.5}$$

Now the first term describes the average power, concentrated in a DC spike, and the second term describes the measurement of the amplitude quadrature at frequency Ω. The variance of the photocurrent at frequency Ω is given by

$$VN_{out}(\Omega) = \alpha_{out}^2 \langle \alpha || \delta\tilde{\mathbf{X1}}_{out}(\Omega)|^2 |\alpha\rangle = \alpha_{out}^2 \, V1_{out}(\Omega) \tag{4.5.6}$$

We have a direct link between the measured values $VN_{out}(\Omega)$ and the property $V1_{out}(\Omega)$ of the state. In particular we see that ultimately the intensity noise of a bright laser is determined by the quantum states of its frequency sidebands. If the laser mode is subjected to amplitude or phase modulations (either deterministic signals or stochastic noise) these can be incorporated into the description as c-number variables. Thus we can modify Equation (4.5.3) to read

$$\tilde{\mathbf{A}}(\Omega) = \alpha\delta(\Omega) + \delta\tilde{\mathbf{A}}(\Omega) + \delta\tilde{S}(\Omega) \tag{4.5.7}$$

where the $\tilde{}$ indicates the Fourier transform of a classical variable. If the power spectra of $\delta\tilde{S}(\Omega)$ are much larger that those associated with the quantum part $(\delta\tilde{\mathbf{A}}(\Omega))$ then the beam will behave classically. However, once again, if all classical noise is suppressed we will still be left with fluctuations due to the quantum noise. If we are trying to measure or send information via small modulations of the beam then the signal to noise associated with this measurement or communication will likewise be determined by the quantum noise.

4.5.2 An example: The propagating coherent state

Let us illustrate the preceding discussion with an idealized, and then more realistic example. In Section 4.2.2 we described a coherent state as the displacement of a single cavity mode initially in a vacuum state. We may think of a propagating coherent state in a similar way as a beam with all its power at a single frequency. This means that the quantum state of the $\delta A(\Omega)$ side-band modes is the vacuum. From Eq. (4.5.6) and using Eq. (4.2.2) and (4.5.4) we immediately find

$$\begin{aligned} VN_{out}(\Omega) &= \alpha_{out}^2 \langle 0 || \delta\tilde{\mathbf{X1}}_{out}|^2 |0\rangle \\ &= \alpha_{out}^2 \langle 0 |(\delta\tilde{\mathbf{A}}^2 + (\delta\tilde{\mathbf{A}}^\dagger)^2 + \delta\tilde{\mathbf{A}}^\dagger\delta\tilde{\mathbf{A}} + \delta\tilde{\mathbf{A}}\delta\tilde{\mathbf{A}}^\dagger |0\rangle \\ &= \alpha_{out}^2 \end{aligned} \tag{4.5.8}$$

where the absolute square is taken because we are in Fourier space. The laser intensity exhibits Poissonian statistics at all frequencies. In a similar way it follows that the quadrature variances are at the quantum noise limit for all angles and all frequencies. These results are as might be expected for a travelling field generalization of a coherent state.

Of course this is an idealization. Real lasers have a finite line width and may have structure in their fluctuation spectra, see Chapter 6. Nevertheless, for lasers oscillating on single longitudinal and spatial cavity modes, so-called *single mode lasers*, at sufficiently high frequencies, the power in the sidebands will become negligible. At or above such frequencies the laser will behave like a coherent state, and it is in this sense that lasers are often referred to as approximately producing a coherent state. A typical solid-state laser will be "coherent" at frequencies above about 10 MHz. Another way of saying this is that on short timescales the laser can be thought of as producing a coherent state.

On the other hand as we go to low sideband frequencies (long timescales) eventually the power in the sidebands will become an appreciable fraction of the central peak power and our linearization assumption will break-down.

4.5.3   Real laser beams

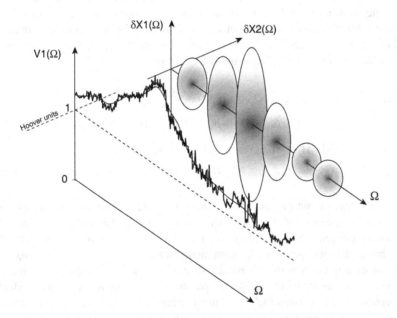

Figure 4.8: The Fourier components of the noise spectrum for a typical real laser. On the right the fluctuations for 6 different detection frequencies Ω are represented by their uncertainty areas. On the left is the corresponding continuous intensity noise spectrum $V1(\Omega)$.

We will find, in Chapter 6, that a real laser beam contains much more noise then a coherent state. Figure (4.8) shows an example of a typical laser which has an intensity noise spectrum $V1(\Omega)$ well above the quantum noise limit $V1(\Omega) = 1$ of the coherent state. This figure shows the link between the values for the quadratures $\delta\tilde{X}1$, $\delta\tilde{X}2$ and the actually measured intensity noise. In addition such a laser beam will contain modulation signals at certain frequencies. We need a technique to derive such spectra and to describe how the noise and the signals will change while the light propagates through the apparatus.

4.5.4   The transfer of operators, signals and noise

We are now in a position to describe a complete optical system and to evaluate the propagation of the noise and the signals through the system. Each optical component of the system can be thought of as having several inputs, described by their coherent amplitudes $\alpha_{in1}, \alpha_{in2}, \ldots$ etc., their fluctuation operators $\delta A_{in1}, \delta A_{in2}, \ldots$ etc. and their states $|\Phi_{in1}\rangle, |\Phi_{in2}\rangle, \ldots$etc. The corresponding spectral variances for each input are

$$
\begin{aligned}
V1_{inj}(\Omega) &= \langle \Phi_{inj}||\tilde{\mathbf{X1}}_{inj}(\Omega)|^2|\Phi_{inj}\rangle \\
V2_{inj}(\Omega) &= \langle \Phi_{inj}||\tilde{\mathbf{X2}}_{inj}(\Omega)|^2|\Phi_{inj}\rangle
\end{aligned}
\tag{4.5.9}
$$

where $j = 1, 2, \ldots$. Inside the optical components the modes are combined. The algebraic equations for the coherent amplitudes are solved to give the output coherent amplitude. We usually choose the arbitrary overall phase of the input fields such that the output coherent amplitude is real. After linearization the equations of motion for the fluctuation operators can be solved in frequency space. The fluctuation operators in the frequency domain for one output mode can then be described by equations of the form

$$
\begin{aligned}
\delta\tilde{\mathbf{A}}_{\text{out}}(\Omega) &= c_{1a}(\Omega)\,\delta\tilde{\mathbf{A}}_{in1}(\Omega) + c_{1c}(\Omega)\,\delta\tilde{\mathbf{A}}_{in1}^{\dagger}(\Omega) \\
&\quad + c_{2a}(\Omega)\,\delta\tilde{\mathbf{A}}_{in2}(\Omega) + c_{2c}(\Omega)\,\delta\tilde{\mathbf{A}}_{in2}^{\dagger}(\Omega) + \cdots \\
\delta\tilde{\mathbf{A}}_{\text{out}}^{\dagger}(\Omega) &= c_{1c}^{*}(-\Omega)\,\delta\tilde{\mathbf{A}}_{in1}(\Omega) + c_{1a}^{*}(-\Omega)\,\delta\tilde{\mathbf{A}}_{in1}^{\dagger}(\Omega) \\
&\quad + c_{2c}^{*}(-\Omega)\,\delta\tilde{\mathbf{A}}_{in2}(\Omega) + c_{2a}^{*}(-\Omega)\,\delta\tilde{\mathbf{A}}_{in2}^{\dagger}(\Omega) + \cdots
\end{aligned}
\tag{4.5.10}
$$

and similarly for all other output modes. The coefficients $c_{kl}(\Omega)$ are complex numbers which describe the frequency response of the component. Note, that the response is not necessarily symmetric about the carrier. The c_{ka} and c_{kc} are linked via commutation requirements. For linear systems the c_{kl} do not depend on the beam intensities and all $c_{kc} = 0$. We can also treat the case of propagation through nonlinear media, such as $\chi^{(2)}$ and $\chi^{(3)}$ media. This includes nonlinear processes such as frequency doubling, parametric amplification, the Kerr effect or nonlinear absorption. The Equations (4.5.10) remain linear, but in these cases the coefficients $c_{1a}, c_{1c}, c_{2a}, c_{2c}, \ldots$ are dependent on the average optical power of the beams. In various examples throughout this book we will show how the coefficients c_{ij} can be derived from the physical properties of the component, such as reflectivities, conversion efficiencies etc.. Once the operator $\delta\tilde{\mathbf{A}}_{\text{out}}(\Omega)$ for the output mode in question has been found the variances can be determined using Equation (4.5.9). Let us consider some general cases:

(1) The simplest case is if firstly all the inputs are independent, i.e.
$\langle \delta\tilde{\mathbf{X1}}_{ink}(\Omega)\delta\tilde{\mathbf{X1}}_{inl}(\Omega)\rangle = \delta_{kl}$. This will be true if all the input beams represent beams from independent sources or different vacuum inputs. Secondly, we require that the coefficients obey $c_{ij}^{*}(-\Omega) = c_{ij}(\Omega)$ i.e. magnitude and phase of the frequency response are symmetric. This will be true if there are no detunings or strong dispersion in the optical system. In this case we obtain the simplest type of transfer function. Indeed, for a linear

system ($c_{1c} = c_{2c} = \ldots = 0$) the transfer functions for both quadratures are identical.

$$
\begin{aligned}
V1_{\text{out}}(\Omega) &= |c_{1a}(\Omega) + c_{1c}(\Omega)|^2 V1_{in1}(\Omega) \\
&\quad + |c_{2a}(\Omega) + c_{2c}(\Omega)|^2 V1_{in2}(\Omega) + \cdots \\
V2_{\text{out}}(\Omega) &= |c_{1a}(\Omega) - c_{1c}(\Omega)|^2 V2_{in1}(\Omega) \\
&\quad + |c_{2a}(\Omega) - c_{2c}(\Omega)|^2 V2_{in2}(\Omega) + \cdots
\end{aligned}
\tag{4.5.11}
$$

(2) The coefficients still obey $c_{ij}^*(-\Omega) = c_{ij}(\Omega)$ but we allow for inputs that are not independent. Now there can be correlations between the quadrature variances of different inputs. However, we can always regain the form of Eq. (4.5.11) by breaking down each of the inputs into a function of its own inputs until we end up with all the inputs being independent.

(3) The coefficients obey $c_{ij}^*(-\Omega) = c_{ij}^*(\Omega)$, i.e. the magnitude of the frequency response is symmetric but not the phase. This will occur if the fluctuations suffer a different phase rotation to that of the coherent amplitude. In this case we get cross-coupling between the transfer functions of $V1_{\text{out}}$ and $V2_{\text{out}}$ such that now

$$
\begin{aligned}
V1_{\text{out}}(\Omega) &= |c_{1a}(\Omega) + c_{1c}(\Omega)|^2 \cos^2(\theta) V1_{in1}(\Omega) \\
&\quad + |c_{1a}(\Omega) - c_{1c}(\Omega)|^2 \sin^2(\theta) V2_{in1}(\Omega) + \cdots \\
V2_{\text{out}}(\Omega) &= |c_{1a}(\Omega) - c_{1c}(\Omega)|^2 \sin^2(\theta) V2_{in1}(\Omega) \\
&\quad + |c_{1a}(\Omega) + c_{1c}(\Omega)|^2 \cos^2(\theta) V1_{in1}(\Omega) + \cdots
\end{aligned}
\tag{4.5.12}
$$

where θ is defined by $c_{ij}(\Omega) = |c_{ij}(\Omega)|e^{i\theta}$. We have assumed that $V1$ and $V2$ are the quadratures of minimum and maximum fluctuations.

(4) The coefficients obey $c_{ij}^*(-\Omega) \neq c_{ij}^*(\Omega)$. This will occur when the frequency response of the optical components is asymmetric around the carrier frequency, for example due to detunings. It is not possible to represent the output as a transfer function of quadrature variances in this case.

The simplest example of a transfer function describes the influence of loss terms only and has the form: $V1_{\text{out}}(\Omega) = |c_{1a}|^2 V1_{in1}(\Omega) + |c_{2a}|^2 + |c_{3a}|^2 + \cdots$, where all the inputs apart from $V1_{in1}(\Omega)$ are vacuum inputs which have the numerical value of 1. Such a transfer function has identical effects on both quadratures. An example of case (2) would be where either a modulation is common or two or more of the input modes have correlated noise. This is the case when the beams originate from one source, such as two beams formed by a beamsplitter, or where two beams are generated by an above threshold OPO. Examples of case (3) can be as benign as mixing beams on a beamsplitter where the coherent amplitudes of the inputs have different phases, or components with strong dispersion where the refractive index changes within a frequency interval of Ω_{det}, or as complex as single ended cavities with optical detuning. The simplest example of case (4) is a double ended optical cavity with detuning.

An appropriate name for Equation (4.5.11) is quantum noise transfer function. This equation is very similar to the classical transfer functions used by communication engineers. The classical function can be derived using a description of the various inputs as classical waves with modulation at Ω. Noise behaves exactly in the same way as the modulations. Quantum noise is not included. The propagation and interference of these waves is determined and the result is the size of the modulation at the output. In contrast, Equation (4.5.11) includes all the quantum effects of the light, in particular the effects of quantum noise. We can directly see how the signals vary in comparison to the quantum noise and can evaluate the coefficient for the transfer of the signal to noise ratio from input to output. This approach has been developed since the 1980s by a number of research groups, very prominently by the groups in France who pioneered squeezing [Fab92].

4.5.5   Sideband modes as quantum states

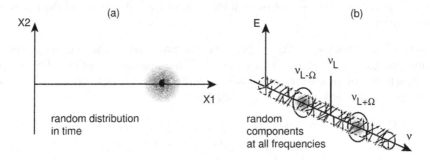

Figure 4.9: Comparison between (a) the phase space diagram and (b) the sideband picture. The quantum properties are encapsulated in noise sidebands at all modulation frequencies above and below the laser frequency ν_L.

We will now outline how we can relate the quantum transfer functions to the detailed quantum state of the field at different frequencies. In particular we will consider the example of modulation of a quantum noise limited beam and see how this can be viewed as producing a coherent state at the modulation frequency. In Section 2.1.6 we saw how small modulations of a classical light beam resulted in coherent power being injected symmetrically into upper and lower sidebands around the carrier. In Section 4.5.1 we described classical modulations in the Heisenberg picture as evolution of the quantum fluctuation operator

$$\delta\tilde{\mathbf{A}}(\Omega) \to \delta\tilde{\mathbf{A}}(\Omega) + \delta\tilde{S}(\Omega) \tag{4.5.13}$$

This Heisenberg evolution is equivalent to the Schrödinger evolution of displacement, see Section 4.2.2, i.e.

$$|\psi\rangle_\Omega \to \tilde{\mathbf{D}}(\delta\tilde{\mathbf{A}}(\Omega))|\psi\rangle_\Omega \tag{4.5.14}$$

Now consider the amplitude quadrature operator in Fourier space. It is easy to show that in general the Fourier transform of a conjugate, e.g. $\tilde{\mathbf{A}}^\dagger(\Omega)$, is related to the conjugate of the Fourier transform, e.g. $\tilde{\mathbf{A}}(\Omega)^\dagger$, via a change in sign of Ω, e.g. $\tilde{\mathbf{A}}^\dagger(\Omega) = \tilde{\mathbf{A}}(-\Omega)^\dagger$. Using this property of the Fourier transform we can observe that

$$
\begin{aligned}
\delta\tilde{\mathbf{X1}}(\Omega) &= \delta\tilde{\mathbf{A}}(\Omega) + \delta\tilde{\mathbf{A}}^\dagger(\Omega) \\
&= \delta\tilde{\mathbf{A}}(\Omega) + \delta\tilde{\mathbf{A}}(-\Omega)^\dagger
\end{aligned}
\tag{4.5.15}
$$

Thus we see that a quadrature measurement actually probes both the upper and lower sidebands. Suppose the initial state of the sidebands is a vacuum. Using Equation (4.5.14) we find modulation produces the coherent sideband state

$$|\psi'\rangle = |\tilde{\delta S}(\Omega)\rangle_\Omega |\tilde{\delta S}(-\Omega)\rangle_{-\Omega} \qquad (4.5.16)$$

where, in response to Equation (4.5.15), we have written the state explicitly in terms of an upper and lower sideband part. We are ignoring the low frequency contributions from the carrier.

For an arbitrary mix of ideal amplitude and phase modulation we have

$$\tilde{\delta S}(\Omega) = \beta\big(\delta(\Omega + \Omega_{\text{mod}}) + \delta(\Omega - \Omega_{\text{mod}})\big) \qquad (4.5.17)$$

The complex number $\beta = M_a/4 + iM_p/4$, where M_a and M_p are the amplitude and phase modulation depths respectively and Ω_{mod} is the modulation frequency, see Section 2.1.6. Averaging over a small range of frequencies around Ω_{mod} we can write our state as

$$|\psi'\rangle = |\beta\rangle_{\Omega_{\text{mod}}} |\beta\rangle_{-\Omega_{\text{mod}}} |0\rangle_{\Omega \neq \pm\Omega_{\text{mod}}} \qquad (4.5.18)$$

A pair of identical coherent states has been created at the upper and lower sideband frequencies. As quadrature measurements probe the sidebands symmetrically, see Equation (4.5.15), the results of quadrature measurements are identical to those for a single frequency coherent state. For example the expectation value of $X1$ in the Schrödinger picture is given by

$$\begin{aligned}
\langle X1 \rangle &= \langle \beta | \tilde{\delta \mathbf{A}}(\Omega_m) | \beta \rangle_{\Omega_m} + \langle \beta | \tilde{\delta \mathbf{A}}(-\Omega_m)^\dagger | \beta \rangle_{-\Omega_m} \\
&= \beta + \beta^* \qquad (4.5.19)
\end{aligned}$$

This is of course the same as the result that would be obtained in the Heisenberg picture. This result generalizes to other states provided the transfer functions remain symmetric about the carrier. Most interesting are states for which the upper and lower sideband modes become entangled as this has no classical analog. Squeezed states, which are described in Chapter 9 are an example of such states.

Using a homodyne detector it is possible to measure the variances $V_\Theta(\Omega)$ at all frequencies (Ω). The properties of the light in a small interval around each frequency can then be described by one Wigner function $W(\Omega)$. The interval is assumed small enough such that the frequency response is flat. In this way the entire beam can be described as a series of quantum states, one for each small frequency interval. This is a Schrödinger picture view of the travelling field which complements very well the actual technology used in optical communication. Here each of these frequency intervals is used as a separate communication channel.

The variance $V_\Theta(\Omega)$ can be determined mathematically by first projecting all parts of the Wigner function onto one axis at angle Θ. This produces a series of one dimensional distribution functions $DF(\Theta)$ which describe in the form of a histogram the time varying photocurrent in a small frequency range. In a second step the variances of these functions are evaluated which provides information about the width of $DF(\Theta)$. In general we would need to evaluate the moments of $DF(\Theta)$ to all orders. However, as we have seen, for the restricted class of Gaussian functions a direct link exists between the Wigner functions and the variances and a complete description is given by $V_{\text{min}}(\Omega)$, $V_{\text{max}}(\Omega)$ and $\Theta_{\text{min}}(\Omega)$, as shown in Section 4.3.2.

It is also possible to obtain the average photon number in the sidebands from the quadrature phase measurements. Consider the sum of the amplitude and phase quadrature variances written in terms of annihilation and creation operators. We have

$$
\begin{aligned}
V1(\Omega) + V2(\Omega) &= 2\langle \delta \tilde{\mathbf{A}}(\Omega)^\dagger \delta \tilde{\mathbf{A}}(\Omega) + \delta \tilde{\mathbf{A}}(-\Omega) \delta \tilde{\mathbf{A}}(-\Omega)^\dagger \rangle \\
&= 2\langle \delta \tilde{\mathbf{A}}(\Omega)^\dagger \delta \tilde{\mathbf{A}}(\Omega) + \delta \tilde{\mathbf{A}}(-\Omega)^\dagger \delta \tilde{\mathbf{A}}(-\Omega) \rangle + 2 \qquad (4.5.20)
\end{aligned}
$$

Hence we can write

$$
\delta N(|\Omega|) = \langle \delta \tilde{\mathbf{A}}(|\Omega|)^\dagger \delta \tilde{\mathbf{A}}(|\Omega|) \rangle = \frac{1}{4}(V1(\Omega) + V2(\Omega) - 2) \qquad (4.5.21)
$$

For a coherent state (of the entire beam) we expect $V1 = V2 = 1$ at all frequencies and so from Equation (4.5.21) there are no photons in the sidebands as discussed in Section 4.5.2.

This discussion requires a caveat: we have assumed throughout that the sidebands are symmetric around the carrier, i.e. our transfer function is not of type (4) as described in the previous section. Asymmetric transfer functions can occur. For example it is possible to produce single sideband modulation. The state produced in such a situation is

$$
|\psi'\rangle = |\beta\rangle_{\Omega_{\mathrm{mod}}} |0\rangle_{\Omega \neq \Omega_{\mathrm{mod}}} \qquad (4.5.22)
$$

Such states are not fully characterized by homodyne quadrature measurements.

4.6 Quantum correlations

A key difference between classical optical fields and their quantum counterparts is that the correlations that can exist between spatially separated beams in quantum mechanics can be stronger than is allowed by classical theory. We now look at examples of how this difference can be quantified both in the single photon and linearized domains.

4.6.1 Photon correlations

In Section 2.2.3 we saw that the intensity correlations between different beams could be characterized by the second order correlation function, $g^{(2)}$, which was defined by

$$
g^{(2)}(\tau) = \frac{\langle I(t) I(t+\tau) \rangle}{\langle I(t) \rangle^2} \qquad (4.6.1)
$$

We also saw that for classical fields $g^{(2)}(0) \geq 1$. In contrast we argued in Section 2.2.3 that $g^{(2)}(0) < 1$ should be possible for quantum mechanical fields. We will now treat this question with the full quantum formalism.

The calculation that led to the restriction on $g^{(2)}$ in the classical case relied on the fact that $\langle A^2 \rangle \geq \langle A \rangle^2$. This relationship still holds if A is an operator, so how is it that we can have $g^{(2)}(0) < 1$ for a quantum mechanical field? To answer this we must look more carefully at what the actual measurement of $g^{(2)}$ involves. As outlined previously we need to divide the field in half at a beamsplitter and then separately measure the intensities of the two halves with a variable time delay, τ, introduced. For a classical field the action of the beamsplitter is $I(t) \rightarrow 1/2I(t)$. Thus by finding the average of the product of the intensities of the two beams and normalizing by their individual averages we arrive at the above expression for $g^{(2)}$.

Now consider the quantum mechanical case. As we have seen in Section 4.4.3 (see Chapter 5 for more details) the correct quantum mechanical expressions for the outputs of a 50:50 beamsplitter are

$$
\begin{aligned}
\hat{a}' &= \sqrt{0.5}\,\hat{a} + \sqrt{0.5}\,\hat{b} \\
\hat{b}' &= \sqrt{0.5}\,\hat{a} - \sqrt{0.5}\,\hat{b}
\end{aligned}
\tag{4.6.2}
$$

where $\tilde{a}$ represents the beam whose intensity we are measuring, whilst $\hat{b}$ is initially in the vacuum state. If we now form the ratio of the average of the product of the intensities of the two beams normalized to their individual averages we obtain

$$
g^{(2)}(\tau) = \frac{\langle (\hat{a}^\dagger(t) + \hat{b}^\dagger(t))(\hat{a}(t) + \hat{b}(t))(\hat{a}^\dagger(t+\tau) - \hat{b}^\dagger(t+\tau))(\hat{a}(t+\tau) - \hat{b}(t+\tau)) \rangle}{\langle (\hat{a}^\dagger(t) + \hat{b}^\dagger(t))(\hat{a}(t) + \hat{b}(t)) \rangle \langle (\hat{a}^\dagger(t+\tau) + \hat{b}^\dagger(t+\tau))(\hat{a}(t) - \hat{b}(t+\tau)) \rangle}
\tag{4.6.3}
$$

Using the field commutation relations and the fact that the $\hat{b}$ mode is initially in the vacuum state (see Section 4.2.1) this expression can be significantly simplified to

$$
g^{(2)}(\tau) = \frac{\langle \hat{a}^\dagger(t)\hat{a}^\dagger(t+\tau)\hat{a}(t)\hat{a}(t+\tau) \rangle}{\langle \hat{a}^\dagger(t)\hat{a}(t) \rangle^2}
\tag{4.6.4}
$$

where we have also assumed the individual beam averages are not a function of time (the system is ergodic).

Equation (4.6.4) is then the correct quantum mechanical expression for the second order correlation function. Notice that the ordering of the operators under the averages is such that all creation operators stand to the left whilst all annihilation operators stand to the right. This is referred to as *normal ordering* and is often a convenient way of expressing results.

We can now consider two simple examples. First suppose the mode $\hat{a}$ was initially in a coherent state. Using Equation (4.6.4) we immediately find that for a coherent state $g^{(2)}(0) = 1$, as expected for an ideal classical field. However, if the mode $\hat{a}$ is in a number state then the numerator is immediately zero, as two applications of the annihilation operator to the one photon state produces zero. Thus for a single photon number state $g^{(2)}(0) = 0$, in stark contrast to the classical limit.

In Chapters 12 and 13 we will see other examples of quantum intensity correlations as functions of different degrees of freedom such as polarization.

4.6.2 Quadrature correlations

With bright beams of light, for which linearization is appropriate, we are more interested in correlations between the quadrature amplitude fluctuations of the spatially separated beams, recorded by separate detectors. These correlations can be quantified by measuring the quadrature difference and sum variances defined by

$$V1_s = \frac{1}{2}\langle(\delta\tilde{\mathbf{X}}1_1 + \delta\tilde{\mathbf{X}}1_2)^2\rangle$$

$$V2_s = \frac{1}{2}\langle(\delta\tilde{\mathbf{X}}2_1 + \delta\tilde{\mathbf{X}}2_2)^2\rangle \tag{4.6.5}$$

for the sum variances of the two quadratures and

$$V1_d = \frac{1}{2}\langle(\delta\tilde{\mathbf{X}}1_1 - \delta\tilde{\mathbf{X}}1_2)^2\rangle$$

$$V2_d = \frac{1}{2}\langle(\delta\tilde{\mathbf{X}}2_1 - \delta\tilde{\mathbf{X}}2_2)^2\rangle \tag{4.6.6}$$

for the difference quadrature variances. Here the subscripts label the beams 1 and 2. If the beams are correlated, say in the amplitude quadrature, we would expect $V1_d < V1_s$. Conversely if the beams were anticorrelated we would find $V1_d > V1_s$. Two coherent beams are found to be uncorrelated, with their noise at the QNL, that is for two coherent beams we have $V1_d = V1_s = V2_d = V2_s = 1$. Beams are said to be quantum correlated if they have sum or difference quadrature variances which fall below 1. Such correlations cannot be consistently accommodated in a classical theory.

More generally we might consider the conditional variance of, say beam 1's amplitude quadrature, given that we have measured beam 2's amplitude quadrature. This tells us how accurately we can infer the value of beam 1's amplitude quadrature given the measurement of beam 2 and is defined as

$$V1_{1|2} = \langle(\delta\tilde{\mathbf{X}}1_1 + g\delta\tilde{\mathbf{X}}1_2)^2\rangle \tag{4.6.7}$$

Here g is a gain which is chosen such as to minimize the value of $V1_{1|2}$. By calculus we can find that the optimum value of g is

$$g = -\frac{\langle\delta\tilde{\mathbf{X}}1_1\delta\tilde{\mathbf{X}}1_2\rangle}{V1_2} \tag{4.6.8}$$

Substituting back into Equation (4.6.7) we obtain

$$V1_{1|2} = V1_1 - \frac{\langle\delta\tilde{\mathbf{X}}1_1\delta\tilde{\mathbf{X}}1_2\rangle^2}{V1_2} \tag{4.6.9}$$

Similarly for the phase quadrature conditional variance we have

$$V2_{1|2} = V2_1 - \frac{\langle\delta\tilde{\mathbf{X}}2_1\delta\tilde{\mathbf{X}}2_2\rangle^2}{V2_2} \tag{4.6.10}$$

For coherent states we find $V1_{1|2} = V2_{1|2} = 1$, that is the conditional variance is no better than the single beam variance indicating no correlation. Quantum correlation occurs when a conditional variance falls below 1. A special situation, which can only happen when the beams are entangled, is when both $V1_{1|2}$ and $V2_{1|2}$ fall below 1 and this will be discussed in Chapter 13.

It is convenient to define the expectation value of the product of the quadratures of the two beams as the amplitude quadrature correlation coefficient, $C1$, i.e.

$$C1 = \langle \delta \tilde{\mathbf{X}} 1_1 \delta \tilde{\mathbf{X}} 1_2 \rangle \qquad (4.6.11)$$

Experimentally we can obtain $C1$ by measuring the sum and difference quadrature variances, i.e.

$$C1 = \frac{1}{2}(V1_s - V1_d) \qquad (4.6.12)$$

In terms of $C1$ we then have that uncorrelated beams have $C1 = 0$, correlated beams have $C1 > 0$ and anti-correlated beams have $C1 < 0$. Similarly for the phase quadrature correlation coefficient we have

$$C2 = \frac{1}{2}(V2_s - V2_d) \qquad (4.6.13)$$

4.7 Summary: The different quantum models

There are three alternative models, or pictures for quantum noise: We can consider light as photons and describe them by quantum states, we can use operators to describe CW beams as modes of the radiation field or we can extend the classical wave model and describe the modulation, and the noise by sidebands. All three models predict the same effects and produce the same results. Which one to use depends on the application.

The photon model

The photon picture is intuitive and obvious for the explanation of emission and detection processes. However, care has to be taken, since photons do not behave like classical particles. Their quantum properties have to be taken into account.

The description in the form of the Schrödinger picture, using photon states, is optimized for situations with low photon numbers and single photon events. A whole separate technology for generating and detecting single photons and photon coincidences has been developed that matches this model well. The special quantum properties of the lights are directly accessible in this regime and they are used extensively for quantum application such as cryptography and quantum information processing, described in Chapter 13.

We will find elsewhere in this guide, Chapter 9, that the photon picture is very useful in explaining the properties of nonlinear processes of CW laser beams, such as second harmonic generation (SHG) or optical parametric oscillation (OPO). While this is a useful and simple illustration of the intensity effects we would lose information about the quadrature phase. The full quantum model, based on the quadrature amplitudes, is indispensable for a full quantitative description of experiments in optics.

Quantum operators

The quantum description based on quantum modes and operators is the most rigorous and complete model for CW laser beams. It is used in all quantitative predictions and, with some practice, can be fairly intuitive. For completeness all commonly used quantum operators, states and functions used in quantum optics are listed in Appendix B.

For all practical applications the simplified linearized model is fully sufficient. The most versatile operators are those for the quadratures, and we can completely describe the beam by the operators $\delta\tilde{\mathbf{X}}1(\Omega)$ and $\delta\tilde{\mathbf{X}}2(\Omega)$. This model is used to directly derive the noise transfer functions for $V1(\Omega)$ and $V2(\Omega)$ describing the different devices. We have one simple recipe to determine the variances via

$$V(\Omega) = \langle|\delta\tilde{\mathbf{X}}(\Omega)|^2\rangle.$$

In this way we can analyse complete optical systems from source to detector by evaluating the operators at the various locations. This model of transfer functions will be extremely useful to derive the properties of light beams which have deeper quantum properties, such as squeezed light, Chapter 9, entanglement between the quadratures and CW quantum processing such as optical teleportation, Chapter 13.

Sidebands

The sideband model is a picture that allows simple predictions of the performance of optical systems and machines. It is mathematically equivalent to the linearized operator model, but the interpretation is chosen to be as close as possible to conventional engineering models, based on Fourier transforms, filters and transfer functions. This model describes individual Fourier components of the field, as is required in most technical applications such as communication, and represents the quantum properties by one single addition term, the quantum noise sideband. The effects of the apparatus, in particular nonlinear media, can frequently be explained more easily by considering the effect on each sideband separately and then how both are combined in the measurement. It is possible to use such simple, classical calculations to describe the propagation of both modulation signals and noise through the apparatus and to incorporate the quantum effects through simple empirical rules. This model should be useful for readers with a technical or engineering background and prefer to use the concept of sidebands rather than the more mathematically rigorous quantum models.

The quantum noise limit

We have found that the coherent state is a limiting case, the quietest light that can be produced with a conventional source, both as thermal light from a lamp, or as coherent light from a laser. Quantum noise is the way the quantum nature of light manifests itself in CW laser beams. However, there is more to come, other types of light states are possible. Within the quantum model it is possible to define light that has reduced fluctuations in one of its quadratures. Such *squeezed states of light* and their generation are the topic of Chapter 9.

We find that for all devices which are sensitive enough to detect quantum noise, and these are getting more and more common due to the advances in technology, it is important to understand the quantum properties of the light. They will influence the performance of the devices. In many cases the quantum effects will set a fundamental limit to the performance, the quantum noise limit (QNL). We will find tricks to avoid these limitations. In other cases the performance will be quite different near the QNL compared to the classical regime well above the QNL. And finally we will find applications which rely completely on the quantum effects.

Further reading

Quantum noise 2nd edition C.W. Gardiner, P. Zoller, Springer, Berlin (2000)
Optical coherence and quantum optics, L. Mandel, E. Wolf, Cambridge University Press (1996)
Tutorial review of quantum optical phase, D.T. Pegg, S.M. Barnett, J.of Mod. Opt. 44, 225 (1997)
Squeezed states of light, C. Fabre, Physics Reports 219, 215 (1992)
Lectures on quantum optics, W. Vogel, D-G. Welsch, Akademie Verlag, Berlin (1994)
Quantum optics in phase space, W.P. Schleich, Wiley-VCH (2001)

Bibliography

[Bac92] Quantum optics experiments with atoms, H-A. Bachor, D.E. McClelland, Phys. Scr. T40, 40 (1992)

[Cah69] Density Operators and quasiprobability distributions, K.E. Cahill, R.J. Glauber, Phys. Rev. 177, 1888 (1969)

[Cav80] C.M. Caves in *Coherence, Cooperation and Fluctuations* eds. F. Haake, L.M. Narducci, D.F. Walls, Cambridge University Press (1980)

[Cas48] The influence of retardation on the London - van der Waals forces, H.B.G. Casimir, D. Polder, Phys. Rev. 73, 360 (1948)

[Col84] Squeezing of intracavity and traveling-wave light fields produced in parametric amplification, M.J. Collett, C.W. Gardiner, Phys. Rev. A 30, 1386 (1984)

[Dru80] Generalised P-representation in quantum optics, P.D. Drummond, C.W. Gardiner, J. Phys. A 13, 2353 (1980)

[Fab92] Squeezed states of light, C. Fabre, Physics Reports 219, 215 (1992)

[Fre95] Two mode quantum phase, M. Freyberger, M. Heni, W.P. Schleich, Quantum. Semiclass. Opt. 7, 187 (1995)

[Gar91] *Quantum noise 2nd edition* C.W. Gardiner, P. Zoller, Springer, Berlin (2000)

[Gla62] Coherent and incoherent states of the radiation field, R.J. Glauber, Phys. Rev. 131, 2766 (1962)

[Gol80] *Classical Mechanics*, H. Goldstein, Addison-Wesley, Singapore (1980).

[Hak70] *Laser Theory* in Encyclopaedia of Physics, Vol. XXV/2c, Light and Matter Interaction, ed. S. Flügge, First ed., Springer, Berlin (1970)

[Hen89] A squeezed state primer, R.W. Henry, S.C. Glotzer, Am. J. Phys. 56, 318 (1988)

[Kit57] Direct Measurement of the long-range van der Waals forces, J.A. Kitchener, A.P. Prosser, Proc. Roy. Soc. A 242, 403 (1957)

[Lou73a] *Quantum theory of light*, R. Loudon, Oxford University Press, Oxford (1973)

[Lou87]   Squeezed light, R. Loudon, P.L. Knight, J. Mod. Opt. 34, 709 (1987)

[Lou73b]   *Quantum statistical properties of radiation*, W.H. Louisell, Wiley, New York (1973)

[Pau30]   *General principles of quantum mechanics*, W. Pauli, Springer (1980)

[Pau95]   *Photonen, eine Einführung in die Quantenoptik*, H. Paul, Teubner, Stuttgart (1995)

[Peg88]   Unitary phase operator in quantum mechanics, D.T. Pegg, S.M. Barnett, Europhys. Lett. 6, 483 (1988)

[Peg89]   Phase properties of the quantized single-mode electromagnetic field, D.T. Pegg, S.M. Barnett, Phys. Rev. A 39,1665 (1989)

[Pla00]   Zur Theorie des Gesetzes der Energieverteilung im Normalspektrum, M. Planck, Verhandlungen der Deutschen Physikalischen Gesellschaft 2, 237 (1900)

[Pow64]   *Introductory Quantum Electrodynamics* by E.A. Power, Longmans (1964)

[Ral96]   Intensity noise of injection locked lasers: quantum theory using a linearised input/output method, T.C. Ralph, C.C. Harb, H-A. Bachor, Phys. Rev. A 54, 4370 (1996)

[Sch93]   Special issue on *Quantum phase and phase dependent measurements* eds. W.P. Schleich, S.M. Barnett, Phys. Scr. T48 (1993)

[Scu97]   *Quantum Optics*, M.O. Scully, M.S. Zubairy, Cambridge University Press (1997)

[Sie86]   *Lasers*, A.E. Siegman, University, Science Books (1986)

[Vog94]   *Lectures on Quantum Optics*, W. Vogel, D-G. Welsch, Akademie Verlag, Berlin (1994)

[Wal94]   *Quantum Optics*, D.F. Walls, G.J. Milburn, Springer (1994)

5 Basic optical components

Real experiments consist of many components – at first sight any quantum optics experiment looks like a bewildering array of lenses, mirrors etc. filling an optical table in a more or less random fashion. However, they all serve a purpose, and their individual characteristics are described in this chapter.

In quantum optics experiments we consider light as laser beams or modes of radiation. Using optical components, we want to change the parameters of the mode, such as beam size, direction, polarisation or frequency. This means we wish to transform one mode into another mode. This can be achieved by a whole range of optical components all of which are *linear*, which means that their output is proportional to their input, independent of the intensity of the beam.

One of the simplest optical processes is attenuation by a beamsplitter, and we will see how it affects both the intensity as well as the fluctuations of the light. While the first is simple, that is, the intensity is reduced, the latter requires more detailed modelling. The complementary process, gain, is similar but requires much more elaborate equipment and theory, which will be covered in Chapter 6, describing lasers. Combining several mirrors allows us to build interferometers and cavities. Several other optical components are used to manipulate the light: spatial parameters can be controlled with lenses and curved mirrors; polarisation can be influenced with various crystals and surface effects; frequency can be shifted using electro-optic modulators.

5.1 Beamsplitters

One of the most common and simplest optical components is a mirror. This is made either of thin layers of metal or multi layer dielectric films deposited on glass. A mirror has two output beams for every input beam, one transmitted the other reflected. Thus it is referred to as a beamsplitter. Let us analyse the properties of a beamsplitter. We can use the different models described in Chapter 4, they will obviously lead to the same conclusion but will provide different interpretations of the effect of a beamsplitter.

5.1.1 Classical description of a beamsplitter

Consider a beamsplitter as shown in Fig. 5.1 with an incoming beam described classically by the complex amplitude $\alpha_{in} \exp(i\phi_{in})$. There are two waves leaving the beamsplitter: the transmitted beam, $\alpha_t \exp(i\phi_t)$, and the reflected beam, $\alpha_r \exp(i\phi_r)$. We describe the properties of

A Guide to Experiments in Quantum Optics, 2nd Edition. Hans-A. Bachor and Timothy C. Ralph
Copyright © 2004 Wiley-VCH Verlag GmbH & Co. KGaA
ISBN: 3-527-40393-0

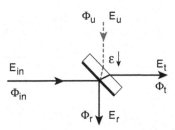

Figure 5.1: The classical beamsplitter

the beamsplitter by the intensity reflection coefficient ϵ

$$
\begin{aligned}
|\alpha_r|^2 &= \epsilon |\alpha_{\text{in}}|^2 \\
|\alpha_t|^2 &= (1 - \epsilon)|\alpha_{\text{in}}|^2
\end{aligned}
\tag{5.1.1}
$$

The special case $\epsilon = \frac{1}{2}$ describes a balanced 50/50 beamsplitter. However, there is also a second input axis defined such that a wave $\alpha_u \exp(i\phi_u)$ would produce at the output a wave in the same place and propagating in the same direction as α_t and α_r. For a simple beamsplitter only α_{in} exists and $\alpha_u = 0$. In other applications two waves α_{in} and α_u are superimposed and, in this way, are made to interfere. The optical frequencies of all input and output beams are identical. All the frequencies, including all the sidebands, are preserved. Assuming it has no losses the mirror will preserve energy. This means

$$
|\alpha_{\text{in}}|^2 + |\alpha_u|^2 = |\alpha_r|^2 + |\alpha_t|^2
\tag{5.1.2}
$$

The intensity response to the other input beam can be written similarly to Equation (5.1.1). The response for amplitude transmission and reflection is not that simple, since the effect of the beamsplitter on the phase of the light has to be considered. Let us arbitrarily set the phase of the two incoming waves to zero, that means $\phi_{\text{in}} = \phi_u = 0$. The relative phase between the two output beams is of interest

$$
\begin{aligned}
\alpha_t \exp(i\phi_t) &= \sqrt{(1 - \epsilon)}\ \alpha_{\text{in}} \exp(i\phi_{\text{in},t}) - \sqrt{\epsilon}\ \alpha_u \exp(i\phi_{u,t}) \\
\alpha_r \exp(i\phi_r) &= \sqrt{\epsilon}\ \alpha_{\text{in}} \exp(i\phi_{\text{in},r}) + \sqrt{(1 - \epsilon)}\ \alpha_u \exp(i\phi_{u,r})
\end{aligned}
$$

There is not sufficient information to determine all four relative phase terms $\phi_{\text{in},t}$; $\phi_{\text{in},r}$; $\phi_{u,t}$; $\phi_{u,r}$ directly. This could be done by solving the electro magnetic boundary conditions explicitly. The solutions, known as Fresnel equations can be found in text books on physical optics. These equations are different for metal and dielectric mirrors and they depend on the polarisation of the input wave. However, we can apply energy conservation and find a general condition for the four waves

$$
\begin{aligned}
|\alpha_{\text{in}}|^2 + |\alpha_u|^2 &= |\alpha_t|^2 + |\alpha_r|^2 \\
&= (1 - \epsilon)|\alpha_{\text{in}}|^2 + \epsilon|\alpha_u|^2 + \epsilon|\alpha_{\text{in}}|^2 + (1 - \epsilon)|\alpha_u|^2 \\
&\quad + \sqrt{\epsilon(1 - \epsilon)}\ \alpha_u \alpha_{\text{in}} \exp(i(\phi_{\text{in},t} - \phi_{u,t})) + c.c. \\
&\quad + \sqrt{\epsilon(1 - \epsilon)}\ \alpha_{\text{in}} \alpha_u \exp(i(\phi_{\text{in},r} - \phi_{u,r})) + c.c.
\end{aligned}
\tag{5.1.3}
$$

which leads to

$$\exp(i(\phi_{in,t} - \phi_{u,t})) + \exp(i(\phi_{in,r} - \phi_{u,r})) = 0 \qquad (5.1.4)$$

The usual choice is to set three of the relative phases to zero: $\phi_{in,t} = \phi_{in,r} = \phi_{u,r} = 0$ and $\phi_{u,t} = \pi$ which means

$$\begin{aligned} \alpha_r &= \sqrt{\epsilon}\alpha_{in} + \sqrt{1-\epsilon}\alpha_u \\ \alpha_t &= \sqrt{1-\epsilon}\alpha_{in} - \sqrt{\epsilon}\alpha_u \end{aligned} \qquad (5.1.5)$$

The normal explanation is that one of the reflected waves has a phase shift of 180 degrees in respect to all other waves. For a simple dielectric layer this would be the reflection of the layer with the higher refractive index. But this is only one possible choice, various mirrors have different properties. All mirrors can be represented by Equation (5.1.5). The above analysis had been discussed as early as 1849 by Stokes [Sto49, Man95] and leads to a more general matrix format

$$\begin{pmatrix} \alpha_r \\ \alpha_t \end{pmatrix} = M \begin{pmatrix} \alpha_{in} \\ \alpha_u \end{pmatrix} = \begin{pmatrix} M_{r,in} & M_{r,u} \\ M_{t,in} & M_{t,u} \end{pmatrix} \begin{pmatrix} \alpha_{in} \\ \alpha_u \end{pmatrix} \qquad (5.1.6)$$

The matrix M has to be a unitary transformation, that means the following conditions apply

$$|M_{r,in}| = |M_{r,u}| \quad \text{and} \quad |M_{t,in}| = |M_{t,u}|$$

$$|M_{r,in}|^2 + |M_{t,in}|^2 = 1$$

$$M_{r,in}M_{t,in}^* + M_{r,u}^*M_{t,u} = 0$$

and our choice of the beamsplitter matrix is

$$\begin{pmatrix} \sqrt{\epsilon} & \sqrt{1-\epsilon} \\ \sqrt{1-\epsilon} & -\sqrt{\epsilon} \end{pmatrix} \qquad (5.1.7)$$

We should note that any real beamsplitter will introduce losses due to absorption and internal scattering. This can be described by a loss term $(1 - A)$ such that

$$|\alpha_t|^2 + |\alpha_r|^2 = (1 - A)\left(|\alpha_{in}|^2 + |\alpha_u|^2\right)$$

In most cases the transmitted and the reflected waves are affected by the same amount. For metal mirrors typical values are $A = 0.05$, for dielectric mirrors A of 10^{-2} to 10^{-4} is common, but losses as low as 5×10^{-6} have been achieved.

Polarisation properties of beamsplitters

In most cases we would want a beamsplitter that acts equally on all polarizations. In the other extreme case one could imagine a polarizing beamsplitter (PBS) that reflects only one polarisation and transmits the orthogonal polarisation. Both these cases can be approximately achieved with different types of beamsplitters. In general, the polarisation response is not trivial, the reflectivity ϵ is polarisation dependent. For this purpose we define two orthogonal polarisation states: S and P linear polarisation. For P polarisation the electric field vector is *parallel* to the surface of the mirror. This is out of the plane of the drawing in Fig. 5.1. For S polarisation the electric field is perpendicular (or *senkrecht* in German) to the P polarisation. This is in the plane of Fig. 5.1. The S polarisation contains a component which is perpendicular to the surface of the mirror and the boundary condition on the mirror surface for the parallel and perpendicular fields are different. Consequently the S and P components of the light are reflected differently, both in amplitude and in phase, and in general ϵ_P and ϵ_S differ from each other. They depend on the way the mirror is manufactured and changes with the input angle and wavelength. For a simple dielectric material, such as uncoated glass, the problem can be solved easily, resulting in the Fresnel reflection coefficients [BW59]. For real multi layer mirrors such calculations are rather involved.

One special solution is $\epsilon_S = 0$ for a dielectric surface under Brewster's angle. Stacks of tilted glass plates, near Brewster's angle, are used as polarizers. As long as we consider only light linearly polarized in the S or P direction the polarisation is maintained and the relative phase shift between the S and the P wave is of no concern. For any other light the polarisation is modified by the mirror. The mirror acts like a combination of an non-polarizing beamsplitter and a wave plate. In summary, to describe a mirror it is not sufficient to state just the reflectivity ϵ, the following properties have to be specified as well: input angle, wavelength, polarisation, and losses.

Finally, the following practical hints are useful for the selection of mirrors: There is a choice of materials for mirrors. For non-polarizing beamsplitters, that means ideally $\epsilon_P = \epsilon_S$. Metal surfaces are very useful and $\epsilon_S = \epsilon_P = 0.48$ can be readily achieved, but the losses are considerable. With dielectric mirrors the difference between ϵ_P and ϵ_S is usually significant, unless they are specifically designed for the specific application. Dielectric mirrors should preferably be used with linearly polarized light aligned in the S or P direction. The losses can be small ($< 0.001\% = 10$ [ppm]). Polarizing beamsplitters with $\epsilon_P = 0.001$ and $\epsilon_S = 0.99$ are available. Note that this property degrades with a change in wavelength or input angle, these polarizers have to be used for specified parameters only. In addition, there is a range of wavelength independent components available which rely on internal total reflection rather than on metallic or dielectric films.

5.1.2   The beamsplitter in the quantum operator model

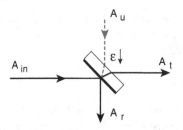

Figure 5.2: The beamsplitter in the quantum operator model.

The properties of a beamsplitter can be described in a quantum mechanical model. We assume that the input beam is a travelling beam described by the operator $\tilde{\mathbf{A}}_{\text{in}}$. A beamsplitter will maintain the mode properties, that means the frequency of the light, the size of the beam, the curvature of the wavefront are all preserved. The direction of propagation is given by geometrical considerations. The polarizing properties of the mirror can be neglected by assuming that the incoming light is either P or S polarized and the calculation for a different polarisation is treated as a linear combination of these special cases. Thus a beamsplitter can be simply described by a transformation from two input mode operators, $\tilde{\mathbf{A}}_{\text{in}}$ and $\tilde{\mathbf{A}}_{u}$, into two output mode operators, $\tilde{\mathbf{A}}_{r}$ and $\tilde{\mathbf{A}}_{t}$ [Lui95].

Thanks to canonical quantization, see Section 4.1, the transformation rules for the quantum operators are equivalent to those of classical complex amplitudes and we can use again the matrix definition of Equation (5.1.7) and obtain

$$\begin{pmatrix} \tilde{\mathbf{A}}_{r} \\ \tilde{\mathbf{A}}_{t} \end{pmatrix} = M \begin{pmatrix} \tilde{\mathbf{A}}_{\text{in}} \\ \tilde{\mathbf{A}}_{u} \end{pmatrix} = \begin{pmatrix} \sqrt{\epsilon} & \sqrt{1-\epsilon} \\ \sqrt{1-\epsilon} & -\sqrt{\epsilon} \end{pmatrix} \begin{pmatrix} \tilde{\mathbf{A}}_{\text{in}} \\ \tilde{\mathbf{A}}_{u} \end{pmatrix} \tag{5.1.8}$$

with the the definition of the matrix as given in Equation (5.1.7). Please note that in many publications in quantum optics the definition of the matrix in Equation (5.1.7) is frequently replaced by alternative definitions. Two examples commonly used are

$$\begin{pmatrix} i\sqrt{\epsilon} & \sqrt{1-\epsilon} \\ \sqrt{1-\epsilon} & i\sqrt{\epsilon} \end{pmatrix} \text{ or } \begin{pmatrix} \sqrt{\epsilon} & i\sqrt{1-\epsilon} \\ \sqrt{1-\epsilon} & -i\sqrt{\epsilon} \end{pmatrix} \tag{5.1.9}$$

They are equivalent and preference to either definition is normally given purely on aesthetic grounds. For consistency this book uses only one definition, as given in Equation (5.1.8).

Suppose there is no light entering the second input. Do we really need to specify both input modes, $\tilde{\mathbf{A}}_{\text{in}}$ and $\tilde{\mathbf{A}}_{u}$? We find a reason for having to include $\tilde{\mathbf{A}}_{u}$ by checking the properties of the output operators. The operators for all modes obey the Boson commutation relations

$$\left[\tilde{\mathbf{A}}_{j}, \tilde{\mathbf{A}}_{j}^{\dagger} \right] = 1 \quad \text{with} \quad j = in, r, t \quad \text{and} \quad \left[\tilde{\mathbf{A}}_{r}, \tilde{\mathbf{A}}_{t}^{\dagger} \right] = 0$$

The last equation describes the fact that both outputs can be measured independently, without affecting each other. If we simply ignored the input mode $\delta\tilde{\mathbf{A}}_{u}$ we obtain on the basis of Equation (5.1.8)

$$\begin{aligned} \tilde{\mathbf{A}}_{r} &= \sqrt{\epsilon}\,\tilde{\mathbf{A}}_{\text{in}} \\ \tilde{\mathbf{A}}_{t} &= \sqrt{1-\epsilon}\,\tilde{\mathbf{A}}_{\text{in}} \end{aligned} \tag{5.1.10}$$

and we find that the commutator equations do not hold. Instead we find

$$\begin{aligned} \left[\tilde{\mathbf{A}}_{r}, \tilde{\mathbf{A}}_{r}^{\dagger} \right] &= \epsilon \left[\tilde{\mathbf{A}}_{\text{in}}, \tilde{\mathbf{A}}_{\text{in}}^{\dagger} \right] &= \epsilon \\ \left[\tilde{\mathbf{A}}_{t}, \tilde{\mathbf{A}}_{t}^{\dagger} \right] &= (1-\epsilon) \left[\tilde{\mathbf{A}}_{\text{in}}, \tilde{\mathbf{A}}_{\text{in}}^{\dagger} \right] &= (1-\epsilon) \\ \left[\tilde{\mathbf{A}}_{r}, \tilde{\mathbf{A}}_{t}^{\dagger} \right] &= \sqrt{\epsilon(1-\epsilon)} \left[\tilde{\mathbf{A}}_{\text{in}}, \tilde{\mathbf{A}}_{\text{in}}^{\dagger} \right] &= \sqrt{\epsilon(1-\epsilon)} \end{aligned}$$

The reason for this discrepancy is that we have ignored the second input port. This was justified in the classical model, since no light enters that way. However, even if no energy is flowing through the second port, in the full quantum model there is zero point energy in the vacuum mode that enters here and contributes to the two output modes. Accordingly, we should use the full description of Equation (5.1.8) with both input beams and find

$$\left[\tilde{\mathbf{A}}_r, \tilde{\mathbf{A}}_r^\dagger\right] = \epsilon \left[\tilde{\mathbf{A}}_{\text{in}}, \tilde{\mathbf{A}}_{\text{in}}^\dagger\right] + (1 - \epsilon) \left[\tilde{\mathbf{A}}_u, \tilde{\mathbf{A}}_u^\dagger\right]$$

$$= \epsilon + (1 - \epsilon) = 1 \tag{5.1.11}$$

and similarly for $\tilde{\mathbf{A}}_t$. This is in agreement with the commutation rules. We see that the vacuum mode plays an important role and is required to satisfy the rules that apply to any quantised field.

5.1.3 The beamsplitter with single photons

On the one hand, the beamsplitter is not splitting individual photons, it is influencing the probabilities for a photon to leave at one or the other of the two outputs. On the other hand, the beamsplitter is splitting the quadratures, dividing the field between the two ports. This ambiguity which makes the action of macroscopic objects on the quantum world dependent on the type of measurements subsequently made is a unique feature of quantum mechanics. This can be described through the quantum states of light, see Chapter 4.

From the photon point of view the beamsplitter acts like a random selector which reflects the photons with a probability ϵ and transmits them with a probability $(1 - \epsilon)$. The effect on any particular photon is unpredictable. If we had an input that always contains a photon, or consider only those intervals when there is a photon, then the input can be approximately described by $|\text{in}\rangle = |1\rangle$. We have the output states $|r\rangle = |1\rangle$ with the probability ϵ, and $|r\rangle = |0\rangle$ with the probability $(1 - \epsilon)$. This means that after the beamsplitter the certainty of finding the input photon in one particular beam is lost. However, for the lossless beamsplitter we have certainty for the correlation between the results for the two outputs. For example with $|r\rangle = |1\rangle$ we have simultaneously $|t\rangle = |0\rangle$ in the same time interval.

Since we have two input ports we can write the total input state in the format of combined states as $|1\rangle_{\text{in}}|0\rangle_u$ for the ports in and u and obtain the two possible output states $|1\rangle_r|0\rangle_t$ and $|0\rangle_r|1\rangle_t$, meaning the photon appears in one or the other output beam. Since the photon number is preserved, the states $|0\rangle_r|0\rangle_t$ and $|1\rangle_r|1\rangle_t$ cannot occur. We see that the beamsplitter expands the basis of states which we have to use. For each beamsplitting process we have to include an extra input and output channel, and that means we increase the basis for the state vectors.

More rigorously we can use the techniques of Section 4.4. Let the unitary operator representing evolution through the beamsplitter be $\tilde{\mathbf{U}}$. We know that $\tilde{\mathbf{A}}_r = \tilde{\mathbf{U}}^\dagger \tilde{\mathbf{A}}_{\text{in}} \tilde{\mathbf{U}}$ and $\tilde{\mathbf{A}}_t = \tilde{\mathbf{U}}^\dagger \tilde{\mathbf{A}}_u \tilde{\mathbf{U}}$. Writing the initial state as $|1\rangle_{\text{in}} |0\rangle_u = \tilde{\mathbf{A}}_{\text{in}}^\dagger |0\rangle_{\text{in}}|0\rangle_u$ we then have

$$\tilde{\mathbf{U}}\tilde{\mathbf{A}}_{\text{in}}^\dagger |0\rangle_{\text{in}}|0\rangle_u = \tilde{\mathbf{U}}\tilde{\mathbf{A}}_{\text{in}}^\dagger \tilde{\mathbf{U}}^\dagger \tilde{\mathbf{U}} |0\rangle_{\text{in}}|0\rangle_u$$

$$= \tilde{\mathbf{U}}\tilde{\mathbf{A}}_{\text{in}}^\dagger \tilde{\mathbf{U}}^\dagger |0\rangle_r|0\rangle_t \tag{5.1.12}$$

Now $\tilde{\mathbf{U}}\tilde{\mathbf{A}}_{\text{in}}^\dagger \tilde{\mathbf{U}}^\dagger$ represents time reversed evolution through the beamsplitter and is obtained by inverting Equation (5.1.8) and taking the Hermitian conjugate. Explicitly

$$\tilde{\mathbf{U}}\tilde{\mathbf{A}}_{\text{in}}^\dagger \tilde{\mathbf{U}}^\dagger = \sqrt{\epsilon}\tilde{\mathbf{A}}_r^\dagger + \sqrt{1 - \epsilon}\ \tilde{\mathbf{A}}_t^\dagger$$

Therefore

$$
\begin{aligned}
\tilde{U}\tilde{A}_{in}^{\dagger} \, |0\rangle_{in}|0\rangle_u &= (\sqrt{\epsilon}\tilde{A}_r^{\dagger} + \sqrt{1-\epsilon}\,\tilde{A}_t^{\dagger})|0\rangle_r|0\rangle_t \\
&= \sqrt{\epsilon}\,|1\rangle_r|0\rangle_t + \sqrt{1-\epsilon}\,|0\rangle_r|1\rangle_t
\end{aligned}
\tag{5.1.13}
$$

As expected we obtain a superposition of the two possibilities, but notice that the "plus" sign between them is not arbitrary but comes directly from the phase convention of the beamsplitter, Eq. (5.1.5). To illustrate this, consider the situation in which the photon arrives at the other port of the beamsplitter. Now our input state is $|0\rangle_{in}|1\rangle_u = \tilde{A}_u^{\dagger}\,|0\rangle_{in}|0\rangle_u$ and our output state is

$$
\begin{aligned}
\tilde{U}\tilde{A}_u^{\dagger} \, |0\rangle_{in}|0\rangle_u &= (\sqrt{1-\epsilon}\,\tilde{A}_r^{\dagger} - \sqrt{\epsilon}\tilde{A}_t^{\dagger})\,|0\rangle_r|0\rangle_t \\
&= \sqrt{1-\epsilon}\,|1\rangle_r|0\rangle_t - \sqrt{\epsilon}\,|0\rangle_r|1\rangle_t
\end{aligned}
\tag{5.1.14}
$$

Keeping track of the sign, or more generally, the phase of the superposition is crucial in understanding interferometers at the single photon level, see Section 5.2.

So far we have assumed only a single photon arrives at the beamsplitter. Now, let us consider a beamsplitter which mixes two single photon inputs. This can produce some surprising results. The combined single photon states can be written as $|1\rangle_{in}|1\rangle_u = \tilde{A}_{in}^{\dagger}\tilde{A}_u^{\dagger}\,|0\rangle_{in}|0\rangle_u$. A particular interesting case is the balanced beamsplitter with $\epsilon = (1-\epsilon) = 0.5$. We obtain the result

$$
\begin{aligned}
\tilde{U}\tilde{A}_{in}^{\dagger}\tilde{A}_u^{\dagger}|0\rangle_{in}|0\rangle_u &= \frac{1}{2}(\tilde{A}_r^{\dagger} + \tilde{A}_t^{\dagger})(\tilde{A}_r^{\dagger} - \tilde{A}_t^{\dagger})|0\rangle_r|0\rangle_t \\
&= \frac{1}{2}\left(\tilde{A}_r^{\dagger}\tilde{A}_r^{\dagger} - \tilde{A}_t^{\dagger}\tilde{A}_t^{\dagger}\right)|0\rangle_r|0\rangle_t \\
&= \frac{1}{\sqrt{2}}\,(|2\rangle_r|0\rangle_t - |0\rangle_r|2\rangle_t)
\end{aligned}
\tag{5.1.15}
$$

If we illuminate the beamsplitter with the states $|1\rangle_{in}|1\rangle_u$, that means we know with certainty that single photons are present at both inputs in the same time interval. The key result from Equation (5.1.15) is that we will have 2 photons either in the reflected or the transmitted beam and the terms $|1\rangle_r|1\rangle_t$ and $|1\rangle_t|1\rangle_r$, with single photons in each output beam, cancel since the input photons are indistinguishable. This means that for the right condition, namely a well balanced beamsplitter and good beam overlap, the events of detecting pairs of single photons, one in each arm, should be rare. This is a quantum interference effect, not a classical interference effect. It can be detected by a coincidence counter, a device which is triggered only when both output beams contain one photon. An experimental demonstration of this effect is discussed in Section 12.2. For a single input beam, $|1\rangle_{in}|0\rangle_u$, or for ordinary laser beams at the input, with a mixture of photon states, there will be no such interference and reduction in coincidence counts. To produce the required light source, which generates many time intervals which are well described by $|1\rangle_{in}|1\rangle_u$, is not trivial (see also Section 3.5). However, with these sources, which can produce pairs of input photons, we can use the beamsplitter and correlator both as a demonstration of quantum interference and as a test of the quality of our photon source.

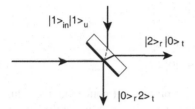

$|1>_{in}|1>_u$

$|2>_r|0>_t$

$|0>_r2>_t$

Figure 5.3: The beamsplitter with two input beams with exactly one photon per time interval. In this hypothetical case we should have only pairs of photons leaving each output and no coincidences between the two output beams.

5.1.4 The beamsplitter and the photon statistics

Let us now consider a beamsplitter with a single input and a single output. That is we consider the other input to be vacuum and assume that the second output is "lost". Such a situation can be used as a generic model for loss due to absorption or scattering. The output state of the single beam will in general be mixed. Using the techniques of Section 4.4 to average (trace) over the possible states of the lost field we find that for a single photon input, $|1\rangle_{in}$, the output state is given by the density operator

$$\tilde{\rho}_t = \epsilon|1\rangle\langle 1| + (1 - \epsilon)|0\rangle\langle 0| \tag{5.1.16}$$

This is just a classical mixture of the possibility that a photon was transmitted or not, weighted by the probability of transmission. Thus in this situation a beamsplitter can be thought of as a random selector of photons, and classical statistics can be applied. In the case of many photons per counting interval we can measure the average photon flux and the width of the counting distribution function. The average photon flux is fully equivalent to the classical intensity. If we have only one input with flux N_{in} and $N_u = 0$ we obtain, based on the probability of the selection of individual photons, $N_r = \epsilon N_{in}$ and $N_t = (1 - \epsilon) N_{in}$.

The photon statistic, however, will be affected differently. We describe the input statistics by the Fano factor f_{in} defined in Equation (4.3.21), where $f_{in} = 1$ corresponds to a Poissonian input statistic and $f_{in} = 0$ to a perfectly regulated beam. The effect of a random selection process is illustrated in Fig. 5.4. Here the time sequence of photons in the input and output beams and their counting distribution functions are sketched. The effects of three beamsplitters with different reflectivities $\epsilon = 0.9, 0.5$ and 0.1 are shown.

In the first case, part (a) of Fig. 5.4, we assume that the input beam is a coherent state, with a Poissonian distribution function. We should note, the general property of any Poissonian distribution is that a random selection yields again a Poissonian distribution. Consequently, the output beam has a Poissonian distribution, with a peak that has moved from N_{in} to $N_r = \epsilon N_{in}$. Similarly, the transmitted beam has a Poissonian distribution peaked at $N_t = (1 - \epsilon)N_{in}$. The normalized width of the distributions, the Fano factors, of all these beams are constant. We can generalist these results and conclude that a coherent state transmitted or reflected by a mirror remains a coherent state (see next section).

The situation is quite different for states with strongly non-Poissonian distributions. In part (b) of Fig. 5.4 the extreme case of an absolutely regular stream of photons is shown.

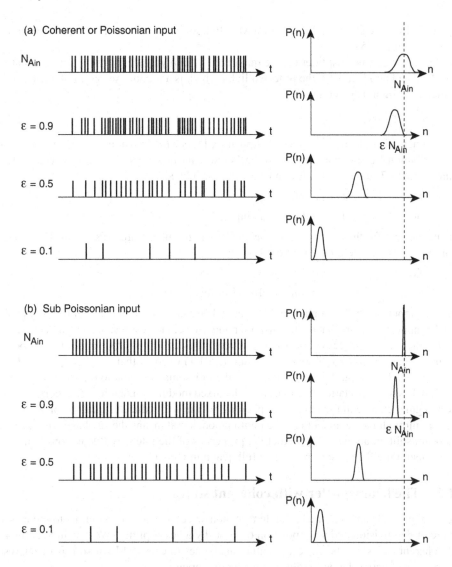

Figure 5.4: The effect of a beamsplitter on a stream of photons for (a) a Poissonian and (b) a Sub-Poissonian input. The left part shows the arrival sequence of photons, the right side the corresponding distribution functions.

The counting distribution for this input beam is a δ-function, that means during each counting interval the same number of photons is available, as shown in the top line of Fig. 5.4. Since only a random selection of these photons is transmitted this regularity is no longer correct for the output beam. There is an uncertainty in the number of photons present in a counting interval. The distribution function is centered at a lower average photon number and is wider. If only a very small fraction is selected, the statistical properties of the selection process will dominate. In this case ($\epsilon \ll 1$) we will have a Poissonian distribution again, as shown in the bottom line of Fig. 5.4.

This effect of a beamsplitter can be quantified by applying the rules of counting statistics (4.3.21). The Fano factor for the reflected light, which is a random selection of a fraction ϵ of the incoming light, is given by

$$f_r = 1 - \epsilon(1 - f_{in}) \tag{5.1.17}$$

The mathematical details are given in Appendix D. For the transmitted, or reflected, beams the distribution functions are closer to the Poissonian limit than for the input beam. Using Equation (5.1.17) the Fano factors can easily be calculated.

$$f_r \;\; = \;\; 1 \quad \text{and} \quad f_t = 1$$

for $\quad f_{in} = 1 \quad$ (Poissonian input)

This means that for the case of a Poissonian input both outputs are Poissonian. However, for a completely noise free input one would obtain

$$f_r \;\; = \;\; 1 - \epsilon \quad \text{and} \quad f_t = \epsilon$$

for $\quad f_{in} = 0 \quad$ (completely noise free input)

When the input is sub-Poissonian, and the second port is not used, the output is always noisier than the input. It appears that the beamsplitter adds noise. However, as we see in Section 5.1.5, it is identical to the effect of the vacuum fluctuations entering through the unused port. If the input beam is super-Poissonian, Equation (5.1.17) shows that $1 < f_r < f_{in}$ and $1 < f_t < f_{in}$. The output beams are closer to the Poissonian distribution than the input beam. This matches the expectation that a beam with a fixed modulation depth, which corresponds to a very regular change in the number of photons, will have a reduced signal to noise ratio after it passed through the beamsplitter. The photon model, that means the stochastic interpretation of a beam splitter, allows us to predict the properties of the intensity fluctuations. The effect on the quadrature fluctuations requires a full quantum model.

5.1.5   The beamsplitter with coherent states

The view of the beamsplitter as a random photon selector is in stark contrast to the classical picture of the division and superposition of fields by a beamsplitter. We saw in Section 4.2.2 that coherent states are the closest quantum analogues to classical light so it is of interest to examine the effect of the beamsplitter on coherent inputs.

We can write our input state as $|\alpha_{in}\rangle_{in}|\alpha_u\rangle_u = \tilde{D}_{in}(\alpha_{in})\tilde{D}_u(\alpha_u)|0\rangle_{in}|0\rangle_u$. Evolution through the beamsplitter then gives

$$\begin{aligned}
\tilde{D}_{in}(\alpha_{in}) \;\; &= \;\; \exp(\alpha_{in}\tilde{A}_{in}^\dagger - \alpha_{in}^*\tilde{A}_{in}) \\
&\rightarrow \;\; \exp\left(\alpha_{in}(\sqrt{\epsilon}\tilde{A}_r^\dagger + \sqrt{1-\epsilon}\tilde{A}_t^\dagger) - \alpha_{in}^*(\sqrt{\epsilon}\tilde{A}_r + \sqrt{1-\epsilon}\tilde{A}_t)\right) \\
&= \;\; \tilde{D}_r(\sqrt{\epsilon}\,\alpha_{in})\tilde{D}_t(\sqrt{1-\epsilon}\,\alpha_{in}) \tag{5.1.18}
\end{aligned}$$

where we have simply transformed the operators in the argument of the exponential in accordance with Equation (5.1.8). Similarly we have

$$\tilde{D}_u(\alpha_u) \rightarrow \tilde{D}_r(\sqrt{1-\epsilon}\alpha_u)\tilde{D}_t(-\sqrt{\epsilon}\alpha_u) \tag{5.1.19}$$

Putting these together we find the output state of the fields is given by

$$|(\sqrt{\epsilon}\,\alpha_{\text{in}} + \sqrt{1-\epsilon}\,\alpha_u))\rangle_r \; |(\sqrt{1-\epsilon}\,\alpha_{\text{in}} - \sqrt{\epsilon}\,\alpha_u))\rangle_t \qquad (5.1.20)$$

The output state remains a pair of coherent states. Their amplitudes are transformed in exactly the same way as the amplitudes of a classical field are transformed by the beamsplitter. As a result the expectation values of the quadrature observables will be identical to the classical predictions. Note also, in contrast to the single photon situation, the output fields are not entangled. That is the output state of Equation (5.1.20) can be factored into a product of a beam r part and a beam t part.

Although the average values of the quadratures are the same as in the classical case we must consider quantum noise in the quantum case. To do this we next discuss the transfer function of the beamsplitter.

Transfer function for a beam splitter

Now we calculate the effect of a beamsplitter on fluctuations and modulations. This is given by a transfer function that transforms $V1_{\text{in}}(\Omega), V2_{\text{in}}(\Omega)$ and $V1_u(\Omega), V2_u(\Omega)$ into $V1_r(\Omega), V2_r(\Omega)$ and $V1_t(\Omega), V2_t(\Omega)$. The beamsplitter is one of the simplest cases of a transfer function. Let us consider a single input beam, that means a photon flux $\bar{N}_{\text{in}}$ and no flux in the other beam, $\bar{N}_u = 0$. The beamsplitter transformation is linear so the fluctuations transform in the same way as the full operators, Eq. (5.1.8). We wish to calculate the variances using the recipe $V1_r(\Omega) = \langle|\delta\tilde{\mathbf{X}}1_r(\Omega)|^2\rangle$ and similarly $V2_r(\Omega) = \langle|\delta\tilde{\mathbf{X}}2_r(\Omega)|^2\rangle$ derived in Section 4.5.1 and thus we convert to the description in terms of the quadrature operators $\delta\tilde{\mathbf{X}}1 = \delta\tilde{a} + \delta\tilde{a}^\dagger$ and $\delta\tilde{\mathbf{X}}2 = i(\delta\tilde{a} - \delta\tilde{a})a^\dagger$. We describe the input beams by the spectra $\delta\tilde{\mathbf{X}}1_{\text{in}}(\Omega)$ and $\delta\tilde{\mathbf{X}}2_u(\Omega)$ and calculate $\delta\tilde{\mathbf{X}}1_r(\Omega)$ and $\delta\tilde{\mathbf{X}}2_r(\Omega)$. Assuming the beamsplitter has a flat frequency response (an excellent approximation over radio frequency ranges) the quadrature operators in frequency space transform in the same way as the operators in the time domain and we can use

$$\begin{pmatrix} \delta\tilde{\mathbf{X}}1_r(\Omega) \\ \delta\tilde{\mathbf{X}}1_t(\Omega) \end{pmatrix} = \begin{pmatrix} \sqrt{\epsilon} & \sqrt{1-\epsilon} \\ \sqrt{1-\epsilon} & -\sqrt{\epsilon} \end{pmatrix} \begin{pmatrix} \delta\tilde{\mathbf{X}}1_{\text{in}}(\Omega) \\ \delta\tilde{\mathbf{X}}1_u(\Omega) \end{pmatrix} \qquad (5.1.21)$$

When we calculate the product $|\delta\tilde{\mathbf{X}}1_r(\Omega)|^2$ we find that for independent inputs all terms of the type $\delta\tilde{\mathbf{X}}1_{\text{in}}(\Omega)\delta\tilde{\mathbf{X}}1_u(\Omega)$ will average to zero. Only the terms $|\delta\tilde{\mathbf{X}}1_{\text{in}}(\Omega)|^2$ and $|\delta\tilde{\mathbf{X}}1_u(\Omega)|^2$ contribute. We obtain

$$\begin{aligned} V1_r(\Omega) &= \epsilon\, V1_{\text{in}}(\Omega) + (1-\epsilon)\, V1_u(\Omega) \\ V1_t(\Omega) &= (1-\epsilon)\, V1_{\text{in}}(\Omega) + \epsilon\, V1_u(\Omega) \end{aligned} \qquad (5.1.22)$$

and the same result for $V2_r(\Omega)$ and $V2_t(\Omega)$. In all these quantum calculations we have to keep track of the fluctuation properties of the operator $\delta\tilde{\mathbf{A}}_u$ for the second input beam that is incident on the beamsplitter. Despite the fact that this mode contains no energy, the fluctuations have to be taken into account as $V1_u(\Omega) = V2_u(\Omega) = 1$. This is a direct consequence of the commutation rules and is the essential difference between the classical and the quantum calculation.

If we have only one input beam with intensity then $V1_u(\Omega)$ can be replaced by 1 and we get $V1_r(\Omega) = \epsilon\, V1_{in}(\Omega) + (1 - \epsilon)$. As a rule the variance of the output light will be moved closer to the QNL. In the limiting case $\epsilon \to 1$, that means a totally reflecting mirror, the output beam is identical to the input. For a balanced 50/50 beamsplitter the output variance of each beam is an equal mixture of the fluctuations from the input state and the empty input port. This result is identical to that for the attenuation of a beam of light, see Fig. 5.4. As mentioned earlier a beamsplitter is fully equivalent to attenuation. Each beamsplitter, or in practical terms each loss mechanism in the experiment, opens one new input channel, and this is affecting the properties of the light for fluctuations or modulation close to the quantum noise level.

Finally, we have the general result again that in the situation where all input beams are coherent, that means $V1_{in}(\Omega) = V2_{in}(\Omega) = 1$, all output beams are coherent as well and we have $V1_r(\Omega) = V2_r(\Omega) = V1_t(\Omega) = V2_t(\Omega) = 1$.

There is a school of thought that wishes to avoid the rigor of quantum mechanical calculations and prefers to use a classical analogue [Edw93]. In simple situations it is possible to interpret a beamsplitter as a beam divider with a noise source built into it. This source generates just the right amount of partition noise and mimics the properties of the stochastic device. All calculations are carried out classically. Obviously, with the correctly chosen noise source, the results for intensity noise are identical to those from a full quantum model. However, complications arise when other quadratures of the light have to be taken into account. Examples are interferometers and cavities, where the phase noise of the light is as important as the amplitude noise; and nonlinear media, where phase and amplitude are linked to each other. In these cases the quantum transfer functions provide a complete and simple description.

5.1.6   The beamsplitter in the noise sideband model

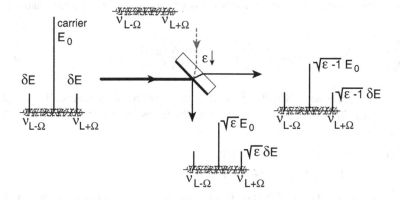

Figure 5.5: The beamsplitter in the sideband model. The carrier is reduced by the amplitude reflectivity $\sqrt{\epsilon}$ and so are the modulation sidebands. However, the quantum noise sidebands remain unchanged.

One of the most elegant descriptions of the beamsplitter is given by the noise sideband model. Here we have to consider the effect of the beamsplitter on the carrier frequency and the modulation sidebands. The measurements show that the intensity of the beam, the absolute size of the modulations, or technical noise all decrease in the same way. The beamsplitter reduces the intensity but retains the relative intensity noise (RIN). Consequently the amplitude of the carrier and the modulation sidebands are both reduced by the amplitude reflectivity $\sqrt{\epsilon}$. In contrast the quantum noise sidebands remain unchanged. That means that the SNR of a quantum noise limited signal is reduced. This can be interpreted as the contribution of the vacuum state entering through the open, unused port of the beamsplitter .

5.1.7 Comparison between a beamsplitter and a classical current junction

It is worthwhile to compare an optical beamsplitter with a junction for electrical currents. These two components have very different, complementary properties. The photons traversing a beamsplitter are non-interacting particles. The beamsplitter acts like a random selector for the photons. It thereby determines the statistical properties of the stream of photons emerging from the beamsplitter and leads to the quantum fluctuations which have been discussed in detail. In the case of electrical currents a strong interaction exists between the particles, a consequence of their electrical charge. The stream of electrons is divided in a completely deterministic way, in accordance with Kirchoff's law. Let us describe the input current $i_{in}(t)$ as a time averaged current i_{in} plus some time dependent fluctuations $\delta i_{in}(t)$. The junction splits off a fraction ζ of the current and produces two output currents $i_t(t) = i_t + \delta i_t(t)$ and $i_r = i_r + \delta i_r(t)$. Both the average current and the fluctuations are affected in the same way

$$
\begin{aligned}
i_r &= \zeta\ i_{in} \\
i_t &= (1 - \zeta)i_{in} \\
\delta i_r(t) &= \zeta\ \delta i_{in}(t) \\
\delta i_t(t) &= (1 - \zeta)\ \delta i_{in}(t)
\end{aligned}
\tag{5.1.23}
$$

The statistical properties of the currents can be described by a normalized variance, defined in the same way as for photons (Equation (4.3.21)). Again $Vi = 1$ represents a Poissonian distribution. The statistics of the output current can be evaluated as

$$
\begin{aligned}
Vi_r &= \frac{\delta i_r(t)E^2}{i_r} = \frac{\zeta \delta i_{in}(t)E^2}{\zeta i_{in}} = \zeta Vi_{in} \\
Vi_t &= (1 - \zeta)Vi_{in}
\end{aligned}
\tag{5.1.24}
$$

This means that, relatively speaking, the two output currents are quieter than the input current and the variance is always reduced. For a Poissonian input current both output currents are sub-Poissonian. The only distribution which remains unchanged would be a delta-function, the distribution for a perfectly quiet current without any fluctuations, $V_i = 0$. The junction does not introduce any noise – it actually reduces it, as measured by the Fano factor. This is a consequence of the way the Fano factor scales. The actual fluctuations have not disappeared – they were redistributed in the same way as the average current. The other consequence of such a fully deterministic junction is that the fluctuations of the two outputs are completely corre-lated to each other and they are correlated to the input. In this case, which can be referred to as the classical case, we could detect $i_r(t)$ and $i_t(t)$ separately and could cancel the fluctuations in one of the signals with the other. Just by measuring one output, all three currents, the input and the two outputs, can be determined with perfect accuracy. The difference between the properties of a random selector (optical beamsplitter) and the deterministic divider (electrical junction) are striking. The beamsplitter will preserve the Poissonian distribution. The output beams generated by a quantum noise limited input beam are completely uncorrelated. In contrast, the junction always reduces the Fano factor and creates perfect correlations between the input and output beams. We clearly have to apply different rules to the two cases. Optical beamsplitters cannot be treated like electrical junctions.

Does the simple description of a junction hold under all conditions? The interaction between the particles (electrons) dominates at the level of fluctuations discussed here, namely relative changes in the current of the order of one part on 10^4 to 10^9. This is the regime where measurements of optical noise are carried out. At the level of much smaller fluctuations the quantum nature of the electrons will contribute some randomness and at such low level of fluctuations Kirchoff's law, Equation (5.1.24), will no longer be correct. In general, quantum effects can be ignored in macroscopic systems such as wires, resistors and other normal electronic components. However, quantum effects have to be considered where the physical dimensions of the components approach the size of the wave function for the electron. These so called mesoscopic systems have some quite unique features with mixed classical and quantum properties. It will probably be possible to build and operate such systems in the near future, resulting in new and interesting ways to control currents.

5.2   Interferometers

One of the most useful and interesting applications of a beamsplitter is to split and later recombine the electro-magnetic waves associated with a beam of light. In the process of recombination the waves interfere, the result is an output beam with an intensity that depends on the relative phase between the two partial waves. Mirrors don't dissipate energy, apart from the (usually very small) scattering losses. Thus the total energy in an interferometer is preserved. For any occurrence of destructive interference in the output there must be a corresponding part of this beam or of another beam leaving the interferometer that shows constructive interference.

The physical layout of interferometers can vary tremendously. But they can all be traced back to one fundamental system, the basic four mirror arrangement, shown in Fig. 5.6, known as the Mach-Zehnder interferometer. This arrangement contains two beamsplitters M_1 and M_4. The other two mirrors (M_2, M_3) are used to redirect the beams and overlap them at the output. For a balanced interferometer reflectivities of the beamsplitters are $\epsilon_1 = \epsilon_4 = 1/2$ and $\epsilon_2 = \epsilon_3 = 1$. The first beamsplitter produces two beams of equal amplitude. These beams travel on different paths. They experience a differential phase shift $\Delta\phi$, which is indicated by the wedge of transparent material with refractive index $n > 1$. The second beamsplitter recombines the beams.

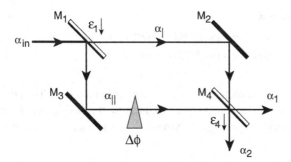

Figure 5.6: The components of a classical interferometer.

5.2.1 Classical description of an interferometer

In the classical wave model a balanced interferometer with 50/50 beamsplitters can be described by the following evolution of the complex amplitudes:

$$
\begin{aligned}
\alpha_I(t) &= \sqrt{1/2}\,\alpha_{\text{in}}(t) \\
\alpha_{II}(t) &= \sqrt{1/2}\,\alpha_{\text{in}}(t) \\
\alpha_1(t) &= \sqrt{1/2}\,\alpha_I(t)\exp(i\Delta\phi) - \sqrt{1/2}\,\alpha_{II}(t) \\
\alpha_2(t) &= \sqrt{1/2}\,\alpha_I(t)\exp(i\Delta\phi) + \sqrt{1/2}\,\alpha_{II}(t)
\end{aligned}
\tag{5.2.1}
$$

where the amplitude reflectivity $\sqrt{\epsilon}$ has been set to $\sqrt{1/2}$. The calculation can easily be generalized to other values of ϵ. Here the convention of Equation (5.1.7) is used. The minus sign in the third equation represents the 180 degree phase shift imposed by one, but not the other, of the reflections. Strictly speaking, this set of equations applies only to plane waves but it can also be applied to spatial Gaussian beams. The output intensity of the interferometer is given by

$$
\begin{aligned}
I_1(\Delta\phi) &= I_{\text{in}}\cos^2(\Delta\phi/2) \\
I_2(\Delta\phi) &= I_{\text{in}}\sin^2(\Delta\phi/2)
\end{aligned}
\tag{5.2.2}
$$

and $I_1 + I_2 = I_{\text{in}}$. There are many applications for interferometers: they can be used to measure physical processes which affect the relative phase difference $\Delta\phi$. One should think about an interferometer as a device which transforms changes in the optical phase into changes of intensity. Equation (5.2.2) holds for any interferometer which is based on the interference of two waves – independent on the layout and the details of the apparatus. This includes Michelson interferometers, Lloyds mirrors, fibre optic interferometers, polarimeters, etc. The main task in describing a specific configuration is to evaluate $\Delta\phi$, once this is done the intensity distribution can be directly evaluated from Equation (5.2.2).

Under practical conditions the interference will not be perfect. The overlap of the two wave fronts will not be perfect or the two waves might not be completely coherent. The spatial or temporal coherence between the two waves, that means the correlation between α_I and α_{II}, will determine the strength of the interference. In the case of perfect correlation the result in Equation (5.2.2) is valid. In the case of no correlation between α_{II} and α_I, or if they are incoherent, the resulting intensity will be

$$I_1(\Delta\phi) = (\epsilon_4|\alpha_{II}|^2 + t_4|\alpha_I|^2)h\Omega = \text{constant}$$

$$I_2(\Delta\phi) = (t_4|\alpha_{II}|^2 + \epsilon_4|\alpha_I|^2)h\Omega = \text{constant}$$

$$I_1(\Delta\phi) + I_2(\Delta\phi) = I_{\text{in}} \tag{5.2.3}$$

where we are now considering an interferometer that is not necessarily balanced. Thus the visibility, or contrast of the interference system can be used to measure the degree of coherence. We can measure the visibility VIS by comparing the extrema I_{max} and I_{min} of the intensity in one output beam and use the definition

$$\text{VIS} = \frac{I_{\text{max}} - I_{\text{min}}}{I_{\text{max}} + I_{\text{min}}} \tag{5.2.4}$$

We can then compare the measured value of VIS with the best possible value $\text{VIS}_{\text{perfect}}$ that we would have achieved for completely correlated beams, where α_{II} and α_I completely interfere.

$$\text{VIS}_{\text{perfect}} = \frac{(\alpha_{II} + \alpha_I)^2 - (\alpha_{II} - \alpha_I)^2}{(\alpha_{II} + \alpha_I)^2 + (\alpha_{II} - \alpha_I)^2} = \frac{\sqrt{I_{II}I_I}}{I_{II} + I_I} \tag{5.2.5}$$

$\text{VIS}_{\text{perfect}}$ can be determined separately, by individually blocking the beams and by recording the intensity of the beams independently. Combining these four measurements of $I_{\text{max}}, I_{\text{min}}, I_{II}$ and I_I in Equation (5.2.5) yields the value

$$\mathcal{V} = \frac{\text{VIS}}{\text{VIS}_{\text{perfect}}} \tag{5.2.6}$$

This is the value for the coherence of the beams, assuming a perfectly aligned interferometer, which is also the value of the correlation between the beams. This measurement cannot produce easily the sign of the correlation since correlated and anti-correlated light produce fringe systems of identical visibility - only shifted with respect to each other. A knowledge of the relative phase between α_{II} and α_I at the maximum interference would allow us to determine the sign. Please note that this technique can be used for any situation, the intensity of the interfering beams do not have to be balanced. However, a simple, direct evaluation is only possible for the balanced situation $|I_{II}| = |I_I|$ since in this case $\text{VIS}_{\text{perfect}} = 1$ and $\mathcal{V} = VIS$. In all cases we have $\mathcal{V} = |C(\alpha_{II}, \alpha_I)|$.

One application of this procedure is to determine the degree of spatial coherence of a beam of light using a double slit experiment. In this particular case α_{II} and α_I are parts of the same wave, but from spatially different sections of the beam. The phase difference $\Delta\phi$ is a purely geometrical term which is given by $\Delta\phi = d\sin(\Theta)$ where d is the separation of the two slits and Θ the angle of detection measured from the axis normal to the plane of the two slits. If the detection occurs at a distance much larger than the slit separation the fringes are intensity minima and maxima which are parallel and equally spaced. The visibility of these fringes are a quantitative measure of the spatial coherence of the light. Similarly, a measurement of the visibility can be used to ascertain the correlation between two electronic signals. For example, the currents from two photon detectors. By adding and subtracting the signals, or by adding them with a variable time delay, interference signals can be generated. The visibility, and thus the correlation coefficient, can be determined. In Section 7.5 this application is discussed in detail.

5.2.2 Quantum model of the interferometer

We now consider a quantum model of the interferometer. As in the classical case we will assume that the instrument is perfectly aligned, that the spatial overlap of the two beams at the recombining beamsplitter is perfect. The effect of spatial mismatch, unmatched wave fronts and partial incoherence between the modes can be added artificially at the end of the calculation.

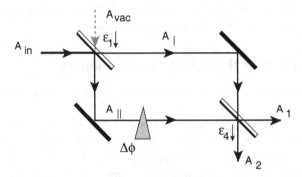

Figure 5.7: Description of an interferometer in the quantum model

The interferometer is illuminated by modes with the operators $\delta \tilde{\mathbf{A}}_{\text{in}}$ and $\delta \tilde{\mathbf{A}}_u$, inside we have the operators $\delta \tilde{\mathbf{A}}_I$ and $\delta \tilde{\mathbf{A}}_{II}$. After recombination we have the two output operators $\delta \tilde{\mathbf{A}}_1$ and $\delta \tilde{\mathbf{A}}_2$ which are detected individually. As with the classical model the case of a Mach-Zehnder interferometer with a phase difference $\Delta\phi$ is treated first. The results can be generalized to any other interferometer and to applications where $\Delta\phi$ is a function of time or space. By canonical quantization the internal operators are linked to the input operators via

$$
\begin{aligned}
\delta \tilde{\mathbf{A}}_I &= \sqrt{1-\epsilon_1}\, \delta \tilde{\mathbf{A}}_{\text{in}} - \sqrt{\epsilon_1}\, \delta \tilde{\mathbf{A}}_u \\
\delta \tilde{\mathbf{A}}_{II} &= \sqrt{\epsilon_1}\, \delta \tilde{\mathbf{A}}_{\text{in}} + \sqrt{1-\epsilon_1}\, \delta \tilde{\mathbf{A}}_u
\end{aligned}
\tag{5.2.7}
$$

The output operators are linked to the internal states via

$$
\begin{aligned}
\delta \tilde{\mathbf{A}}_1 &= \exp(i\Delta\phi)\sqrt{1-\epsilon_4}\, \delta \tilde{\mathbf{A}}_{II} - \sqrt{\epsilon_4}\, \delta \tilde{\mathbf{A}}_I \\
\delta \tilde{\mathbf{A}}_2 &= \exp(i\Delta\phi)\sqrt{\epsilon_4}\, \delta \tilde{\mathbf{A}}_{II} + \sqrt{1-\epsilon_4}\, \delta \tilde{\mathbf{A}}_I
\end{aligned}
\tag{5.2.8}
$$

For the balanced interferometer we can set $\epsilon_1 = \epsilon_4 = \frac{1}{2}$ and the equations can be combined to give

$$
\begin{aligned}
\delta \tilde{\mathbf{A}}_1 &= \exp(i\Delta\phi/2)\left(\cos(\Delta\phi/2)\, \delta \tilde{\mathbf{A}}_{\text{in}} + i\sin(\Delta\phi/2)\, \delta \tilde{\mathbf{A}}_u\right) \\
\delta \tilde{\mathbf{A}}_2 &= \exp(i\Delta\phi/2)\left(i\sin(\Delta\phi/2)\, \delta \tilde{\mathbf{A}}_{\text{in}} + \cos(\Delta\phi/2)\, \delta \tilde{\mathbf{A}}_u\right)
\end{aligned}
\tag{5.2.9}
$$

5.2.3 The single photon interferometer

The output state of a single photon incident on a balanced interferometer can be calculated by inverting Equation (5.2.9). For the input state $|1\rangle_{\text{in}}|0\rangle_u = \tilde{\mathbf{A}}_{\text{in}}^\dagger |0\rangle_{\text{in}}|0\rangle_u$ we obtain the output state

$$\cos(\Delta\phi/2) \ |1\rangle_1|0\rangle_2 + i \sin(\Delta\phi/2) \ |0\rangle_1|1\rangle_2 \qquad\qquad (5.2.10)$$

Depending on the phase shift in the interferometer the photons can be made to exit from one or other output ports with certainty, or be in a superposition of both ports. This effect is clearly a wave effect and cannot be understood by thinking of the beamsplitters as random photon partitioners. The behaviour of the photons depends on path length differences between the arms of the interferometer on the order of fractions of the optical wavelength.

5.2.4 Transfer of intensity noise through the interferometer

We now consider the intensity noise observed at the interferometer output when a bright beam is incident.

In order to determine the photon flux, for example at output A_2, we have to evaluate the operator $\tilde{N}_2 = \tilde{A}_2^\dagger \tilde{A}_2$. However, for a bright beam we can consider this as a classical part α for the average values and the operator $\delta\tilde{A}$ for the fluctuations, e.g. $\tilde{A}_2 = \alpha_2 + \delta\tilde{A}_2$. To calculate the average intensities we only have to consider the terms in α, all terms with $\delta\tilde{A}$ make no contribution and we obtain

$$\begin{aligned}
\bar{N}_1 &= \bar{N}_{\text{in}} \cos^2(\Delta\phi/2) \\
\bar{N}_2 &= \bar{N}_{\text{in}} \sin^2(\Delta\phi/2).
\end{aligned} \qquad\qquad (5.2.11)$$

Not surprisingly, this is the same result as for the classical calculation. The interferometer is like a variable beamsplitter that distributes the energy between the two output ports. The ratio of the distribution can be adjusted by the differential phase $\Delta\phi$.

Now we come to the transfer function for the fluctuations. The direct calculation of the fluctuations would require the evaluation of $VN_1 = \langle \delta\tilde{A}_1^\dagger \delta\tilde{A}_1 \delta\tilde{A}_1^\dagger \delta\tilde{A}_1 \rangle - \langle \delta\tilde{A}_1^\dagger \delta\tilde{A}_1 \rangle^2$. This would be very laborious, but for bright beams we can linearise, see Section 4.5, and use $VN_1 = \alpha_1^2 \ V1_1(\Omega) = \alpha_1^2 \ \langle |\delta\tilde{X}1_1(\Omega)|^2 \rangle$. Consequently, turning Equation (5.2.9) into a transfer function for the amplitude quadratures, the result is

$$\begin{aligned}
V1_1(\Omega) &= \cos^2(\Delta\phi/2) \ V1_{\text{in}}(\Omega) + \sin^2(\Delta\phi/2) \ V2_u(\Omega) \\
V1_2(\Omega) &= \sin^2(\Delta\phi/2) \ V1_{\text{in}}(\Omega) + \cos^2(\Delta\phi/2) \ V2_u(\Omega)
\end{aligned} \qquad (5.2.12)$$

Again, this is not a surprising result. The fluctuations of the two outputs are proportional to the average photon fluxes. The size of the fluctuations reflects the division of the flux between the two outputs. However, it is important to note that the fluctuations of the phase quadrature of the unused port enter these calculations, not the phase fluctuations of the input laser beam, which cancel completely. It was first pointed out by C. Caves [Cav81] that the noise in the output can be interpreted as a beat between the coherent input beam and the vacuum state with $V2_u = 1$. In this balanced interferometer the phase noise of the input beam cancels, it does not contribute.

If we have only one input beam we can make the assumptions that: (i) the second port is not used, the other input beam is the vacuum mode and $V1_u = 1$, and (ii) the photon flux of the input beam is dominant ($\bar{N}_{in} \gg 1$). Now we obtain

$$
\begin{aligned}
V1_1(\Omega) &= \cos^2(\Delta\phi/2) \, V1_{in}(\Omega) + \sin^2(\Delta\phi/2) \\
V1_2(\Omega) &= \sin^2(\Delta\phi/2) \, V1_{in}(\Omega) + \cos^2(\Delta\phi/2)
\end{aligned}
\tag{5.2.13}
$$

This is identical to the transfer function of a single beam with a beamsplitter of variable reflectivity, determined by the phase shift difference $\Delta\phi$ in the interferometer. Again, the spectrum of the fluctuations remains unchanged.

5.2.5 Sensitivity limit of an interferometer

How should we operate the interferometer to get the best sensitivity? What is the limit in sensitivity? It is necessary to determine the signal that can be created by modulating the phase difference. For a small change $\delta\phi$ of the phase shift ($\delta\phi \ll \pi$) the change in the optical flux at the detector is equal to the first derivative of the output photon flux in regard to the phase multiplied with $\delta\phi$

$$
\delta N_2 = \delta\phi \frac{d(N_2)}{d\phi} = \delta\phi 2 \cos(\Delta\phi/2) \sin(\Delta\phi/2) \frac{1}{2} N_{in} \approx \frac{\delta\phi}{2} \sin(\Delta\phi) N_{in}
\tag{5.2.14}
$$

An AC detector measures the signal power p_{sig} which is proportional to the square of the change in optical power, see Section 2.1, and includes τ, the inverse of the detection bandwidth

$$
p_{sig} = \left(\frac{\delta\phi}{2} \sin(\Delta\phi) \, N_{in}\tau \right)^2
\tag{5.2.15}
$$

As illustrated in Fig. 5.8, the signal is largest when the change in power is most rapid, that is at $\Delta\phi = \pi/2$, in the middle between the maximum and minimum of a fringe. The signal can be compared with the variance of the photon number, or the quantum noise on the detector

$$
p_{qnoise} = V N_2 = \sin^2(\Delta\phi/2) \, N_{in}\tau
\tag{5.2.16}
$$

and we can evaluate the signal to noise ratio

$$
\text{SNR} = \frac{\sin^2(\Delta\phi)}{\sin^2(\Delta\phi/2)} \, N_{in} \, \tau \, \frac{\delta\phi^2}{4}
\tag{5.2.17}
$$

which has the largest value when $\Delta\phi = 0, 2\pi$, where the output optical power is at a minimum. In the jargon of the trade: we operate the interferometer at a *dark fringe*. This is a very different operating point than one would use for a classical interferometer with intensity independent noise. The reason for this result is that while both signal and noise disappear at $\Delta\phi = 0$, the quantum noise disappears more rapidly than the signal and the ratio of the two reaches a maximum, see the SNR trace (c) in Fig. 5.8.

Note that this is a somewhat academic result. In practice the perfect dark fringe is not an optimum since there are always some additional intensity independent noise terms such as electronic noise. The optimum will be at a point where the quantum noise component in the current fluctuations is larger than the additional noise. Details are given in Section 10.4. We

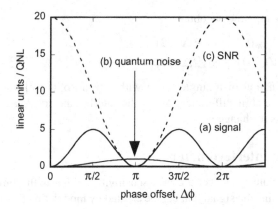

Figure 5.8: Results for an interferometer which is limited by quantum noise. Shown as a function of the detuning $\Delta\phi$ are: (a) the optical power which also corresponds to a signal with fixed modulation depth, (b) the quantum noise and (c) the signal to noise ratio.

define a quantum noise limited (QNL) instrument as a device where the quantum noise is the dominant noise source. The minimum detectable phase change $\delta\phi_{\text{QNL}}$ in a QNL interferometer corresponds to a signal to noise ratio of $SNR = 1$. Near the optimum dark fringe we can approximate

$$\frac{\sin^2(\Delta\phi)}{\sin^2(\Delta\phi/2)} \approx \frac{(\Delta\phi)^2}{(\Delta\phi/2)^2} = 4$$

$$\delta\phi_{\text{QNL}} = \frac{1}{\sqrt{N_{in}\tau}} \qquad (5.2.18)$$

for a coherent input state. This is the standard quantum noise limit for interferometry. It is the smallest achievable measurement of a change in phase using coherent light. A comment on units: $\delta\phi_{\text{QNL}}$ is measured in [rad]. N_{in} is the number of photons travelling from the interferometer to the front face of a detector in one measurement interval. Thus N_{in} is a number with no dimensions. The value of $\delta\phi_{\text{QNL}}$ is not fixed, it depends on the integration time τ. If we spend more time on the measurement, integrate for longer, the value would improve. But the integration time cannot be longer than the period of the signal. As an example the QNL for an interferometer operated with 1 [mW] of light at $\lambda = 1[\mu\text{m}]$ and a detection interval of 1 [ms] is 4.4×10^{-7} [rad].

This quantum description is not complete. Equation (5.2.18) shows that the effect of the quantum noise decreases with more optical power. It seems that it would be possible to reach any arbitrary precision, this must be impossible. It is one of the deep truths of quantum mechanics that we need to *complement* (using an expression coined by Niels Bohr) the concept of quantum noise of the light with a description of the measurement process. We have to consider both the system we investigate and the process of measurement. In straight forward analogy to the Gedanken experiment of Heisenberg's microscope there is a back effect of the measurement process onto the object we measure.

In an interferometer we investigate the position of the mirrors, which defines the optical path length of the interferometer. The measurement process is the interference of the two reflected modes of light. The reflection of a beam of light with power P_{refl} exerts a force on the mirrors

$$F_{RP} = \frac{2P_{\text{refl}}}{c}$$

The fluctuations of this force are due to the quantum noise fluctuations in the photon flux, as discussed above, and the amplitude spectral density of the force will be

$$\delta F_{RP}^2(\Omega) = \frac{2\,\hbar\,P_{\text{refl}}}{c\lambda} \tag{5.2.19}$$

which is independent of the detection frequency. Such a stochastic force will produce fluctuations in the mirror position and results in a position noise spectrum [Sau94]

$$\delta x^2(\Omega) = \frac{1}{m(2\pi\Omega)^2}\delta\ F_{RP}^2(\Omega) = \frac{1}{m\Omega^2}\sqrt{\frac{\hbar P_{\text{refl}}}{8\pi^3 c\lambda}} \tag{5.2.20}$$

where m is the mass of the mirror. This displacement occurs in both arms of the interferometer, but with opposite sign. The total effect gives a spectrum of the angular fluctuations

$$\delta\phi_{RP}(\Omega) = \frac{\hbar}{m\Omega^2\lambda^2}\sqrt{N_{\text{in}}\,\tau}$$

This noise contribution increases with the input power. It also rises at lower detection frequencies. The total noise will be

$$\delta\phi_{QM}^2 = \delta\phi_{\text{QNL}}^2 + \delta\phi_{RP}^2 \tag{5.2.21}$$

and the balance between the two terms depends on detection frequency and the power level chosen. Only at low frequencies and high powers does the radiation pressure dominate. At higher frequencies and lower powers the system is limited by intensity quantum noise alone (QNL). Very special arrangements, such as high power lasers and very light weight mirrors are required to make the light pressure visible. An example for this balance is shown in Fig. 10.10 in Chapter 10. For each frequency there is an minimum noise level, also called the *standard quantum limit* or *QNL*.

This limit can be reached for an optimum power of $P_{\text{opt}} = \pi c\lambda m\Omega^2$ which is typically quite large. For example with $m = 1$ [kg], $\lambda = 1$ [μm], $\Omega = 100$ [Hz] one obtains $P_{\text{opt}} = 9.4$ [MW]. At the optimum condition the system reaches its limit in sensitivity as dictated by the quantum physics of the measurement process. The QNL described in Equation (5.2.21) cannot be avoided with any measurement process based on normal coherent light but it can be surpassed by using correlated photon states and quantum non-demolition techniques which are described in Chapters 9, 11 and 10.

5.3   Cavities

One component of widespread use in quantum optics experiments is the optical cavity. In the simplest case this consists of two mirrors which are aligned on one optical axis such that they

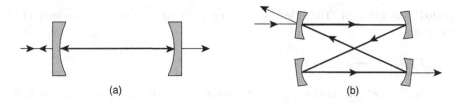

Figure 5.9: Schematic diagrams of linear and ring cavities

exactly retro-reflect the incident light beam. In the more general case it is a group of mirrors aligned in such a way that the beam of light is reflected in a closed path, such that after one round trip it interferes perfectly with the incident wave. The curvature of the mirrors is chosen that they match the curvature of the internal wave fronts at all reflection points. The optical path length for one round trip through the cavity is the perimeter p, which takes into account all the refractive indices inside the cavity. The light is reflected many times around the cavity. Each round trip generates one partial wave, all these waves will interfere.

For the case that p is an integer multiple of the wavelength of the incident light a strong field will be built up inside the cavity due to constructive interference, the cavity is on resonance. The resonance frequencies satisfy the condition

$$\nu_{cav} = \frac{jc}{p}; \quad j = \text{integer} \tag{5.3.1}$$

Note that for a linear cavity the optical length is $p = 2nd_{12}$, where d_{12} is the physical separation between the two mirrors and n the refractive index. If either the cavity length or the wavelength of the light is scanned the cavity will go in and out of resonance. The cavity resonance frequencies are equally spaced and the difference between two adjacent resonance frequencies is called the *free spectral range* (FSR)

$$FSR = c/p \tag{5.3.2}$$

The magnitude of the enhancement of the field and the width of the resonance depend on the reflectivities of the mirror and losses inside the cavity. Cavities are used for a number of purposes: The laser itself is a cavity with an active medium inside which provides the gain. The cavity provides the optical feedback required to achieve oscillation. Hence the name: Laser resonator. The cavity determines the spatial properties of the laser beam. In addition, it selects the frequencies of the laser modes. A cavity can be used to investigate the frequency distribution of the incident light. For this purpose, the length of the cavity is scanned and the transmission is plotted as a function of the length. In this mode of operation a cavity is called an optical spectrum analyser.

On the other hand, a cavity can be used as a very stable reference for the frequency of the light. The detuning of the incident light from the resonance frequency can be measured and the frequency of the laser light can be controlled via electronic feedback. In this way the laser is locked to the cavity and the frequency of the laser light is stabilised. For details see Section 8.4. The light reflected from the cavity provides optical feedback. This is very undesirable for many laser systems, and sometimes even destructive. However, feedback can be used with some lasers, such as diode lasers, to form an external cavity and thereby control the laser frequency. The high optical field inside the cavity can be used to enhance nonlinear optical effects. Such buildup cavities are used in many of the CW squeezing experiments described in Chapter 9.

On resonance, the cavity will set up internal spatial modes. In the case of a laser resonator this determines the shape of the laser mode, its waist position and waist size. Only certain combinations of mirror curvatures will lead to a stable resonance condition [Sie86]. A passive cavity will only be resonant with that component of the incident field which corresponds to its internal spatial mode. All other light will be directly reflected. In other words, the incident light has to be mode-matched to the cavity in order to get into the cavity. In return, the transmitted light has a very well defined, clean spatial mode. For this reason cavities can be used as spatial mode cleaners.

There are several applications of cavities for influencing the fluctuations of the light. The buildup of the internal field is not instantaneous. It requires the interference of many waves which have to travel through the entire cavity. A steady state can be achieved after a time t_{cav}. This is also the time it takes for the internal field to decay in the absence of an incident light field. Thus the cavity acts like an integrator on the fluctuations with a time constant t_{cav}. The high frequency components of the incident light, at frequencies larger than $\gamma = 1/t_{cav}$, will not enter the cavity, they will be reflected. These frequency components will be strongly attenuated in the transmitted light. On resonance, the cavity acts like a low pass filter on the intensity fluctuations.

Finally, there are interesting applications of a cavity which is close to, but not exactly on, resonance with the input field. The phase of the reflected amplitude, compared to the phase of the incident light, varies strongly, by about π, when scanned across one cavity linewidth. Thus the sidebands of the light, which have different cavity detunings to the carrier frequency, are reflected with different optical phases. As a consequence, the quadrature of the reflected light is rotated in respect to that of the incident light. We will find, in Section 9.3, that cavities can be used to analyse the quadratures of light, coherent or squeezed, and as a device to change the squeezing quadrature.

5.3.1   Classical description of a linear cavity

The intensity of the internal, reflected and transmitted light and the relative phase of these beams can be evaluated with a purely classical model. Let us consider a simple cavity with input mirror M_1 and output mirror M_2. This could be a standing wave cavity with two mirrors only or a ring cavity where all other mirrors are near perfect reflectors. The cavity can be described by the following parameters and the corresponding total cavity properties:

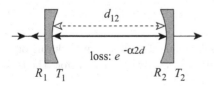

Figure 5.10: Properties of a linear cavity

Table 5.1: Parameters and properties of a linear cavity

Parameters of a linear cavity	
T_1	= intensity transmission of the input mirror
T_2	= intensity transmission of the output mirror
R_1	= intensity reflectivity of the input mirror
R_2	= intensity reflectivity of the output mirror
p	= optical path length for one round trip
d_{12}	= distance from input to output mirror
α	= the loss coefficient for internal absorption or scattering
$\exp(-\alpha p)$	= round trip intensity loss of the cavity
Properties of a linear cavity	
g_m	= loss parameter
P_{in}	= power of the incident beam
P_{out}	= power of the transmitted beam
P_{refl}	= power of the reflected beam
γ	= cavity linewidth
Δ	= normalized cavity detuning = $(\nu_L - \nu_{\text{cav}})/\gamma$
FSR	= free spectral range = p/c
F	= Finesse = FSR/γ

These properties can be derived using the classical equations for multiple interference of waves. Here we will use the standard textbook approach of adding partial waves and combine it with the more elegant approach of solving self-consistency equations at the boundaries of the cavity. The amplitude α_{in} of the incident field will be partially transmitted through the first mirror, the internal field just past the first mirror field will be $\alpha_0 = \sqrt{T_1}\alpha_{\text{in}}$. It will circulate inside the cavity and during each round-trip it will be reduced in amplitude by a factor [Sie86]

$$g_m = \sqrt{R_1 R_2 \exp(-\alpha p)} \tag{5.3.3}$$

which includes the amplitude reflectivities at the mirrors and the internal losses. The wave will also pick up a phase difference $\delta\phi = 2\pi(\nu_L - \nu_{\text{cav}})p/c$ for every round trip.

Let us describe the field amplitude α_{cav} inside the cavity as a sum of all circulating terms

$$\alpha_{\text{cav}} = \alpha_0 + \alpha_1 + \alpha_2 + \alpha_3 + \alpha_4 + \cdots = \sum_{j=0}^{\infty} \alpha_j$$

where

$$\alpha_{j+1} = g_m \exp(-i\delta\phi(\nu))\alpha_j = g(\nu)\alpha_j$$

Here $g(\nu)$ is a complex number which describes the total effect of the cavity during one round trip. The amplitude of α_2 is related to the amplitude of α_1 by the same factor. The total field is described by

$$\alpha_{\text{cav}} = \alpha_0 \left(1 + g(\nu) + g(\nu)^2 + g(\nu)^3 + g(\nu)^4 + \cdots\right) = \frac{\alpha_0}{1 - g(\nu)} \tag{5.3.4}$$

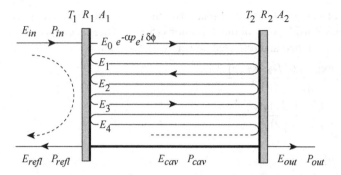

Figure 5.11: Multiple amplitude interference inside a linear cavity.

Note that this is equivalent to the effect of a closed feedback system which adds a component $g(\nu)\alpha_0$ inside the feedback loop. The intensity, and thus the power of the light inside the cavity is built up to

$$
\begin{aligned}
P_{\text{cav}} \propto |\alpha_{\text{cav}}|^2 &= \frac{|\alpha_0|^2}{|1 - g(\nu)|^2} \\
&= \frac{|\alpha_0|^2}{|1 - g_m \exp(-i\phi)|^2} \\
&= \frac{|\alpha_0|^2}{(1 - g_m)^2 + 4g_m \sin^2(\delta\phi/2)}
\end{aligned}
\tag{5.3.5}
$$

Thus we get

$$
P_{\text{cav}} = P_{\text{max}} \frac{1}{1 + (2F/\pi)^2 \sin^2(\delta\phi/2)}
\tag{5.3.6}
$$

with $P_{\text{max}} = \left(\frac{\alpha_0}{1-g_m}\right)^2$ and

$$
F = \text{Finesse} = \frac{\pi\sqrt{g_m}}{(1 - g_m)}
$$

The finesse is generally used as a measure for the quality of cavity. It is the number most frequently quoted in experimental publications. We can now link the amplitudes of the various fields through the following boundary conditions, where we take α_{cav} as the internal field just on the inside of mirror M_1

$$
\begin{aligned}
\alpha_0 &= \sqrt{T_1}\alpha_{\text{in}} \\
\alpha_{\text{out}} &= \sqrt{T_2}\exp(-\alpha d_{12} + i\delta\phi/2)\alpha_{\text{cav}} = \frac{\sqrt{T_2}\,g(\nu)}{\sqrt{R_1 R_2}}\alpha_{\text{cav}} \\
\alpha_{\text{refl}} &= -\sqrt{R_1}\alpha_{\text{in}} + \sqrt{T_1}\sqrt{R_1}\alpha_{\text{cav}}
\end{aligned}
\tag{5.3.7}
$$

The second condition takes into account the internal losses on the path d_{12} from mirror M_1 to mirror M_2 and the transmission through mirror M_2. The last condition takes into account that the wave at the outside of mirror M_1 is out of phase (minus sign) with the wave transmitted through this mirror from the inside. Combining Equations (5.3.7) with Equations (5.3.4) leads to the link between the transmitted and reflected powers with the input power

$$\frac{P_{\text{out}}}{P_{\text{in}}} = T_1 T_2 \frac{\exp(-2\alpha d_{12} + i\delta\phi)}{|1 - g(\nu)|^2}$$

$$\frac{P_{\text{refl}}}{P_{\text{in}}} = \frac{(R_1 - (R_1 + T_1)g(\nu))^2}{R_1 |1 - g(\nu)|^2}$$

$$\frac{P_{\text{cav}}}{P_{\text{in}}} = \frac{T_1}{|1 - g(\nu)|^2} \tag{5.3.8}$$

Consider the special case of a linear cavity with no losses in the mirrors, that means $p = 2d_{12}$ and $R_1 + T_1 = 1$, $R_2 + T_2 = 1$. This is commonly discussed in textbooks as the case for an optical spectrum analyser or a Fabry-Perot etalon. In this case we obtain the result

$$\frac{P_{\text{out}}}{P_{\text{in}}} = \frac{T_1 T_2 g(\nu)}{\sqrt{R_1 R_2} |1 - g(\nu)|^2}$$

$$\frac{P_{\text{refl}}}{P_{\text{in}}} = \frac{|R_1 - g(\nu)|^2}{R_1 |1 - g(\nu)|^2}$$

$$\frac{P_{\text{cav}}}{P_{\text{in}}} = \frac{T_1}{|1 - g(\nu)|^2} \tag{5.3.9}$$

The cavity buildup can be dramatic and the transmission of a cavity can be much larger than the individual transmissions of the mirrors. Note that these equations hold for all detunings, on resonance simply replace $g(\nu)$ by g_m. For illustration it is worthwhile to consider some special cases:

i) A symmetric, loss less cavity. That means $T_1 = T_2 = T$ and $\alpha = 0$. The loss parameter for this case is $g_m = R = (1 - T)$ and we obtain from Equation (5.3.8) $P_{\text{out}}/P_{\text{in}} = T^2 R / R (1 - R)^2 = 1$ and $P_{\text{refl}}/P_{\text{in}} = (1 - R)^2 / R (1 - R)^2 = 0$. Such a cavity is perfectly transmitting and has no reflected power at all. This clearly illustrates the dramatic effect of the constructive interference. The internal power is $P_{\text{cav}} = P_{\text{in}} T / (1 - R)^2 = P_{\text{in}}/T$. Which means that mirrors of transmission $T = 0.01$, corresponding to $R = 0.99$, are required for a power buildup by a factor of one hundred. The finesse of such a cavity can be approximated as $F = \pi/T = 300$.

ii) A cavity with loss. All cavities will have some losses. An interesting case is a cavity where the input transmission T_1 just balances the sum of all losses. That means $R_1 = g_m$, which is equivalent to $R_1 = R_2 \exp(-\alpha p)$. For this case we get $P_{\text{refl}} = 0$, there is no reflection. The leakage of the intra-cavity power out through the input mirror is equal to the directly reflected amplitude. They cancel each other. This case is known as the impedance matched case, in analogy to the matching of electronic transmission lines.

$$\begin{array}{lll}
\text{under-coupled :} & R_1 < R_2 \exp(-\alpha p) \\
\text{impendance matched:} & R_1 = R_2 \exp(-\alpha p) \\
\text{over-coupled :} & R_1 > R_2 \exp(-\alpha p)
\end{array}$$

The two other cases, under-coupled and over-coupled lead to some reflection and the transmission is not optimum. While both cases result in similar reflected power, the sign of the reflected wave changes from the under-coupled to the over-coupled case. However, in most experiments this sign cannot be detected. All these cases are illustrated in Fig. 5.12 by the traces (i), (ii) and (iii).

iii) A single ported cavity ($T_2 = 0$). This obviously has no transmission. However, it is still useful. It shows internal power buildup and all the leakage back out is through the input mirror. The cavity will have a partial power reflection. This light will contain information about the internal field. The reflected beam can be separated and detected by using a linear cavity and an optical isolator or alternatively a ring cavity where the light is reflected into a different direction.

The loss parameter g_m describes the reduction in the intensity of the internal field on resonance with the cavity after one round trip. The fraction $(1 - g_m)$ leaves the cavity during each round trip. One round trip takes the time p/c. Thus the energy loss from the cavity is $(1 - g_m)P_{cav}c/p$ measured in [J/s]. Since this loss is proportional to the energy stored in the cavity the power P_{cav} will decay exponentially in time in the absence of an incident field.

$$P_{cav}(t) = P_{cav}(0) \exp(-t/t_{cav})$$
$$\frac{1}{t_{cav}} = \frac{c}{p}(1 - g_m)$$

The amplitude of the field decays with twice the time constant and we can define a coupling rate κ for the amplitude

$$\alpha_{cav}(t) = \alpha_{cav}(t = 0) \exp(-t/2t_{cav}) = \alpha_{cav}(t = 0) \exp(-\kappa t)$$

with $\kappa = (c/p)(1 - g_m)$. This decay constant is the appropriate term for the temporal description of the cavity response. It can be linked to the cavity linewidth γ. The Fourier transform of the intensity decay gives the spectral response function of the cavity. It results in a function proportional to $1/(1 + i4\pi t_{cav})$. The resonance of the spectral response has a full width half maximum FWHM of $\gamma = 1/(2\pi t_{cav})$ and is measured in units of [Hz]. The same result could have been obtained by determining the FWHM of the cavity response function given in Equation (5.3.6).

5.3.2 The special case of high reflectivities

It is useful to treat the transmission through the mirrors and the loss inside the cavity in a similar way. They are all processes for the leakage of the field out of the cavity. For this purpose one can describe the properties of the mirrors and the losses with coupling constants δ [Sie86]. These replace the exponential terms which are required for cavities with large losses. We convert the mirror reflectivity into a mirror coupling constant via

$$R_i = \exp(-\delta_{mi}) \quad \text{or} \quad \delta_{mi} = \ln(1/R_i) = -\ln(R_i)$$

and define an equivalent coupling constant for the intra-cavity losses as $\delta_{lo} = \alpha p$. In this way the loss parameter g_m, defined in Equation (5.3.3) for a cavity with two partially reflecting mirrors turns into

$$g_m = \exp(-(\delta_{m1} + \delta_{m2} + \delta_{lo})) = \exp(-\delta)$$

An important case in quantum optics is the case of a high finesse cavity. Here the round trip losses are small. That means $\delta_{mi} \ll 1$, $\delta_{lo} \ll 1$ and thus $\delta \ll 1$. This effectively means that the amplitude is changing so little per round trip that spatial effects along the propagation axis can be neglected. We assume a uniform internal amplitude. We also assume that losses inside the mirrors can be neglected, that means $(R_i + T_i) = 1$. This leads to the following approximations:

$$\begin{aligned}
g_m = \exp(-\delta) &\approx 1 - \delta \\
\delta_{mi} = -\ln(R_i) &\approx 1 - R_i = T_i \\
\kappa = \frac{(1 - g_m)c}{2p} &\approx \frac{\delta c}{2p} = \frac{(\delta_{m1} + \delta_{m2} + \delta_{lo})c}{2p}
\end{aligned} \tag{5.3.10}$$

We can define the loss rates for the total cavity and for the individual processes as

$$\kappa = \kappa_{m1} + \kappa_{m2} + \kappa_{lo} \quad \text{with} \quad \kappa_{mi} = \frac{T_i c}{2p} \quad \text{and} \quad \kappa_{lo} = \frac{\alpha c}{2} \tag{5.3.11}$$

These definitions are commonly used in describing high finesse cavities. We find them for example in laser models, Chapter 6, and many papers on nonlinear processes inside cavities. They are very useful for most quantitative models. However, the limitations of this approximation should not be forgotten. There are many experiments where intra- cavity losses as large as several percent are unavoidable. Or we might have a strong nonlinear conversion out of the internal mode into another mode. One example would be efficient second harmonic generation. In these cases the approximations of the high Finesse limit might not hold and the exact equations should be used.

5.3.3   The phase response

The cavity introduces a phase shift between the incoming, the intra-cavity, the reflected and the transmitted fields. The boundary conditions for the electric field, given in Equation (5.3.7) can be used to evaluate these phase shifts. Let us define all phase shifts in respect to the phase of the incoming amplitude. That means $\phi_{cav} = \arg(\alpha_{cav}) - \arg(\alpha_{in}) = \arg(\alpha_{cav}/\alpha_{in})$, where the function $\arg(z)$ determines the phase angle ϕ of the complex number z. Using the results of Equation (5.3.7) and some algebraic manipulation the following results are obtained:

$$\begin{aligned}
\phi_{cav} &= \arctan\left(\frac{g_m \sin(\delta\phi)}{1 - g_m \cos(\delta\phi)}\right) \\
\phi_{out} &= \arctan\left(\frac{-\sin(\delta\phi)}{g_m - \cos(\delta\phi)}\right) \\
\phi_{refl} &= \arctan\left(\frac{-T_1\sqrt{R_1}\sin(\delta\phi)}{\sqrt{R_1} + T_1\sqrt{R_1}(1 - g_m\cos(\delta\phi))}\right)
\end{aligned} \tag{5.3.12}$$

A graphical evaluation of these equations for three special cases of under-coupled, over-coupled and impedance matched are shown in Fig. 5.12(i),(ii),(iii) alongside the results for

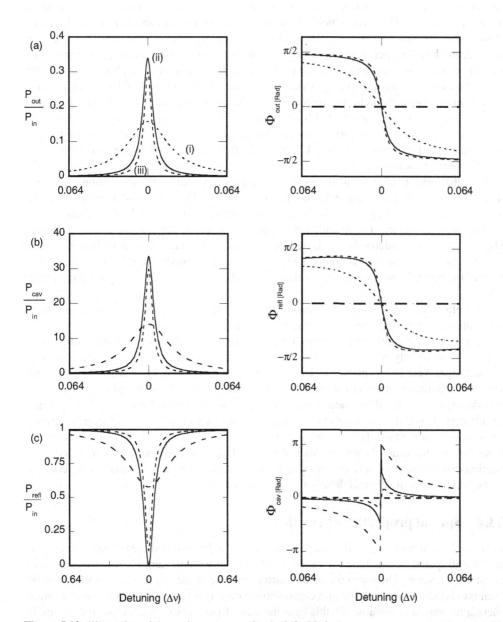

Figure 5.12: Illustration of the cavity response. On the left side is the power response, on the right side the phase response. Shown are the three results: (a) transmitted light, (b) the internal field and (c) the reflected light. For each situation three different cases are shown. (i) the largely under-coupled case (long dashes), (ii) the impedance matched case (solid line) and (iii) the over-coupled case (dotted line).

the transmitted powers. On resonance, that means $\delta\phi = 0$, all three phase terms are equal to zero, modulus π. The electric field of the incoming, intra-cavity and transmitted light are all in phase. The reflected light is either in phase or out of phase. Around the resonance the phase terms show a dispersive behaviour. The width of the dispersion curve (right hand side) is comparable to the cavity linewidth γ (left hand side). There is a 2π phase change right across the cavity resonance. The intra-cavity field and the transmitted field have a phase advance over the incoming field that increases with detuning. For the reflected light the phase can be advanced (+ sign) or retarded (- sign), depending on the parameters of the cavity. If the directly reflected light dominates (under-coupled case) the phase difference is 180 degrees on resonance and the phase is retarded. If the light from the cavity dominates (over-coupled case) the phase is advanced.

Note that sidebands of the incoming wave can experience different attenuation and phase shifts. For the limiting case where the modulation frequency Ω_{mod} of the sidebands is less than the cavity linewidth ($\Omega_{\text{mod}} \ll \gamma$) this effect is negligible. In this case the sidebands remain in phase with the cavity. For modulation frequencies much larger than the cavity linewidth ($\Omega_{\text{mod}} \gg \gamma$) the two sidebands are shifted by $+\pi$ and $-\pi$ respectively and again they appear to be in phase with the central component. If, however, the modulation frequency is of the order of the cavity linewidth the two sidebands at $\pm\Omega_{\text{mod}}$ will experience very different phase shifts.

This effect can be exploited to measure the detuning of the cavity. The incoming light is modulated and the transmitted or reflected light intensity is detected. At Ω_{mod} we obtain a signal which is the sum of two beat signals between the sidebands and the central laser frequency. Commonly, phase modulation is used with the consequence that on resonance these two beat signals are out of phase and cancel each other. On resonance, there is no detectable intensity modulation on either the incoming, the transmitted or the reflected light. However, for a detuned cavity the difference in the phase shifts for the two sidebands will prevent this cancellation. A modulation signal at frequency Ω_{mod} appears and phase modulation is partially turned into intensity modulation. The signal strength can be measured by demodulating the photo-current, for example with a mixer driven at Ω_{mod}. The shape of this response as a function of detuning is closely linked to the dispersion curve for the individual phase terms. It can be derived analytically and details are given in Section 8.4.

5.3.4 Spatial properties of cavities

So far we have treated the cavity and the light in the cavity as one dimensional systems, we have lost all spatial information. However, not all cavities with the same mirror reflectivity and length are the same. The curvature of the mirrors and the shape of the wave front have to be taken into account. There are certain combinations of mirror separation and mirror curvature where a resonance is possible. In this case the wave front, returning after one round trip, is keeping its size and is not expanding significantly. A spatial resonance will build up and as a consequence the curvature wave front of the resulting resonant mode will be identical to the curvature of the mirror.

As the simplest example let us consider a linear, two mirror cavity. Here the resonator is defined by the separation d_{12} of the mirrors and the mirror curvatures $\mathcal{R}_1$ and $\mathcal{R}_2$. The focal length of the mirrors is given by $f_1 = \mathcal{R}_1/2$ and $f_2 = \mathcal{R}_2/2$. In the standard notation we define the mirror parameters [Sie86]

$$g_1 = 1 - \frac{d_{12}}{2f_1} \quad \text{and} \quad g_2 = 1 - \frac{d_{12}}{2f_2} \tag{5.3.13}$$

The condition for a stable resonator, that means a resonator which builds up a resonant wave, can be derived [Sie86] and results in the stability criterion

$$0 \le g_1 g_2 \le 1. \tag{5.3.14}$$

Based on this criterion we find that for a given set of mirror curvatures (focal lengths) resonance only occurs for a restricted range of cavity lengths. We have to choose the length and the curvatures together. Some examples of stable resonators are:

- The (near) concentric cavity: $d_{12} = 2 \, \mathcal{R}_1 = \mathcal{R}_2$ that means $g_1 = g_2 = -1$ and $g_1 g_2 = 1$.

- The (near) planar cavity: $\mathcal{R}_1 = \mathcal{R}_2 = \infty$, that means $g_1 = g_2 = 1$ and $g_1 g_2 = 1$.

- The confocal cavity : $d_{12} = \mathcal{R}_1 = \mathcal{R}_2$ that means $g_1 = g_2 = 0$ and $g_1 g_2 = 0$.

The concentric and the planar cavities are right at the limit of the stability criterion. They can be stable but even minor changes in length or small misalignments tend to affect the performance. These cases are not recommended. It is better to satisfy the stability criterion with some margin. The confocal cavity is in the middle of the stability region and thus very useful as a robust design for a resonator.

Cavities can support higher order $TEM_{i,j}$ spatial modes, and if the cavity is not confocal this will occur at cavity lengths different to the resonance length for the $TEM_{0,0}$ mode. This feature can be used to devise techniques for locking a cavity to the frequency of a laser beam [Sha00] by analysing the spatial properties of the transmitted and/or reflected beam, see Section 8.4.

Mode matching

The cavity has its own internal geometric mode, determined by the curvature of the mirrors. The wave front of the mode of light illuminating the cavity might have a different wave front than the internal mode. The incoming wave can only fully interact with the cavity if the incoming wave front and the internal wave front are matched. This is known as the *mode matched* case. The spatial properties of the incoming wave are matched, the intensity distribution (mode size) is identical to the internal mode, the wave front curvature is identical to the curvature of the mirror, which in turn is the curvature of the internal mode. In this particular case, which is highly desirable but in practice requires extensive alignment work, the description of the cavity and the coupled light is simple, the one dimensional model in the previous section is correct. All the light interacts with the cavity. Most theoretical models assume mode matching.

What will happen to light incident on a cavity which has a different geometric mode? The cavity will support one fundamental mode and a whole series of higher order spatial modes. In general, these modes will not be resonant simultaneously – different detunings, that means small shifts in the mirror separations, will be required for the resonance condition of the higher order modes. For any given mirror separation the cavity will only support one mode. The one special case, where all spatial modes are resonant simultaneously, is the confocal cavity. For all other cases the resonators are non degenerate – the resonances occur at different cavity detunings. On resonance, the cavity will select the part of the external field which corresponds to the resonant internal mode, an eigenmode of the cavity. In order to calculate the reflection and transmission of a cavity it is useful to describe the external field in terms of all the internal eigenmodes of the cavity. These spatial eigenmodes are a useful basis set for a mathematical expansion of the external field.

Let us classify the spatial modes as Hermite-Gaussian modes (see Section 2.1) with the classification TEM$_{mn}$ describing the possible modes. We can calculate the fraction P_{mn} of the total power that corresponds to the resonant internal mode TEM$_{mn}$. In most cases we align the cavity to resonate in the fundamental mode TEM$_{00}$. Only this light will enter the cavity and contribute to the buildup of the internal power. The remaining power will be reflected, assuming that the front mirror has a high reflectivity. The equations for the transmitted and reflected power (Equations (5.3.8)) have to be modified to include the fraction P_{mm}. For example, the transmission will then be $P_{mn} \times P_{\text{trans}}/P_{\text{in}}$, the reflectivity will be $(1 - P_{mn}) + P_{mn} \times P_{\text{refl}}/P_{\text{in}}$. Obviously, $P_{00} = 1$ corresponds to the perfectly mode matched situation. To achieve $P_{00} > 0.95$ requires considerable patience and good quality optical components. Once aligned to the fundamental mode, both the internal and the transmitted fields will have the spatial intensity distribution of a Gaussian beam. The cavity selects the Gaussian component from the incident field, it acts like a spatial filter. Such a cavity is therefore also referred to as a spatial mode cleaner.

How can the mode matched condition be achieved, how can the mode mismatch be quantified? By detuning the cavity, that means by changing the mirror separation a small fraction of a wavelength, the cavity can be set on resonance with the higher order spatial modes. The exact resonance frequencies can be evaluated by taking into account the phase differences between the various higher order modes and the fundamental mode, known as Guoy phase shifts. These phase shifts depend on the order numbers m and n and can be expanded as $\cos^{-1}(\pm\sqrt{g_1 g_2})$ where g_1 and g_2 are the mirror parameters defined in Equation (5.3.13). The changes in the resonance frequencies $\Delta\nu_{mn}$ for the modes TEM$_{mn}$ away from the resonance of the fundamental mode are given by [Sie86]

$$\Delta\nu_{mn} = \frac{(n + m + 1)\cos^{-1}(\pm\sqrt{g_1 g_2})}{\pi 2 d_{12}} \; c \tag{5.3.15}$$

with the following special cases:

$$\Delta\nu_{mn} \quad \begin{aligned} &= 0 &&\text{for confocal cavity} \\ &= \tfrac{1}{2}FSR &&\text{for near planar cavity} \\ &= FSR &&\text{for near concentric cavity} \end{aligned}$$

If the cavity is not confocal, the higher order spatial modes map out a whole series of resonances as shown in Fig. 5.13(b). These can be made visible by scanning the cavity length, or the frequency of the laser. At each sidepeak the cavity is in resonance with one of the higher order spatial modes and the transmitted light will show the corresponding spatial patterns of this mode. In most practical situations we want to achieve the mode matched case. That

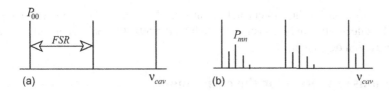

Figure 5.13: The frequency response of a scanning cavity which is longer than confocal to (a) a mode matched input beam and (b) a mismatched beam. Improved alignment will get us from case (b) to the optimum case (a).

means we want to minimise the higher order modes, as shown in Fig. 5.13. By using lenses to change the wave-front curvature and by optimizing the position of the lenses systematically mode matching can be achieved. In the same way the alignment of the beam to the axis of the cavity is improved. This is largely an iterative process, since in most situation there are six degrees of freedom that have to be optimized: two degrees for the tilt of the input beam in regard to the cavity, two degrees for a lateral shift, and two degrees for the wave front curvature. Measuring the size of the high order peaks P_{mn} is far more sensitive than measuring the reflected power P_{refl}. Clever strategies have been devised to detect misalignment and mode matching separately. There have been several approaches how alignment can be automised [And84, Sam90, Mor94], but the problem of automatic mode matching is still largely unresolved.

Polarisation

Polarisation was not included in the above discussion. For a completely isotropic cavity the polarisation does not matter; the description using a single, arbitrary polarisation state of light would be correct. However, in most practical cases at least one of the components of the cavity, or inside the cavity, is anisotropic and induces some birefringence or dichroism. In that case one has to find the polarisation eigenstates of the cavity. Light in these states will keep its polarisation while propagating. It will return after one complete round trip with the same polarisation. Note that at any individual point inside the cavity these eigenstates can be quite different. There are always two such orthogonal states. Losses, detuning and thus the finesse can be different for the two states and the response of the cavity is described by two independent sets of equations, one for each polarisation eigenstate. The best known example is the unidirectional device inside a ring laser which uses polarisation and losses to suppress laser oscillation in one of the two counter-propagating directions.

Tunable mirrors

In many practical situations it would be very desirable to be able to tune the reflectivity of a mirror in a cavity. This can be achieved by making one of the mirrors a cavity itself or by using a trick called *frustrated internal reflection*. Here the two uncoated surfaces are used with a small gap to form a mirror. The light is aligned such that each surface would totally reflect the light. However, when the surfaces have a separation of a wavelength or less the evanescent field, on the outside of the totally reflecting surfaces, reaches to the next surface and light is transmitted. The magnitude of this effect can be controlled by changing the separation, thus a mirror of tunable reflectivity can be constructed [Sch92, BW59]. It should be noted that such mirrors are polarisation dependent.

5.3.5   Equations of motion for the cavity mode

For the rest of our discussion of cavities we will assume that the technical requirements for obtaining high finesse operation have been satisfied. Under such conditions the equations of motion of a single cavity mode can be obtained in a straightforward way. Conceptually it is easiest to consider the case of a ring cavity as depicted in Fig. 5.9. The cavity has two partially transmitting mirrors with transmissions T_1 and T_2 and internal loss with a transmission of T_l. External fields, α_{in1} and α_{in2} enter at the first and second mirrors respectively. We calculate the change in the circulating mode amplitude, α_c, incurred in making one round trip of the cavity under the assumption that that change is small. We have

$$\alpha_c(t+\tau) = \alpha_c(t)e^{i\phi}\sqrt{1-T_1}\sqrt{1-T_2}\sqrt{1-T_l}$$
$$+ \alpha_{in1}(t)e^{i\phi}\sqrt{T_1}\sqrt{1-T_2} + \alpha_{in2}(t)e^{i\phi}\sqrt{T_2} \qquad (5.3.16)$$

where τ is the cavity round trip time and ϕ is the differential phase shift acquired by the fields in that time. If $T_1, T_2, \phi \ll 1$ then the change in the field in one round trip will be small. As such we expand to first order the LHS of Equation (5.3.16) in a Taylor series and make small T approximations on the RHS. We obtain

$$\alpha_c(t) + \tau\dot{\alpha}_c(t) = \left(\alpha_c(t)\left(1 - \frac{1}{2}T_1 - \frac{1}{2}T_2 - \frac{1}{2}T_l \right) + \alpha_{in1}\sqrt{T_1} + \alpha_{in2}\sqrt{T_2} \right)e^{i\phi}$$

Making a small ϕ (modulo 2π) approximation, i.e. assuming we are close to a cavity resonance, and rearranging we obtain

$$\sqrt{\tau}\dot{\alpha}_c = \sqrt{\tau}\alpha_c\frac{(i\phi - \frac{1}{2}T_1 - \frac{1}{2}T_2 - \frac{1}{2}T_l)}{\tau} + \alpha_{in1}\sqrt{\frac{T_1}{\tau}} + \alpha_{in2}\sqrt{\frac{T_2}{\tau}} \qquad (5.3.17)$$

Equation (5.3.17) is normally presented with the following substitutions: $\alpha = \sqrt{\tau}\alpha_c$ is the amplitude of the standing mode of the cavity; $\Delta = \phi/\tau$ is the detuning of the cavity; $\kappa_i = T_i/(2\tau)$ are the decay rates of the cavity. With these substitutions we obtain the following equation of motion for the cavity mode:

$$\dot{\alpha} = \alpha(i\Delta - \kappa_1 - \kappa_2 - \kappa_l) + \alpha_{in1}\sqrt{2\kappa_1} + \alpha_{in2}\sqrt{2\kappa_2} \qquad (5.3.18)$$

This equation of motion will be the starting point for our quantum description. The field exiting from the first mirror is

$$\alpha_{out1} = \sqrt{T_1}\alpha_c - \sqrt{1-T_1}\alpha_{in1}$$
$$\approx \sqrt{2\kappa_1}\alpha - \alpha_{in1} \qquad (5.3.19)$$

and similarly the field exiting from the second mirror is

$$\alpha_{out2} = \sqrt{T_2}\alpha_c - \sqrt{1-T_2}\alpha_{in2}$$
$$\approx \sqrt{2\kappa_2}\alpha - \alpha_{in2} \qquad (5.3.20)$$

Equations (5.3.19), (5.3.20) are referred to as the boundary conditions for the cavity.

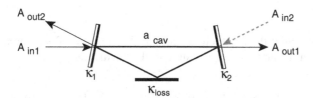

Figure 5.14: A cavity described in the quantum model

5.3.6 The quantum equations of motion for a cavity

The fundamental object of the quantum model of light is the cavity mode. The cavity is the experimental equivalent to the abstract concept of a single frequency mode and its subsequent identification with the quantum model of a harmonic oscillator. In general, a description of the quantum evolution of a cavity requires the specification of the Hamiltonian describing the interactions within the cavity as well as those between the cavity mode and the continuum of field modes outside the cavity. Equations of motion for the field can be obtained either in the Heisenberg picture (operator Langevin equations) or in the Schrödinger picture (Master equation) [Gar00]. Combining this with quantum boundary conditions [Col84] allows calculations of the properties of fields leaving the cavity. It was techniques such as these which were used to rigorously establish the behaviour of quantum fields that we have described in this and the previous chapters. In later chapters we will use such approaches. However, here we proceed via the simpler approach of canonical quantization of the classical equations of motion (Equation (5.3.18)). This leads directly to

$$\dot{\tilde{a}} = -(\kappa - i\Delta)\hat{\tilde{a}} + \sqrt{2\kappa_1}\hat{\tilde{A}}_{in1} + \sqrt{2\kappa_2}\hat{\tilde{A}}_{in2} + \sqrt{2\kappa_l}\hat{\tilde{A}}_l \qquad (5.3.21)$$

where $\tilde{a}$ is the standing cavity mode while the $\tilde{A}_j$ are travelling modes and $\kappa = \kappa_1 + \kappa_2 + \kappa_l$ is the total decay rate of the cavity. As for the beamsplitter and the interferometer the essential differences between the quantum and classical equations is the transformation to operators and the explicit inclusion of an operator representing the vacuum mode introduced by loss. Equation (5.3.21) is an example of an operator quantum Langevin equation. The boundary condition for the transmitted field

$$\hat{\tilde{A}}_{out2} = \sqrt{2\kappa_2}\hat{\tilde{a}} - \hat{\tilde{A}}_{in2} \qquad (5.3.22)$$

and the reflected field

$$\hat{\tilde{A}}_{out1} = \sqrt{2\kappa_1}\hat{\tilde{a}} - \hat{\tilde{A}}_{in1} \qquad (5.3.23)$$

can also be obtained directly from the classical boundary conditions, see Eq. (5.3.19), (5.3.20)).

5.3.7 The propagation of fluctuations through the cavity

In order to analyse the effect of a cavity on the noise of an incident propagating mode of light we consider the transfer functions for the cavity. We use the model described in Chapter 4. We approach the problem step by step:

1. The fluctuations of the input field are described as the spectrum of the amplitude variance $V1_{\mathbf{A}in}(\Omega)$.

2. The cavity is described through the equation of motion. The resulting fluctuations are analysed. We first consider the amplitude quadrature of the fluctuations.

3. The variances $V1_{out}$ of the various output fields are determined. They are given by a linear combination, the transfer function, of the variances of all the input fields.

The variances of the fluctuations of the internal field are of academic interest only. They cannot be measured by conventional means. We find that, for a cavity on resonance, the output amplitude fluctuations are linked directly to the input amplitude fluctuations. For a cavity with detuning, both quadrature amplitudes of the input contribute to the output amplitude fluctuations. The output noise spectrum can be written as a linear function of the variance of the input mode and of contributions from the various unused input ports, which is the major difference to the classical model. While the unused ports do not contribute to the spectrum of the output power, they contribute to the noise spectra of the light. In a way, the effect of the cavity on the fluctuations becomes similar to that of an electronic filter. The response is frequency dependent, it is linear and it can be described by a transfer function.

We consider the system shown in Fig. 5.14 and include the fluctuations. As usual each operator is written as a classical complex value for the amplitude and an operator for the fluctuations: $\tilde{a} = \alpha + \delta a$, $\tilde{\mathbf{A}}_{in1} = \alpha_{in1} + \delta \mathbf{A}_{in1}$ etc. The classical parts obey Equations (5.3.18), (5.3.19), (5.3.20), whilst the operator parts obey Equations (5.3.21), (5.3.22), (5.3.23). Considering just the quantum fluctuations we find

$$\dot{\hat{\delta a}} = -\kappa \hat{\delta a} + \sqrt{2\kappa_1}\delta \hat{\mathbf{A}}_{in1} + \sqrt{2\kappa_2}\delta \hat{\mathbf{A}}_{in2} + \sqrt{2\kappa_l}\delta \hat{\mathbf{A}}_l$$

By adding the complex conjugate an equation for the amplitude quadrature can be formed:

$$\delta \dot{\hat{\mathbf{X}}}1 = -\kappa \delta \hat{\mathbf{X}}1 + \sqrt{2\kappa_1}\delta \hat{\mathbf{X}}1_{in1} + \sqrt{2\kappa_2}\delta \hat{\mathbf{X}}1_{in2} + \sqrt{2\kappa_l}\delta \hat{\mathbf{X}}1_l \qquad (5.3.24)$$

The boundary conditions of Eq. (5.3.19), (5.3.20) are rewritten for the fluctuations

$$\begin{aligned} \delta \hat{\mathbf{X}}1_{out2} &= \sqrt{2\kappa_2}\delta \hat{\mathbf{X}}1 - \delta \hat{\mathbf{X}}1_{in2} \\ \delta \hat{\mathbf{X}}1_{out1} &= \sqrt{2\kappa_1}\delta \hat{\mathbf{X}}1 - \delta \hat{\mathbf{X}}1_{in1} \end{aligned} \qquad (5.3.25)$$

The next step is to consider the Fourier components of the fluctuation. Using the property of Fourier transforms $\mathcal{F}(df(t)/dt) = -i2\pi\Omega\mathcal{F}(f(t))$, we can combine Equation (5.3.24) with Equation (5.3.25) and obtain

$$\delta \tilde{\mathbf{X}}1_{out2} = \frac{\sqrt{4\kappa_1\kappa_2}\,\delta \tilde{\mathbf{X}}1_{in1} + (2\kappa_2 - \kappa + i2\pi\Omega)\,\delta \tilde{\mathbf{X}}1_{in2} + \sqrt{4\kappa_1\kappa_l}\,\delta \tilde{\mathbf{X}}1_l}{\kappa - i(2\pi\Omega)}$$

In order to obtain the spectrum of the fluctuations transmitted through the cavity we determine the variance

$$
\begin{aligned}
V1_{out2}(\Omega) &= \langle |\delta \tilde{\mathbf{X}} 1_{out2}|^2 \rangle \\
&= \{ \quad 4\kappa_2\kappa_1 \quad V1_{in1}(\Omega) \\
&+ \quad ((2\kappa_2 - \kappa)^2 + (2\pi\Omega)^2) \quad V1_{in2}(\Omega) \\
&+ \quad 4\kappa_2\kappa_l \quad V1_l(\Omega) \} \\
&\times \quad \frac{1}{\kappa^2 + (2\pi\Omega)^2}
\end{aligned}
\tag{5.3.26}
$$

This is the desired set of transfer functions. It shows how the different input variances are linked and how the fluctuations propagate through the system. We can repeat the above calculations and obtain for the reflected beam:

$$
\begin{aligned}
V1_{out1}(\Omega) &= \{ ((2\kappa_1 - \kappa)^2 + (2\pi\Omega)^2) \, V1_{in1}(\Omega) \\
&+ 4\kappa_2\kappa_1 \, V1_{in2}(\Omega) \\
&+ 4\kappa_1\kappa_l \, V1_l(\Omega) \} \\
&\times \frac{1}{\kappa^2 + (2\pi\Omega)^2}
\end{aligned}
\tag{5.3.27}
$$

The first useful check is to consider the special case where all inputs are quantum noise limited (QNL), that means they are either not used or illuminated by a coherent state. Using Equation (5.3.11) this results in

$$
\begin{aligned}
V1_{out2}(\Omega) &= \frac{4\kappa_2\kappa_1 + (2\kappa_2 - \kappa)^2 + (2\pi\Omega)^2 + 4\kappa_2\kappa_l}{\kappa^2 + (2\pi\Omega)^2} \\
&= V1_{out1}(\Omega) = 1
\end{aligned}
$$

As expected, the two output noise spectra for this special case are also QNL. This is an important result that can be applied to all passive systems no matter how complicated the instrument is: *if all the input beams are at the QNL than also all the output states are at the QNL.*

But note, that even when all output variances have the variances $V1_{out} = 1$, the origin of this noise can still vary with the frequency since the components of the transfer function itself are frequency dependent. This becomes important when at least one of the input beams has fluctuations above or below the QNL, and either noise or squeezing is propagating through the cavity. In general, the transfer function is frequency dependent.

Consider first the special case of a perfect ring cavity, with $\kappa_1 = \kappa_2 = \kappa/2$. Then the set of transfer functions will become

$$
V1_{out2}(\Omega) = \frac{\kappa^2 \, V1_{in1} + (2\pi\Omega)^2 \, V1_{in2}}{\kappa^2 + (2\pi\Omega)^2}
$$

$$
V1_{out1}(\Omega) = \frac{(2\pi\Omega)^2 \, V1_{in1} + \kappa^2 \, V1_{in2}}{\kappa^2 + (2\pi\Omega)^2}
$$

We can now examine the low and high frequency performances of the cavity, which are distinctly different. For low frequencies, we see that in the limit $\Omega \to 0$ the spectra approach

$$
\begin{aligned}
V1_{out2}(0) &= V1_{in1} \\
V1_{out1}(0) &= V1_{in2}
\end{aligned}
$$

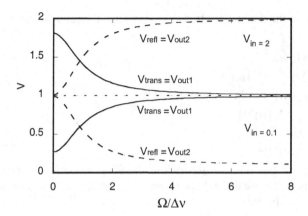

Figure 5.15: Noise response of a cavity as a function of the detection frequency. Both the transmitted and the reflected noise are shown. Two cases: (a) a cavity with noisy input ($V1_{in1} = 2$), (b) a cavity with an input noise below the QNL ($V1_{in1} = 0.1$).

In other words, the input noise passes directly through to the output, and the noise of the unused port, normally vacuum noise, passes straight through into the reflected beam. The cavity is transparent to noise in either direction. More interestingly, at high frequencies, we see that in the limit of $\Omega \rightarrow \infty$ these spectra approach

$$V1_{out2}(\infty) = V1_{in2}$$
$$V1_{out1}(\infty) = V1_{in1} \tag{5.3.28}$$

This is the opposite situation to that above. The cavity is completely opaque to noise at high frequencies. The input noise is reflected into the reflected field and the noise of the unused port is reflected into the output field. This result is true for all cavities, as described by Equations (5.3.26) and (5.3.27). For high frequencies ($(2\pi\Omega)^2 \gg \kappa^2$) the noise from the input source will not be able to propagate through the cavity. Only low frequency noise, within the linewidth of the cavity, is transmitted. The cavity acts like an integrator and smoothes out fast fluctuations. However, these fluctuations are not lost, they appear in the reflected light. For low frequencies, $(2\pi\Omega)^2 \ll \kappa^2$, the noise is transmitted. For frequencies around $(2\pi\Omega) = \kappa$ there is a smooth transition. This response is shown in Fig. 5.15 for the two cases of a noisy input ($V1_{in1} = 2$) and a squeezed input ($V1_{in1} = 0.1$).

In practice, it means that for a laser beam with modulation at Ω_{mod} incident on a cavity the transmission function for the laser carrier frequency ν_L and for the modulation sidebands is not the same, compare Equations (5.3.8) and (5.3.26). In particular, for high frequencies the modulation sidebands cannot be transmitted, they are reflected. A cavity can be used as a filter for noise. The cavity is placed after the output coupler of the laser, all high frequency classical noise is rejected and the output beam is quantum noise limited at these frequencies.

A second, very important consequence of Equation (5.3.27) is the case where the internal fluctuations are not at the quantum noise limit, either below due to some internal nonlinear process, see Chapter 9, or above due to an additional noise or modulation. The above formalism can be used to evaluate the noise spectrum at the output. For example, consider the following simplified case: a single ended cavity with an internal noise spectrum $V1_{cav}(\Omega)$, losses are neglected. That means $\sqrt{2\kappa_l} = \sqrt{2\kappa_2} = 0$ and we obtain an expression which links the external with the internal spectrum.

$$V1_{out1}(\Omega) = \frac{\kappa^2 + (2\pi\Omega)^2}{2\kappa} \ V1_{cav}(\Omega)$$

This is important when we modify the internal noise spectrum through nonlinear processes but we can only observe the external noise spectrum.

Note that all of the above equations have treated the resonant case. The transfer functions for the phase quadrature, $V2$, or indeed any arbitrary quadrature angle, are equivalent to those we have calculated for the amplitude quadrature, $V1$, for this case. It is straightforward to include detunings from cavity resonance, but this generates significantly more complex equations. The main difference is that in the detuned case both quadrature amplitudes, $X1$ and $X2$, have to be taken into account. A detuned cavity couples the fluctuations of the two quadratures – it links amplitude and phase fluctuations. This effect can be exploited for very sensitive measurements of phase fluctuations, for cavity locking (see Section 8.4) and for the rotation of the squeezing axis (see Section 9.3).

5.3.8   Single photons through a cavity

We now look at the effect of the cavity on a single photon field. As for previous elements we will use the techniques of Section 4.4 to generate the state evolution. However, in contrast to the beamsplitter and interferometer, here we need to explicitly work in the frequency picture. Let us consider the case of a resonant, lossless, two-sided cavity with the input and output mirrors of equal reflectivity. Using Equations (5.3.21), (5.3.22), (5.3.23) with $\kappa_1 = \kappa_2 = \kappa/2$ and $\kappa_l, \Delta = 0$ we obtain the following operator evolutions through the cavity in frequency space

$$\tilde{\mathbf{A}}_{out2} = \frac{\kappa \, \tilde{\mathbf{A}}_{in1} + i2\pi\Omega \, \tilde{\mathbf{A}}_{in2}}{\kappa - i(2\pi\Omega)} \tag{5.3.29}$$

and

$$\tilde{\mathbf{A}}_{out1} = \frac{\kappa \, \tilde{\mathbf{A}}_{in2} + i2\pi\Omega \, \tilde{\mathbf{A}}_{in1}}{\kappa - i(2\pi\Omega)} \tag{5.3.30}$$

Let us consider a single photon input at the first input port and vacuum at the other. Up till this point we have written our single photon states in a generic way as $|1\rangle = \tilde{\mathbf{A}}^\dagger|0\rangle$. Now we need to make the frequency dependence of the state explicit. We write

$$|\phi(\Omega)\rangle = \int_{-\infty}^{\infty} d\Omega' f(\Omega') \tilde{\mathbf{A}}^\dagger_{in1,\Omega'} |0\rangle_{in1} |0\rangle_{in2}$$

where $\tilde{\mathbf{A}}^\dagger_{in1,\Omega'}$ is explicitly the photon creation operator at frequency Ω' in the input mode to the first port, such that $\tilde{\mathbf{A}}^\dagger_{in1,\Omega'} |0\rangle_{in1} = |1\rangle_{in1,\Omega'}$. Also $|f(\Omega')|^2 d\Omega'$ is the probability to find the photon in the small frequency interval $d\Omega'$ around Ω'.

Using the inverted forms of Equations (5.3.29), (5.3.30) we can evolve the state through the cavity giving

$$\tilde{U}(\Omega)|\phi(\Omega)\rangle = \int_{-\infty}^{\infty} d\Omega' f(\Omega') \frac{\kappa \, \tilde{A}^{\dagger}_{out2} + i2\pi\Omega \, \tilde{A}^{\dagger}_{out1}}{\kappa + i(2\pi\Omega)} |0\rangle_{out1}|0\rangle_{out2} \qquad (5.3.31)$$

As expected, the effect of the cavity in frequency space is to act like a frequency dependent beamsplitter, transmitting the low frequency components of the single photon state but reflecting the high frequency components. For example suppose the input photon has a very narrow frequency range centred at $\Omega' = 0$. We can write approximately $f(\Omega') = \delta(\Omega')$ and so the input state is just $|1\rangle_{in1,0}|0\rangle_{in2}$ whilst the output state is $|0\rangle_{out1}|1\rangle_{out2,0}$, that is the photon is definitely transmitted through the cavity. On the other hand if the photon has a narrow frequency range centred on some frequency $\Omega \gg \kappa$ then the output state will be $|1\rangle_{out1,\Omega}|0\rangle_{out2}$, that is the photon will be completely reflected by the cavity.

For an input state with a very broad frequency distribution, which we can approximate by supposing $f(\Omega') = K$, where K is a constant, the output state will be in a superposition of the two output modes

$$\tilde{U}(\Omega)|\phi(\Omega)\rangle = K \int_{-\infty}^{\infty} d\Omega' \frac{\kappa \, \tilde{A}^{\dagger}_{out2} + i2\pi\Omega \, \tilde{A}^{\dagger}_{out1}}{\kappa + i(2\pi\Omega)} |0\rangle_{out1}|0\rangle_{out2} \qquad (5.3.32)$$

If we consider the reflected field to be "lost" then the transmitted field will be a mixed state of the vacuum and a single photon state with the Lorentzian frequency distribution

$$\frac{\kappa^2}{\kappa^2 + (2\pi\Omega)^2} \qquad (5.3.33)$$

In this way cavities select specific components of the radiation field, even in the limit of single photons.

5.4   Other optical components

5.4.1   Lenses

We require optical components to change the spatial properties of a laser beam, or mode of light. Actually, in most of optics imaging is the main concern. The ability of lenses and

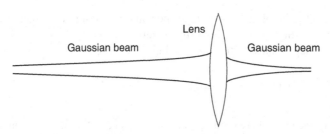

Figure 5.16: Spatial mode transformation by a lens

mirrors to form and reproduce images is the reason for using these components. The effect of lenses is a change of the wave-front curvature and consequently a transformation from one Gaussian input beam to another Gaussian beam with different waist position and size. In practice, lenses are used to transform the spatial mode coming from the laser into the mode required in the experiment, to match it to any cavity that exists in the experiment and finally, to focus the mode onto the detector.

The quality of a lens or mirror is determined by the shape of the lens, its surface finish, and the uniformity of the material. Most imaging lenses are multi element lenses and the details of the lens design are of crucial importance to the quality of the image, the spatial resolution that can be achieved. For practical details consult one of the handbooks on optical instruments [OHB95]. For most cases the paraxial transformation will hold and we can use the results discussed in Section 2.1. It assumes that the input is in a single spatial mode and that the lens is well represented by a parabolic phase term. Consequently, the output will be in a single spatial mode. Imperfections of the lens will result in deviations from a Gaussian beam and aspherical aberrations are fairly common. In this case the light has no longer a parabolic wave front, it is distributed over a variety of spatial modes. This will make any interference experiment, including homodyne detection, very difficult. It will reduce the visibility of the interference fringes and thereby make it an inefficient process. In the context of quantum optics the spatial aspect of the light is almost completely left out. The quantum nature of light is of little concern to imaging – except that it forms a general noise background. So far, spatial properties of the light have been of no importance to quantum optics. Spatial fluctuations have been ignored as unimportant. They could possibly form a new set of interesting experiments – but little has been done. Thus we need to describe lenses mainly in terms of their losses.

The loss of light due to reflection on the lens surfaces is a major experimental problem. As an example, for glass lenses, $n = 1.5$, the intensity reflection per surface is typically 4 % for every surface. This can quickly add up in a whole experiment. Thus most surfaces have an anti-reflection coating that can reduce the reflection to less than 0.5 %. Scattering losses inside the components are usually even smaller than 0.1 % and will only be of concern if the optical component is placed inside a cavity or is extremely sensitive to scattered light.

While losses are one concern in quantum experiments, the other is the need to maintain one single mode. Assume we have a mode of light with an interesting intensity noise statistics which is sub-Poissonian. It passes a lens with distortions and turns into a non-Gaussian mode. If we detect all of this light we still find a sub-Poissonian result. If, however, the detector is sensitive to only the Gaussian mode, because it contains a cavity or it is a homodyne detector that requires the interference and beating with an optical local oscillator, the results will be different. If the partition between the modes is random then the other modes will appear like a power loss and the statistics will be closer to Poissonian. If the partition is somehow deterministic then additional noise could be introduced. Fortunately all passive components, such as mirrors and lenses, result in a random partition. The other modes can be treated as loss terms. In contrast, active systems such as lasers are different and can introduce additional partition noise, as shown in Chapter 6.

A problem that exists with high intensity experiments, and in particular pulsed experiments, is thermal lensing. Glasses will absorb some small fraction (typically 0.1 [%/cm]) of the light and the absorbed power can lead to local heating of the glass which results in local changes of the refractive index. As a consequence self focussing can occur. The refractive

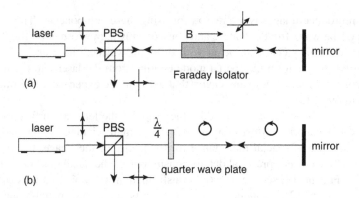

Figure 5.17: Optical isolators using either a Faraday rotator (a) or a quarter wave plate (b).

index effectively forms a lens, formed by the beam. This tends to focus, increasing the effect in a positive feedback. This can have disastrous effects – and many lenses have been destroyed with holes and channels burned into the glass.

5.4.2 Crystals and polarizers

Glass and crystals vary in two very important points: crystals are non isotropic, they can show an electro optics response and they can show nonlinear effects. The anisotropy of the crystals results in properties such as birefringence, optical activity and dichroism. The details obviously vary with different types of crystals, their composition and geometrical structure. It also varies with the alignment of the light beam with the crystal axis. All these properties can be used to control the polarisation properties of the light. The idea here is to introduce a relative phase change between two different polarisation components of the field. For example, introduce a $\pi/2$ phase change of one linear polarisation and not the other and thereby turn linearly polarized light, aligned with the axis of the crystal, into circularly polarized light. This would be called a quarter wave plate. The results of any such component can be analysed using the rules of classical optics, in particular the formalism of Jones matrixes [BW59]. Birefringent crystals can be manufactured into wave plates and polarizers. In this regard quantum optics experiments are no different to classical optics experiments. All tricks available to control and influence the polarisation will be used in the various experiments.

One particularly useful component is the optical isolator. This device is equivalent to a diode in electronics, it provides a different effect in the forward and backward direction. The light propagating in the reverse direction is redirected and does not go back to the source. This avoids interference between the incoming beam and the retro-reflected light. This effect is described as optical feedback which can lead to very noisy or even unstable performance of a laser. At the same time an optical isolator will give access to the retro-reflected beam, which is important in experiments where all of the light has to be detected. It is also useful for situations where locking of a cavity to a laser is important. In this case the redirected light contains the information about the cavity detuning. Two types of optical isolators are commonly used: the Faraday isolator and wave plates, see Fig. 5.17. The Faraday isolator uses a material, usually a special glass or crystal, with a high Verdet constant. This means a large rotation of the plane of polarisation for linearly polarized light can be induced by a static magnetic field along the crystal. The isolator is tuned to the wavelength used in the particular experiment by adjusting the magnetic field strength to give a 45° rotation. The return beam is rotated again by 45° in the same direction, that means both beams are linearly polarized at 90° to each other and can be separated by a polarizing beamsplitter (PBS). The second design uses wave plates, see Fig. 5.17(b). The source generates linearly polarized light. A quarter wave plate is used to turn this into left hand circularly polarized light ($\sigma+$ light). Upon reflection in the experiment the returning light is right hand circularly polarized ($\sigma-$ light) and the wave plate turns this into a counter propagating beam with orthogonal linear polarisation. This light can be separated with a polarizing beam splitter.

Crystals can also contribute entirely new effects. They can have a nonlinear response which allows us to generate new frequencies, such as twice the frequency (second harmonic generation), half the frequency (optical parametric gain) or sums and different frequencies (four wave mixing). Crystals can show a Kerr effect, that means an intensity dependent refractive index, or nonlinear absorption. These effects will influence not only the intensities but also the fluctuations of the light and consequently they are discussed in the Chapter 9 on squeezing experiments.

5.4.3 Modulators

Many crystals have significant electro-optic coefficients. That means the refractive index can be manipulated with an external electric field. The direction of the electric field which affects the refractive index along a particular optical axis depends on the type of crystal used. In some cases the transverse effect is large, that means the electric field is orthogonal to the optical beam. In other crystals a longitudinal field, parallel to the optical beam, is required. The refractive index will follow the field up to very high frequencies, several [GHz]. Thus these devices can be used as electro-optic modulators (EOM). Large changes in refractive index will require electric fields of the order of 100–1000 [V/cm]. The limit in speed is normally set by the electric power requirements. Fast response and large degrees of modulation can be achieved by using resonant electronic circuits which provide a large electric field in a very narrow frequency range.

This effect can be used to make a phase modulator, see Fig. 5.18(a). Either a crystal with isotropic response is chosen or the axis of linear polarisation is carefully aligned parallel to the crystal axis. For a single modulation frequency Ω_{mod} the output light will get one set of sidebands, Ω_{mod} and its higher harmonics, as described in Equation (2.1.21). In the ideal case phase modulated light does not produce any modulation of the photo-current. In practice, considerable care has to be taken to avoid amplitude modulation. FM sidebands are frequently used to probe the detuning of cavities and to lock lasers to cavities.

Alternatively, a birefringent crystal can be used and aligned such that the linear input polarisation is at 45° to the optical axis of the crystal, see Fig. 5.18(b). In this case the two orthogonal components of the linear polarisation are affected differently. The voltage induces a difference in the change of the refractive index and the modulator is acting like the AC equivalent of a wave plate. The polarisation is rotated synchronously with the applied electric field. After transmission through a polarizer the light has a mixture of amplitude modulation, from the polarisation rotation, and phase modulation.

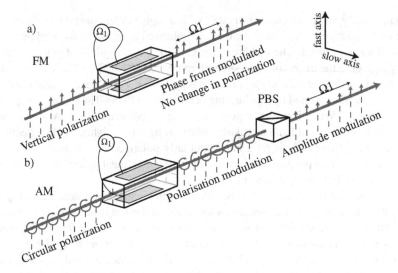

Figure 5.18: Schematic operation of an electro-optical modulator EOM used for (a) Phase or frequency (FM) and (b) amplitude (AM) modulation.

One common difficulty in using modulators is that their performance changes critically with temperature. The alignment and the crystal parameters can change when the crystal is heated, either by the RF energy of the driving field or the laser beam. It is common that the crystal has to be realigned when the transmitted optical power or the RF power has been changed, and I have seen modulators which gave pure phase modulation at low optical powers but could not be realigned for large intensities.

An entirely different effect is used in acousto-optical (AOM) modulators. These utilize the fact that an acoustic standing wave can be generated inside crystals through an RF electric field. Light transmitted through such a standing wave sees a periodic density grating and is diffracted. The diffracted beam is shifted in frequency by the modulation frequency.

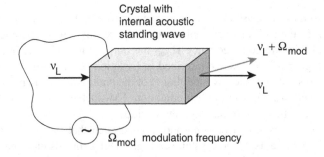

Figure 5.19: Acousto-optical modulator

The common application of an AOM is as fast scanners, such as laser printers. Since the diffraction angle is linked to the modulation frequency, we can redirect a beam by simply changing the frequency of the RF signal. We can change the intensity of the diffracted beam by changing the RF power. Both RF frequency and power can be controlled rapidly and the output beam is thus scanned in one dimension. It is the ideal component for a laser printer. In our experiments the main application is the shift in optical frequency. For an ideal AOM the diffracted beam has the same mode properties as the input beam, apart from a change in frequency and direction. The fluctuations of the light are simply transmitted. There are many applications, such as tuning to an atomic or cavity resonance or the probing of atoms with a frequency offset. AO modulators can also be used as isolators. The beam which is retro reflected will have twice the frequency shift imposed by the modulators and is less likely to interfere with the laser.

5.4.4 Optical fibres

Optical fibres are a convenient component to deliver light from one place to another. The idea is simple: the fibres have a refractive index structure that guides the light. For sufficiently small core sizes, a few times the optical wavelength, the field inside the fibre has one single mode. Larger diameter fibres show many modes. The mode properties of fibres are complex and are discussed in great detail in the specialized literature. It is possible to design a lens that will couple very efficiently a Gaussian laser beam into a single mode fibre. This requires a lens of short focal length, such as a microscope objective, that has a numerical aperture matched to that of the fibre. When the beam parameters are correctly chosen up to 80% of the power can be coupled into the fibre. The losses inside the fibre are so exceedingly small that the actual fibre length is normally of no consequence.

It should be noted that simple symmetric fibres completely scramble the polarisation of the light. Any polarisation dependence of the optics after the fibre can lead to enormous intensity noise introduced by the fibre. There are special, non symmetric polarisation preserving fibres. They have to be carefully aligned. And even these are very sensitive to mechanical movements of the fibre. The output mode from a single mode fibre is very close to a Gaussian beam. Fibres can be used as mode cleaners, they will transform an input beam with complex wave front into a very nice Gaussian beam. The light that is not matched will be lost by scattering. The higher losses of a fibre mode cleaner are frequently compensated by its simplicity and the fact that, unlike a mode cleaner cavity, it requires no active locking. The high internal intensity inside a fibre make it also very attractive for nonlinear interactions. Some of the first and some of the best squeezing experiments have been carried out in fibres, see Section 9.7.

5.4.5 Optical noise sources

Optical materials can contribute noise. This is equivalent to saying that additional light can be scattered into the mode under investigation. One example is atoms which are excited by one of the laser beams in the experiment and which spontaneously emit at the frequency and into the spatial mode which is detected. In a similar way stimulated processes can add light to the mode. These could be Raman transitions or other nonlinear processes, such as four wave mixing or second harmonic generation. We have to be very careful in experiments which contain strong optical fields, basically any process that can generate light at the frequency of

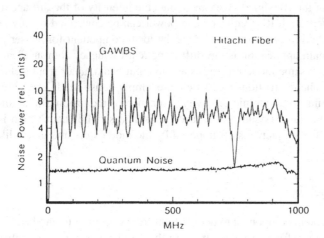

Figure 5.20: An example of optical noise: The noise introduced in a single mode optical fibre by guided acoustic wave Brillouin scattering (GAWBS) measured with a CW laser, as reported by R. Shelby et al. [She86].

the mode we wish to detect is potentially a source of competing noise that could swamp an effect as small as quantum noise.

These types of noise add to the intensity fluctuations in the mode. In addition, the refractive index of the optical material can be affected by fluctuations which will generate noise in the phase quadrature of the transmitted light. For example, a common source of noise in optical fibres is guided acoustic wave Brillouin scattering, abbreviated as GAWBS. This noise occurs in any system where the wave is guided by geometrical structures such as a fibre or planar wave guide. The noise spectrum shows very pronounced maxima, which can be related to the resonances of acoustic waves which are reflected by the boundaries of the fibre. It covers such a wide spectrum that the detection of QNL light becomes very difficult. Nonlinear crystals have other additional sources of noise [Kim92]. In all experiments a careful characterization of the noise spectrum has to be carried out, before we can assume that quantum noise is actually observed. Chapter 8 describes these techniques in detail.

5.4.6 Nonlinear processes

In summary, all linear components have a very simple effect on the fluctuations: *a coherent state of light will remain a coherent state*. Even for light which is not a coherent state the effects are simple, we will find that any losses make the fluctuations look more like those of a coherent state. In the limit of large losses any input light will approach a coherent state.

Quantum optics would not be much fun if there were only linear processes, but fortunately there is more. Gain combined with saturation allows us to make a laser, the all important light source, see Chapter 6. One of the main features of nonlinear processes is that they can generate additional new frequencies. In particular, second harmonic generation (SHG) and degenerate optical parametric oscillation (OPO) are used to generate beams at twice or half the original frequency. The nonlinear Kerr effect transfers changes in the intensity to the refractive index. Four wave mixing (4WM) links modulation sidebands. The nonlinear processes are too numerous to list here explicitly and instead the extensive literature should be consulted, for example the textbooks by M. Levenson & S.S. Kato [Lev88] and R.W. Boyd [Boy92]. These nonlinear effects allow us to change the statistics of the light, to manipulate quantum noise and to make *measurements better than allowed by the QNL*. Nonlinear media are essential for the generation of squeezed and entangled light and these experiments will be discussed, one by one, in Chapter 9.

Bibliography

[And84]   Alignment of resonant optical cavities, D.Z. Anderson, Appl. Opt. 23, 2944 (1984)

[BW59]   *The principles of Optics*, M. Born, E. Wolf, Pergamon Press (1999)

[Boy92]   *Nonlinear Optics*, R.W. Boyd, Academic Press (1992)

[Cav81]   Quantum-mechanical noise in an interferometer, C.M. Caves, Phys. Rev. D 23, 1693 (1981)

[Col84]   Squeezing of intracavity and traveling-wave light fields produced in parametric amplification, M.J. Collett, C.W. Gardiner, Phys. Rev. A 30, 1386 (1984)

[Edw93]   Reduction of optical shot noise from light emitting diodes, P.J. Edwards, IEEE J. Qu. Electronics 29, 2302 (1993)

[Gar00]   *Quantum noise*, 2nd edition, C.W. Gardiner, P. Zoller, Springer, Berlin (2000)

[Kim92]   Squeezed states of light: an (incomplete) survey of experimental progress and prospects, H.J. Kimble, Physics Reports 219, 227 (1992)

[Lev88]   *Introduction to nonlinear laser spectroscopy*, M.D. Levenson, S.S. Kano, Academic Press (1988)

[Lui95]   A quantum description of the beamsplitter, A. Luis, L.L. Sanchez-Soto, Quantum Semiclass. Opt. 7, 153 (1995)

[Man95]   *Optical coherence and quantum optics*, L. Mandel, E. Wolf, Cambridge University Press (1995)

[Mil87]   Optical-fibre media for squeezed-state generation, G.J. Milburn, M.D. Levenson, R.M. Shelby, S.H. Perlmutter, R.G. DeVoe, D.F. Walls, J. Opt. Soc. Am. B 4, 1476 (1987)

[Mor94]   Experimental demonstration of an automatic alignment system for optical interferometers, E. Morrison, B.J. Meers, D.I. Robertson, H. Ward, Appl. Opt. 33, 5041 (1994)

[OHB95]   *Optics Handbook*, American Optical Society (1995)

[Ral95]   Noiseless amplification of the coherent amplitude of bright squeezed light using a standard laser amplifier, T.C. Ralph, H-A. Bachor, Opt. Comm. 110, 301 (1995)

[Sam90]   Stabilisation of laser beam alignment to an optical resonator by heterodyne detection of off-axis modes, N.M. Sampas, D.Z. Anderson, Appl. Opt. 29, 394 (1990)

[Sau94]   *Fundamentals of interferometric gravitational wave detection*, P.R. Saulson, World Scientific (1994)

[Sch92]   A fused silica monolithic total internal reflection resonator, S. Schiller, I.I. Yu, M.M. Fejer, R.L. Byer, Opt. Lett. 17, 378 (1992)

[Sha00]   Modulation free control of a continuous-wave second harmonic generator, D.A. Shaddock, B.C. Buchler, W.P. Bowen, M.B. Gray, P.K. Lam, J. Opt. A 2, 400 (2000)

[She86]   Quantum optics and nonlinear interactions in optical fibres, R.M. Shelby, M.D. Levenson, IBM Res. Report (1986)

[Sie86]   *Lasers*, A.E. Siegman, University Science Books (1986)

[Sto49]   G.G. Stokes, Cambridge and Dublin Math. J. 4, 1 (1849)

6 Lasers and Amplifiers

6.1 The laser concept

The LASER is the key component in any quantum optics experiment. It was the invention of the laser that gave us access to the world of coherent optics, nonlinear optics and to the quantum features of light. Without the laser we would be restricted to thermal light – missing out on most modern applications and on almost all the experiments described in this guide.

In quantum optics we are concerned with the characteristics of the light generated by the laser. In particular, we will investigate the fluctuations in the intensity of the light. In theoretical terms the laser is the system of choice for the realization of many quantum optics ideas: the laser light itself is close to a pure coherent state. In many cases the laser can be regarded as the source for a single mode of radiation as described in the theory. If we consider a continuous wave (CW) laser the mode will have constant properties and even pulsed lasers can be considered as constant in the sense that we describe them as a steady stream of pulses – each pulse like any other.

Many books have been written about the laser, one of the most comprehensive introductions has been given by Siegman [Sie86] who provides all the technical details required to understand and operate lasers. In simple technical terms the laser is an optical oscillator, see Fig. 6.1, that relies on feedback, much like an electronic circuit. It consists of a narrow band optical amplifier, which generates a well defined output beam of light, and an optical arrangement which feeds part of the light, in phase, back to the input. The amplification is provided by a gain medium, atoms or molecules which are excited by an external source of energy, the pump. The realization that such a situation is possible was one of the early results of atomic quantization [Ein17]. The feedback is provided by a cavity surrounding the active medium. The mirrors of the cavity generate a large internal optical field, which is amplified during each roundtrip. Of course there are also losses inside the cavity and in order to achieve sufficient feedback for oscillation to occur the gain has to be larger than the losses.

Without any input there would be no output, no optical feedback signal and no laser beam. However, the presence of even a small amount of light, some small optical amplitude within the bandwidth of the resonant amplifier system, will initiate the process and lead to a rapid growth of the output. Spontaneous emission from atoms in their excited state can provide such a seed. The amplitude increases until the gain of the amplifier saturates at a level which just balances the losses due to the mirrors and other sources. The output power will stabilize at a steady state value and the laser achieves constant CW operation.

The conditions for lasing are twofold: Firstly, the amplifier gain must be greater than the loss in the feedback system so that a net gain is achieved in each round trip. All lasers show

A Guide to Experiments in Quantum Optics, 2nd Edition. Hans-A. Bachor and Timothy C. Ralph
Copyright © 2004 Wiley-VCH Verlag GmbH & Co. KGaA
ISBN: 3-527-40393-0

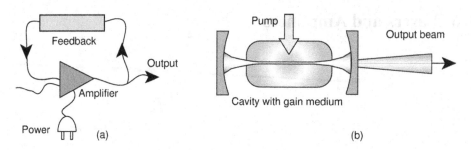

Figure 6.1: Comparison between an (a) electronic amplifier/oscillator and (b) a laser.

a threshold, that means they require a minimum pump power P_{thr} to start steady oscillation and generate an output beam. Secondly, the total phase accumulated in a single round trip must be an integer multiple of 2π to provide phase matching to the feedback signal, i.e. the laser frequency must correspond to a cavity mode. Ignoring small frequency pulling effects, which can be somewhat larger in diode lasers, the laser mode(s) will correspond with the cavity mode(s) closest to the peak of the gain profile. In the simplest case this is a single fixed frequency.

Above threshold the laser produces a steady output beam. The geometrical properties of the output beam are those of the spatial modes of the cavity, in the simplest case the fundamental Gaussian TEM_{00} beam, see Eq. (2.1.8). The combination of these two cases is known as *single mode operation*, both transversely (spatially) as well as longitudinally (spectrally). This is the ideal case for the study of the quantum properties of light. The intra-cavity field can be well described as one quantum mechanical mode of the radiation field that extends throughout the cavity. The output beam can be described by a propagating mode, or coherent state as described in Section 4.5. As we will see, the present technology gets us extremely close to the realization of such a perfect laser.

6.1.1 Technical specifications of a laser

For CW lasers we can use a wide range of lasing materials, excitation mechanisms and cavity designs. CW lasers range from gas lasers, such as the widely used He-Ne lasers and the powerful Ar Ion and CO_2 lasers, to the complex tunable dye lasers and the technically almost simplistic solid state diode lasers. There is no space here to describe all the technical details and numerous engineering tricks that are used in the design of these lasers. A range of specialized literature is available to provide this information [Sie86, Tei93]. However, the trend of recent years is clearly towards a much wider application of solid state lasers, either directly as diode lasers, as diode pumped solid state lasers, or as a fibre lasers [Koe96]. These lasers are starting to dominate the technological applications. Even in the area of atomic spectroscopy, where tunability and narrow bandwidth is required, diode lasers and Ti-sapphire rapidly replace the traditional dye lasers. As a consequence most of the practical examples of laser performance in this book refer to solid state lasers.

The performance of a laser can be described by the following parameters: optical power, spatial mode structure, output wavelength, frequency spectrum and the noise spectrum. The

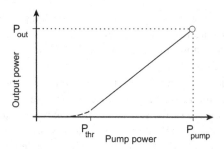

Figure 6.2: Optical output power versus pump power. The laser has a threshold above which the output increases linearly with the pump power.

power is quoted as the total optical power P_{out} of the output beam. Of practical interest is the total efficiency of the laser $\eta_{total} = P_{out}/P_{pump}$, where P_{pump} is the pump power required to achieve an output power P_{out}. Well above threshold the power scales linearly and the slope efficiency $\eta_{slope} = dP_{out}/dP_{pump}$ is constant. This is shown in Fig. 6.2. Note that operation near threshold is not reliable. At that point the power is not stable, it tends to be noisy and P_{thr} may not be easy to measure.

The *spatial properties* of the laser beams are determined by the properties of the optical resonator. In most cases a stable resonator is used and the mode analysis in terms of Hermite Gaussian $TEM_{m,n}$ modes, as discussed in Chapter 2, can be used. There are a few exceptional cases, particularly for the design of high power lasers where more complicated unstable resonators are used. However, many lasers operate in a single spatial mode, the lowest order $TEM_{0,0}$ mode, which is particularly simple to describe. In practice the resonator is never a perfect optical cavity, there will always be spatial some distortions in the output beam and a parameter $M_{m,n}$ is normally quoted which describes how closely the beam is approximated by a $TEM_{m,n}$ mode. The perfect match corresponds to $M_{m,n} = 1$. Some lasers produce beams which have elliptical rather than circular beam profiles. For these lasers it is important to know the aspect ratio of the ellipse and whether the spatial mode is simply a stretched version of a Gaussian beam or a more complex wave front.

The *output wavelength* is determined by the properties of the gain medium, and can vary from the UV (excimer laser 169 [nm]) to the far Infrared NH_3 laser (440 [μm]). Many laser materials have gain for several output wavelengths. For example, the Ar Ion laser has several green/blue laser transitions. The desired wavelength can be selected by optimizing the mirror coatings and by using selective intra- cavity components, such as a prism or a grating. Some lasers are tunable, that is a range of laser wavelengths can be chosen. Some materials with a tuning range have a broad gain profile – prominent examples are dye lasers and the Ti-Sapphire laser, in other materials the gain profile is temperature dependent. For example, in a diode laser the wavelength can be tuned by changing the temperature of the device, controlling the current and using an extended cavity for precise control of the laser frequency.

The *frequency distribution* of the laser output is determined by the cavity properties and the gain profile. In addition, it is important whether the gain profile is an in-homogeneous or homogeneous profile. In the first case different cavity modes obtain their gain from different active atoms and the various modes can oscillate independently. The result is a series of spectral output modes, separated by the free spectral range of the laser resonator. In the latter case all active atoms in the laser contribute to the laser in the same way. There will be competition between the spectral modes and normally only one will win and the output spectrum will be dominated by one single spectral mode.

Such a single mode laser still has a frequency bandwidth. The absolute frequency can change with time. The bandwidth is given as the range of frequencies which are present in a set observation time. The bandwidth is a measure of the size of the frequency fluctuations. For example, suppose only frequencies within a 1 [MHz] interval are measured within 1 [sec] interval. This information tells us that the laser could be used to resolve a detailed feature of a given spectrum with a resolution of 1 [MHz] and no less. However, short term laser bandwidth and long term frequency drift can vary widely. The same laser might drift by 100 [MHZ] in 10 [minutes]. We would have to carry out the experiment very quickly. A more complete description of the frequency properties is given by the Allan variance, a statistical analysis of the frequency changes for a range of different observation times [All87]. We will find that the frequency fluctuations are commonly a consequence of technical imperfections, such as vibrations, electrical and acoustical noise. The quantum mechanical fluctuations play hardly a role in the frequency noise. The phase of a laser is unconstrained and will drift. This is known as phase diffusion. This means that the variance of the phase quadrature, see Section 4, will grow unlimited over long time periods. On the other hand for short times, or high frequencies, the phase can be quite stable and will tend to the coherent state level at sufficiently high frequencies.

Finally, there are fluctuations of the intensity of the output beam. These are described as intensity noise spectra, as the relative changes of the laser power, $\delta P_{out}/P_{out}$, measured with a bandwidth RBW at a frequency Ω. The measured variance of the optical power fluctuations can be directly compared with the modelled variance of the amplitude quadrature $V1_{las}(\Omega)$, see Chapter 4, which is normalized by the quantum noise associated with a coherent state of the same optical power.

In summary, lasers are available that produce output beams which are very close to the theoretical limit. Their light can be considered as the real world analogue of a coherent state. But not all lasers do this. Some lasers are not built well enough and produce beams that are distorted or noisy. Some other lasers have noise that is intrinsic to the lasing process. Finally, we will find some lasers that actually perform better than a coherent state.

6.1.2 Rate equations

Although different laser systems differ widely in their details it is possible to model many aspects of laser behaviour from a quite general rate equation approach. The main assumption in such an approach is that coherent atomic effects are negligible. Coherent effects can be neglected in the presence of strong dephasing effects, such as inter-atomic collisions in a gas or lattice vibrations in a solid. This is an excellent approximation for all but a few lasers, such as far-infrared gas lasers, which we will not consider in this discussion.

In a rate equation model we consider the gain medium to be made up of M atoms. These atoms have internal states of excitation, with energies E_1, E_2 and E_3 as shown in Fig. 6.3. We assume lasing occurs between the states with energies $E3$ and $E2$. The frequency of the laser

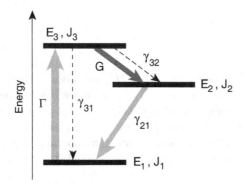

Figure 6.3: The semi-classical model of a 3 level laser system

light is centered around $\nu_L = (E_3 - E_2)/h$. The actual laser frequency will be determined by the cavity resonance, taking into account all the nonlinear effects inside the laser resonator.

The excitation of the medium is described by the populations M_1, M_2, M_3 of the individual states. These populations should really be interpreted as the probabilities J_1, J_2, J_3 of finding any individual atom in one of the three states. The total population is preserved, thus

$$M = M_1 + M_2 + M_3 = M(J_1 + J_2 + J_3) \tag{6.1.1}$$

A number of processes link the states. Pumping is modelled as a uni-directional transfer of population from state $|E_1\rangle$ to state $|E_3\rangle$. Normally any pumping process will also produce de-excitation of the upper pump level at a rate proportional to its population. The uni-directional pumping is to be understood as the limit of pumping to a higher state which then decays very rapidly to state $|E_3\rangle$. The population of all excited states decay through spontaneous emission. Stimulated emission and absorption of photons in the laser field couples the states $|E_3\rangle$ and $|E_2\rangle$. All these processes are described by rates.

As an example, the spontaneous emission from state $|E_2\rangle$ to state $|E_1\rangle$ results in $M_2\gamma_{21} = MJ_2\gamma_{21}$ photons emitted per second, which are leaving the laser in random directions, and a corresponding reduction in the population of state $|E_2\rangle$ and increase of population of state $|E_1\rangle$. Similarly the rate of spontaneous emission from state $|E_3\rangle$ is described by rate constant γ_{32}. The pumping rate depends on the pump power, the excitation cross-section and the pumping efficiency. It is described by one rate $M_1\Gamma$.

For simplicity all we neglect spatial effects and consider one spatial mode only. Similarly only one spectral mode, at frequency ν_L, is considered. All optical effects are described by the gain and loss of light from this single mode. For a high finesse cavity the total loss of the resonator is just the sum of the individual losses, see Section 5.3.21. We write the total loss κ as the sum of losses due to the output mirror κ_m and those due to all other mechanisms κ_l. If the mode inside the laser contains n_t photons, then the number of photons leaving the cavity per second will be $2\kappa n_t$. Of these only the fraction κ_m/κ exit in the output beam, the remainder are dissipated internally. The energy difference between the atomic states $|E_3\rangle$ and $|E_2\rangle$ corresponds to a frequency which is resonant with the mode. Stimulated emission occurs at a rate $G_{32}J_3n_t$ and stimulated absorption at a rate $G_{32}J_2n_t$. The rate constant G_{32}

is defined as

$$G_{32} = \sigma_s \rho c / n_r \tag{6.1.2}$$

where σ_s is the stimulated emission cross-section for the $|E_3\rangle$ to $|E_2\rangle$ transition, ρ the density of lasing atoms in the medium, c the speed of light and n_r is the refractive index. The two processes produce an overall change of the photon number of $G_{32}(J_3 - J_2)n_t$ [*photons per second*]. The inversion $(J_3 - J_2)$ drives this amplification process. At this point we ignore the small contribution to the photon number due to spontaneous emission into the lasing mode. Combining all these individual rates we can write a system of four coupled rate equations which describe the dynamics of the laser;

$$
\begin{aligned}
\frac{dJ_1}{dt} &= -\Gamma J_1 + \gamma_{21} J_2 \\
\frac{dJ_2}{dt} &= G_{32}(J_3 - J_2)n + \gamma_{32} J_3 - \gamma_{21} J_2 \\
\frac{dJ_3}{dt} &= -G_{32}(J_3 - J_2)n - \gamma_{32} J_3 + \Gamma J_1 \\
\frac{dn}{dt} &= G_{32}(J_3 - J_2)n - 2\kappa n \tag{6.1.3}
\end{aligned}
$$

where n is the photon number per atom such that $nM = n_t$. We can also write the equation of motion of the cavity field (per atom) as

$$\frac{d\alpha_c}{dt} = G_{32}(J_3 - J_2)\alpha_c - \kappa\alpha_c \tag{6.1.4}$$

where $\alpha_c^*\alpha_c = n$. The steady state solution for the photon number per atom, n, can be obtained by setting the time derivatives to zero. We find either n=0 or

$$n = \frac{\Gamma(\gamma_{21} - \gamma_{32})}{2\kappa(2\Gamma + \gamma_{21})} - \frac{\kappa}{G_{32}} \frac{\Gamma\gamma_{21} + \gamma_{21}\gamma_{32} + \Gamma\gamma_{32}}{2\kappa(2\Gamma + \gamma_{21})}$$

Provided we have $\gamma_{21} > \gamma_{32}$, there is a threshold pump rate, Γ_t, above which the non-zero solution for n will be positive and stable. This represents laser oscillation while the zero solution is unstable. For pump rates below the threshold the non-zero solution is negative and it is the zero solution that is stable. That is, the laser does not oscillate. Physically that means the losses in the resonator, per round trip, are larger than the gain so that any light introduced in to the cavity is quickly damped away.

Generally the lifetime of the lower lasing level will be very short so we can set $\gamma_{21} \gg \gamma_{32}, \Gamma$. Under this condition the population of the second level can be neglected and the ground state can be considered undepleted ($J_2 = 0$, $J_1 = 1$). Our equations of motion reduce to

$$
\begin{aligned}
\frac{dJ_3}{dt} &= -G_{32}J_3\alpha_c^*\alpha_c - \gamma_{32}J_3 + \Gamma \\
\frac{d\alpha_c}{dt} &= G_{32}J_3\alpha_c - \kappa\alpha_c \tag{6.1.5}
\end{aligned}
$$

and we obtain the following simpler solution for the output photon flux per atom of the laser, n_{out}

$$n_{\text{out}} = 2\kappa_m n = \left(\Gamma - \frac{2\kappa}{G_{32}} \gamma_{32} \right) \frac{\kappa_m}{\kappa} \tag{6.1.6}$$

The pump threshold is $\Gamma_t = \frac{2\kappa}{G_{32}} \gamma_{32}$. We also have that above threshold, in the steady state

$$J_3 = \frac{2\kappa}{G_{32}} \tag{6.1.7}$$

and the following alternative form of Equation (6.1.6) can be written

$$n_{\text{out}} = (\Gamma - \gamma_{32} J_3) \frac{\kappa_m}{\kappa} \tag{6.1.8}$$

That is, the rate at which laser photons leave the cavity is simply the rate at which atoms are placed in the upper lasing level minus the rate at which photons are lost through spontaneous emission. Threshold occurs when pumping exceeds the loss due to spontaneous emission. The useful output is scaled by the fraction of photons which leave via the cavity mirror and hence depends on the internal losses. The total efficiency of the laser, η_{total}, depends on this fraction, the relative pump power P_{pump}/P_{thr} as well as the pumping efficiency.

Very close to threshold the performance is obviously very dependent on fluctuations. The laser can switch on and off and our simple theoretical model would have to be extended to include such effects. For reliable operation the actual pump rate Γ should be considerably higher than Γ_t, that means $P_{\text{pump}} > P_{thr}$. Here the threshold the output power increases linearly with the pump rate, see Fig. 6.2. Hence the slope efficiency, η_{slope}, is constant above threshold, depending only on the internal losses and pump efficiency. At even higher pump powers the output power can level off. Theoretically this occurs when the pump rate begins to deplete the ground state. In practice this tends to happen at lower pump powers, than a simple estimate would suggest, since a number of additional nonlinear effects might occur that can have power limiting back effects. For example, laser materials show thermal expansion, due to absorption, that can lead to a nonuniform refractive index and thus thermal lensing. This and other power induced wave front distortions introduce additional losses in the resonator which limit the internal power.

An alternative possible configuration of the laser is where the lasing transition occurs between state $|E_2\rangle$ and the ground state, state $|E_1\rangle$, with a stimulated emission cross-section proportional to G_{21}. Important examples of this type of laser are the solid-state Ruby and Erbium lasers. The analysis proceeds as before. Making the assumption now that state $|E_3\rangle$ decays very rapidly we get the following solution for the output photon flux per atom;

$$n_{\text{out}} = 2\kappa_m n = \frac{1}{2} \left(\Gamma - \gamma_{21} - \frac{2\kappa}{G_{21}} (\Gamma - \gamma_{21}) \right) \frac{\kappa_m}{\kappa} \tag{6.1.9}$$

The threshold pump rate is given by

$$\Gamma_t = \frac{1 + \frac{2\kappa}{G_{21}}}{1 - \frac{2\kappa}{G_{21}}} \gamma_{21} \tag{6.1.10}$$

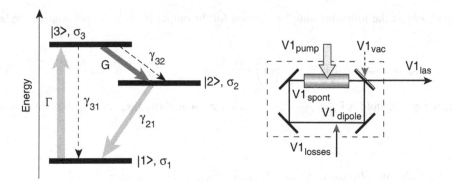

Figure 6.4: The quantum model of a 3-level laser

The major difference between this laser system and the previous one is the threshold behaviours. Whilst in the previous case the pump threshold could in principle be made arbitrarily small by reducing the ratio $\frac{2\kappa}{G_{32}}$, that is increasing the cavity finesse or active atom density, in this case the threshold cannot become less than the spontaneous emission rate between the laser levels, γ_{21}. Physically this is because we have to at least half empty the ground level before we can create an inversion.

6.1.3   Quantum model of a laser

In order to determine the dynamics of the fluctuations of the laser system a quantum model is required. In the quantum description the laser consists of a laser medium of M atoms with internal states $|E_1\rangle$, $|E_2\rangle$, $|E_3\rangle$, a pump source with pump rate Γ and noise spectrum $V1_P(\Omega)$ and an internal optical mode $\hat{a}$ with an expectation value of α_c. Operators $\hat{J}_1, \hat{J}_2, \hat{J}_3$ describe the collective excitation of the atoms. As in the semi-classical model, given in the previous section, the expectation values of these operators, J_1, J_2, J_3 respectively, give the probability of finding any one atom in the respective state. Figure 6.4 shows the laser system in this notation.

Making the simplifying assumptions which led to Equation (6.1.5) (i.e. negligible second level population and an undepleted ground state), the quantum operator Langevin equations of motion can be written as

$$
\frac{d\hat{J}_3}{dt} = -G_{32}\hat{J}_3\hat{a}^\dagger\hat{a} - \gamma_{32}\hat{J}_3 + \Gamma + \sqrt{\gamma_{32}}\,\delta\hat{X}_\gamma + \sqrt{G_{32}\hat{a}^\dagger\hat{a}}\,\delta\hat{X}_p - \sqrt{\Gamma}\,\delta\hat{X}_P
$$

$$
\frac{d\hat{a}}{dt} = \frac{G_{32}}{2}\hat{J}_3\hat{a} - \kappa\hat{a} + \sqrt{2\kappa_l}\,\delta\hat{A}_l + \sqrt{2\kappa_m}\,\delta\hat{A}_m - \sqrt{G_{32}\hat{J}_3}\,\delta\hat{A}_p^\dagger \qquad (6.1.11)
$$

where all terms with "δ" prefixes are vacuum fields, with zero expectation values. The equations of motion of the expectation values correspond to the semi-classical rate equations, Equation (6.1.5). All the processes are described in a similar way – now by using operators to describe the states of the atoms and the single mode of the light. In addition, the full quantum model has to contain operators describing all possible noise sources. These include the pump noise, characterized by the operator $\delta\hat{X}_P$, and with an intensity noise spectrum

$V1_P(\Omega) = \langle|\delta\tilde{\mathbf{X}}_P|^2\rangle$, the vacuum input through the cavity mirror, $\delta\hat{\mathbf{A}}_m$ and the vacuum input due to other intra-cavity losses, $\delta\hat{\mathbf{A}}_l$. The rapid phase decay of the atomic coherence due to collisions with atoms or, in the case of solids, lattice induced disturbances, introduces dipole fluctuations which can be characterized by the operator $\delta\hat{\mathbf{X}}_p = \delta\hat{\mathbf{A}}_p + \delta\hat{\mathbf{A}}_p^\dagger$. Finally, noise in the populations due to spontaneous emission has to be included via the operator $\delta\hat{\mathbf{X}}_\gamma$. The effects of all these noise sources appear explicitly in the fluctuations of the output operator $\tilde{\mathbf{A}}_{\mathrm{las}}$. The derivation of Equation (6.1.11) is quite involved and beyond the scope of this book. Details of the derivation and the identification of the noise operators with their various physical origins can be found in reference [Ral03].

The steady state solutions for the probabilities J_3 and photon number per atom, n, in the quantum model, obtained when all the time derivatives are set to zero, are identical to those obtained from the semi classical Equations (6.1.5). We proceed by linearizing around these stable solutions, see Section 4.5.1. Writing $\hat{\mathbf{a}} = \alpha_c + \delta\hat{\mathbf{a}}$ and $\hat{\mathbf{J}}_3 = J_3 + \delta\hat{\mathbf{J}}_3$ and only keeping terms to first order in the vacuum fields we obtain the following linearized equations for the quantum fluctuations

$$
\frac{d\delta\hat{\mathbf{J}}_3}{dt} = -G_{32}J_3\alpha_c\delta\hat{\mathbf{X}}_a - (\gamma_{32} + G_{32}\alpha_c^2)\delta\hat{\mathbf{J}}_3 + \sqrt{\gamma_{32}}\,\delta\hat{\mathbf{X}}_\gamma
$$
$$
+ \sqrt{G_{32}\alpha_c^2}\,\delta\hat{\mathbf{X}}_p - \sqrt{\Gamma}\,\delta\hat{\mathbf{X}}_P
$$

$$
\frac{d\delta\hat{\mathbf{a}}}{dt} = \frac{G_{32}}{2}\alpha_c\delta\hat{\mathbf{J}}_3 + \sqrt{2\kappa_l}\,\delta\hat{\mathbf{A}}_l + \sqrt{2\kappa_m}\,\delta\hat{\mathbf{A}}_m - \sqrt{G_{32}J_3}\,\delta\hat{\mathbf{A}}_p^\dagger \qquad (6.1.12)
$$

where we have taken $\alpha_c = \sqrt{n}$ real without loss of generality (see Section 6.1.5). These equations can be solved in Fourier space. The internal mode is coupled to the external field (see Section 5.3) – resulting in the frequency space operator, $\tilde{\mathbf{A}}_{\mathrm{las}} = \sqrt{2\kappa_m}\tilde{\mathbf{a}} - \delta\hat{\mathbf{A}}_{\mathrm{in}}$, describing the fluctuations of the output field. The result of interest is the noise spectrum $V1_{\mathrm{las}}(\Omega)$ of the photon number, or intensity, of this mode. Under our linearization condition this is given by the amplitude quadrature spectrum of the output mode: $V1_{\mathrm{las}}(\Omega) = \langle|\delta\tilde{\mathbf{X}}_A|^2\rangle$ where $\delta\tilde{\mathbf{X}}_A = \delta\tilde{\mathbf{A}}_{\mathrm{las}} + \delta\tilde{\mathbf{A}}_{\mathrm{las}}^\dagger$. We obtain the following solution for the system of transfer functions:

$$
V1_{\mathrm{las}}(\Omega) = \frac{1 + 4\kappa_m^2((2\pi\Omega)^2 + \gamma_L^2) - 8\kappa_m\kappa G_{32}n\gamma_L}{((2\pi\Omega_{RRO})^2 - (2\pi\Omega)^2)^2 + (2\pi\Omega)^2\gamma_L^2} V_{\mathrm{vac}}
$$
$$
+ \frac{2\kappa_m G_{32}^2 n\Gamma}{((2\pi\Omega_{RRO})^2 - (2\pi\Omega)^2)^2 + (2\pi\Omega)^2\gamma_L^2} V_p(\Omega)
$$
$$
+ \frac{(2\kappa_m G_{32}^2 n\gamma_{32}J_3)}{((2\pi\Omega_{RRO})^2 - (2\pi\Omega)^2)^2 + (2\pi\Omega)^2\gamma_L^2} V_{\mathrm{spont}}
$$
$$
+ \frac{2\kappa_m G_{32}((\gamma_{32} + \Gamma)^2 + (2\pi\Omega)^2)}{((2\pi\Omega_{RRO})^2 - (2\pi\Omega)^2)^2 + (2\pi\Omega)^2\gamma_L^2} V_{\mathrm{dipole}}
$$
$$
+ \frac{4\kappa_m\kappa_l(\gamma_L^2 + (2\pi\Omega)^2))}{((2\pi\Omega_{RRO})^2 - (2\pi\Omega)^2)^2 + (2\pi\Omega)^2\gamma_L^2} V_{\mathrm{losses}} \qquad (6.1.13)
$$

where $\gamma_L = G_{32}n + \gamma_{32} + \Gamma$, $(2\pi\,\Omega_{RRO})^2 = 2G_{32}n\kappa$ and the solutions for n and J_3 are given by Equations (6.1.6), (6.1.7). It should be noted that the parameters have to be chosen separately for each laser. The material constants σ_s, ρ, n_r, γ_{32} and γ_{21} are specific to the type of laser and reasonably well known. In particular the rapid decays from the highly excited states and the lower lasing level for most lasers justify the use of this simplified 3 level model. Other parameters, such as the resonator parameters κ_l and κ_m and the total number of lasing atoms vary with the individual laser. How close to a coherent state with $V1_{\text{las}}(\Omega) = 1$ can this solution be? The general answer is: many lasers approach this limit if they are operated well above threshold. However, there are important differences for the various types of lasers. We will now discuss a number of typical cases focusing especially on solid-state and diode lasers.

6.1.4   Examples of lasers

Classes of lasers

The finesse of the laser resonator, determined by the magnitude of κ, is generally chosen to achieve optimum laser efficiency. The finesse is controlled by changing the transmission of the output mirror (κ_m). Reducing the mirror transmission lowers the laser threshold and hence acts to increase efficiency. However, it also reduces the ratio $\frac{\kappa_m}{\kappa}$ due to the presence of internal losses (κ_l), which decreases the efficiency. Thus for a particular laser material there will be an optimum value of κ. Three classes of lasers with distinctly different dynamical behaviour can be identified on the basis of the relationship between the sizes of κ and the decay rate of the upper lasing level (γ_{32}). They are

- Class 1: $\gamma_{32} \gg \kappa$ – e.g. dye lasers

- Class 2: $\gamma_{32} \approx \kappa$ – e.g. visible gas lasers

- Class 3: $\gamma_{32} \ll \kappa$ – e.g. solid state and semiconductor lasers.

We will now examine the predicted behaviour of examples from each of these classes, focusing particularly on the class 3 solid-state and semi-conductor lasers. A fourth class of lasers, mentioned earlier, in which the photon decay rate is more rapid than the coherence dephasing rate can not be described via rate equations and will not be discussed here.

Dye lasers and Argon ion lasers

There is a wide range of possible lasers which are used in quantum optics experiments. In particular when the nonlinearity is a resonant atomic or molecular medium it is important to have a laser which can be tuned on or arbitrarily close to the resonance. While even that can be achieved nowadays with diode lasers, using external cavities, the traditional work horse in these situations is the Argon ion laser pumped dye laser. In these devices the active medium is a dye solution, made up of large, complex dye molecules which can provide gain continuously over a wide range of optical frequencies. The actual laser frequency is selected with a number of frequency dependent components inside the laser resonator.

Typically a dye laser will have parameters of $\kappa \approx 1.5 \times 10^7$ [1/sec] and $\gamma_{32} \approx 3 \times 10^8$[1/sec] and hence is a class 1 laser. Provided we are operating more than about five times above threshold we can approximate the noise spectrum (6.1.13) by

$$V1_{\text{las}}(\Omega) = 1 + \frac{(2\kappa)^2 (V1_P - 1)}{(2\kappa)^2 + (2\pi\Omega)^2} \tag{6.1.14}$$

This is a Lorentzian spectrum with a half width equal to the cavity decay rate. At low frequencies the spectrum will be dominated by the noise of the pump laser. At high frequencies ($2\pi\Omega \gg 2\kappa$) the spectrum will be quantum noise limited. In order to obtain quantum noise limited operation from a dye laser at frequencies in the tens of MHz a quantum noise limited pump laser is needed.

A dye laser is pumped by an Ar ion laser, providing large optical powers (more than 10 [W]) at wavelengths in the blue end of the visible spectrum. Typically an argon laser will have $\kappa \approx 1.0 \times 10^6$ [1/sec] and $\gamma_{32} \approx 2 \times 10^6$ [1/sec] and hence is a class 2 laser. Sufficiently far above threshold a class 2 laser has a noise spectrum very similar to that of a class 1 laser, see Equation (6.1.14). Due to the slower cavity decay rate and hence narrower pump (electric discharge) noise region we would expect the noise spectrum of the argon laser to be quantum noise limited at frequencies of tens of MHz. However, due to the small gain that the Ar plasma provides, these lasers have long cavities and thus in the presence of inhomogeneous broadening frequently run on several longitudinal modes. The output spectrum of such a pump laser would hence be dominated by strong spectral features with a spacing of typically 100 [MHz]. This noise from an Ar ion laser will appear on the intensity noise spectrum of the dye laser. It is possible to suppress the higher order modes using intracavity etalons, but only with a large power penalty.

For quantum optics experiments, which require a laser beam which is as close as possible to the quantum limit the noise properties of the Ar ion and dye lasers have posed considerable problems. Most experiments were carried out at specific detection frequencies, in between the large noise spikes, where the noise spectrum happened to be QNL. In recent years the progress in the development of high power low noise diode lasers and diode laser pumped solid state lasers has been so remarkable that now dye lasers are avoided as much as possible.

The CW Nd:YAG laser

We will now discuss the most popular solid state laser, the Nd:YAG laser. Gain is provided by excited Neodymium ions in a Yttrium Aluminium Garnet glass substrate. Traditionally the neodymium ions were excited by a flash lamp but it is becoming more common these days to use diode laser arrays as the pump source. In a miniature monolithic ring configuration these lasers can be made to perform very close to the ideal single mode laser described by our rate Equations (6.1.3). Thus a careful quantitative comparison between theory and experimental results is warranted here. Typically solid state lasers have $\kappa \approx 10^7 - 10^8$ [1/s] depending on their size and $\gamma_{32} \approx 10^2 - 10^3$ [1/s] and hence are class 3 lasers. For a class 3 laser all the terms in the noise spectrum, Eq. (6.1.13), remain important even when operating up to 10 times above threshold and the spectrum is quite complex. It contains contributions from the pump noise $V1_P(\Omega)$, spontaneous emission, dipole fluctuations and from vacuum fluctuations. The mixture of these contributions varies with the detection frequency.

It is instructive to plot these various contributions separately. This is done in Fig. 6.5(a), where the various contributions are described in the figure caption.

It is instructive to plot these various contributions separately. This is done in Fig. 6.5(a), where the various contributions are described in the figure caption. There are some very clear features and the spectrum is dominated by a resonant feature, called the resonant relaxation

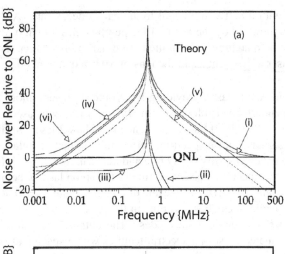

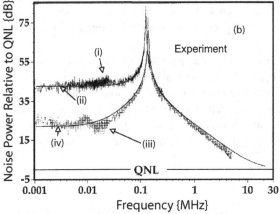

Figure 6.5: Noise spectrum of a free running Nd:YAG laser. (a) The different contributions to Equation (6.1.13) are drawn separately. Trace (i) is the contribution of the vacuum noise, (ii) is the effect of the pump noise, (iii) is the noise due to spontaneous emission, (iv) is the noise associated with the dipole fluctuations, (v) is the noise due to intracavity losses and (vi) is the total noise for a QNL pump. Part (b) shows experimental results for such a Nd:YAG laser. In this realistic case (i) the pump noise level was about 40 [dB] above the QNL. For comparison the predicted noise performance is shown in (ii). Trace (iii) was obtained with a pump noise of 20 [dB]. Trace (iv) is the corresponding theoretical prediction. Adapted from C.C. Harb et al.[Har97].

oscillation which occurs at the frequency

$$2\pi\, \Omega_{RRO} = \sqrt{2G_{23}n\kappa} \tag{6.1.15}$$

In this particular case $\Omega_{RRO} = 400$ [kHz]. This oscillation is due to the interaction of the atomic state inversion $(J_3 - J_2) \approx J_3$ with the excitation of the internal optical mode. In

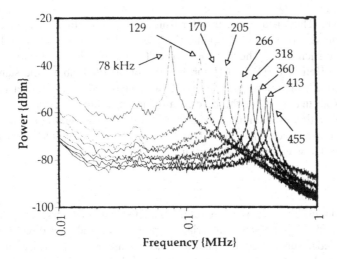

Figure 6.6: Experimental noise trace for a Nd:YAG laser with various pump powers. The relaxation oscillation frequency Ω_R is stated for each case in units of [KHz]. It increases with pump power as stated in Equation (6.1.16). After C.C. Harb

physical terms, the energy in the laser is moving back and forth between the laser medium and the cavity mode resulting in a strong modulation of the output power. This oscillation is a prominent feature and corresponds to an under-damped oscillation which is continually excited by the noise sources within the laser [Yar70]. Note that this oscillation does not occur in class 1 and 2 lasers because for them $\gamma_L \gg \Omega_{RRO}$, making the oscillation overdamped, as can be seen from the denominator in Equation (6.1.13). An alternative expression for Ω_{RRO} in terms of the pump rate is

$$2\pi\, \Omega_{RRO} = \sqrt{2\kappa\gamma_{32}}\ \sqrt{\frac{\Gamma - \Gamma_t}{\Gamma_t}} \tag{6.1.16}$$

That means, the harder the laser is pumped, increasing Γ, the higher is the frequency of the relaxation oscillation and the less pronounced it is. This can be seen by comparing the traces in Fig. 6.6 which are plotted for the same laser, but with different pump powers. In some situations fitting the measured frequencies Ω_{RRO} for a range of pump powers is a more reliable way to determine Γ_t than an attempt to measure the power at which the laser turns on.

Above Ω_{RRO} the laser has a long, drawn out noise tail. For very high frequencies the laser approaches the quantum noise limit. The origin of the quantum noise can be traced back to the various vacuum inputs in the model. It should be noted that the noise of the pump source has no influence on the laser noise for frequencies $\Omega \gg \Omega_{RRO}$. A noisy pump will not affect the laser at such high frequencies. However, it also means that in this frequency regime the laser output power cannot be controlled by modulating the pump power. This is certainly a disadvantage for communication systems, since pump power modulation is normally the simplest and cheapest modulation scheme.

At low frequencies, $\Omega \ll \Omega_{RRO}$, the noise spectrum of the laser is dominated by the noise of the pump source. There is a strong transfer from the pump to the laser output. By carefully choosing the noise properties of the pump the laser output can be quietened down. Figure(6.5(b)) shows two such cases. The first case, traces (i) and (ii), is realistic for a laser with large pump noise. Here $V1_p(\Omega)$ is a factor $10,000$, or 40 [dB], above the standard quantum limit. The second case, traces (iii) and (iv), were measured with a quieter pump, $V1_p(\Omega) = 20$ [dB]. Note that even for the case of a quantum limited pump $(V1_p(\Omega) = 0$ [dB]), the resonant relaxation oscillation would survive. Only for the impossible case of a completely noise free pump source $(V1_P(\Omega) = 0)$ and no cavity losses or dipole fluctuations would the relaxation oscillation completely disappear.

Diode lasers

Perhaps the simplest and most efficient lasers are semi-conductor or diode lasers. Diode lasers use the band-gap in doped semi-conductor materials to produce optical gain directly from an electrical current. Under suitable conditions diode lasers can be described by our rate Equations (6.1.3) as class 3 lasers. Major differences with solid-state lasers arise from their small size and high gain. Typically a diode laser will have $\kappa \approx 10^{11}$ [1/sec] and will be run 5–10 times above threshold. Under these conditions the RRO will be pushed out to frequencies of tens of [GHz] and will be quite small. Most detection schemes are only sensitive up to frequencies of a few [GHz] thus for all practical purposes the spectrum can be written

$$V1_{\text{las}}(\Omega) = \frac{\kappa_m}{\kappa}[1 + \zeta\,(V1_P(\Omega) - 1)] \tag{6.1.17}$$

where it is understood that $2\pi\Omega \ll \kappa$ and ζ is the pump efficiency. Moreover, unlike the situation of the dye laser where the pump source is a potentially very noisy Ar ion laser, the pump source of the diode laser is an electric current which can be made very quiet. Indeed we can have $V1_{\text{las}}(\Omega) = 1$ if the pump current is given by $V1_P(\Omega) = 1$. This means that for a quiet, Poissonian noise limited pump current a diode laser has a quantum noise limited output. Equation (6.1.17) also means the laser output responds directly to modulation of the pump source. This is important for communication applications where we want to use the current driving the diode laser as the input of modulation. In fact there is no fundamental reason why we cannot have $V1_P(\Omega) < 1$, that is a sub-Poissonian current. Unlike light, electric currents can easily be made sub-Poissonian. This would imply $V1_{\text{las}}(\Omega) < 1$, the laser output would be sub-Poissonian. Diode lasers have successfully been made to produce squeezed light using this method and technical details are discussed in Chapter 9. From Equation (6.1.17) we see that, at least in theory, the squeezing is limited only by the internal losses and the pump efficiency ζ.

Limits of the single mode approximation in diode lasers

From the previous discussion of diode lasers it would appear that, provided sufficient care was taken to suppress noise in the pump current, these lasers would always run at or below the QNL. For a number of reasons, in practice this is not the case. In particular, multi-mode behaviour can lead to increased noise. Normally one expects single mode operation from a homogeneously broadened medium, such as that of a diode laser, due to gain competition. However, this is not the whole story, if there is strong spatial dependence in the laser mode some regions of the gain medium can become gain depleted whilst others do not, allowing modes with different spatial dependence to access the undepleted gain and lase. Competition between these modes can lead to elevated noise levels. An example is the high power diode array, consisting of a strip of diode gain medium sharing a common cavity. The cavity can support a number of transverse modes many of which lase simultaneously due to spatial hole-burning, thus introducing excess noise. Further, the far field spatial patterns of the different modes are different, such that different spatial regions of the beam can have different noise properties.

Even in a purely homogeneously broadened medium, spontaneous emission into the lasing mode, which is neglected in our rate Equations (6.1.3), can allow for the existence of weakly lasing longitudinal side-modes [Ino92]. The homogeneity of the medium means that the production of a photon in the main mode implies no photon is concurrently emitted in the side modes and vice versa. Hence there is a strong anti-correlation between the side modes and the main mode. If just the main mode is observed it is found to be very noisy due to the deletion of photons to the side modes [Mar95]. However, the anti correlation means that the noise tends to cancel when all the light is observed. Unfortunately, differences in the losses experienced by the main mode and the side modes can lead to a small amount of excess noise which does not cancel. This can be sufficient to destroy squeezing. This problem has been solved using injection locking or external cavities to suppress the side modes. In this way reliable squeezing has been obtained – the details of these experiments are given in Chapter 9.

In summary we find that lasers themselves are interesting but in many cases complex light sources. Only the single mode laser operating well above threshold will be QNL at all detection frequencies. For all other cases we have to explicitly derive the intensity noise spectrum, or measure it directly.

6.1.5   Laser phase noise

In Section 6.1.2 we solved for the steady state values of the intra-cavity photon number, n, and the upper laser level population, J_3. We also introduced the intracavity complex amplitude α_c. Notice however, that it is not possible to solve for the phase of α_c. Physically this indicates that the laser phase is meta-stable and will tend to drift in time. This effect is known as phase diffusion.

When we linearized the quantum equations of motion to obtain the equations of motion for the fluctuation operators, Eq. (6.1.12), we took the phase of α_c to be constant and real. This is of no consequence for the intensity spectrum because intensity measurements $(\hat{A}^\dagger \hat{A})$ are phase insensitive. However, we need to be more careful if considering phase sensitive measurements. We can calculate the phase fluctuation spectrum of the laser output from

$$V2_{\mathrm{las}}(\Omega) = <|\delta\tilde{\mathbf{X}}\mathbf{2}_{\mathrm{las}}|^2> \qquad (6.1.18)$$

where as usual $\delta\tilde{\mathbf{X}}\mathbf{2}_{\mathrm{las}} = i(\delta\tilde{\mathbf{A}}_{\mathrm{las}} - \delta\tilde{\mathbf{A}}_{\mathrm{las}}^\dagger)$. The phase fluctuation spectrum can be obtained by making a phase sensitive measurement of the field with respect to a stable phase reference, that means using homodyne detection described in Section 8.1.4. Our assumption of constant classical phase (α_c real) will only be valid for the phase spectrum if the fluctuations around that constant phase are small. The solution is given by

$$V2_{\text{las}}(\Omega) = \frac{(2\pi\Omega)^2 + 2\,(2\kappa)^2}{(2\pi\Omega)^2} \qquad\qquad (6.1.19)$$

At very high frequencies the phase spectrum tends to the quantum noise level, $V2_{\text{las}}(\Omega) = 1$. As we move to lower and lower frequencies the phase noise becomes larger and larger, eventually diverging at zero frequency. The divergence of the phase noise spectrum at zero frequency indicates that our linearization breaks down for sufficiently small Fourier frequencies. On long enough time scales, or low enough frequencies, the phase drifts significantly, thus invalidating the linear approximation. This is the signature of laser phase diffusion. In the absence of technical noise, this phase diffusion determines the laser linewidth.

In practice phase diffusion does not present a major problem as generally the same laser is used as a phase reference as is used to produce the signal. The phase drift is then common to both modes and so is not observed. Thus provided we stay within the coherence length of our laser, see Section 2.2.3, and work at frequencies for which the laser's intensity noise is at the QNL, a laser beam can represent an excellent realization of an ideal coherent state.

6.2   Amplification of optical signals

It is possible to use an optical medium to amplify the power of a laser beam. For example, we can imagine an input power of 1 [mW] and an output power of 1 [W], that means a power gain G of a factor of 1000. A medium that can achieve this purpose is called an active optical medium. Normally it is just a laser medium without optical feedback from a cavity. Such a device would obviously be very useful for practical applications where high power is required. Some of the biggest and most expensive optical systems are the Nd:glass amplifiers which have been built to generate powerful laser pulses for fusion studies using inertial confinement [Fus]. Another application is in optical communications where the beam is attenuated over distance, for example by losses inside an optical fibre. In this case it is necessary to amplify not only the power but also the modulation sidebands which contain the information that is communicated. In a fibre system an in-line amplifier, such as an Erbium doped fibre amplifier, is a very useful and attractive device. The gain bandwidth should extend to the largest modulation sideband that is used. Typically this is a few [GHz], which is not difficult to achieve. The next question is how does an optical amplifier affect both the signal and the noise? Does the signal to noise ratio vary?

It might be easiest to understand the properties of optical amplifiers by first describing the simpler and more obvious case of an electronic amplifier. Consider a circuit with a linear amplifier which has a power gain $G_{\text{el}}(\Omega)$. The gain is frequency dependent, usually $G_{\text{el}}(\Omega)$ drops to less than unity at high frequencies. The input into the amplifier is a current with modulation at Ω_m. In linear systems we can consider each Fourier component, each modulation frequency, separately and independently of all the others. Ideally the input and output are related by

$$\delta i_{\text{out}}(\Omega) = \sqrt{G_{\text{el}}(\Omega)}\,\delta i_{\text{in}}(\Omega)$$

The variances of modulations, or fluctuations, of the input and the output are linked by

$$\text{Var}(i_{\text{out}}) = G_{\text{el}} \ \text{Var}(i_{\text{in}}) \tag{6.2.1}$$

This amplifier affects both the signal and the noise in the same way. In principle the signal to noise ratio will be maintained – in practice a small additional noise component will be added by the amplifier. However there is no fundamental limit to how small this added noise can be. As long as the added noise is much smaller than the noise floor of the input the signal to noise ratio is maintained. The performance of this system could be improved further if the amplifier is given a reference signal which is at the same frequency and phase locked to the modulation signal. In this case a significant part of the noise, which has random phase, can be suppressed by this apparatus, normally called a lock-in amplifier.

The performance of an optical amplifier is quite different [Hau62], [Cav82], [Hau00]. The input and the output can be described in the Heisenberg picture by the operators $\tilde{\mathbf{A}}_{\text{in}}$ and $\tilde{\mathbf{A}}_{\text{out}}$. Suppose our input state is a coherent state, $|\alpha\rangle$. We expect the coherent amplitude to be amplified such that

$$\alpha_{\text{out}} = \langle \tilde{\mathbf{A}}_{\text{out}} \rangle_\alpha = \sqrt{G} \ \alpha_{\text{in}}$$

This suggests the following operator description

$$\tilde{\mathbf{A}}_{\text{out}} = \sqrt{G} \ \tilde{\mathbf{A}}_{\text{in}}$$

However, this violates the commutation rule

$$[\tilde{\mathbf{A}}_{\text{out}}, \tilde{\mathbf{A}}_{\text{out}}^\dagger] = G \ [\tilde{\mathbf{A}}_{\text{in}}, \tilde{\mathbf{A}}_{\text{in}}^\dagger] = G \neq 1$$

This problem is similar to the quantum theory of damping, where a careful consideration of the role of quantum noise has to be made. The vacuum state entering through any unused port has to be taken into account. See the description of a beamsplitter in Section 5.1 for details. In a similar way we have to include the vacuum state in this amplification process. The correct description is [VW94]

$$\tilde{\mathbf{A}}_{\text{out}} = \sqrt{G} \ \tilde{\mathbf{A}}_{\text{in}} + \sqrt{G-1} \ \delta\tilde{\mathbf{A}}_{\text{vac}}^\dagger \tag{6.2.2}$$

where $\delta\tilde{\mathbf{A}}_{\text{vac}}$ is an operator describing an input mode originally in the vacuum state that is mixed in by the amplifier. It is completely independent of $\tilde{\mathbf{A}}_{\text{in}}$. The result is that the light, after amplification, is considerably noisier than before. The amplifier is increasing not only the power but also the noise level. The variance of the intensity of the amplified light is

$$V1_{\text{out}} = G \ V1_{\text{in}} + (G-1) \ V1_0 \tag{6.2.3}$$

Here $V1_0 = 1$ is at the QNL. For large gains ($G \gg 1$) the second term results in an increase of an extra 3 [dB] compared to the classical case, Eq. (6.2.1). As shown in Fig. 6.7 both the average amplitude and the fluctuations increase.

To see that this is indeed the effect of amplification by a laser medium consider again the quantum mechanical equations of motion for the laser, Eq. (6.1.11). Suppose that we are

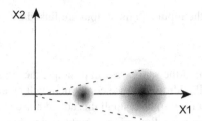

Figure 6.7: Effect of optical gain on a quantum state

below threshold such that $\kappa > G_{32}J_3/2$ but that instead of vacuum entering the output mirror a field, $\hat{A}_{in}$, is injected. Further let us assume that we are sufficiently below threshold that there is very little depletion of the upper lasing level and that we can therefore replace the operator $\hat{J}_3$ with its expectation value J_3. We obtain the following equation of motion

$$\frac{d\hat{a}}{dt} = \frac{G_{32}}{2}J_3\hat{a} - \kappa\hat{a} + \sqrt{2\kappa_l}\,\delta\hat{A}_l + \sqrt{2\kappa_m}\,\hat{A}_{in} - \sqrt{G_{32}J_3}\,\delta\hat{A}_p^\dagger \tag{6.2.4}$$

Equation (6.2.4) is now linear and can be solved exactly in Fourier space. Neglecting internal losses ($\kappa_l = 0$) leads to the following solution for the output from the amplifier

$$\tilde{A}_{out} = \frac{\kappa + G_{32}J_3/2 + i2\pi\Omega}{\kappa - G_{32}J_3/2 + i2\pi\Omega}\,\tilde{A}_{in} - \frac{\sqrt{2\kappa G_{32}J_3}}{\kappa - G_{32}J_3/2 + i2\pi\Omega}\,\delta\tilde{A}_p^\dagger \tag{6.2.5}$$

The spectrum for either quadrature is then

$$V i_{out}(\Omega) = \frac{(\kappa + G_{32}J_3/2)^2 + (2\pi\Omega)^2}{(\kappa - G_{32}J_3/2)^2 + (2\pi\Omega)^2}\,V i_{in} + \frac{2\kappa G_{32}J_3}{(\kappa - G_{32}J_3/2)^2 + (2\pi\Omega)^2}$$

which is of the same form as Equation (6.2.3) with

$$G(\Omega) = \frac{(\kappa + G_{32}J_3/2)^2 + (2\pi\Omega)^2}{(\kappa - G_{32}J_3/2)^2 + (2\pi\Omega)^2}$$

We see that in this case the additional quantum noise is introduced via the dipole fluctuations of the gain medium.

The *3dB penalty* is imposed by the optical amplifier when it acts equally on all quadratures. Such an amplifier is called *phase insensitive*. The penalty can only be avoided by using an amplifier that is *phase sensitive*, as will be discussed in the next section.

6.3   Parametric amplifiers and oscillators

Phase insensitive amplification is of limited utility when dealing with quantum fields because of the excess noise it introduces. We would really like to be able to amplify our quantum limited light beam without adding additional noise. Let us consider the fundamental commutation problem again. If we write the annihilation operator in terms of the amplitude and phase quadratures we see the problem clearly:

$$\tilde{\mathbf{A}} = \frac{1}{2}(\tilde{\mathbf{X}}\mathbf{1} + i\tilde{\mathbf{X}}\mathbf{2}) \tag{6.3.1}$$

If both quadratures are amplified equally then $\tilde{\mathbf{A}}$ is just multiplied by some gain factor and via the arguments of the last section it is clear that an additional field, inevitably introducing noise, must be mixed in by the amplifier. However, consider the situation in which the amplitude quadrature is amplified by some amount, $\sqrt{K}$, but the phase quadrature is *de-amplified by the same amount*. The output operator would now look like

$$\tilde{\mathbf{A}}_{\text{out}} = \frac{1}{2}\left(\sqrt{K}\,\tilde{\mathbf{X}}\mathbf{1} + \frac{i}{\sqrt{K}}\,\tilde{\mathbf{X}}\mathbf{2}\right) \tag{6.3.2}$$

It is straightforward to confirm that the commutation relations still hold in this case and that this is a legitimate transformation and no additional noise operators are required. Thus, provided the product of the gains of the two quadratures is unity, noiseless amplification of one of the quadratures can be achieved. This is a very important quantum optical interaction. In analogy with the linear amplifier we can also write Equation (6.3.2) in the form

$$\tilde{\mathbf{A}}_{\text{out}} = \sqrt{G}\,\tilde{\mathbf{A}}_{\text{in}} + \sqrt{G-1}\,\tilde{\mathbf{A}}_{\text{in}}^{\dagger} \tag{6.3.3}$$

where the gain is

$$G = \frac{(K+1)^2}{4K}$$

For historical reasons a phase sensitive amplifier is referred to as a *parametric amplifier*.

6.3.1 The second-order non-linearity

Classically the response of an atomic medium to an incident optical beam can be described by the induced dipole polarisation $P(E)$ caused by the beam's electric field E:

$$P(E) = \chi E + \chi^{(2)} E^2 + \chi^{(3)} E^3 + \cdots \tag{6.3.4}$$

A non-linear optical material is one in which, at the available field strength, higher than linear terms in the expansion become significant. With the invention of the laser, incident field strengths became sufficiently high that many materials could be made to show non-linear behaviour. We will find that phase sensitive amplification and other interesting quantum optical phenomena can be produced by materials exhibiting a second-order $\chi^{(2)}$ non-linearity.

The key feature of the second-order non-linearity is its ability to couple a fundamental field (oscillating at ν) to a harmonic field (oscillating at 2ν). The Hamiltonian, or energy function, for such an interaction can be written

$$H = i\hbar\chi^{(2)}(\beta^*\alpha^2 - \alpha^{*2}\beta)$$

where α is the fundamental whilst β is the harmonic field. Canonical quantization then leads to the quantum mechanical Hamiltonian

$$\tilde{\mathbf{H}} = i\hbar\chi^{(2)}(\hat{\mathbf{b}}^\dagger\hat{\mathbf{a}}^2 - \tilde{\mathbf{a}}^{\dagger 2}\hat{\mathbf{b}})$$

This Hamiltonian has an elegant interpretation in terms of photons. The first term represents the annihilation of two photons at the fundamental frequency and the creation of one at the harmonic, whilst the second represents the reverse process of a single harmonic photon being annihilated and two fundamental photons being created. The Heisenberg equations of motion, see Section 4.4, are then given by

$$
\begin{aligned}
\dot{\hat{b}} &= \frac{1}{i\hbar}[\hat{\mathbf{b}}, \hat{\mathbf{H}}] = \chi^{(2)}\,\hat{\mathbf{a}}^2 \\
\dot{\hat{a}} &= -2\,\chi^{(2)}\,\hat{\mathbf{a}}^\dagger\hat{\mathbf{b}}
\end{aligned}
\tag{6.3.5}
$$

Notice that because the process is energy conserving (and coherent), i.e. two photons of energy $h\nu$ equals one of energy $h\,2\nu$, then no additional vacuum modes are required to be coupled into the fields.

In order that either the up-conversion (fundamental to harmonic) or the down-conversion (harmonic to fundamental) process takes place with some reasonable efficiency we require not only energy conservation but also that momentum is conserved for the converted beam in some particular direction(s). In this way the emission from many dipoles can coherently add to produce a single output beam. This condition is known as phase-matching.

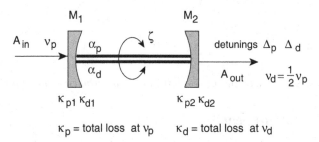

Figure 6.8: A parametric medium inside a cavity. The various input, output and intra-cavity modes are labelled. This system has a threshold and can operate below threshold as a phase sensitive amplifier of the vacuum modes and above threshold as an oscillator. The subscript A refers to the fundamental frequency and the subscript B refers to the subharmonic frequency.

Let us consider the situation shown schematically in Fig. 6.8. A non-linear crystal is placed inside a pair of cavities, one resonant at the fundamental and the other resonant at the harmonic, which are oriented such that there is good phase matching between the two spatial modes defined by the cavities. By combining the cavity equations of motion derived in Section 5.3.6 with those of Equation (6.3.5) we obtain the equations of motion for the internal cavity modes as

$$\dot{\hat{b}} = \chi^{(2)}\,\hat{\mathbf{a}}^2 - \kappa_b\,\hat{\mathbf{b}} + \sqrt{2\,\kappa_B}\,\hat{\mathbf{B}}_{in} + \sqrt{2\kappa_{lb}}\,\delta\hat{\mathbf{B}}_l \tag{6.3.6}$$

$$\dot{\hat{a}} = -2\,\chi^{(2)}\,\hat{a}^\dagger\hat{b} - \kappa_a\hat{a} + \sqrt{2\kappa_A}\,\hat{A}_{\text{in}} + \sqrt{2\kappa_{la}}\,\delta\hat{A}_l \tag{6.3.7}$$

where κ_A and κ_B are the loss rates of the input/output mirrors for the $\hat{a}$ and $\hat{b}$ modes respectively. Similarly κ_{la} and κ_{lb} are the respective internal loss rates for the two modes and $\kappa_a = \kappa_A + \kappa_{la}$, $\kappa_b = \kappa_B + \kappa_{lb}$. We also have the respective input modes $\hat{A}_{\text{in}}$ and $\hat{B}_{\text{in}}$ and the respective vacuum modes which couple in due to the internal losses $\delta\hat{A}_l$ and $\delta\hat{B}_l$. Equations (6.3.6) and (6.3.7) can describe a large range of important quantum optical experiments.

6.3.2 Parametric amplification

Let us now consider parametric amplification. We will assume that the harmonic field is an intense field which is virtually undepleted by its interaction with the non-linear crystal, whilst the fundamental field is "small". We expect down-conversion under these conditions with energy being transferred from the harmonic to the fundamental field. We shall also assume that $\kappa_a \ll \kappa_b$ such that the dynamics of the harmonic mode can be neglected. Under these conditions we can replace the operator $\hat{b}$ with the c-number β. We are then left with just a single, linear equation for the $\hat{a}$ mode

$$\dot{\hat{a}} = -\chi\hat{a}^\dagger - \kappa_a\hat{a} + \sqrt{2\kappa_A}\,\hat{A}_{\text{in}} + \sqrt{2\kappa_{la}}\,\delta\hat{A}_l \tag{6.3.8}$$

where

$$\chi = 2\chi^{(2)}\,\beta = 2\chi^{(2)}\sqrt{\frac{2}{\kappa_b}}\,\beta_{\text{in}} \tag{6.3.9}$$

and $\beta_{\text{in}} = \langle\hat{B}_{\text{in}}\rangle$. The parameter χ is intensity dependent. Equation (6.3.8) can be solved in Fourier space to give the output mode at the fundamental frequency

$$\tilde{A}_{\text{out}} = \frac{\chi^2 + \kappa_a^2 + (2\pi\Omega)^2}{(\kappa_a + i2\pi\Omega)^2 - \chi^2}\,\tilde{A}_{\text{in}} + \frac{2\chi\kappa_a}{(\kappa_a + i2\pi\Omega)^2 - \chi^2}\,\tilde{A}_{\text{in}}^\dagger \tag{6.3.10}$$

Here we have neglected internal losses to the fundamental mode ($\kappa_{la} = 0$) for simplicity. Equation (6.3.10) is formally equivalent to Equation (6.3.3). The variance of the amplitude quadrature is given by

$$V1_{\text{out}}(\Omega) = \frac{(\kappa_a^2 + 2\chi\kappa_a + \chi^2 + (2\pi\Omega)^2)^2}{\chi^4 + (\kappa_a^2 + (2\pi\Omega)^2)^2 - 2\chi^2(\kappa_a^2 - (2\pi\Omega)^2)}\,V1_{\text{in}}(\Omega) \tag{6.3.11}$$

where $V1_{\text{in}}$ is the amplitude quadrature variance of the input at the fundamental. The output is an amplified version of the input amplitude quadrature with no added noise as hoped. This can be seen more clearly at zero frequency where the expression reduces to

$$V1_{\text{out}}(0) = \frac{(\kappa_a + \chi)^2}{(\kappa_a - \chi)^2}\,V1_{\text{in}}(0) \tag{6.3.12}$$

As the harmonic intensity is increased χ becomes larger, the denominator of Equation (6.3.12) becomes smaller and hence the amplification increases. On the other hand the phase quadrature (at zero frequency) is given by

$$V2_{\text{out}}(0) = \frac{(\kappa_a - \chi)^2}{(\kappa_a + \chi)^2}\,V2_{\text{in}}(0) \tag{6.3.13}$$

which suffers increasing deamplification as χ increases. Notice that if $V2_{\text{in}}$ is at the QNL then $V2_{\text{out}}$ will be *below the QNL*, that means $V2_{\text{out}} < 1$. This is termed *squeezing* and is discussed in detail in Chapter 9.

6.3.3   Optical parametric oscillator

Examining Equation (6.3.12) we note that the spectrum diverges as the value of χ approaches that of κ_a. This indicates that an approximation is failing, in this case that of negligible pump depletion. As the amplification of the harmonic field becomes larger and larger, eventually it starts to significantly deplete the harmonic. To see what happens in this case we must go back to Equations (6.3.6) and (6.3.7) and linearize.

We continue to consider the experimental situation where the cavity for the harmonic mode is either only weakly resonant or even non-existent, that is we effectively have $\kappa_a \ll \kappa_b$. As a result the evolution of the mode $\hat{b}$ occurs on a much faster time scale than that of the $\hat{a}$ mode and we can adiabatically eliminate it by setting its time derivative to zero. We also continue to assume that $\kappa_{lb} \ll \kappa_b$. That means the losses for the mode $\hat{b}$ can be neglected and Eq. (6.3.6) and (6.3.7) reduce to

$$\dot{\hat{\delta a}} = -\frac{2\chi^{(2)^2}}{\kappa_b}\hat{\delta a}^\dagger \hat{a}^2 - 2\chi^{(2)}\sqrt{\frac{2}{\kappa_b}}\hat{\delta a}^\dagger \hat{B}_{\text{in}} - \kappa_a\,\hat{\delta a} + \sqrt{2\kappa_A}\,\hat{A}_{\text{in}} + \sqrt{2\kappa_{la}}\,\hat{\delta A}_l$$

To proceed we first consider the semi-classical steady-state solution of this equation. By taking expectation values, and making the approximation that they may be factored, we obtain

$$\dot{\alpha} = -\frac{2\chi^{(2)^2}}{\kappa_b}\alpha^*\alpha^2 - 2\chi^{(2)}\sqrt{\frac{2}{\kappa_b}}\alpha^*\beta_{\text{in}} - \kappa_a\alpha + \sqrt{2\kappa_A}\alpha_{\text{in}} \tag{6.3.14}$$

The steady-state solution is obtained by setting the time derivative to zero. If we assume that the first term on the RHS of Equation (6.3.14) is small and can be neglected then we obtain the parametric amplification considered in the previous section. Instead we now retain that term but consider the situation in which there is no input field at the fundamental ($\alpha_{\text{in}} = 0$). The equation that needs to be solved is

$$0 = -\frac{2\chi^{(2)^2}}{\kappa_b}\alpha^*\alpha^2 - 2\chi^{(2)}\sqrt{\frac{2}{\kappa_b}}\alpha^*\beta_{\text{in}} - \kappa_a\alpha \tag{6.3.15}$$

Equation (6.3.15) has two solutions: either $\alpha = 0$ or $\alpha \neq 0$. We recognize this as threshold behaviour similar to that of the laser. At a sufficiently high value of the harmonic pump field the system will start to oscillate and produce coherent output at the fundamental. Under these conditions the device is called an optical parametric oscillator. If $\alpha \neq 0$ we have

$$\alpha^*\alpha = -\frac{\sqrt{2\kappa_b}}{\chi^{(2)}}\beta_{\text{in}}e^{-i2\phi_\alpha} - \frac{\kappa_a\kappa_b}{2\,\chi^{(2)^2}} \tag{6.3.16}$$

where ϕ_α is the phase of α. To be a valid stable solution $\alpha^*\alpha$ must be real and positive. This is only possible if $\phi_\beta - 2\phi_\alpha = \pi$. For simplicity we take β_{in} real and negative ($\phi_\beta = \pi$) then α will be real, and the solution for the output photon number is

$$n_{\text{out}} = 2\kappa_a\alpha^2 = \frac{\kappa_a\kappa_b}{\chi^{(2)^2}}(\chi - \kappa_a) \tag{6.3.17}$$

where we have used the definition of χ from Equation (6.3.9). When $\chi < \kappa_a$ the $\alpha = 0$ solution will be stable, when $\chi > \kappa_a$ then Equation (6.3.17) gives the photon flux of the output field at the fundamental. Notice that the threshold condition is the same as the point of divergence of the noise spectrum in the previous section.

Noise spectrum of the parametric oscillator

We can now investigate the fluctuations of the above threshold fundamental output by linearizing around the semi-classical solution. By linearizing Equation (6.3.10) we obtain the following equation of motion for the fluctuations:

$$\dot{\delta a} = -\frac{2\,\chi^{(2)^2}}{\kappa_b}\alpha^2\left(\delta\hat{a}^\dagger + 2\delta\hat{a}\right) - 2\chi^{(2)}\sqrt{\frac{2}{\kappa_b}}\left(\delta\hat{a}^\dagger\beta_{in} + \alpha\,\delta\hat{B}_{in}\right)$$
$$- \kappa_a\delta\hat{a} + \sqrt{2\kappa_A}\delta\hat{A}_{in} + \sqrt{2\kappa_{la}}\delta\hat{A}_l \qquad (6.3.18)$$

This equation can be solved in Fourier space to give the quadrature fluctuations of the output field:

$$\delta\tilde{X}1_{out} = \frac{4\kappa_a - 2\chi - i2\pi\Omega}{2\chi - 2\kappa_a + i2\pi\Omega}\delta\tilde{X}1_{Ain} + \frac{2\sqrt{2\kappa_a}(\chi - \kappa_a)}{2\chi - 2\kappa_a + i2\pi\Omega}\delta\tilde{X}1_{Bin}$$

$$\delta\tilde{X}2_{out} = \frac{2\kappa_a - 2\chi - i2\pi\Omega}{2\chi + i2\pi\Omega}\delta\tilde{X}2_{Ain} + \frac{2\sqrt{2\kappa_a}(\chi - \kappa_a)}{2\chi + i2\pi\Omega}\delta\tilde{X}2_{Bin} \qquad (6.3.19)$$

where we have neglected internal losses of the fundamental for simplicity and $\delta\tilde{X}1_{Ain}$ and $\delta\tilde{X}2_{Ain}$ are the amplitude and phase quadrature fluctuations entering at the fundamental input/output mirrors respectively. Notice also that non-negligible pump depletion now couples fluctuations from the pump harmonic field, $\delta\tilde{X}1_{Bin}$ and $\delta\tilde{X}2_{Bin}$, into the fundamental field. The quadrature variances at zero frequency are

$$V1(0) = 1 + \frac{\kappa_a^2}{(\chi - \kappa_a)^2}$$

$$V2(0) = 1 - \frac{\kappa_a^2}{\chi^2} \qquad (6.3.20)$$

where we have assumed that the harmonic pump fluctuations are at the QNL. The coupling of the harmonic fluctuations means that the amplification of the amplitude quadrature is no longer noiseless above threshold. Strong noised reduction, that means $V2 \ll 1$, is still present on the phase quadrature close to threshold. However, it becomes negligible far above threshold where the output tends towards a coherent state.

6.3.4 Pair production

Finally let us return to the parametric amplifier operating far below threshold for another important application of the χ_2 interaction. Far below threshold, with no input at the fundamental, we may form the following physical picture: the harmonic beam is mostly transmitted without change through the non-linear medium, however every now and then a harmonic photon is converted into a pair of fundamental photons. This picture suggests that if we counted photons at the fundamental output under these conditions we should only ever see them arrive as pairs.

To test this proposition we consider the Schrödinger evolution of the fundamental far below threshold where its initial state is the vacuum. The evolution of the fundamental mode is not passive (energy conserving) as energy is being added, however it is unitary (reversible) as no noise is being added. This means we cannot directly use the techniques described in Chapter 4 but we can use a modified version. We use the fact that we can write the vacuum state in the following way: $\tilde{\mathbf{A}}\tilde{\mathbf{A}}^\dagger|0\rangle$ which is clearly still the vacuum state as the action of creation and then annihilation nullify each other. This motivates us to write the output state as

$$
\begin{aligned}
|\phi\rangle &= \tilde{\mathbf{U}}|0\rangle = \tilde{\mathbf{U}}\tilde{\mathbf{A}}\tilde{\mathbf{A}}^\dagger|0\rangle \\
&= (\tilde{\mathbf{U}}\tilde{\mathbf{A}}\tilde{\mathbf{U}}^\dagger)(\tilde{\mathbf{U}}\tilde{\mathbf{A}}^\dagger\tilde{\mathbf{U}}^\dagger)(\tilde{\mathbf{U}}|0\rangle) \\
&= \tilde{\mathbf{A}}'\tilde{\mathbf{A}}'^\dagger|\phi\rangle
\end{aligned}
\tag{6.3.21}
$$

Equation (6.3.21) shows that the output state is the eigenstate of the operator $\tilde{\mathbf{Y}} = \tilde{\mathbf{A}}'\tilde{\mathbf{A}}'^\dagger$ with eigenvalue "one". Here the dashes indicate reverse time Heisenberg evolution (see Section 4.4). Solving this eigenvalue equation leads to (at worst) a second order recurrence relation for linear or linearized systems. Inverting Equation (6.3.10) and its corresponding conjugate equation we obtain

$$
\begin{aligned}
\tilde{\mathbf{Y}}(\Omega) &= |p|^2\tilde{\mathbf{A}}(\Omega)\tilde{\mathbf{A}}(\Omega)^\dagger - pq^*\tilde{\mathbf{A}}(\Omega)\tilde{\mathbf{A}}(-\Omega) \\
&\quad -p^*q\tilde{\mathbf{A}}(-\Omega)^\dagger\tilde{\mathbf{A}}(\Omega)^\dagger + |q|^2\tilde{\mathbf{A}}(-\Omega)^\dagger\tilde{\mathbf{A}}(-\Omega)
\end{aligned}
\tag{6.3.22}
$$

where

$$
\begin{aligned}
p &= \frac{\chi^2 + \kappa_a^2 + (2\pi\Omega)^2}{(\kappa_a + i2\pi\Omega)^2 - \chi^2} \\
q &= \frac{2\chi\kappa_a}{(\kappa_a + i2\pi\Omega)^2 - \chi^2}
\end{aligned}
\tag{6.3.23}
$$

Note the presence of creation and annihilation operators for both upper and lower sidebands in the result, see Section 4.5.5. Solving the eigenvalue equation under the assumption that $\chi \ll \kappa_a$ leads to the following output state for a particular frequency component

$$
|\phi\rangle_\Omega = |0\rangle + \frac{2\chi\kappa_a}{\kappa_a^2 + (2\pi\Omega)^2}|1\rangle_\Omega|1\rangle_{-\Omega}
\tag{6.3.24}
$$

As expected, most of the time "vacuum in" results in "vacuum out". However, occasionally a pair of photons is produced; one in the upper sideband, one in the lower sideband, as required for energy conservation. Given that the input state is a superposition of all frequencies we may write the total output state as

$$
|\phi\rangle_T = |0\rangle + \int d\Omega \frac{2\chi\kappa_a}{\kappa_a^2 + (2\pi\Omega)^2}|1\rangle_\Omega|1\rangle_{-\Omega}
\tag{6.3.25}
$$

Operating in this configuration the system is often referred to as a *down-converter* or pair generator and this and similar systems are presently the workhorses of single photon experiments, as described in Chapter 12 and 13.

6.4 Summary

In this chapter we have examined the theory of laser oscillators, the conditions under which real lasers behave like the simplified theory, and the characteristics of the laser light under those conditions. We conclude that well built lasers can produce output beams that behave approximately like coherent states. We call these quantum noise limited laser, or QNL, beams.

Finally, we considered the theoretical limits to amplification of quantum limited light and how they could be circumvented. This led us to consider the theory of the non-linear χ_2 interaction. We found a general description of this process and considered the specific examples of parametric amplification, parametric oscillation and pair production. Experiments based on these examples and many variations on this theme will feature strongly in the following chapters.

Bibliography

[All87] Time and frequency (time-domain) characterization, estimation, and prediction of precision clocks and oscillators, D.W. Allan, ICEE Transact. on Ultrasonics, Ferroelectrics and frequency control, Vol. UFFC, No. 6, 647 (1987)

[Cav82] Quantum limits on noise in linear amplifier, C. Caves, Phys. Rev. D 26, 1817 (1982)

[Ein17] Zur Quantentheorie der Strahlung, A. Einstein, Physik. Zeitschrift 18, 121 (1917)

[Fus] details of one of the largest laser can be found at http://www.llnl.gov/nif/milestone/world.record.html

[Hau62] Quantum noise in linear amplifiers, H.A. Haus, J.A. Mullen, Phys. Rev. 128, 2407 (1962)

[Hau84] Quantum theory of injection locking, H.A. Haus, Y. Yamamoto, Phys. Rev. A 29, 1261 (1984)

[Hau00] *Electromagnetic noise and quantum optical measurements*, H.A. Haus, Springer (2000)

[Har97] Intensity-noise dependence of Nd:YAG lasers on their diode-laser pump source, C.C. Harb, T.C. Ralph, E.H. Huntington, D.E. McClelland, J. Opt. Soc. Am. B 14, 2936 (1997)

[Koe96] *Solid state laser engineering*, W. Koechner, 4th ed., Springer Verlag(1996)

[Mar95] Squeezing and inter-mode correlations in laser diodes, F. Marin, A. Bramati, E. Giacobino, T-C. Zhang, J-Ph. Poizat, J-F. Roch, P. Grangier, Phys. Rev. Lett. 75, 4606 (1995)

[Ino92] Quantum Correlations between longitudinal-mode intensities in a multi-mode squeezed semi-conductor laser, S. Inoue, H. Ohzu, S. Machida, Y. Yamamoto, Phys. Rev. A 46, 2757 (1992).

[Ral96] Intensity noise of injection locked lasers: quantum theory using a linearised input/output method, T.C. Ralph, C.C. Harb, H-A. Bachor, Phys. Rev. A 54, 4370 (1996)

[Ral03] Squeezing from Lasers, T.C. Ralph, in *Quantum Squeezing*, Edited by P. Drummond and Z. Ficek, Springer (2003).

[Sie86]   *Lasers*, A.E. Siegman, University Science Books (1986)

[Tei93]   *Fundamentals of Photonics*, B.E.A. Saleh, M.C. Teich, Wiley (1991)

[VW94]   *Lectures on quantum optics*, W. Vogel, D-G. Welsch, Akademie Verlag, Berlin (1994)

[Yar70]   *Quantum Electronics*, A. Yariv, Wiley, (1989)

7 Photodetection techniques

Light is not directly detected in the laboratory: it is nearly always converted to electrical signals and it is these that are detected and analysed. Light is a high frequency electromagnetic field and, as with all such fields, there are two methods of detection: *field detection*, where both amplitude and phase information can be measured and *intensity detection*, which yields no phase information. In the radio, microwave and mid-infrared (mid-IR) regions of the spectrum, field detection can be done directly. In the near-infrared (near-IR), visible, and ultra-violet (UV) regions, field detection is achieved indirectly via homodyne or heterodyne techniques: the light field is interfered with a reference field of known amplitude and phase and the resultant field is detected with intensity detectors, as described in Section 8.1.4. Thus in the optical region of the spectrum, intensity detectors are necessary for both field and intensity detection.

Broadly speaking there are two different ways of recording light intensity, we can detect individual photons or we can measure currents. Both the time response of the photodetector and the magnitude of the photon flux determine which mode of operation is suitable. If the photon flux is low enough and the detector is fast enough we can record and count the individual events. Alternatively, we can integrate the individual events and record a photo-current $i(t)$ which varies in time.

7.1 Photodetector characteristics

Whether used to count individual photons or record photocurrents, there are key physical characteristics of any photodetector.

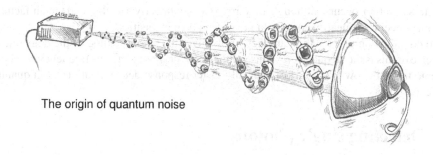

The origin of quantum noise

Figure 7.1: The origin of quantum noise

A Guide to Experiments in Quantum Optics, 2nd Edition. Hans-A. Bachor and Timothy C. Ralph
Copyright © 2004 Wiley-VCH Verlag GmbH & Co. KGaA
ISBN: 3-527-40393-0

Quantum efficiency

The quantum efficiency, η, is the probability of the incoming light being converted to measurable signal, that is,

$$\text{rate of signal pulses} = \eta_{\text{det}} \text{ rate of photons.} \qquad (7.1.1)$$

It is a key parameter in many quantum optics experiments, for example, if $\eta < 1$ then a perfectly regulated stream of photons will be seen as a somewhat randomized stream of signal pulses, where the lower the quantum efficiency, the more random the stream appears. It is thus desirable that photodetectors have high quantum efficiency so that they accurately monitor the optical field. The quantum efficiency depends on the material, the geometry of the detector, and the wavelength of the light – the variation in the latter is often known as the *spectral response* of the detector.

Dark noise

Even in the absence of light, photodetectors typically produce some level of signal. This is the dark noise, which is due to various effects, ranging from thermal effects, to material impurities, to stray fields. In room temperature photodetectors, thermal effects are often the primary source of dark noise, and so are often cooled to reduce it: for sealed detectors, Peltier cooling can be used to achieve temperatures as low as -35°C; for open detectors, where the glass has been removed to reduce optical losses, cooling is limited to the frost point.

Speed and saturation

Photodetectors have a variety of response times: chiefly the excitation time, τ_ε which is the time for incident light to excite a detectable signal, and the dead-time, τ_Δ, which is the time the detector takes to recover before it can be excited again. Through a combination of intrinsic physical processes and engineering, the excitation time is very fast in most detectors and is often taken to be instantaneous for all practical purposes. The dead-time is significantly longer: two, or more, photons arriving at the detector at time intervals less than τ_Δ will generate one electrical pulse only. If the input intensity is such that the count rate is larger than τ_Δ, detector saturation occurs. Photodetectors are still usable over a range of intensities before complete saturation occurs, although they respond nonlinearly: to this end, manufacturers sometimes specific a linearity correction factor for given intensities.

The combination of excitation and dead-time gives the response time of the detector, τ – the inverse of this is the frequency response of the detector. Photodetector frequency responses can range from the low Hz to GHz – typically faster response detectors are used in quantum optics experiments.

7.2   Detecting single photons

The photons in the field carry an energy proportional to the field frequency, ν, specifically, $E = h\nu$, where h is Planck's constant. For visible light, frequencies are on the order of 10^{15} Hz, meaning the energies of photons in this region are on the order of 10^{-19} J – a very small amount of energy indeed. To detect single photons a mechanism is needed to convert this small amount of energy into a macroscopic signal that can be detected in the laboratory. Three mechanisms are used in current photon counters: *photochemical*, where the incoming photon triggers a measurable chemical change; *photoelectric* where the photon is converted to an electron, and this then produces a measurable current or voltage; and *photo thermal*, where the photon is converted to a phonon, and this produces a measurable temperature change, which in turn may influence secondary characteristics such as electrical properties. A fourth mechanism has been proposed but not yet demonstrated as a photon counter: *photo-optical*, where a single photon is converted to a many-photon cascade, which is then detected via one of the above mechanisms [Jam02].

Photochemical detectors

The earliest single photon detectors were photochemical: the energy of a single photon is sufficient to lead to a chemical change in a single grain of photographic film [Tay09]. Film has the advantages of high sensitivity, extremely low noise (spontaneous changes in the grains are very rare) and good spatial resolution. The chief disadvantage of film is that it does not produce an electronic signal output – it must be chemically and mechanically processed, and this does not allow for real-time monitoring of an experiment. Thus it is rarely used in modern quantum optics experiments except as an inexpensive imaging device, e.g. to measure the spatial variation of a photon source [Kwi95].

Photoelectric detectors

In contrast to film, fast electrical signals are provided by a variety of photoelectric detectors. The most ubiquitous is the Photo-Multiplier Tube or PMT. A PMT works via the *photoelectric effect*, where light incident on a metal surface in a vacuum liberates electrons from the surface (as long as the energy of the photon is greater than the binding energy of electrons to the metal – the work function). Typically a single electron is liberated by a single photon, and would be equally difficult to detect. However, in a PMT there is a series of metal plates placed by the photosensitive surface. Each plate is at an increasingly positive voltage with respect to the one before it, such that when a single electron strikes a plate it liberates several electrons. A relatively small cascade of plates thus leads to a large number of electrons, which can be detected as a macroscopic current pulse (this is the origin of the Multiplier part of the name PMT). This mode of operation is sometimes known as Geiger mode, as early Geiger counters worked in a similar fashion. Current commercial PMTs can operate quickly (with linear outputs up to 1.5 million counts per second), suffer moderate dark noise (>10 dark counts per second with moderate Peltier cooling) and have a high spectral bandwidth, detecting light from 185 to 900 [nm]. Unfortunately, they are not very efficient at turning photons into electrons: η ranges from fractions of a percent at high wavelengths (600–900 [nm]), to a few percent at low wavelengths (185–300 [nm]), to tens of percent near the peak wavelengths. Currently the best commercial PMT achieves $\eta \sim 40\%$ at 550 [nm].

A photon counter with higher efficiencies than the PMT is the Avalanche Photo Diode (APD). An APD works via the internal photoelectric effect, where light incident on a semiconductor surface promotes an electron from the valence to the conduction band, as illustrated in Fig. 7.2 Of course, this requires a material with a band gap smaller than the photon energy,

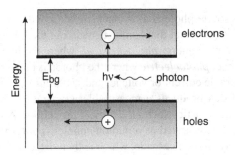

Figure 7.2: Semiconductor and band gap.

$h\nu$ – a difficult task in certain regions of the spectrum. As with a PMT, a single electron is typically liberated by a single photon, so APDs are also run in Geiger mode to amplify the single electrons to macroscopic current pulses. Commercially available APDs are often packaged with an amplifier, discriminator and TTL output driver, which turns the current pulses into pulses with well defined height and rise time, so called Transistor-Transistor Logic or TTL voltage pulses (defined as Logic $1 = 2 - 5$ [V] and Logic $0 = 0 - 0.8$ [V]), which is making them very user-friendly. These detectors are also reasonably fast (with maximum counting rates of ~ 10 million photons per second, linear to ~ 1 million photons per second) and suffer moderate dark noise (>25 dark counts per second). The wavelength range is smaller than PMTs but decidedly more efficient: 5% at 400 [nm], 70% at 650 [nm], 50% at 830 [nm] and 2% at 1060 [nm].

An even more efficient device, but one not yet in widespread commercial distribution, is the Solid State Photomultiplier (SSPM). This is a semiconductor material doped with neutral atoms where the energy levels of the dopant atoms are very close to the conduction band of the semiconductor (e.g. Si doped with As has a 54 meV gap). Briefly, at an operating temperature of 6~7 K the electrons are frozen onto the impurity atoms, as they lack enough thermal energy to be excited into the conduction band. When an incoming photon is absorbed it generates an electron-hole pair: the electron is accelerated to the anode, impact ionizing other impurity As atoms on the way, and so generating an electron cascade within the device. Since little energy is needed to photoexcite an impurity atom so close to the conduction band, SSPMs are sensitive well into the infra-red: from 30 [μm] down to the visible (except for wavelengths in the Silicon absorption band, which unfortunately includes the common telecom wavelengths). The broad IR sensitivity means that SSPMs must be carefully screened so that only light at the desired wavelength is coupled to the detector – else ambient IR will quickly lead to detector saturation. So screened, and correcting for optical losses, the quantum efficiency of an SSPM in the visible has been measured to be 96±3% [Ebe94].

In an SSPM, photoabsorption and gain occur throughout the bulk of the detector: in a variant known as the Visible Light Photon Counter (VLPC) the absorption and gain regions are separated. This reduces the sensitivity to 1–30 [μm] radiation to ~2% whilst maintaining the visible spectral response. At 694 [nm], uncorrected quantum efficiencies of 88±5% have been measured [Tak99], corresponding to corrected efficiencies of 95±5%. Unlike APDs and PMTs, SSPMs and VLPCs can distinguish between one and several photons hitting the detector at the same time: an SSPM has been shown to distinguish two photons with 47% quantum efficiency [Kim99]. Apart from requiring cryogenic operation, the other disadvantages to these systems are their cost and relatively high dark noise, typically 2×10^4 dark counts per second.

Spatial images can be obtained photoelectrically, although typically the response time is orders of magnitude slower than that possible with a single detection element device like an APD. There are several technologies for achieving this, perhaps the most common being the photon counting CCD. A traditional configuration is the intensified CCD (ICCD), where an image intensifier (e.g. a photoelectric photocathode, followed by an array of multichannel plates for electronic amplification, followed by a phosphor screen) is imaged, via free-space or fibre-bundle, to a low-noise CCD camera. More recent devices dispense with the image intensifier, utilizing high voltage to produce an on-chip electron cascade, and thus gain. Current CCDs have a quantum efficiency in excess of 92% (with resolutions up to 1024x1024 pixels and an associated readout rate of 30 frames per second), and will doubtlessly improve. Figure (7.3) shows the output of a spontaneous down conversion photon source (see next section) imaged with an image intensified photon counting CCD.

The photoabsorbing surface in most photoelectric devices is a metal or semi-conductor: high refractive-index materials that intrinsically reflect a significant proportion of incident light (e.g. see Table 7.1). The quantum efficiency of photoelectric detectors can be substantially improved via the application of anti-reflection (AR) coatings that allow more light to be coupled into the bulk of the material, where the intrinsic quantum efficiency may be quite high, e.g. $\eta = 99\%$ for Si [Zal80]. In commercial single element devices, the AR coatings are relatively simple (e.g. a single layer of TiO_2) and the improvement in quantum efficiency is modest. In commercial CCDs the situation is somewhat better: mid-band visible AR coatings are available that improve the efficiency from $\sim$61% to $\sim$92%. It is worth remembering that for a specific wavelength application a custom coating can lead to a significant improvement, up to the intrinsic efficiency of the device.

Photo thermal detectors

Bolometers, which absorb light and convert it to heat, have long been used for accurate measurement of optical intensity. Traditionally, however, they have been regarded as effectively DC devices, with response speed limited by the bulk thermal response of the absorbing material. Perhaps surprisingly, it is possible to produce extremely fast bolometric photon counters – indeed faster than possible via photoelectric effects. In a *superconducting bolometer*, the photoabsorbing material is operated near its superconducting transition temperature. The absorption of a single photon then pushes the material into the normal conduction regime, and the resulting large change in impedance can be measured [Cab98]. Bolometers can also accurately measure the energy deposited, albeit at the expense of response time, meaning that bolometric photon counters can identify the number of photons incident within a given detection period [Mil03]. Although still at the experimental stage, the early results are encouraging, two examples being: a NbN bolometer operating at 4.2 [K] with a quantum efficiency of $\sim$70% at 400 [nm] and a GHz counting rate [Ver02]; and a photon number resolving tungsten bolometer operating at 100 [mK] with a quantum efficiency of 18% and a time constant of 10 [μs] [Mil03].

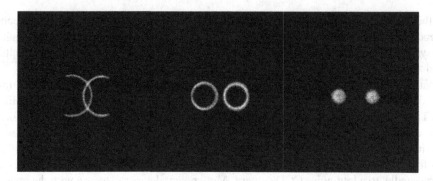

Figure 7.3: Images acquired with a photon-counting CCD camera: output of a Type II sponta-neous downconverter. [Pry02].

7.3   Photon sources and analysis

Photons are detected as discrete events, and so, unlike with photocurrents, experimental anal-ysis is about registering the place and time of arrival of these events. In arguably the first quantum optics experiment, G.I. Taylor attenuated a thermal light source to the single pho-ton level, passed the light through a double slit, and recorded the photon arrival positions on a piece of film, as discussed in Chapter 3. The important information in this case is the recorded intensity at a given point, the effect occurs regardless of the statistics of the light source. The photons can arrive with any temporal distribution: thermal (e.g. a natural light source); Poissonian (a coherent light source, e.g. a laser); or regularly spaced in time (a Fock state, e.g. a single photon source). Indeed, many quantum optics phenomena based on self-interference are insensitive to photon arrival statistics, with the general caveat that the light is sufficiently attenuated so that the chance of a multiple photon count in any given interval is very low. Examples include quantum cryptography, and "interaction-free" measurements, which are discussed in Chapters 12 and 13.

Analysis in these experiments is relatively straightforward, since the important parameter is the intensity, or rate of photons at a given output port, measured over some integration time. If film is used, the integration time is set by the exposure; for photodetectors, the integration time is set by the count time. In the latter case, if the photodetector output is digitized, e.g. TTL pulses, it can be counted with a commercial pulse counter (e.g. a Perkin Elmer Ortec 974); otherwise it can be digitized using a commercial discriminator, fast oscilloscope, or similar device, and then counted. In some protocols, such as quantum cryptography, it is important that the photons arrive in a well-defined time period. Even in these cases, any source, thermal, coherent or Fock state source, can be used, as long as some combination of the source, the optical path, or the photodetector detection time are appropriately modulated. Such pulses will be regular, with the source statistics dominant within the pulse length. Photon time-of-arrival information is a useful diagnostic in these experiments, and may be obtained by using a Time-to-Amplitude Converter (TAC, e.g. a Perkin Elmer Ortec 567) to measure the difference between the expected time, as provided by the modulator, and the actual time, as provided by the photodetector. The modulator provides a periodic start signal, which is ended by a photodetection event if one occurs – the time difference between the start and stop pulses then sets the amplitude of a regular pulse. The amplitude distribution of these pulses can be measured with a Multi-Channel Analyser (MCA), providing information on the distribution of photon arrivals.

Another class of quantum optics experiments, including non-classical interference, teleportation and quantum computation as they are discussed in Chapter 13 on single photon experiments and quantum information, are multiple-mode interference experiments. In these experiments it is critical that the various modes are indistinguishable – that is that they have identical distributions in energy (photon number *and* frequency); momentum (spatial modes); and time. In general, the modes must also have states of definite photon number, i.e. be Fock states. In the case of pulsed modes, this requirement is best satisfied if the pulses are identical pulse to pulse. In practical terms, it is not easy to build a source that meets all these requirements. Approaches tried to date have been numerous and include single atoms [Kuh02], colour centres [Kur00, Bro00, Bev02], molecules [DeM96, Lou00], magneto-optic traps [Kuz03], quantum wells, and quantum dots [Gér99, Mic00, Zwi01, Mor01, Yua02, San02]. Many of these approaches are of interest because they hold the promise of a *deterministic* photon source, or photon gun, where photon arrivals occur at pre-determined times, and examples are discussed in Chapter 13.

However, the most common technology used in photonic quantum optics experiments to date has been non-deterministic – spontaneous parametric down-conversion (SPDC). SPDC sources are bright (high photon count rates), can produce good optical modes with well defined frequency and spatial characteristics [Tak01, Kur01], and can produce tunably entangled modes [Kwi99a, Whi99]. In SPDC, high frequency photons are passed through a nonlinear crystal, where with low probability ($\sim 10^{-8} - 10^{-10}$) they are converted down into two lower frequency daughter photons. The daughter photons are born within 10's of femtoseconds of one another, and so are tightly correlated in time; their energy and spatial distributions are a common function of the pump mode and crystal mode-matching characteristics and are also highly correlated – in fact, they are entangled [Rar90, Bre91].

The tight time correlation means that if a daughter photon is detected in one mode, then any photon detected within a short time interval in the other arm is likely to be its twin – given the source brightness, it is unlikely that there are unrelated photons in that time. Thus the information gathered by one mode can be used to condition the other, so that it is a good approximation to a true single photon source – albeit one with non-deterministic arrival times, determined by the pump laser statistics. This conditioning can be a fast-forward circuit that unblocks the optical path on detection of a photon in the twin mode, or, more commonly a coincidence-circuit that post-selects events from one mode based on results from the other. Indeed, if two single photon sources are required, it is not necessary to set up two SPDC sources – instead both modes of one source can be used, where only events detected in coincidence are post-selected. Such photon coincidences can be registered with commercial photon coincidence counters (e.g. Stanford Research Instruments Model SR400), which are limited to a 5 ns coincidence window, or with fast counting instrumentation used in particle physics (e.g. the Perkin Elmer Ortec 567 and 974 described above), where the coincidence window can be sub-nanosecond. The advantage of a short coincidence window is that it allows screening of background counts due to extraneous light sources, particularly absorption-induced fluorescence which can be a problem at short wavelengths. Currently the lower limit to coincidence window length is the combined jitter of the photodetector digitization circuit and the coincidence circuit – which is in the order of [ns]. Coincidence windows lower than this can result in lost coincidence counts.

7.4   Detecting photocurrents

The technology used with CW laser beams is quite different from that used with single photons, as described in the previous Section 7.2. Here we analyse the properties of the photocurrent $i(t)$. The results we want for a laser beam are a value for the average optical power, which is obtained with a power meter, and a spectrum of the variance of the optical power, such as shown in Fig. 7.9, which is obtained with a electronic spectrum analyser.

Photocurrent

For CW detection the electronic signal from the photodetector is integrated over many detection events for a time interval Δt and this generates a current $i(t)$ which is proportional to the power $P(t)$, measured in [W], of the light beam on the detector or, in other words, the rate at which photons reach the detector

$$i(t) = \frac{n_e(t)e}{\Delta t} = \frac{P(t)e}{h\nu\eta_{\text{det}}} \tag{7.4.1}$$

where e is the charge of a single electron and η_{det} is the efficiency of the conversion process. We will be interested in the average value of the current $\bar{i} = \langle i(t) \rangle$. This value is proportional to the average optical power P_{opt}, thus a device which measures $\bar{i}$ is generally referred to as a power-meter. Each such apparatus will have to be calibrated and care should be taken that the response is linear, since the device will saturate if the optical power is too high.

Beat measurements

The photodetector provides not only an average current $\bar{i}$, the DC current, but also a whole spectrum of time varying signals, the AC current. Any AC component at detection frequency Ω can be traced back to a beat between two components of the optical spectrum with a separation Ω. These beat signals are very useful to identify the mode structure of the light source. Multi-mode laser operation will lead to strong beat signals, which can be detected and processed with high accuracy and sensitivity using conventional electronic radio frequency equipment. The measurement of beat signals between two lasers for example will establish the frequency difference and the convolution of the linewidth of both lasers. Such observations are extremely useful for the establishment of optical frequency standards and also the measurement of laser linewidth, as discussed in Section 8.1.5. In addition, the AC signal also contains information about the noise and the signal carried by the laser beam, since the fluctuations and modulation of the light intensity corresponds directly to the size of the amplitude sidebands. The presence of these sidebands lead to beat signals, which are similar but much weaker and noisier than the beat signals from two lasers.

Intensity noise and the shot noise level

For the quantitative measurement of the fluctuations we are interested in the variance $\Delta i(t)^2$ of the photocurrent. This is linked to the variance of the number of single photon detection events via

$$\Delta i(t)^2 = \Delta \left(\frac{n_e(t)e}{\Delta t} \right)^2 = \frac{\Delta(n_e(t))^2}{\Delta t^2} \, e^2 \qquad (7.4.2)$$

If we consider the generation of all the photo-electrons as independent processes, we can apply the statistics for independent particles. The counting statistics for the electrons should be a Poissonian distribution with variance $\Delta(n_e(t))^2 = n_e = \bar{i}\Delta t/e$. For large numbers of electrons in each counting interval this should be a Gaussian distribution resulting in

$$\Delta i(t)^2 = \frac{\bar{i}\ e}{\Delta t}$$

We will be investigating the spectrum of the current fluctuations. In this case the detection interval Δt is related to the bandwidth B of the apparatus via $B = 1/2\Delta t$ which leads to

$$\Delta i(t)^2 = 2\,\bar{i}\,e\,B \qquad (7.4.3)$$

assuming that $\bar{i}$ does not vary during the measurement. This is the well known *shot noise* formula in electronics that also applies to photodetection. It should be noted that Equation (7.4.3) can be derived for any stream of electrons that are generated independently. However, this is a rather special case for a current and generally a current can be noisier or quieter. There are many practical situations, such as the use of electronic current controllers, which can generate currents with less noise than the shot noise limit.

It is important to note that this *shot noise is not a property of the detector*. The noise reflects the fluctuations of the light intensity and we can use detectors to probe the statistics of the light fluctuations. This will be correct as long as the quantum efficiency of the detector is high. For $\eta_{det} = 1$ the statistics of the photons and electrons will be identical, the current fluctuations are directly linked to those of the light and we can investigate the statistical properties of the light by studying the current fluctuations. It should be noted that after the detection of the light any quantum properties of the fluctuations no longer exist. Current noise can be treated classically, as already discussed in Section 5.1. For any practical purpose there is no quantum limit for the noise of a current. However, at the same time the *current fluctuations are a reliable measure of the quantum properties of the light*. This distinction is important in the context of measurement theory, the generation of nonclassical light, Chapter 9, and the applications of electro-optic feedback described in Section 8.3.

How closely the current noise reflects the light fluctuations depends on the efficiency η_{det}. A quantum efficiency much less than unity corresponds to a random selection of a fraction η_{det} of the light. This is fully equivalent to having a perfect detector behind a beamsplitter which selects a small fraction of the photons. This results in a variance of the current that is closer to the shot noise limit than to the variance of the light intensity. To describe this quantitatively, it is useful to consider normalized variances for individual Fourier components of the current

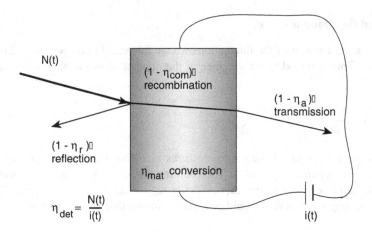

Figure 7.4: The different loss mechanisms present in a photodetector

fluctuations

$$V_i(\Omega) = \mathcal{F}\left(\frac{\Delta i(t)^2}{2\,\bar{i}\,e\,B}\right) \tag{7.4.4}$$

The corresponding statistics of the light is described by the normalized spectrum $V1_{out}(\Omega)$ of the photon flux, as was shown in Chapter 4. The effect of the detection efficiency η_{det} is given by

$$V_i(\Omega) = 1 + \eta_{det}(V1_{out}(\Omega) - 1)$$

This is identical to treating the inefficient detector as a perfect detector that sees only the light reflected by a beamsplitter with reflectivity $\epsilon = \eta_{det}$ as calculated in Section 5.1. The measured spectrum $V_i(\Omega)$ varies with the fraction of the light detected. For a very inefficient detector the normalized variance of the current approaches the shot noise limit $V_i(\Omega) = 1$.

Quantum efficiency

The detection efficiency η_{det} contains all processes that reduce the conversion from light into a current. In practice, there are a number of such processes, as shown in Fig. 7.4: Only a fraction η_r of the light reaches the photodetector material, the rest is reflected or scattered at the surface of the detector. The detector absorbs a fraction η_a of the light, the rest is scattered or transmitted through the detector. The absorption scales exponentially with the thickness of the material and depends on the doping of the material. The material has a quantum efficiency η_{mat} which depends strongly on the wavelength of the light. This efficiency can be very close to unity as long as the wavelength of the light is close to but shorter than the cutoff wavelength set by the band gap energy of the material. Finally, there is a chance for the electron – hole pair to combine before it can leave the detector. This recombination happens more likely at the surface of the material then inside. At shorter wavelengths the light tends to be absorbed more strongly, it does not penetrate as deeply. Thus more detection processes occur close to a surface and the losses due to recombination are larger. Only the fraction η_{com} will actually contribute to the photocurrent. All these losses are independent and can be combined to the total detector efficiency η_{det}

$$\eta_{det} = \eta_{mat} \times \eta_a \times \eta_r \times \eta_{com}$$

The efficiency η_{det} is generally independent of the intensity and the detection frequency Ω. However, it has been noted that η_{com} will get worse at higher intensities and that some photodetectors get slower at higher intensities, when the detector material is either getting hot or there are insufficient fast charge carriers to drive the AC current. The efficiency η_{det} has to be determined for the parameters of the light used in the specific experiment, in particular its wavelength, total power and focusing geometry. Techniques for increasing η_{det} in order to approach the ideal case of $\eta_{det} = 1$ are very important in all experiments with non-classical light.

Photodetector materials

Photo-detection relies on the interaction of light with materials which results in the absorption of photons and the generation of electron-hole pairs. These pairs change the conductance of the material, they contribute to the photocurrent. The material for the photodetector has to be selected for a given frequency of the light. The energy of the photons $h\nu$ has to be matched to the band gap separation E_{bg} of the material. Semi-conducting solids, such as germanium (Ge), silicon (Si) or indium-gallium-arsenide (InGaAs) are well suited for this application. These materials work in the visible and near infrared. The properties of these materials are listed in Table 7.1 and the wavelength dependence of the efficiency is shown in Fig. 7.5. Note that the properties of primary semiconductors (Si, Ge) are fixed but that the properties of tertiary materials (such as In GaAs) depend on the relative composition, that means the concentration of the individual elements.

The material quantum efficiency η_{mat} of a very pure and selected material can be very high, up to $\eta_{mat} = 0.98$. This is the limit of the efficiency of the detector and can be achieved in practice provided the sample of material used is thick enough to absorb all the light and reflection losses are avoided. Since most photodetector materials have very high refractive indices their reflectivity is also very high; for example, for Si with a refractive index of 3.5 at the peak detection wavelength of 800 [nm], the reflectivity under normal incidence is 31% of the intensity. To this the reflection loss at the cover glass (typically 8% from both surfaces of the glass) has to be added. In order to increase the quantum efficiency it is common to remove the cover glass and to either tilt the detector closer to Brewster's angle, which requires large detectors, or to use an anti-reflection coating on the detector. In addition, the light reflected at the surface can be retro-reflected on to the same detector. Using these techniques typical quantum efficiencies are 65% at 553 [nm] (EG&G FND 100 SI photo-diode) and 95% at 1064 [nm] (Epitax 500 InGaAs photo-diode) . Alternatively, several detectors can be cascaded, each subsequent detector receiving the light reflected by the previous detector and all the individual photocurrents are combined. Here electronic delays of the different currents have to be carefully avoided in order to keep the signals synchronized with an accuracy better than the inverse of the detection frequency. Using cascaded detectors efficiencies close to

100% have been reported for detection frequencies of up to 100 kHz using Si detectors [Gar94] and up to 10 MHz using InGaAs detectors.

Table 7.1: Properties of photo-diode materials

	Si	Ge	InGaAs
Band gap E_{bg}	1.11 eV	0.66 eV	≈ 1.0 eV
Cut off wavelength	1150 [nm]	1880 [nm]	≈ 1300 [nm]
Peak detection wavelength λ_{peak}	800 [nm]	1600 [nm]	≈ 1000 [nm]
Material quantum efficiency η_{mat} at peak wavelength	0.99	0.88	0.98
λ_{min} and λ_{max} for $\eta_{mat} = 0.5$	600–900 [nm]	900–1600 [nm]	≈ 900–1200 [nm]
Refractive index n at λ_{peak}	3.5	4.0	3.7
Reflectivity at normal incidence	31 %	36 %	33 %
Brewster's angle	74°	75°	76°

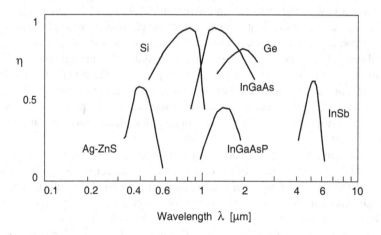

Figure 7.5: Quantum efficiencies of various photo-diode materials.

7.4.1   The detector circuit

Photodiodes

The photo-detection process leads to a change in the conductivity of the semiconductor material. A simple photo-resistor could be used to produce a photocurrent. However, the most common configuration is a PIN photo-diode, which is a composite structure made out of P and N semiconductors with an intermediate layer. In this device an external voltage, the bias voltage, U_{bias}, creates an electrical field in the region between the P and the N layer. Here electron-hole pairs can be separated and they contribute to the photocurrent.

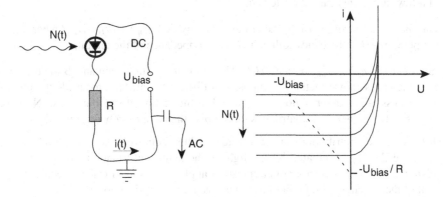

Figure 7.6: Circuit of a photo-diode with reverse bias voltage U_{bias} and the typical diagram that shows the photocurrent $i(t)$ as a function of the bias voltage for various values of the photon flux $N(t)$.

There are three modes of operation for a photo-diode. The diode can be operated with an open circuit. This will produce a voltage proportional to the photon flux, no current will flow. Alternatively, a low resistance is switched across the diode and the circuit is effectively a short circuit. In this configuration a DC current is generated and the photo-diode operates like a solar cell. Both these modes have slow responses. It is advantageous to apply a voltage, the reverse bias voltage U_{bias}, to the detector which accelerates the charge carriers. In this mode the photodetector is used like a diode to generate a voltage signal across an external resistor. The diode current is proportional to the photon flux $N(t)$. For a sufficiently large bias voltage the photocurrent is proportional to $N(t)$. A schematic circuit and a typical response curve is shown in Fig. 7.6. It is common to use the highest possible bias voltage ($U_{bias} = 5 - 80$ [V] depending on the diode) since this will increase the drift velocity of the charge carriers and thereby the speed of the detector.

The photocurrent contains both a DC and an AC component which can be directed to different receivers using simple capacitors or more complicated electronic networks. The AC components contain all the information about modulation and the fluctuations.

Amplifiers and electronic noise

The typical AC noise component is of the order of 1 part in 10^7 of the DC current. It requires amplification. This can be achieved by various types of amplifiers, as described in [Gra98]. The most obvious solution is to terminate the photo-diode into a 50Ω resistor and to use a commercial RF amplifier. Such an amplifier can be obtained with wide bandwidth (1–200 MHz), large gain (10 to 60 [dB]) and low noise performance. Please note that the gain is normally stated as the power amplification in dB, where [dB] is $10 \log P_{out}/P_{in}$, see Eq. (7.5.5). The amplifier will increase both signal and noise at the input. Figure 7.7 shows the combination of noise and signal that is generated. The different noise contributions are:

- *Dark noise* of the detector $\Delta i(t)^2_{\text{dark}}$. Note that this is not the shot noise, it is only the noise due to false detection events. For most of our measurements with optical powers of a few [mW], this can be neglected.

- *Thermal, or Johnson, noise* of the circuit given by $\Delta i(t)^2_{th} \approx 4k_B T B/R$ where T is the temperature, B is the bandwidth and R is the impedance of the circuit.

- *Amplifier noise* $\Delta i(t)^2_{\text{amp}} \approx 2e\, G\, B\, F$. Here G is the voltage gain and B the bandwidth. F is the excess noise factor, a composite of different contributions inside the amplifier. The excess noise factor is normally quoted in the technical specifications. Note that a large bandwidth detector contributes more noise than a narrow band amplifier.

All these noise contributions are independent, thus their variances can be added. Note that the noise $\Delta i(t)^2_{\text{optical}}$ created by the light is the actual information we wish to detect. It corresponds to the signal in a communication application and in order to achieve a clear detection of the signal the total noise has to be less than the optical noise

$$\Delta i(t)^2_{\text{el}} = \Delta i(t)^2_{\text{dark}} + \Delta i(t)^2_{\text{amp}} + \Delta i(t)^2_{th} \ll \Delta i(t)^2_{\text{optical}} \qquad (7.4.5)$$

One of the largest contributions is normally the thermal noise. Consider the following example: for a DC current of 1 [mA], equivalent to 1 [mW] of power at $\lambda = 1000$ [[nm]], the thermal noise at room temperature for an impedance of 50 [Ohm] and a detection bandwidth of 100 [kHz] is $\Delta i(t)^2_{\text{thermal}}/\bar{i}^2 = 4k_B\, T\, B/\, R\, \bar{i}^2 = 4 * 1.38\ 10^{-23} * 300 * 10^5/(50 * 10^{-6}) = 3\ 10^{-11}$. That means a relative current noise $\Delta i(t)^2_{\text{thermal}}/\bar{i}^2$ of about $5\ 10^{-6}$. The relative noise which corresponds to the shot noise for this bandwidth and average power is $\Delta i(t)^2/\bar{i}^2 = 2\, e\, B/\bar{i} = 2 * 1.6\ 10^{-19} * 10^5/10^{-3} = 3\ 10^{-11}$.

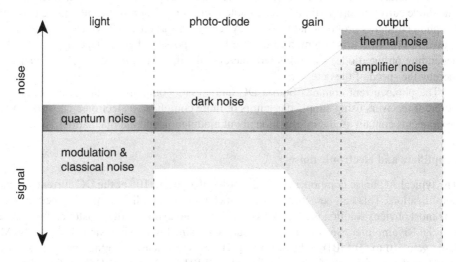

Figure 7.7: Signal and noise contributions in a photodetector. Noise terms (up) and signals (down) propagate differently through the stages of the detectors circuit.

That means, the thermal noise is as large as the shot noise at this optical power and such a detector could only show quantum noise at much higher powers. As a consequence many custom made photodetectors contain a so called transimpedance amplifier, which is a circuit that converts the AC photocurrent into an AC voltage, is fast and still has a high impedance. In such a circuit the thermal noise will be much lower, about 10^{-8} at 1 [mA] DC current and we can clearly resolve quantum noise. This is the reason why photodetectors for quantum optics experiments are frequently custom made. For illustration, an example of a complete circuit diagram of such an amplifier is shown in Fig. 7.8. It shows the photo-diode, the transimpedance amplifier (IC1), the driver amplifier (IC2) and all the filter components required for a practical low noise circuit.

Saturation

Frequently the light has strong modulations at specific detection frequencies. This could be due to laser noise, such as beats with other laser modes, or due to the sidebands introduced for the purpose of locking the laser to resonators, as discussed in sec.(8.4). We expect the photodetector to handle both the very small fluctuations and the modulation without distortions and consequently we require a very large dynamic range, for some applications as large as 60 [dB]. If the dynamic range of the amplifiers is not chosen large enough the circuit will be saturated. This frequently leads to a decrease of all AC components, even those at other than the modulation frequency that originally caused the saturation.

This saturation can be confusing and can mimic false results. It is particularly worrisome if several amplifiers are used, a first amplification stage with wide bandwidth and a later stage with narrow bandwidth. In this situation the first stage can saturate from a modulation at frequencies outside the bandwidth of the final stage. Thus the modulation is not visible at the output and the noise level decreases when the signal is turned on – mimicking a noise suppression. In more than one case this malfunctioning of the detector has fooled researchers and was taken as the demonstration of some quantum optical effect. As a precaution one should always record spectra for the entire bandwidth of the detector and each amplifier, and any low pass filter should be placed directly after the photo-diode. In addition the scaling of the noise traces with the input optical power should be checked, as outlined in Section 8.1.1.

7.5 Spectral analysis of photocurrents

An electronic spectrum analyser (ESA) is used in most experiments to record the Fourier spectrum of the photocurrent. This is much faster and more reliable than to record the photocurrent directly and to perform a numerical fast Fourier transform (FFT). A typical spectrum displayed by such an instrument is shown in Fig. 7.9. The principle function of an ESA is to determine and display the power spectrum of the electrical input signal. We consider in detail the transfer function from the spectrum of the input signal to the display. It is important that we understand the many options and settings of the analyser in order to obtain reliable data that can be interpreted as the properties of the light beam. A state of the art analyser is a very powerful instrument with many options, menus and knobs. An ESA contains a combination of analog and digital electronics which produce a string of data which will be recorded alongside

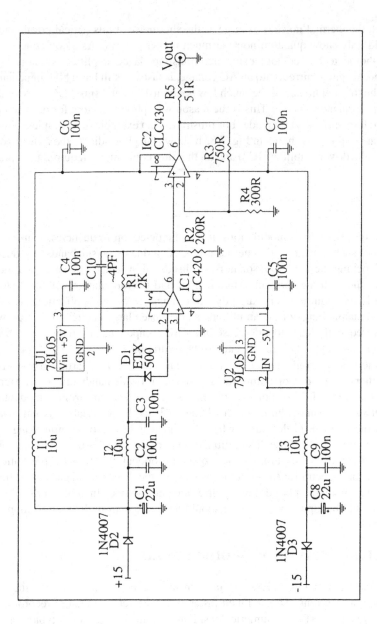

Figure 7.8: Example of an actual circuit diagram of a photodetector. For other examples see
M. Gray et al. [Gra98]

its actual settings. These machines are designed for the recording of signals, such as radio
signals. In this section we will discuss the technical details of the ESA, such as the various
bandwidth settings, sensitivity scales, sweep speeds etc. and the tricks for using the analyser

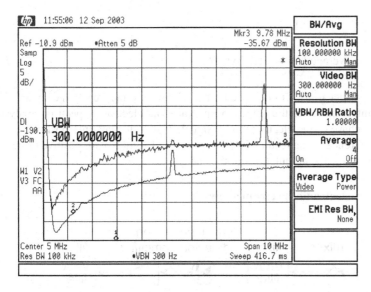

Figure 7.9: A typical intensity spectrum as displayed by an electronic spectrum analyser. Here the settings of the instrument are shown together with the data

for the recording of noise and signals. Finally, we will discuss alternatives such as electronic mixers, which are useful for more advanced experiments such as tomography.

From optical sidebands to the current spectrum

We are interested in the sidebands of the optical field which contain all the information about the fluctuations and the statistics as well as all the information that is transmitted with the light in the form of modulation. Typical sideband frequencies range from [kHz] to [GHz]. These frequency intervals are too small for ordinary spectroscopic instruments, such as grating spectrometers or optical spectrum analysers (OSA) and are just about accessible with optical cavities. However, the most reliable and best separation of the sidebands is by the Fourier analysis of the photocurrents.

The information is transferred from the optical domain($\nu \approx 10^{15}$ [Hz]) into the domain of electronic instruments, that means detection frequencies $\Omega < 10$ [GHz]. For this purpose several light beams are combined on a photodetector. The resulting beat signals contain information about the quadrature of the light. These techniques, homodyne and heterodyne detection, are described in Section 8.1.4. In all cases the information is contained in the temporal development of the current $i(t)$ generated by the photodetector. The purpose of an electronic spectrum analyser is to measure and display the spectral composition $p_i(\Omega)$ of the electrical power of an electrical signal $i(t)$. Mathematically this corresponds to a determination of the power spectrum, or the square of the Fourier transform, of the input voltage. The instrument measures the power $p_i(\Omega)$ of the oscillating field in a spectral interval around the *detection frequency* Ω. The spectral response of the analyser is given by the function res(Ω) which has a width given by the *resolution bandwidth* (*RBW*). For a fixed input resistance R, typically

50 [Ohm], the measured power is

$$p_i(\Omega) = \text{time average} \left(\int \text{res}(\Omega) \, i(\Omega) \, R \, d\Omega \right)^2 \tag{7.5.1}$$

where $i(\Omega)$ is the Fourier transform of the current $i(t)$. We are measuring the square of the AC photocurrent integrated over the detection interval. The electronic spectrum analyser is not sensitive to the phase of the electrical input signal. It will display only positive frequencies, there is no difference in the result of a measurement of signals with positive or negative frequencies, of sine or cosine modulations of the current.

The operation of an electronic spectrum analyser

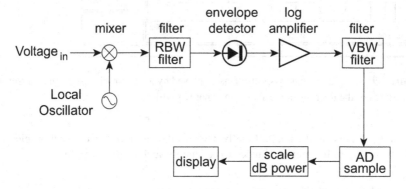

Figure 7.10: Block diagram of a typical spectrum analyser.

Technically, the electronic spectrum analyser consists of a mixer which selects a certain frequency, an envelope detector, in most cases a logarithmic detector and a circuitry for averaging the signal, as shown in Fig. 7.10. In most modern analysers the signal is digitized and displayed with a wide choice of scales and other options. In the following we analyse the transfer function between the electronic input signal and the display. The ESA is built around the concept of electronic heterodyne detection. A reference signal with fixed amplitude and variable frequency, the local oscillator, is generated internally. The frequency Ω_{det} of the local oscillator determines the frequency under detection. The input signal is amplified and mixed with the local oscillator. The output from the mixer is filtered by a narrow band pass filter with bandwidth RBW. The result of the measurement, at a given Ω, can still fluctuate in time and we can describe the output of the ESA by a distribution function with the mean value μ_{disp} and it is this value from which we infer the spectral density $p_i(\Omega)$ of the photocurrent. In our optical experiments we make relative measurements, that means we are not interested in the absolute values of μ_{disp}, but in relative values, such as the ratio of the μ_{disp} to a calibrated value which represents quantum noise.

The output signal is integrated using a low pass filter with a cut-off frequency known as the *video bandwidth VBW*. The video bandwidth is then inverse of the time constant Δt_{ave} with which the measured signals is averaged, $\Delta t_{\text{ave}} = 1/VBW$. In the limit of small VBW, large integration times, the output is one single, steady value $\mu_{\mathrm{disp}}(\Omega)$ for each detection frequency. In many practical situations we want to see the spectrum rapidly and a lower value for the *VBW* is chosen, this does not vary $\mu_{\mathrm{disp}}(\Omega)$ but produces a more noisy trace.

Modern radio frequency (RF) spectrum analysers cover a wide range of input frequencies, from about 1 [kHz] to 10 [GHz]. They can achieve this wide span through a sophisticated combination of several internal mixing processes. Nevertheless, the idea outlined above of a narrow band scanning filter followed by an integrating power detector is appropriate. The instrument covers a total frequency interval, or *span*, *SPA*, and scans it with a set scan speed, measured in [Hz/s]. The instrument requires at least a time of $1/RBW$ for each measurement, after that it moves on by a frequency interval *RBW*. Consequently the maximum scan speed is RBW^2. For typical values of *RBW* of 10^4 [Hz] the fastest possible speed is 10^8 [Hz/s] = 100 [MHz/s]. Faster scans would lose detailed information. The highest possible *VBW* is equal to *RBW*, but usually the *VBW* is chosen lower. In particular for noise measurements it is necessary to select a low *VBW* in order to obtain a well defined result. Frequently, measurements with fixed detection frequency Ω_{det} are carried out. In this case the scan represents a time history of the power of the single Fourier components.

Detection of signal and noise

The response of the spectrum analyser to a single frequency harmonic input voltage signal is straight forward, if we can assume that the *RBW* is larger than the spectral width of the signal. In this case the function displayed will have the width of the *RBW* and the maximum value of μ_{disp} will be proportional to the square of the amplitude S_{sig}^2 of the single harmonic input. This peak value displayed is the only information of interest and is independent of *RBW*. However, the response of the ESA to a signal with a spectral distribution that is wider than *RBW* is different. In this case $\mu_{\mathrm{disp}}(\Omega)$ corresponds to the convolution stated in Eq. (7.5.1) for a fixed value of $\Omega = \Omega_{\mathrm{det}}$. Consequently, details in the spectrum of the input signal narrower than *RBW* will not be detected. The exact shape of the function $res(\Omega)$ depends on the specific analyser used, a Gaussian shape is generally assumed, but you should check the manual of your analyser.

A second, and very common, application is the detection of noise with a bandwidth much larger than *RBW*. Both harmonic signal and broadband noise will be displayed, however the way the two influence μ_{disp} is quite different. We need to consider the details of the processing occurring inside the ESA to predict the distribution function of the displayed values and to interpret μ_{disp}. The input signal to the spectrum analyser is bipolar, that means positive and negative voltages are present. A single frequency harmonic signal will oscillate between the two extreme values $\pm S_{\mathrm{sig}}$. These maxima have no uncertainty during the time of measurement, that means the signal distribution function has effectively a negligible width, $\sigma_{\mathrm{sig}} = 0$. In contrast, the noise has a Gaussian distribution, centered at zero, $\mu_{\mathrm{noise}} = 0$, with the width of the distribution given by σ_{noise}.

The analyser produces the power spectrum, that means it is taking the square of the input signal and it also performs an averaging process. Consequently it shows only positive values and the distribution at the output is asymmetric, similar to a Rayleigh function, as shown in

Fig. 7.11. The mean value μ_{disp} of this distribution is close to but not exactly at the maximum value of the distribution and it has a width σ_{disp} measured from full maximum to half maximum.

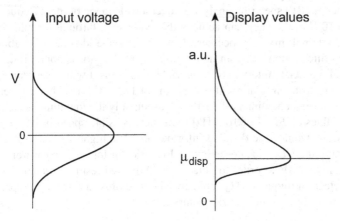

Figure 7.11: Input and display distribution functions for a spectrum analyser. The width σ_{noise} of the distribution at the input determines the peak value μ_{disp} of the distribution on the display.

We measure μ_{disp} and expect it to represent both the variance of the noise, given by σ^2_{noise}, and the power of the input signal given by S^2_{sig}. The actual details are a little more complicated, since we have to evaluate the full distribution function. The full mathematical details of calculating the Rayleigh distribution are given by D.A. Hill and D.P. Haworth [Hil90] and for signals which are substantially larger than the noise the result is given by

$$\mu_{\text{disp}} = \sigma^2_{\text{noise}} + C_S \, S^2_{\text{sig}} \qquad (7.5.2)$$

This is frequently referred to *as adding the signal and the noise in quadrature*, in the same way as the variance of two noisy signals with Gaussian distribution functions can simply be added. This is a remarkably simple result since the actual mathematical background is quite complex. In conclusion, we can add the square of the signal amplitude and the square of the RMS of the noise. The scaling factor C_S is a consequence of the actual sampling process inside the ESA and this is only important if we want to measure absolute signal and noise levels or determine absolute signal to noise ratios (SNR).

With increasing signal the output value μ_{disp} rises gradually, since the ESA always measures *(signal + noise)* and not *signal* alone. Very small modulations cannot be detected, they disappear in the noise and Hill and Haworth showed that Eq.7.5.2 does not necessarily hold for all ESAs in the limit of a small signal with SNR$\simeq$1. The most interesting application in quantum optics will be exactly in this regime of small signals and thus we have to be aware of these properties of the ESA. However, once the signal is dominant the display is proportional to the signal strength alone.

Please note that most instruments have been calibrated accurately for signal rather than noise measurements and an absolute comparison between signal and noise, or determination of C_S, can be difficult. This is well known to the designers of the ESA and the details are given in their manuals, for example [HPmanual]. These details are further elaborated in reference [Hil90], but it is sufficient in most cases to work with a correction factor C_S which corresponds to about 2.5 [dB].

It is not possible to distinguish between a pure noise source and a mixture of signal and noise by just looking at the trace of the ESA at one frequency. In this case we could not say if we have a laser beam with quantum noise alone, a laser beam with quantum and classical noise or laser beam with noise and modulation. We need additional information. For example a scan of frequencies would reveal the peak that the signal is narrowband and the noise is broadband. Or a measurement of the orthogonal quadrature, using a homodyne detector, would show the characteristics of a signal and of noise, or squeezing and anti-squeezing. From these traces we can isolate *signal* and *noise*. We can determine the SNR as

$$SNR = \frac{\mu_{\text{disp}}(\text{signal} + \text{noise}) - \mu_{\text{disp}}(\text{noise})}{\mu_{\text{disp}}(\text{noise})} \tag{7.5.3}$$

What would happen if we had more than one noise source and more than one signal present? As long as the noise terms have Gaussian distributions with widths σ_{n1} and σ_{n2} we can combine them into one value $\sigma_{\text{noise}}^2 = \sigma_{n1}^2 + \sigma_{n1}^2$. In a similar way we can combine two signals with S_1 and S_2, taking into account that they have either a fixed or a random phase relationship, as detailed below. This results in one combined signal intensity S_{signal}^2. Consequently, the calculation in Eq. (7.5.3) for one noise and one signal term is generally sufficient for all cases that we will encounter.

Note that it is possible to reduce σ_{noise} for a given noise source by decreasing the resolution bandwidth, effectively detecting less noise and the same amount of signal. This will improve the balance in Eq. (7.5.2) and allow smaller signals to be measured. However, this will make the detection process slower. We can also determine the measured value μ_{disp} better by averaging the fluctuations for a sufficiently long time, that means using a low VBW. In the limit of $VBW \rightarrow 0$, or extremely long integration times, the response of the analyser to a noise input with constant RMS will be a constant reading. The entire noise distribution would now be reduced to one piece of information. In most practical situations the integration time will be limited to about 0.1 [s] ($VBW = 10$ [Hz]) since systematic drifts influence the performance of the experiment and do not allow excessively long scans. When the noise is white, that means the fluctuations are independent of the frequency, the spectrum will be a flat horizontal line on the display, μ_{disp} is constant.

The decibel scale

Some examples of actual measurements of laser intensity noise are given in Chapter 8. The display of the spectrum analyser is normally logarithmic. The units used are logarithmic power units, called [dBm]. The conversion of [Watt] into [dBm] is given by

$$p_i[\text{dBm}] = 10 \log \left(\frac{P[\text{W}]}{1[\text{mW}]} \right) \tag{7.5.4}$$

The unit [dBm] is an absolute unit of power, for example the RF power of 1 [mW] corresponds to 0 [dBm], 0.001 [mW] = 1 [μW] corresponds to -30 [dBm], etc. While these numbers appear to be rather small it should be noted that they refer to the power of the AC components only, not the optical power of the beam that generated the photocurrent. It would require a strong modulation of the intensity of the beam to get a significant fraction of the energy into the sidebands. For example, a beam with average optical power P_{opt} and single frequency harmonic modulation of the intensity by 1% has both a positive and a negative sideband at the modulation frequency, $\pm\Omega_{mod}$. Each contains 1% of the amplitude of the optical field, which corresponds to $i(\Omega_{mod}) = 0.02$ of $\bar{i}$ and $p_i(\Omega_{mod}) = 4 \times 10^{-4} P_{opt}$.

In most situations the relative change of the signal, or the noise, is of interest and a relative unit of power, the decibel or [dB], is used. It describes the ratio of two power levels. On a logarithmic scale this corresponds to a difference $\Delta P(\Omega)$ given by

$$\Delta P_{1,2}(\Omega)[dB] = 10 \ \log\left(\frac{P_1(\Omega)}{P_2(\Omega)}\right) \tag{7.5.5}$$

An expression such as 'the signal was amplified by 3 [dB]' means that the power of the signal increased by a factor of $10^{3/10} = 1.995 \approx 2$. Note that for such an increase the actual AC current would grow by only a factor $\sqrt{2}$. Similarly 6 [dB] correspond to a factor 4 increase in power. The ratios in Eq. (7.5.5) are independent of the units, we record them from the display as ratios between values of μ_{disp}, we think of them as the ratio of two noise powers P and interpret them as the ratio of quadrature variances $V1$. All results of noise increase or noise suppression are quoted in [dB], which gives directly the factor of change in the quadrature variance $V1$ of the light.

Adding electronic AC signals

In many situations more than one signal or one noise component are present at the same detection frequency. Only the total power of all the contributions will be displayed. It is important to be able to determine the contributions from the individual terms and we need rules for the adding of several AC currents. These rules depend on the nature and the correlation of the individual currents. Clearly, there is a distinction between currents that are coming from the same source or from different, independent sources. First consider the extreme case of two currents, $i_1(\Omega)$ and $i_2(\Omega)$, which are generated by the same source. They are perfectly correlated, their cross correlation coefficient as defined in Chapter 2, is equal to unity: $C(i_1, i_2) = 1$. In practical terms that means that the two signals have identical time histories. For single harmonic functions this means that they have the same frequency and are locked in phase to each other. For these signals the combined power is simply the sum of the amplitudes:

$$P_{i_1+i_2}(\Omega) = \int \text{res}(\Omega)[i_1(\Omega') + i_2(\Omega')]^2 R \ d\Omega' \tag{7.5.6}$$

If the ratio of the amplitudes of the two currents is given by the number $c = i_2(\Omega)/i_1(\Omega)$, the measured power will change by a factor

$$\frac{P_{i_1+i_2}}{P_{i_1}} = (1+c)^2 \tag{7.5.7}$$

or in [dB] by

$$\Delta P_{i_1+i_2}[dB] = 10 \ \log\left((1+c)^2\right) \tag{7.5.8}$$

For the case of two identical signals, $i_1(\Omega) = i_2(\Omega)$ or $c = 1$, the power increase is 6 [dB]. The other extreme case is that of two signals which have a random phase difference to each other, that means they are uncorrelated. In this case the two signals add in quadrature, that means

$$\Delta P_{i_1+i_2}[\mathrm{dB}] = 10 \ \log(1 + c^2) \tag{7.5.9}$$

and for the case of two signals with equal amplitude the power increase is 3 [dB]. The same technique can be applied to noise terms, which are only different from signals in so far as the power of the noise at detection frequency Ω varies in time while the power of a signal remains fixed. In particular, signal and noise terms are always uncorrelated to each other. Two noise terms are uncorrelated when they come from different sources. They are correlated when they can be traced back to the same source. Correspondingly, these noise signals are added in the respective way, either using Eq. (7.5.8) or Eq. (7.5.9). In the first case, uncorrelated terms, the variances of the two noise terms are added in quadrature. Note that this procedure is strictly speaking only correct if both noise signals have the same distribution functions, eg. Gaussian distribution, which we can assume for all practical situations.

Measuring the cross correlation function with a spectrum analyser

As a practical example of how spectrum analysers are used, consider the case where we wish to measure the correlation of two fluctuating currents $i_1(t)$ and $i_2(t)$. The two currents are combined in a device called a RF splitter/combiner which has two outputs providing the sum and the difference of the two input signals and has the appropriate electrical impedance which matches the impedance of the cables. With the spectrum analyser we can measure the four values $i_1(t)^2$, $i_2(t)^2$, $[i_1(t) + i_2(t)]^2$ and $[i_1(t) - i_2(t)]^2$ by using the two output ports of the splitter/combiner or by terminating the inputs individually. All four values have to be measured separately and are combined to calculate $\mathbf{C}$. According to Equation (4.6.11) the cross correlation coefficient can be expressed as

$$\mathbf{C}(i_1, i_2) = \frac{[i_1(t) + i_2(t)]^2 - [i_1(t) - i_2(t)]^2}{[i_1(t) + i_2(t)]^2 + [i_1(t) - i_2(t)]^2} \times \frac{i_1(t)^2 + i_2(t)^2}{\sqrt{2i_1(t)^2 i_2(t)^2}} \tag{7.5.10}$$

One extreme case is the balanced situation where the two currents have the same long term average, that means $i_1(t)^2 = i_2(t)^2$. Here Eq. (7.5.10) is greatly simplified to

$$\mathbf{C} = \mathcal{V} = \frac{[i_1(t) + i_2(t)]^2 - [i_1(t) - i_2(t)]^2}{[i_1(t) + i_2(t)]^2 + [i_1(t) - i_2(t)]^2} \tag{7.5.11}$$

and in this special case only two measurements are required. For perfect cross correlation, $\mathbf{C} = 1$, we have $i_1(t)^2 = i_2(t)^2$ and the minimum of the signal is equal to zero which is easily detectable. For $\mathbf{C} = 0$, that means completely uncorrelated signals, the cross terms $i_1(t)i_2(t)$ all cancel and we get the condition that $[i_1(t) + i_2(t)]^2 = [i_1(t) - i_2(t)]^2$, which means there is no difference between these two measurements. Actually, the two signals could be added with any phase difference and the result remains constant. This is an important test: the total power of two uncorrelated signals is the sum of the individual powers and is the same for addition and subtraction.

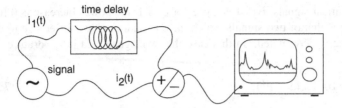

Figure 7.12: Layout of experiment with splitter/combiner.

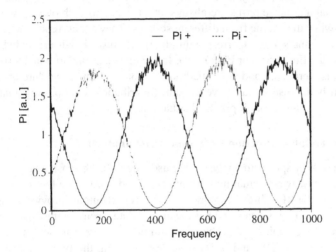

Figure 7.13: Noise correlation for two classical noise sources added with a delay line.

As mentioned above, this applies to noise signals as well as modulation signals and the spectrum analyser can be used to evaluate the correlation of two noise sources. In other words, it can test the independence of the noise sources. If they are independent, the variances should simply add in quadrature, the variance of the sum or the difference should be equal to the sum of the variances. In practice to change from sum to difference can be more technically involved than it appears from the equations. The change in sign can be experimentally achieved by swapping the outputs of the electronic splitter/combiner, provided that there is no phase- lag, or delay between the two signals. This is difficult to achieve ar higher frequencies Ω since the delay has to be much smaller than the inverse of Ω. A reliable test is to deliberately introduce a common modulation signal to the two currents. This provides a strong output and we know that these signals are perfectly correlated and by measuring C in the way described above the entire apparatus can be tested.

An alternative technique is to systematically vary the phase-lag Φ between the two signals. This can be done by introducing a fixed time delay $\Delta\tau$, for example by sending one signal through an extra length of cable or through an RC time delay line, and by recording the spectra of both outputs of the splitter/combiner for range of frequencies, Ω_1 to Ω_2. The delay $\Delta\tau$ should be such that $\Delta\tau > 1/\Omega_1$. The phase lag for each detection frequency is

$\Phi(\Omega) = 2\pi\Delta\tau/\Omega$ and $\Delta\tau$ should be chosen sufficiently long that $\Phi(\Omega)$ varies by more than 2π in the range of detection frequencies used. In this way a spectrum is recorded which contains information of both $[i_1(t) + i_2(t)]^2$ and $[i_1(t) - i_2(t)]^2$ on one trace. The spectrum looks much like an interference fringe, the visibility of these "fringes" can be measured and the value of the correlation coefficient can be determined, as discussed in Section 2.2. If the signals are correlated there will be a strong modulation of the trace, the minimum signal is close to zero, the visibility is close to 1 and the correlation function $\mathbf{C}$ is close to unity. Figure(7.12) shows the layout for this technique with the results given in Fig. 7.13. Two signals from a classical noise source are fed into a splitter/combiner, one signal is delayed by 1 [μs] in regard to the other. They are combined with a second splitter/combiner generating the sum P_{i+} and difference P_{i-} signals. Ideally the trace should show perfect modulation. However, Fig.(7.13) shows that the two currents P_{i+} and P_{i-} are almost out of phase, but not exactly. The splitter/combiner components are not perfect. In this case they provide a shift of about $170°$, not the optimum $180°$.

These techniques are used extensively in the investigation of sub-Poissonian light sources, Section 9.1, in the QND experiments, Chapter 11, and in the investigation of optical entanglement as described in Chapter 13.

Further reading

Detection of Light: from the Ultraviolet to the Submillimeter, G. Rieke, Cambridge University Press (2002)
Introduction to Photonics, B. E. A. Saleh, M. C. Teich, Wiley (1991)
Optical sources, detectors and systems, R. H. Kingston, Academic Press (1995)

Bibliography

[Bev02]   Room temperature stable single-photon source, A. Beveratos, S. Kuhn, R. Brouri, T. Gacoin, J-P. Poizat, P. Grangier Eur. Phys. J. D 18, 191 (2002)

[Bou97]   Experimental qunatum teleportation, D. Bouwmeester, J.-W. Pan, K. Mattle, M. Eibl, H. Weinfurter, A. Zeilinger, Nature 390, 575 (1997)

[Bre91]   Time resolved dual-beam two-photon interference with high visibility, J. Brendel, E. Mohler, W. Martiensen, Phys. Rev. Lett. 66, 1142 (1991)

[Bro00]   Photon antibunching in the fluorescence of individual color centres in diamond, R. Brouri, A. Beveratos, J-Ph. Poizat, P. Grangier, Opt. Lett. 25, 1294 (2000)

[Cab98]   Detection of single infrared, optical, and ultraviolet photons using superconducting transition edge sensors, B. Cabrera, et al., Appl. Phys. Lett. 73, 735 (1998)

[DeM96]   Single-mode generation of quantum photon states by excited single molecules in a microcavity trap, F. De Martini, G. Di Giuseppe, M. Marrocco, Phys. Rev. Lett. 76, 900 (1996)

[Ebe94]   Detection efficiency and dark pulse rate of Rockwell (SSPM) single photon counters, P.H. Eberhard, P.G. Kwiat, M.D. Petroff, M.G. Stapelbroek,

H.H. Hogue, in *Proceedings of the IEEE International Conference on Applications of Photonics Technology (ICAPT '94)*, Toronto, Canada, June 22 (1994)

[Gar94]   Transmission trap detectors, J.L. Gardner, Appl. Optics 33, 5914 (1994)

[Gér99]   Strong Purcell effect for InAs quantum boxes in three-dimensional solidstate microcavities, J.M. Gérard, B. Gayral, J. Lightwave Technol. 17, 2089 (1999)

[Gra98]   Photodetector designs for low-noise, broadband and high power applications, M.B. Gray, D.A. Shaddock, C.C. Harb, H-A. Bachor, Rev. of Sci. Instr. 69, 3755 (1998)

[Gis02]   Quantum cryptography, N. Gisin, G. Ribordy, W. Tittel, H. Zbinden, Rev. Mod. Phys. 74, 145 (2002)

[Hil90]   Accurate measurement of low signal-to-noise ratios using superheterodyne spectrum analysers. D.A. Hill, D.P. Haworth, IEEE transactions on instrumentation and measurement, 19, 432 (1990)

[HPmanual] for example manual of Hewlett Packard Spectrum Analyser HP B6457 or manuals of similar electronic spectrum analysers

[Jam02]   Atomic vapor-based high efficiency optical detectors with photon number resolution, D.F.V. James, P.G. Kwiat, Phys. Rev. Lett. 89, 183601 (2002)

[Kim99]   Multiphoton detection using visible light photon counter, J. Kim, S. Takeuchi, Y. Yamamoto, H.H. Hogue, Appl. Phys. Lett. 74, 902 (1999)

[Kuh02]   Deterministic single-photon source for distributed quantum networking, A. Kuhn, M. Hennrich, G. Rempe, Phys. Rev. Lett. 89, 067901 (2002)

[Kur00]   Stable solid-state source of single photons, C. Kurtsiefer, S. Mayer, P. Zarda, H. Weinfurter, Phys. Rev. Lett. 89, 290 (2000)

[Kur01]   High-efficiency entangled photon pair collection in type-II parametric fluorescence, C. Kurtsiefer, M. Oberparleiter, H. Weinfurter, Phys. Rev. A 64, 023802 (2001)

[Kuz03]   Generation of nonclassical photon pairs for scalable quantum communication with atomic ensembles, A. Kuzmich, W.P. Bowen, A.D. Boozer, A. Boca, C.W. Chou, L.-M. Duan, H.J. Kimble, Nature 423, 731 (2003)

[Kwi95]   New high-intensity source of polarization-entangled photons, P.G. Kwiat, K. Mattle, H. Weinfurter, A. Zeilinger, A. Sergienko, Y. Shih, Phys. Rev. Lett. 75, 4337 (1995)

[Kwi99]   High-efficiency quantum interrogation measurements via the quantum Zeno effect, P.G. Kwiat, A.G. White, J.R. Mitchell, O. Nairz, G. Weihs, H. Weinfurter, A. Zeilinger, Phys. Rev. Lett. 83, 4726 (1999)

[Kwi99a]   Ultrabright source of polarization-entangled photons, P.G. Kwiat, E. Waks, A.G. White, I. Appelbaum, P.H. Eberhard, Phys. Rev. A 60, R773 (1999)

[Lou00]   Single photons on demand from a single molecule at room temperature, B. Lounis, W.E. Moerner, Nature 407, 491 (2000)

[Mic00]   A quantum dot single-photon turnstile device, P. Michler, A. Kiraz, C. Becher, W.V. Schoenfeld, P.M. Petroff, L. Zhang, A. Imamoglu Science 290, 2282 (2000)

[Mil03]   A.J. Miller, S.W. Nam, J.M. Martinis, A. Sergienko, private communication (2003)

[Mor01]   Single mode solid-state single photon source based on isolated quantum dots in pillar microcavities, E. Moreau, I. Robert, J.M. Gerard,IAbram, L. Manin, V. Thierry-Mieg Appl. Phys. Lett. 79, 2865 (2001)

[Pry02]   G.J. Pryde, J.L. O'Brien, A.G. White, private communication (2002)

[Rar90]   Experimental violation of Bell's inequality based on phase and momentum, J. Rarity, P. Tapster, Phys. Rev. Lett. 64, 2495 (1990)

[Rie02]   *Detection of Light: from the Ultraviolet to the Submillimeter*, G. Rieke, Cambridge University Press, UK, 2nd ed. (2002)

[Sal91]   *Introduction to photonics*, B.E.A. Saleh, M.C. Teich, Wiley (1991)

[San02]   Indistinguishable photons from a single-photon device, C. Santori, D. Fattal, J. Vučković, G.S. Solomon, Y. Yamamoto, Nature 419, 594 (2002)

[Tak99]   Development of a high-quantum-efficiency single-photon counting system, S. Takeuchi, J. Kim, Y. Yamamoto, H.H. Hogue, Appl. Phys. Lett. 74, 1063 (1999)

[Tak01]   Beamlike twin-photon generation by use of type II parametric downconversion, S. Takeuchi, Opt. Lett. 26, 843 (2001)

[Tay09]   Interference fringes with feeble light, G.I. Taylor, Proc. of the Cambridge Philosophical Society 15, 114 (1909)

[Ver02]   Detection efficiency of large-active-area NbN single-photon superconducting detectors in the ultraviolet to near-infrared range, A. Verevkin, J. Zhang, R. Sobolewski, A. Lipatov, O. Okunev, G. Chulkova, A. Komeev, Smirnov, G.N. Gol'tsman, A. Semenov, Appl. Phys. Lett. 80, 4687 (2002)

[Whi99]   Nonmaximally entangled states: production, characterisation, and utilization, A.G. White, D.F.V. James, P.H. Eberhard, P.G. Kwiat, Phys. Rev. Lett. 83, 3103 (1999)

[Yua02]   Electrically driven single photon source, C. Yuan, B.E. Kardynal, R.M. Stevenson, A.J. Shields, C.J. Lobo, K. Cooper, N.S. Beattie, D.A. Ritchie, M. Pepper, Science 295, 102 (2002)

[Zal80]   Silicon photodiode absolute spectral response self calibration, E.P. Zalewski, J. Geist, Applied Optics 19, 1214 (1980)

[Zwi01]   Single quantum dots emit single photons at a time: antibunching experiments V. Zwiller, H. Blom, P. Jonsson, N. Panev, S. Jeppesen, T. Tsegaye, E. Goobar, M-E. Pistol, L. Samuelson, G. Björk, Appl. Phys. Lett. 78, 2476 (2001)

8 Quantum noise: Basic measurements and techniques

We now turn our attention to the measurement of quantum noise. Several theoretical models have already been described in Chapters 2, 4 and the experimental components have been discussed in Chapters 5, 7. Based on this let us build up, step by step, the experiments for measuring and manipulating quantum noise.

8.1 Detection and calibration of quantum noise

We will concentrate on CW laser beams and the noise associated with the intensity, amplitude and phase of the light. We have to devise ways of interpreting the electrical noise spectra. How do we distinguish quantum noise from ordinary technical noise? How can we can calibrate our measurement? Three different techniques are commonly used, direct detection (8.1.1), balanced detection (8.1.2) and homodyne detection (8.1.4). The choice depends on the properties of laser beam.

8.1.1 Direct detection and calibration

The obvious way of detecting quantum noise is to *directly detect* the light with a single photo-detector and to analyse the resulting photo-current using an electronic spectrum analyser. This is the same arrangement as one would use to measure an amplitude modulation of the laser beam and in most applications we will be interested in quantum noise as the ultimately limiting effect in the signal to noise ratio. Direct detection gives us an electrical noise power $p_i(\Omega)$, which for a perfect laser mode and detection system should be entirely due to the quantum noise of the light. There are obviously some technical pitfalls that need to be avoided and we need a reliable procedure that convinces us, and anybody else, that we are really measuring quantum noise.

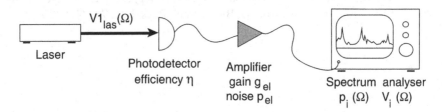

Figure 8.1: Schematic diagram of direct detection

A Guide to Experiments in Quantum Optics, 2nd Edition. Hans-A. Bachor and Timothy C. Ralph
Copyright © 2004 Wiley-VCH Verlag GmbH & Co. KGaA
ISBN: 3-527-40393-0

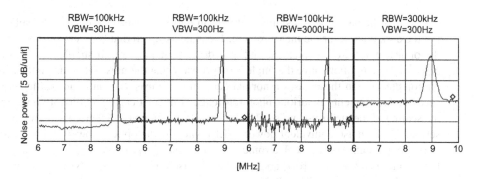

Figure 8.2: Experimental noise spectra for different combination of resolution and video bandwidth. These are four separate experimental traces, with settings as shown in Fig. 7.9.

The noise power $p_i(\Omega)$ displayed by the spectrum analyser is a measure of fluctuations of the photo-current, the amplification and the noise contributions of the electronics. For a quantum noise limited source we obtain the value $p_{i,\mathrm{QNL}}(\Omega)$. All measurements will be normalized to this value, resulting in the normalized variance

$$V_i(\Omega) = \frac{p_i(\Omega)}{p_{i,\mathrm{QNL}}(\Omega)} \qquad (8.1.1)$$

Once the recorded noise traces have been calibrated it is obvious that a new unit should be used. A creative suggestion was to label the standard quantum noise limit as 1 [Hoover], in recognition of the role of the vacuum noise [Lev85]. All noise diagrams in this guide with linear scales use this unit, most however use a logarithmic scale with units of decibels [dB] defined on page 193.

We need to know the total quantum efficiency of the detection system to link the variance $V_i(\Omega)$ to the fluctuations of the light $V1(\Omega)$, see Equation (4.3.21). All values for the noise power are long term values obtained from the display as shown in Fig. 8.2(a). A quantum noise limited light source with modulation at 9 [MHz] illuminates the detector. The left hand side traces shows the display for a spectrum from 6 to 10 [MHz] with constant RBW of 100 [kHz] and variable VBW of 30, 300, 3000 [Hz]. Note that the average noise remains constant while the actual time fluctuations of the noise increase with VBW. The signal remains unchanged. The width of the signal is given by RBW. The right hand side trace shows the results for the RBW of 300 [kHz]. The average noise level increases. The signal height remains the same since the signal is much larger then the noise and the signal width increases with RBW. Clearly we gain by selecting a larger RBW and get further away from the electronic noise. However, there are limitations, in particular if we have not a flat but a structured noise spectrum and there are a number of technical complications:

1. *Electronic noise.* The photo-detector generates independent electronic noise, $p_{\mathrm{el}}(\Omega)$, which adds to the optical noise spectrum. We can avoid this complication by designing special detectors with low electronic noise, see Section 7.4.1 and by independently measuring the *dark noise* of the detector. This is usually done by simply blocking the

detector. If the dark noise is significant we can correct the measured signal by subtracting the dark noise after the measurement. It can be assumed that the electronic noise and
the optical noise are independent and are added in quadrature. It should be noted that
some exotic amplifiers have the unpleasant property that the noise depends on the actual
photon current. In these cases a normal dark current measurement is not sufficient and
several measurements at various photo-currents, or intensities, are required to investigate
the scaling of the electronic noise. The performance of a given detector is commonly described by the noise equivalent power (NEP) which is the optical power on the detector
at which quantum noise and electronic noise are equal $(p_{i,\text{QNL}}(\Omega) = p_{\text{el}}(\Omega)$. That means
the recombined trace is 3 [dB] above the dark noise trace. Values of NEP = 1 [mW] are
typical. Values as low as 10 [μW] have been achieved.

2. *AC response.* Most detector circuits will have a frequency dependent gain, given by
   $g_{\text{el}}(\Omega)$. For all detectors there will be a high frequency roll off, depending on the type
   of photo-diode and amplifier used. Frequently, there is also some electronic mismatch
   between the components leading to a sinusoidal frequency response curve. This is more
   a nuisance than a complication. It is taken into account by normalizing the traces, Equation (8.1.1), and $g_{\text{el}}(\Omega)$ does not effect $V_i(\Omega)$ directly.

3. *Efficiencies.* The detector efficiency is not perfect. A low efficiency has the same effect
   as a beamsplitter which diverts only a fraction of the light into the detector. For measurements of the quantum noise alone this is not crucial. For example, we can measure
   the quantum noise level (QNL) even from only a fraction of the beam and compensate
   with more electronic gain. However, knowledge of the efficiency is crucial to calculate
   the SNR of a signal or the degree of squeezing in the light, that means differences to the
   QNL. These effects disappear with low quantum efficiency. Note that electronic gain and
   efficiency play different roles. The golden rule for any squeezing experiment is that all
   of the light should be detected and that quantum efficiencies have to be close to unity.

4. *Power saturation.* Detectors can only handle a limited amount of optical power. The
   actual limit varies from [mW] for small, fast photo-diodes up to [W] for large detectors.
   Once the limit in power is exceeded the detector can actually be damaged due to heating
   processes. The absorbed light will heat the diode material and, more importantly, the bias
   current will lead to further ohmic heating. Even below this damage limit the performance
   can become nonlinear. The number of available carriers inside the detection material is
   limited and the response drops if the local intensity is too high. Thus the detector has to
   be uniformly illuminated even at intensities below the recommended power limit.

5. *Saturation of the amplifiers.* Several cases have been reported where a detector showed
   a peculiar response and the reason was the presence of a strong modulation signal which
   saturated the entire circuit. Saturation at one RF frequency can affect the performance of
   the detector at many other frequencies. Care should be taken that no such strong signal is
   present by recording large bandwidth spectra up to the roll off frequency of the detector.
   A particular pitfall is to use a series of amplifiers where the bandwidth of the earlier
   stages is broader than that of later stages, for example by using a low pass filter at the
   output of an amplifier. In this case the large high frequency signal would not be visible

in the output, but it is saturating the amplifiers ahead of the filter. Saturation can change the noise spectrum over a wide bandwidth, it can even lower the noise level and in some experiments saturated amplifiers have generated noise traces that mimicked, and were mistaken for, squeezing.

After all precautions have been taken we can trust the detector. But in order to measure the QNL we still need to calibrate it. One obvious technique is to calculate the expected noise power $p_{i,QNL}(\Omega)$ from the formula for shot noise, Equation (3.3.6), and using the appropriate constants we get

$$p_{i,QNL}(\Omega) \text{ [dBm]} = -138.0 + 10 \log(i \times \text{RBW})\text{[dBm]} + p_{el}(\Omega)\text{[dBm]} + g_{el}(\Omega)\text{[dB]}$$

where RBW is measured in [Hz], i in [A] and an input resistance of 50 [Ohm] is assumed. $p_{i,QNL}(\Omega)$ can be determined once the average photo-current i_{ave} is known. This in turn should be done best by a direct measurement of the DC current from the photo-diode. It also requires a knowledge of the detection bandwidth, which is normally the resolution bandwidth of the spectrum analyser. The AC gain of the various amplifying stages has to be included. Finally, the actual response characteristics of the spectrum analyser to noise, rather than to a signal, has to be taken into account, see Section 7.5, which could require corrections as large as 2 [dB]. After all these measurements and calculations one could expect to get a value for the noise level with an accuracy of typically 20% (1 [dB]). This is good enough to test the apparatus for inconsistencies but insufficient to calibrate the traces accurately. A useful check is to investigate the scaling of the detected electrical noise power with the optical power, as shown in Fig. 8.3. Quantum noise, electronic noise and classical photon noise, or modulations, all scale differently. Electronic, or dark noise, is usually independent of the optical power P_{opt}. The noise power due to quantum noise is proportional to $P_{opt}^0.5$. The noise due to classical optical noise is proportional to P_{opt}. Figure 8.3 shows four cases of different mixtures of noise and their scaling. A reliable measurement of the quantum noise should only be attempted for situations which show a linear dependence of the measured noise level on P_{opt}, in this case near $P_{opt} = 1$.

White light calibration

An alternative is to calibrate the detector with a light source which we know is quantum noise limited. This is usually a thermal white light source, which has Poissonian photon statistics, see Section 2.2. In principle all that is required is to illuminate the detector with white light that generates the same DC photo-current – and to record the RF noise spectrum. However, care has to be taken: The wavelength should be filtered to the same wavelength (about ±20–50 [nm]) as used in the actual experiment, since the response of photo-diodes is known to be wavelength dependent. The absorption depth depends on the wavelength. In addition, the illumination pattern has to be similar, since localized illumination can lead to charge carrier depletion. Finally, the diode, particularly the very small InGaAs diodes, can easily be geometrically over filled. Once a large fraction of the light misses the photo-sensitive area it can heat up the entire chip and distort the results. From experience, large photo-detectors can be calibrated using white light to about ±0.1 [dB], while small InGaAs detectors can be calibrated to an accuracy of about ±0.4 [dB].

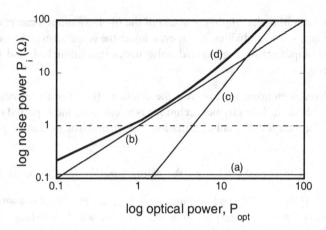

Figure 8.3: Scaling of different types of noise for a beam of light with fixed classical relative intensity noise (RIN = constant) and variable optical power P_{opt} The noise terms scale differently with P_{opt}. (a) electronic noise. (b) quantum noise. (c) classical noise. (d) total noise as seen in an experiment. For convenience all terms are normalized by the quantum noise present at $P_{opt} = 1$.

8.1.2   Balanced detection

An alternative technique for calibration is to use a balanced detection system. Here the incoming light is divided into two beams of equal optical power which are detected by two matched photo-detectors, see Fig. 8.4. The two photo-currents are either added or subtracted. Two noise spectra are created by the electronic spectrum analyser and are recorded as the sum and difference spectrum. This technique can distinguish between optical quantum noise and classical optical noise. The beamsplitter sends classical noise onto both detectors - the classical optical noise in both beams and thus the photo-currents are correlated to each other. That means they can be subtracted and cancelled. When added, the full noise power is recovered. For quantum noise the effect of the beam splitter is different – the two resulting photo-currents are not correlated. That means the two channels add in quadrature for both the sum and the difference. In the quantum noise limited case the two spectra are identical. This argument is based purely on the correlation properties of the sidebands containing the information about the noise, as described in Section 4.5. A more formal derivation of these results is given in Appendix D, which describes the balanced detector in terms of the propagation of coherent states through a beamsplitter.

Electronic noise from the two detectors is obviously not correlated – and adds in quadrature. It cannot easily be distinguished from quantum noise. Consequently, the apparatus can suppress all classical, technical noise on this mode of light and leaves only the optical

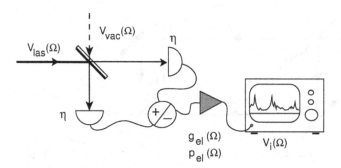

Figure 8.4: Balanced detector

quantum and the electronic noise. The description is

$$
\begin{aligned}
p_+(\Omega) &= g_{\text{el}}(\Omega)\, p_i(\Omega)\, V_i(\Omega) + p_{\text{el}}(\Omega) \\
p_-(\Omega) &= g_{\text{el}}(\Omega)\, p_i(\Omega) + p_{\text{el}}(\Omega) \\
V_i(\Omega) &= \frac{p_+(\Omega) - p_{\text{el}}(\Omega)}{p_-(\Omega) - p_{\text{el}}(\Omega)}
\end{aligned}
\tag{8.1.2}
$$

In the case of sub-Poissonian intensity noise the plus trace displays the noise suppression, the minus trace shows quantum noise only. The technical difficulties with this detection scheme is the balancing of the detectors. The two photo-detectors have to have the same electronic RF response. We have effectively built an electronic interferometer and both the gain and the phase have to be matched at all RF frequencies which are used. This can be extraordinarily time consuming since the response frequently depends not only on the notional values of the components used but also on the actual physical layout of the circuit. Obviously cable lengths and delay times in the circuits have to be matched to achieve a proper balance. Passive splitter & combiners are used as adders & subtractors at high frequencies ($\geq$ 50 [MHz]) while active circuits are useful for low and medium frequencies. The optical balancing can be tricky since it ideally requires a precise 50/50 beamsplitter. This can be achieved by careful adjustment, since the splitting ratio for most dielectric beamsplitters depends on the input angle. Alternatively, a polarizing beam splitter can be used. For a linearly polarized output beam the ratio can be continuously adjusted by rotating the polarisation direction ahead of the beamsplitter with a half wave plate.

In all these arrangements optical losses have to be avoided if squeezing is to be detected since they would effectively decrease the quantum efficiency. It is tempting to compensate a mismatch in the electronic gain with an adjustment in the optical power. But again, this would decrease the quantum efficiency. Thus many of the common tricks used in detecting ordinary coherent light are not allowed for squeezed light.

8.1.3   Detection of intensity modulation and SNR

The size of a modulation can be measured directly. After all, electronic spectrum analysers were designed for this purpose. The height of the spectral peak is proportional to the power

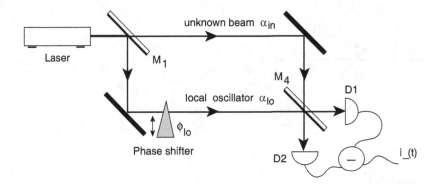

Figure 8.5: Homodyne detector

of the Fourier component. The relative intensity noise is linked to the measured power by

$$p_i(\Omega) \;\;=\;\; \left(\bar{\imath} \, \text{RIN}(\Omega) \right)^2 \times \text{RBW}$$

$$\text{RIN}(\Omega) \;\;=\;\; 10^{\frac{1}{2} p_i(\Omega)} [\text{dBm}] \tag{8.1.3}$$

For example, a beam which generates an average photo-current of 1 [mA] with a modulation of 1 part in 10^4 (RIN $= 10^{-4}$) produces a noise peak of -10 [dBm] when measured with a RBW of 100 [kHz]. For large modulations, the photo-current SNR, as defined in Eq. (7.5.3), can be measured directly as the difference, in [dB], between the signal peak and the noise floor. For small modulations we have to take into account that the signal peak contains both the signal and the noise and have to calculate the value SNR = (signal peak - noise) / noise. For very small SNR (< 2) the exact response of the spectrum analyser has to be considered, see Appendix E. Electronic gain will not affect the SNR. Finally, the total quantum efficiency, including any form of attenuation of the light, has to be taken into account when using Eq. (7.5.3) to evaluate the optical SNR from the measurements of the photon current.

8.1.4   Homodyne detection

The direct detection schemes described in Section 8.1 are used to measure the intensity as well as the intensity noise of a laser beam. For a single mode field the combination of two detectors with a beamsplitter can be used to evaluate the contribution of classical modulation and quantum noise. However, the direct detection cannot differentiate between the quadratures. For this purpose we need a reference beam, commonly called the local oscillator. This beam has to be phase locked to the input, otherwise it cannot provide a phase reference to distinguish between the quadratures. If the local oscillator has the same frequency as the input, we have a *homodyne* detector. Alternatively, if the local oscillator is frequency shifted the system is known as a *heterodyne detector*.

The homodyne detector for classical waves

For a homodyne detector we can assume that the local oscillator is created by the same source as the input beam we want to measure. In practice the link could be much more indirect. The entire system can be represented by one large interferometer, see Fig. 8.5. The input beam α_{in} and the local oscillator beam α_{lo} are combined by the mirror M_4 which has a balanced 50/50 reflectivity. The whole apparatus has to be aligned in the same way and with the same precision as a conventional interferometer. The two detectors D_1 and D_2 will receive the interference signal. Firstly, let us classically analyse the performance of the system. The two beams can be represented by the amplitudes

$$
\begin{aligned}
\alpha_{in}(t) &= \alpha_{in} + \delta X1_{in}(t) + i\delta X2_{in}(t) \\
\alpha_{lo}(t) &= [\alpha_{lo} + \delta X1_{lo}(t) + i\delta X2_{lo}(t)] \, e^{i\phi_{lo}}
\end{aligned} \tag{8.1.4}
$$

Here $e^{i\phi_{lo}}$ represents the relative phase difference between the input amplitude and the local oscillator amplitude. It can be controlled by changing the length of the one arm of the interferometer designated as the local oscillator. In practice, this involves the fine control of a mirror position by a piezo or the rotation of a glass plate in the local oscillator beam. The local oscillator should dominate the output signal. We consider the case that the local oscillator beam is far more intense than the input beam: $\alpha_{lo}^2 \gg \alpha_{in}^2$. In this case we can assume that all the intensity is from the local oscillator and that the intensities on both detectors are the same

$$
|\alpha_{D1}|^2 = |\alpha_{D2}|^2 = \frac{1}{2}|\alpha_{lo}|^2 \tag{8.1.5}
$$

All terms proportional to α_{in} can be dropped in comparison to those proportional to α_{lo}. The signal at the two detectors can be described using the standard convention for the phase shift in a single beamsplitter (Equation (5.1.1)), including a 180 degree phase shift for one of the reflected beams.

$$
\begin{aligned}
\alpha_{D1} &= \sqrt{1/2}\,\alpha_{lo}(t) + \sqrt{1/2}\,\alpha_{in}(t) \\
\alpha_{D2} &= \sqrt{1/2}\,\alpha_{lo}(t) - \sqrt{1/2}\,\alpha_{in}(t)
\end{aligned} \tag{8.1.6}
$$

The intensities and the photo-currents from the two detectors are proportional to $|\alpha_{D1}|^2$ and $|\alpha_{D2}|^2$ respectively.

$$
|\alpha_{D1}|^2 = 1/2 \left[|\alpha_{lo}(t)|^2 + \alpha_{lo}(t)\alpha_{in}^*(t) + \alpha_{in}(t)\alpha_{lo}^*(t) + |\alpha_{in}(t)|^2 \right]
$$

which can be approximated as

$$
\begin{aligned}
|\alpha_{D1}|^2 &= 1/2 \left[|\alpha_{lo}|^2 + 2\alpha_{lo}\delta X1_{lo}(t) \right. \\
&\quad \left. + 2\alpha_{lo} \left(\delta X1_{in}(t)\cos(\phi_{lo}) + i\delta X2_{in}(t)\sin(\phi_{lo}) \right) \right]
\end{aligned} \tag{8.1.7}
$$

by neglecting terms such as $\alpha_{lo}\alpha_{in}^*$ compared to $|\alpha_{lo}|^2$ and $\alpha_{lo}\delta X^*$ compared to $\alpha_{in}\delta X^*$. In addition all higher order terms in δX have been left out. Similarly, the result for the other detector can be found. We can now define the difference current $i_-(t)$ from the two detectors.

$$
i_-(t) \approx 2\alpha_{lo}(\delta X1_{in}(t)\cos(\phi_{lo}) + i\delta X2_{in}(t)\sin(\phi_{lo})) \tag{8.1.8}
$$

This is a remarkable result. The current $i_-(t)$ scales with the amplitude of the local oscillator, but the fluctuations, or noise, of the local oscillator are completely suppressed. Its noise level could be well above the QNL and the result would not be affected. On the other hand the power of the input beam has no influence, as long as it is small compared to the power of the local

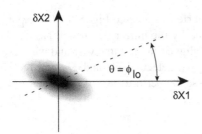

Figure 8.6: Projection through the uncertainty area.

oscillator. This is an extremely useful and somewhat magic device. The difference current $i_-(t)$ contains only higher frequency, or AC, terms. Obviously the DC currents cancel each other. In the experiment we could use a splitter / combiner or a differential AC amplifier to generate this fluctuating current. Finally, the variance of this current is evaluated. We should note that in this calculation all cross terms of the type $\delta X1\delta X2$ cancel out in the averaging process. We obtain

$$\Delta i_-^2 \approx 4\alpha_{lo} \left(\delta X 1_{in}^2 \cos^2(\phi_{lo}) + \delta X 2_{in}^2 \sin^2(\phi_{lo}) \right) \tag{8.1.9}$$

This means the measured variance is a weighted combination of the variance of the fluctuations in the two quadratures. In particular for the phase $\phi_{lo} = 0$ the variance in the quadrature amplitude is measured while for $\phi_{lo} = 90 = \pi/2$ the variance in the quadrature phase is measured. An alternative interpretation of Equation (8.1.9) is that the homodyne detector measures the variance of the fluctuating state under a variable projection angle $\theta = \phi_{lo}$. By scanning the local oscillator phase the projection angle can be selected.

The sideband interpretation says that the different types of modulation differ by the relative phase θ between the two beats between the two sidebands and the carrier. The homodyne detector selects one such phase. The modulation with the other phase ($\theta + 90$) is not detected at all. The homodyne detector filters out specific quadrature of the modulation sideband. With the fixed phase shift ϕ_{lo} for the local oscillator the detector will have best sensitivity for one particular type of modulation. For example for $\phi_{lo} = 0$ only amplitude modulation is detected, for $\phi_{lo} = 90$ only phase modulation is detected. The sensitivity of the detector to this quadrature changes with $\cos^2(\phi_{lo})$.

The homodyne detector is very popular in radio applications and is a key technology in our present communication systems. Signals are detected in phase with a local oscillator, in the case of radio waves an electronic oscillator inside the receiver, which can be controlled in frequency very well, for example using quartz crystal oscillators or could even be locked to the incoming radio frequency (known as phase locked loop or *PLL*). Radio homodyne and all the later heterodyne detection techniques have allowed us to place different carriers very close to each other in the radio spectrum and to detect very weak signals. This increases dramatically the carrying capacity compared to a simple intensity coded system and allowed applications such as the mobile phone, satellite communications etc.

Quantum mechanical calculations are required to check the validity of the above classical results for the quantum properties of the light. The balanced homodyne detector is a special case of a Mach Zehnder interferometer, Section 5.2, with a 50/50 output mirror. The two incident fields on the beam splitter M_4 can be described by the operators $\delta\tilde{\mathbf{A}}_{\text{in}}$ and $\delta\tilde{\mathbf{A}}_{\text{lo}}e^{i\phi_{\text{lo}}}$. Using the quantum mechanical description of the beamsplitter we can describe the beams in front of the detectors by

$$\delta\tilde{\mathbf{A}}_{D1} = \sqrt{1/2} \ \left(\delta\tilde{\mathbf{A}}_{\text{lo}}e^{i\phi_{\text{lo}}} + \delta\tilde{\mathbf{A}}_{\text{in}}\right)$$

$$\delta\tilde{\mathbf{A}}_{D2} = \sqrt{1/2} \ \left(\delta\tilde{\mathbf{A}}_{\text{lo}}e^{i\phi_{\text{lo}}} - \delta\tilde{\mathbf{A}}_{\text{in}}\right) \tag{8.1.10}$$

The frequency components of the current generated by the entire apparatus is given by

$$V1_{HD}(\Omega) = \alpha_{\text{lo}}^2 \, V\Theta_{\text{in}}(\Omega) \tag{8.1.11}$$

In complete analogy to Eq. (8.1.9) but now for the full quantum operators. The derivation of Eq. (8.1.11) holds for the approximation that $\alpha_{\text{lo}} \gg \alpha_{\text{in}}$.

In this simple interferometer we have a beautiful device for the measurement of the variance for a single quadrature, including all quantum properties. We can select the quadrature by simply varying the phase angle of the local oscillator. For a coherent state, where all quadratures are identical, this technique has no particular advantage over direct detection. However, once the variances of the quadratures change, as in a squeezed state, the homodyne detector is the essential measuring system. It was first proposed by H.P. Yuen and J.H. Shapiro [Yue80] and D.F. Walls and P. Zoller [Wal81] and has since been a key component in all experiments with quadrature squeezed states, as discussed in Chapter 9.

Finally, a few practical hints: It should be noted that the homodyne detector can be operated even when the intensity of the input is small. As long as $|\alpha_{\text{in}}|^2$ is smaller than the intensity $|\alpha_{\text{lo}}|^2$ of the local oscillator all the results given above are correct. Thus the detector can be used to measure squeezed vacuum states as well as bright squeezed light. If $|\alpha_{\text{in}}|^2$ is comparable to $|\alpha_{\text{lo}}|^2$ the output intensity will change periodically with ϕ_{lo} and the variance will be a mixture of variances from both orthogonal quadratures. Balancing the homodyne detector is the most difficult part of this experiment. Both the amplitude and the wave front curvature have to be matched for both outputs. This generally means that both the local oscillator and the input beam should travel comparable distances and should experience similar optical components such as lenses and mirrors. In squeezing experiments we cannot afford any losses, otherwise some vacuum noise would be added, the noise suppression would deteriorate. That means some of the tricks quite common with normal, classical interferometers cannot be used here. For example, we have to detect all of the beam, using an aperture is not allowed. We cannot balance the two beam splitter outputs by attenuating one of the beams. It is also important that all the electrical components are balanced, they should have the same gain and phase difference at all detection frequencies used.

The quality of the alignment can be judged by measuring the fringe visibility VIS of the system, see definition in Eq. (5.2.4). This can be done in some experiments by turning up the power of the input beam to be the same as that of the local oscillator. As long as some input intensity is available the normalized visibility $\mathcal{V} = VIS/VIS_{\text{perfect}}$ can be measured,

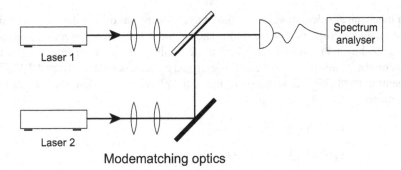

Figure 8.7: Example of heterodyne detection of the bandwidth of two lasers

where VIS is the directly measured value and VIS_perfect is the visibility predicted from the measurement of the intensities. For $V = 1$ the homodyne detector records all the correlations accurately. The measurement of the extrema is reduced when $V < 1$, the correlations between the input and the local oscillator beams appear to be smaller than they really are. The quality of the homodyne detector is described by an overall efficiency

$$\eta_{hd} = \eta_\text{det} V^2 \tag{8.1.12}$$

where η_det is the normal efficiency of the photo-detector. In most situations η_det is smaller than V^2. Values as good as $V^2 = 0.98$ have been achieved. These details will be discussed in the description of the individual experiments. The homodyne detector is the most common technique to measure the noise and signal properties of both quadratures. It is more complex, requires more patience than a local oscillator beam, but it provides the full information.

8.1.5   Heterodyne detection

While in homodyne detection we are mixing two beams with the same optical frequency on one detector, with a beamsplitter we can also mix beams with different optical frequencies ν_1 and ν_2. This is known as heterodyne detection. As long as the difference frequency $\Omega_\text{HT} = \nu_1 - \nu_2$ is within the detection range of the photodetector, and in practice that is limited typically to about 1–2 [GHz], we will record a strong, narrow signal at Ω_HT which is surrounded by beat signals created by all the sidebands of either laser beam beating with the respective carriers of the other beam. Note that a spectrum analyser cannot distinguish between positive and negative beat frequencies and these two will be folded together, see Fig. 8.7.

Heterodyne detection is frequently used to determine the linewidth $\Delta\nu_2$ of an unknown laser. The spectrum around the detection frequency $\Omega_\text{HT} = \nu_1 - \nu_2$ contains information about both beams. The lineshape of the heterodyne signal is a convolution of the lineshapes of the two lasers. If one of the laser beams has a narrow linewidth compared to the other, e.g. $\Delta\nu_1 \ll \Delta\nu_2$, the width $\Delta\Omega_\text{HT}$ of the heterodyne signal corresponds to the broader linewidth, in this case $\Delta\nu_2$. The exact link between $\Delta\Omega_\text{HT}$ and $\Delta\nu$ depends on the lineshape. For example if both lasers have Lorentzian lineshapes, which are quite common, and $\Delta\nu_1 \approx \Delta\nu_2$

the link is $\Omega_{HT} = \Delta\nu_1 + \Delta\nu_2$. Furthermore, the heterodyne spectrum will show any additional laser modes in the input beams. These will appear as strong, narrow spectral features around Ω_{HT}.

In heterodyne detection we are measuring signals at high detection frequencies, outside the prevailing technical noise at low frequencies, the heterodyne technique is very sensitive in the amplitude quadrature. Ideally it is only limited by the quantum noise level of the stronger of the two input beams and in this case the detection is at the QNL, or commonly called *shot noise limited*. This technique has been used to detect with very high sensitivity the fluorescence spectrum of individual ions in traps or individual atoms and to determine such properties as the ion trap frequency (as low as hundreds of [Hz]) and the excitation temperature. In this regime heterodyne detection competes with single photon detection in sensitivity.

Measuring other properties

The detection schemes discussed so far concentrate on the measurement of the intensity and the quadratures of the laser light. However, there are other properties of the light, in particular the position and the polarisation of the light which are worth a consideration. Many applications rely on the modulation and detection of these properties, as can be seen from Fig. 8.8. Detector (c) measures the beam position by comparing the difference in the intensity falling onto the two halves of a split detector. This could also be done in both transverse directions using a quadrant detector. Detector (d) uses a polarizing beam splitter (PBS) to analyse the degree of polarisation, in this case horizontal and vertical linear polarisation. Other states of polarisation can be determined by including a waveplate ahead of the PBS.

This comparison shows that in all cases we are using two photodetectors and we are processing the difference of the photocurrents. In all these cases we will find that for a coherent and single mode beam, the perfect laser beam, each detector produces a photocurrent which includes quantum noise and which simply adds to the quantum noise from the other detector. This means we have a quantum noise for the detection of all these properties: intensity, quadrature position and polarisation. The last one is normalized differently, the uncertainty can be expressed in terms of the Stokes parameters which have a cyclic commutation relationship [Rob74], [Sch03], but is still a QNL. We will find in chapters(9)and (10) that for all these properties we can generate non-classical states of light with reduced noise below the QNL.

8.2 Intensity noise

Laser noise

Lasers are the light sources in all quantum optics experiments, but they are far from ideal; most lasers have intensity noise well above the QNL as discussed in Chapter 6, page 157. With careful preparation some lasers can be QNL in restricted parts of the spectrum. Good examples are the diode pumped, monolithic Nd:YAG laser which is one of the quietest and most reliable lasers available [Kan85, Koe96, Innolight]. These lasers are QNL for frequencies well above 5 MHz, as shown in Fig. 6.5, and thus are rather useful light sources for quantum optics experiments.

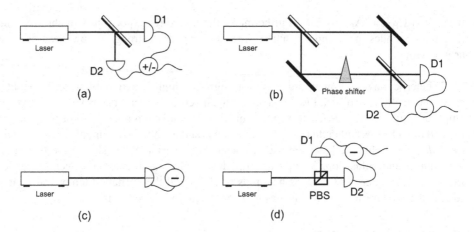

Figure 8.8: Comparison of detection schemes: (a) balanced detector for $V1(\Omega)$, (b) homodyne detector for the quadratures, (c) split detector for the position of the beam and (d) polarisation detector.

Diode lasers themselves can be excellent low noise light sources at frequencies up to several hundred [MHz], as discussed on page 160. In particular the low power (< 200 [mW]) single element diode lasers can be QNL at a wide range of frequencies. The cavities are very short (< 1 [mm]) but the gain bandwidth can be very large and thus single mode operation is usually only achieved with either an extended cavity design or an internal Distributed Bragg reflector (DBR). Even without these additional features the total photon flux can indeed be very quiet, basically representing the fluctuations of the input current which can easily be controlled to the QNL. However, in practice these lasers tend to excite other modes near threshold, which results in large intensity noise, as discussed on page 160.

8.3   The intensity noise eater

It is an obvious idea that the intensity noise of a laser can be reduced using an electro-optic modulator and a feedback controller. In the jargon of the experimentalists such a device is called a *noise eater*. Practical applications are the suppression of the large modulations associated with the relaxation oscillation, as shown in Fig. 6.5, and the suppression of technical noise due to mechanical imperfections of the laser. The schematic layout of a noise eater is shown in Fig. 8.9. It consists of the intensity modulator, a beam splitter to generate a beam that is detected by the in-loop photo-detector D_{il} and control electronics, with a combination of proportional, integral and differential (PID) gain, that drives the modulator. The laser has a noise spectrum $V1_{\text{las}}(\Omega)$ which is reduced by the feedback system to a much quieter spectrum $V1_{\text{out}}(\Omega)$. For large intensity noise, well above the QNL, the feedback system is entirely classical and we can use conventional control theory to describe it. However, once we approach the QNL the properties change. We will find that for a QNL laser input the noise actually increase and completely reversing the classical predictions.

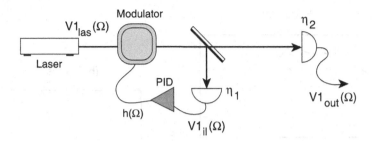

Figure 8.9: Schematic diagram of a noise eater.

For the control of the intensity any one of the different intensity modulators, described in Section 5.4.3 can be used. A fixed voltage is applied to the modulator which reduces the power P_{las} of the laser to a working point $(1 - D)P_{\text{las}}$. We assume $D \ll 1$, but even the largest fluctuation of the laser is less than DP_{las}. The voltage to the modulator is controlled by the feedback circuit to counteract any changes in the intensity coming from the laser. In principle this could be achieved for all frequencies Ω. As a result, the noise at the output, $V1_{\text{out}}(\Omega)$, is quiet. We want to derive the transfer function for $V1_{\text{out}}(\Omega)$.

8.3.1 Classical intensity control

Using standard control theory [Dor92] the feedback system can be described by the open loop amplitude gain $h(\Omega)$. This gain can be visualized as the action of the entire feedback system, including beamsplitter, detector, amplifier and modulator, on an intensity modulation. Here $h(\Omega) = 1$ means the effect of the modulator would be equal to the signal on the laser beam and $h(\Omega) = 0$ means that there is no feedback. For good noise suppression we need $h(\Omega) > 1$. The complex gain $h(\Omega)$ changes the magnitude and the phase of the modulation. The phase shifts are due to the delay in the loop, introduced by the physical layout of the system, and the phase shifts inside the electronic components. The magnitude $h(\Omega)$ is affected by the conversion efficiency of the detector, the gain of the electronic amplifier and the characteristics of the modulator. Figure 8.10 shows one typical example of the gain spectrum $h(\Omega)$. This is usually plotted in separate diagrams for the magnitude $|h(\Omega)|$, Fig. 8.10(a trace i), and the phase $\arg(h(\Omega))$ as functions of Ω, Fig. 8.10(a trace ii). This pair of diagrams is known as the Bode plot. A more useful alternative is the Nyquist plot, Fig. 8.10(b), in which $h(\Omega)$ is shown in a polar plot. The individual frequencies are marked along this plot. It should be noted that in all practical cases the gain will be low, $|h(\Omega)| < 1$ at very low frequencies and very high frequencies. In addition any such a system has to have a phase shift from DC to the highest frequency of at least 2π.

We can derive the effect of the feedback system by tracing the amplitude of the modulation at different places within the system, see Fig. 8.9. For a classical system we can ignore the effect of the vacuum fluctuations. We have the amplitude $\delta \tilde{A}_{il}$ directly after the modulator, $\delta \tilde{A}_f$ after the beamsplitter, $\delta \tilde{A}_e$ inside the electronic feedback system, $\delta \tilde{A}_{\text{out}}$ for the laser beam leaving the noise eater. For the classical system the different amplitudes are linked via

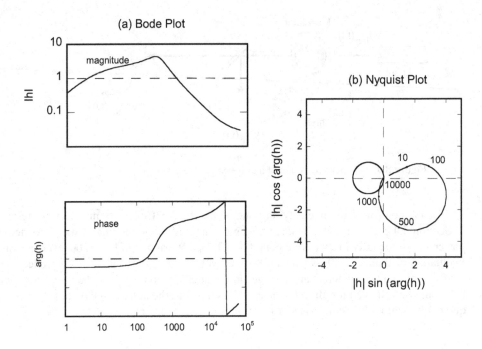

Figure 8.10: (a) Bode plot of a typical feedback loop roundtrip gain, (b) Nyquist plot of the same gain

the equations

$$\delta\tilde{\mathbf{A}}_{il} = \delta\tilde{\mathbf{A}}_{las} + h(\Omega)\,\delta\tilde{\mathbf{A}}_{las}$$
$$\delta\tilde{\mathbf{A}}_{il} = \delta\tilde{\mathbf{A}}_f = \delta\tilde{\mathbf{A}}_e \tag{8.3.1}$$

this results in

$$\delta\tilde{\mathbf{A}}_{il} = \delta\tilde{\mathbf{A}}_{las} + h(\Omega)\delta\tilde{\mathbf{A}}_{il} \tag{8.3.2}$$

and consequently, still ignoring the effect of the vacuum fluctuations and with $h(\Omega)$ compensating the reduction in amplitude caused by the beamsplitter, we obtain

$$\delta\tilde{\mathbf{A}}_{il} = \frac{\delta\tilde{\mathbf{A}}_{las}}{|1 - h(\Omega)|^2} \tag{8.3.3}$$

and we get the well known result from classical control theory

$$V1_{il}(\Omega) = \frac{V1_{las}(\Omega)}{|1 - h(\Omega)|^2} \tag{8.3.4}$$

which is also detected in the beam leaving the noise eater. Without feedback the noise of the light in front of the in-loop detectors is $V1_{il}(\Omega) = V1_{las}(\Omega)$. Once the feedback loop is

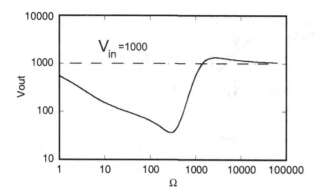

Figure 8.11: The action of the noise eater on a large classical signal using the feedback gain shown in the previous diagram. Input noise is $V1_{las}(\Omega) = 1000$, very large compared to the QNL. The resulting output $V1_{out}(\Omega)$ remains well above the QNL of $V1_{out}(\Omega) = 1$.

closed the fluctuations will be suppressed, the tendency is to make the noise of the in-loop current $V1_e(\Omega)$ as low as possible. This noise suppression is best for large gain. Naively one might expect that the noise suppression depends simply on the sign of $h(\Omega)$; that for negative feedback, $h(\Omega) < 0$, the noise is suppressed and for positive feedback, $h(\Omega) > 0$, the noise is increased. This is not correct. The suppression depends on the value $|1 - h(\Omega)|^2$. This is best explained with a Nyquist diagram (Fig. 8.10(b)). Equation (8.3.4) means that the feedback loop decreases the variance at all those frequencies where $|1 - h(\Omega)| > 1$, that is where $h(\Omega)$ is outside a circle of radius 1 drawn around the point $(1,0)$. The reduction in the variance is proportional to the square of the distance between $h(\Omega)$ and $(1,0)$. If the line that traces $h(\Omega)$ is inside the circle the variance is actually increased, not reduced. If $h(\Omega)$ gets close to $(1,0)$ the resulting noise is extremely large.

The stability of the systems can be described by one simple rule: If the line that traces $h(\Omega)$ in the Nyquist diagram encloses the point $(1,0)$ the system is unstable. It will oscillate madly. This includes the case where $h(\Omega) = 1$. All other systems, where $h(\Omega)$ does not enclose $(1,0)$, are stable. But note that even in a stable system there can still be frequencies where the noise is increased, as demonstrated in the example shown in Fig. 8.11. Finally, we find the variance of the modulation of the beam leaving the classical system is scaled by the transmission $1 - \epsilon$ of the beamsplitter and the quantum efficiency η of the detector, in the same way as the intensity of the output beam, resulting in the transfer function

$$
\begin{aligned}
V1_{out}(\Omega) &= \eta_2(1 - \epsilon)\, V1_{il}(\Omega) \\
&= \eta_2(1 - \epsilon)\, \frac{V1_{las}(\Omega)}{|1 - h(\Omega)|^2}
\end{aligned}
\tag{8.3.5}
$$

where η_2 is the quantum efficiency of the output detector.

To illustrate Equation (8.3.5) consider the output noise spectrum shown in Fig. 8.11 which is calculated for a large constant laser noise $V1_{las}(\Omega) = 1000$ very much larger than the QNL. The feedback system has the gain $h(\Omega)$ shown in Fig. 8.10. We see the effect of a feedback

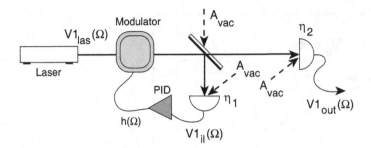

Figure 8.12: Schematic diagram of a noise eater with all vacuum terms.

amplifier with a maximum gain at 300 [kHz] and unity gain points at 6 [kHz] and 1000 [kHz]. The gain drops below 1 at frequencies below 6 [kHz] and above 1000 [kHz]. This feedback loop has an internal delay time of about 4 [μs]. The phase response is determined by the somewhat resonant nature of this circuit. In the Nyquist diagram this corresponds to a line that starts with straight negative feedback until $h(\Omega)$ changes sign at about 400 [kHz]. However, the noise is reduced even at higher frequencies up to 1000 [kHz] since the noise reduction is proportional to the square of the distance to the point $(1, 0)$. The complete spectrum of the output variance is shown in Fig. 8.11. The noise is suppressed from very low frequencies to 1000 [kHz]. The maximum gain results in the best noise suppression at 300 [kHz]. But when the line $h(\Omega)$ enters the circle around $(1, 0)$ the noise increases. A clear hump is visible from 1000 [kHz] to 40 [MHz]. This feedback loop functions well at medium frequencies with high gain. But at higher frequencies the phase has increased too much and excess noise is created. This result is not atypical for a realistic feedback system. In practice one has to work very hard to find a design that reduces the gain quickly enough and simultaneously avoids too much electronic phase change.

8.3.2   Quantum noise control

The behaviour of the feedback system near the standard quantum limit (QNL) is quite different. In this case we have to include the vacuum fluctuations. from the unused input port of the beamsplitter. We can also no longer assume that the variance inside and outside the feedback loop will remain the same. Figure(8.12) shows the apparatus including all vacuum fields. The input into the noise eater is described by $\delta\tilde{\mathbf{A}}_{\text{las}}$ We distinguish between the in-loop field $\delta\tilde{\mathbf{A}}_{il}$ in front of the beamsplitter, the field in front of the in-loop detector $\delta\tilde{\mathbf{A}}_f$, the field which is leaving the apparatus $\delta\tilde{\mathbf{A}}_o$ and the two electrical signals, one inside the feedback loop $\delta\tilde{\mathbf{A}}_{\text{el}}$, which is fed back to the modulator, and $\delta\tilde{\mathbf{A}}_{\text{out}}$ from the detector which measures the beam leaving the apparatus.

It is now necessary to define a quantum description of the feedback. This problem was addressed as early as 1980 and was approached in various ways. The most obvious and direct approach was pioneered by Wiseman [Wis95]. The main feature is that the current signal at time t is not only related to the instantaneous value of the photon flux N_{il} but is the convolution with a response function which takes the history of the signal into account and simulates the gain of the feedback loop. For frequencies which are smaller than the inverse of the roundtrip time of the feedbackloop ($\Omega < 1/t_{loop}$), the effect is similar to the effect in a classical feedback system and we can describe the action of the modulator simply by

$$\delta \tilde{\mathbf{A}}_{il} = \delta \tilde{\mathbf{A}}_{las} + g\sqrt{\epsilon\eta}(\Omega)\,\delta \tilde{\mathbf{A}}_{el} \tag{8.3.6}$$

Here the first line describes the action of the modulator which combines the input field and the electrical signal. The very small absorption of this device, which also corresponds to a vacuum input, has been ignored. The gain of the in-loop electrical amplifier is given by the value g. The various fields are linked by

$$
\begin{aligned}
\delta \tilde{\mathbf{A}}_f &= \sqrt{\epsilon}\delta \tilde{\mathbf{A}}_{il} + \sqrt{1-\epsilon}\,\delta \tilde{\mathbf{A}}_{vac} \\
\delta \tilde{\mathbf{A}}_{el} &= \sqrt{\eta}\delta \tilde{\mathbf{A}}_f - \sqrt{1-\eta}\,\delta \tilde{\mathbf{A}}_{vac} \\
\delta \tilde{\mathbf{A}}_o &= \sqrt{1-\epsilon}\delta \tilde{\mathbf{A}}_{il} + \sqrt{\epsilon}\,\delta \tilde{\mathbf{A}}_{vac} \\
\delta \tilde{\mathbf{A}}_{out} &= \sqrt{\eta}\delta \tilde{\mathbf{A}}_o - \sqrt{1-\eta}\,\delta \tilde{\mathbf{A}}_{vac}
\end{aligned}
\tag{8.3.7}
$$

where we have assumed that the two detectors have the same quantum efficiency η. We can now combine the equations in (8.3.7) to provide explicit expressions for $\delta \tilde{\mathbf{A}}_{el}$ and $\delta \tilde{\mathbf{A}}_{out}$, convert these into the quadrature operators $\delta \tilde{\mathbf{X}} 1_{el}$ and $\delta \tilde{\mathbf{X}} 1_{out}$ and using our standard recipe to calculate the noise spectra $V1_{el}(\Omega)$ for the current inside the loop and $V1_{out}(\Omega)$ for the current from the external detector. The details of the calculation are given in Appendix K and the resulting transfer functions are

$$V1_{el}(\Omega) = \frac{\eta\epsilon(V1_{las}(\Omega) - 1) + 1}{|1 - h(\Omega)|^2} \tag{8.3.8}$$

and

$$V1_{out}(\Omega) = 1 + \frac{1-\epsilon}{\epsilon}\,\frac{\eta\epsilon(V1_{las}(\Omega) - 1) + |h(\Omega)|^2}{|1 - h(\Omega)|^2} \tag{8.3.9}$$

where $h(\Omega) = \eta\epsilon\, g(\Omega)$ is the open loop roundtrip gain, which includes both the electronic gain $g(\Omega)$ and the optical attenuation. Note that our model of the modulator requires the intensity attenuation to be combined with the voltage gain.

These results show that the properties of the light inside the controller and the beam leaving the controller are entirely different. Inside, Eq. (8.3.8), the modulation and the noise are strongly suppressed and for a large gain the variance $V1_{el}(\Omega)$ can be well below the QNL. The noise eater works well inside, which is possible, since the laser beam inside is a conditional field. It only exists inside the feedback system and the conditions for a freely propagating beam do not apply. The standard commutation rules do not apply and allow this violation of the normal quantum noise rule. Such light is sometimes referred to as *squashed light* [Wis00]. For high gain, $|h(\Omega)| > 1$, and a low level of laser noise, $V1_{las}(\Omega) \approx 1$, the variance $V1_{el}(\Omega)$ of the control current can be substantially less than the QNL. This was observed by A.V. Masalov et al. [Mas94], [Tro91] and others. It was concluded that the light inside the loop was sub-Poissonian and could be used for measurements with a better signal to noise ratio than the QNL would allow. However, it was overlooked that any signal, for example a modulated absorption, would be controlled by the feedback loop in a similar way as the classical noise. A detailed analysis has shown that even with sub-Poissonian control currents it is not possible to improve the signal to noise ratio. Any apparent advantage of the low noise of the control current cannot be used [Tau95].

The properties of the freely propagating output beam, Eq. (8.3.9), are quite different. This beam obeys the standard commutation rules and conditions for the quantum noise and the spectrum $V1_{out}(\Omega)$ in most cases above the QNL. We find that this light has properties which are quite different from the predictions of the classical model, see Eq. (8.3.5), and contain unique quantum features. In many situations the effect of the *noise eater* are actually counter-productive, the machine increases the noise level. This is one of the striking examples where the properties of an optical instrument with small signals, near the QNL, is considerably different to the conventional classical predictions, and standard engineering solutions will actually fail. In the area of laser noise control this lesson had to be learned and quantum optics rules are important for the technical progress.

It is worthwhile to consider the following special cases:

(1) For very low gain ($h(\Omega) \rightarrow 0$) this result shows simply the effect of attenuation by the combination of beamsplitter transmission ($1-\epsilon$) and quantum efficiency η with $V1_{out}(\Omega) = 1 + (1-\epsilon)\eta \; (V1_{las}(\Omega) - 1)$. This is identical to a simple beamsplitter. Note that the effect of the quantum noise contribution is included.

(2) For a QNL input laser ($V1_{las}(\Omega) = 1$) the feedback loop creates extra noise which increases with the gain $h(\Omega)$ with $V1_{out}(\Omega) = 1 + \frac{1-\epsilon}{\epsilon} \; \frac{|h(\Omega)|^2}{|1-h(\Omega)|^2}$. This case is interesting for engineering applications, since it means that in systems with very small signals, close to the QNL, the real feedback system shows a behaviour that is quite different from the classical approximation. The higher the gain and the smaller the fraction ϵ of the light that is used for control, the more excess noise is generated. The physical explanation is that the vacuum noise at the beamsplitter is added and amplified. For example, if only 10% of the light is used ($\epsilon = 0.1$) and the open loop gain is $h = -5$ (negative feedback) the output variance will be $1 + ((1 - 0.1)/0.1) \; 5^2/(1+5)^2 = 7.25$. It could be significantly larger if the feedback phase is not optimum. In order to avoid this noise penalty completely all the light would have to be detected ($\epsilon = 1$) which is clearly impractical. One can minimise the excess noise by detecting most of the light, leaving only a small intensity at the output. In the limit of very large gain we obtain the limiting value $V1_{out}(\Omega) = 1 + \frac{1-\epsilon}{\epsilon}$. This value is independent of the gain, a kind of saturation effect producing a fixed noise level, as shown in Fig. 8.13.

(3) For very large signals ($V1_{out}(\Omega) > \frac{h(\Omega)}{\eta\epsilon}$), the system shows the classical properties with the noise suppression given by $V1_{out}(\Omega) = \frac{\eta\epsilon V1_{las}}{|1-h(\Omega)|^2}$. Quantum effects no longer play any role.

To illustrate this quantum effect the performance of the noise eater previously described in Fig. 8.10(b) has been evaluated for a wide range of input laser noise levels. They range from $V1_{las} = 1000$ to $V1_{las} = 1$. All traces in Fig. 8.14 are normalized to the QNL. At the

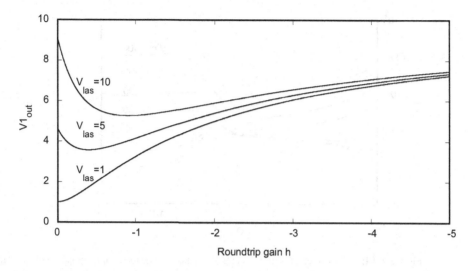

Figure 8.13: The effect of the feedback system with $\epsilon = 0.1$, for different modulations $V1_{\text{las}} = 10, 5, 1.$, as a function of the open loop gain h. At small gains the larger signals are suppressed, at larger gains ($h = -5$) all signals approach the same value, in this case $V1_{\text{out}} = 7.3$

highest frequencies where the gain is much less than one the noise spectrum is unaffected. The top trace, with the large noise level, $V1_{\text{las}} = 1000$, clearly shows the same response as the classical feedback loop. Biggest suppression is achieved close to the maximum gain 300 [kHz] and some excess noise at higher frequencies, 2000 [kHz]. However, the lowest trace for a QNL laser, clearly demonstrates the noise penalty. In this particular case the feedback system produces about 8 [dB] of excess noise. At laser noise levels of about 10 times the QNL the feedback loop has very little effect indeed since in this case noise suppression and noise penalty balance each other.

One explanation for the noise penalty is the fact that the beamsplitter acts as a random selector of photons, not a deterministic junction, (see the discussion in Section 5.1). The noise measured by the in-loop detector is a random selection of the laser noise which is not correlated to the fluctuations transmitted to the output. This small fraction of the noise is detected, converted into a current, amplified, and added by the modulator to the remaining fluctuations of the light from the laser. Obviously all this has to happen within a time less than $1/\Omega$. Unlike in a classical feedback circuit, the noise which was detected cannot cancel the remaining noise since there is no correlation between them. Instead, they are added in quadrature leading to the excess noise. The more gain we apply the worse it gets. This phenomenon has been observed and described by a number of researchers. One particularly instructive description was given by Mertz et al. who used a Monte Carlo technique to describe the properties of the beamsplitter [Mer93].

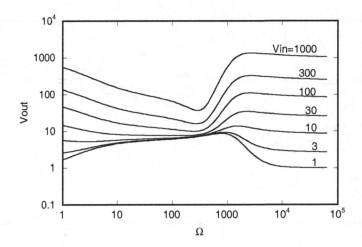

Figure 8.14: The effect of a noise eater on a laser. Various input noise levels ($V1_{\mathrm{las}}$ = $1, 3, 10, \ldots, 1000$) are shown. At low noise levels this device creates rather than eats noise.

Practical consequences

The consequence of this performance at the QNL is that we have to reduce the gain of the feedback loop to much less than unity at all detection frequencies where the laser should be quantum noise limited. Most lasers are QNL at large frequencies. Consequently, the task is to develop special electronic amplifiers that combine high gain at low frequencies, where classical noise dominates, with low gain ($|h| < 10^{-2}$) at the frequencies where we need QNL performance. The trick is to achieve this very steep roll off in the gain without too much phase lag being incorporated [Rob86] and this requires specialized electronic designs.

There are different options for intensity control. As discussed so far, an external intensity modulator can be added to the laser. It is also possible to control the pump power to the laser by feedback. In the case of diode lasers the injection current can be directly controlled, which is a very elegant approach. The performance of this system can be predicted by incorporating the laser noise transfer function for pump noise into the description of the feedback loop. The solution is not trivial. The practical difficulty is that most current control units are not suitable for this since their external inputs include filters designed to avoid noise transients. This produces a phase lag which is far too large for a good feedback performance. In many cases it is better to leave the current source untouched and to add a small control current. This was demonstrated successfully by C.C. Harb et al. [Har94] for a diode laser pumped CW Nd:YAG ring laser. The relaxation oscillation was suppressed by more than 40 [dB] and the noise at low frequencies was largely eliminated. Similar circuits are available in commercial lasers [Kan90].

In conclusion, a feedback intensity controller, or noise eater, is extremely useful to reduce technical noise in the intensity of a laser. With careful design the QNL can be achieved. However, it cannot be used to either generate an output beam that is sub-Poissonian (squeezed) or to measure any effect with a signal to noise ratio better than the standard quantum limit that can already be reached with a coherent state. A complementary design, electro-optic feedforward has quite different properties which can be utilized for noiseless amplification, Chapter 10 or quantum information processing, see Chapter 13.

8.4   Frequency stabilization, locking of cavities

In many experiments it is very important to operate lasers and cavities which are locked to each other, that means they are tracking each other in the frequency domain. Active feedback techniques are required to achieve this aim. The length of either the cavity or the laser resonator is constantly adjusted to keep both of them in resonance at one frequency.

Applications for such a system are buildup cavities for non-linear processes, mode cleaner cavities and the stabilization of the laser frequency. In the latter case, a very quiet reference cavity is chosen, which is protected as much as possible from external influences, such as mechanical vibrations and temperature fluctuations. The laser is locked to this reference and will have a much narrower linewidth than a free running laser. The performance of lasers can best be described using the concept of the Allan variance and quoting the timescale on which the smallest variance is observed [All87]. Frequency stabilization has been used since the very early days of laser design. Many of the milestones in this technology have been achieved by J. Hall and his colleagues in a continuous series of technical developments since the 1970's. Two current examples illustrate the degree of improvement that can be achieved. A free running dye laser has a linewidth of many [MHz], measured over an interval of 1 [sec]. Using active control techniques, the linewidth can be reduced to about 1 [kHz]. A free running diode laser pumped CW Nd:YAG laser has already a narrow linewidth, 10 [kHz] measured over 100 [sec], due to its simple mechanical structure. Using active control techniques linewidths of less than 1 [Hz] measured in 100 [sec] can be achieved. And this is not a fundamental limit [Day92].

Many tricks are used in the design of the special reference cavities, such as low expansion materials, special mounting systems, cryogenic control. Extraordinary precision has been achieved through painstaking engineering and some ingenious design. Here only the principles of the locking technique will be outlined, as an introduction to the techniques commonly used. In most CW quantum experiments at least one cavity locking system is required. Frequently several are used simultaneously. In most cases the locking system can be described by the schematic block diagram shown in Fig. 8.15(i). The aim of the device is to dynamically control the detuning $\Delta\nu$, the difference between the cavity and the laser frequency and to maintain resonance ($\Delta\nu = 0$).

The light from the laser is mode matched, using a lens system, into the reference cavity. An error signal is generated which is a measure of the detuning $\Delta\nu$. This signal is processed using an amplifier of suitable gain, bandwidth and phase response. Finally, an actuator is used, such as a piezo crystal mounted on a mirror, to control either the laser frequency or the cavity frequency. Let us discuss the most important features of these components. The cavity parameters are chosen to produce a suitable control range. Typically the control range is as wide as the cavity linewidth. For lasers with large frequency excursions, such as dye lasers, a low finesse cavity is required. For narrow band lasers, such as CW Nd:YAG lasers, a high Finesse (up to 200 00 or more) is chosen to provide a very steep frequency response.

The cavities can be confocal, this reduces the difficulties of mode matching, or can have a conventional, stable resonator configuration.

Pound-Drever-Hall locking

The error signal has to be responsive to very small changes in $\Delta\nu$. We require a technique that measures the detuning accurately and with low noise. The most popular technique is to measure the response of the cavity to frequency modulation sidebands. This technique was first used by Pound for microwave systems and later refined by several scientists for the optical domain. It is frequently known as the Pound-Drever-Hall technique. In this case a phase modulator, driven by a frequency Ω_{PM}, is placed between the laser and the cavity. The light reflected from the cavity is detected. Since optical feedback has to be avoided an optical isolator, Section 5.4, is normally placed between the cavities, and the reflected beam is directly accessible. The reflected light has modulation sidebands at frequency Ω_{PM}. These are detected using a mixer and, as local oscillator, a part of the modulator driving signal. The amplitude of the resulting low frequency signal provides a suitable error signal.

If the cavity is on resonance with the laser beam, that means $\Delta\nu = 0$, the reflected light will be purely phase modulated. After the mixing process no signal is available, since the two sidebands balance each other exactly. The error signal for $\Delta\nu = 0$ is zero. However, once the cavity is detuned, both PM sidebands will experience a different phase shift, the two beat signals will have a phase difference $\Delta\phi_{out}(\Delta\nu)$ with respect to each other. They will no longer cancel, there will be some AM modulation, a detectable error signal after the mixer.

The shape of the error signal, as a function of cavity detuning, is easiest explained for the case that the modulation frequency Ω_{pm} is much larger than the cavity linewidth Γ_{cav}. The relative phase of the two beat signals is determined by the phase difference between the central component ϕ_c and the two sidebands ϕ_-, ϕ_+. The three components have a distinctively different phase, as discussed in Section 5.3. A typical phase response is shown in Fig. 8.15 (i). The corresponding power response, in this case for an impedance matched cavity, is given in Fig. 8.15(iii)b. We can arbitrarily define $\phi(0) = 0$ at resonance. That means for large modulation frequencies we obtain

$$\phi(-\Omega_{PM}) = \phi(+\Omega_{PM}) = 0$$

The phase difference at the output is

$$\begin{aligned}
\Delta\phi_{out}(\Delta\nu) &= (\phi_+ - \phi_c) - (\phi_c - \phi_-) \\
&= \phi(\Delta\nu + \Omega_{pm}) - \phi(\Delta\nu) - (\phi(\Delta\nu) - \phi(\Delta\nu - \Omega_{pm}))
\end{aligned} \tag{8.4.1}$$

This output phase difference is shown in Fig. 8.15(c).

On resonance both terms are equal, $\phi(\Delta\nu) = 0$ and the two beat signals keep their original phase, $\Delta\phi_{out}(0) = \Delta\phi_{in}(0)$. The modulation remains pure phase modulation. The error signal is zero. For any other detuning we have to evaluate the phase response explicitly. It crosses zero on resonance, $\Delta\nu = 0$, and rises steeply on either side. Close to resonance is has a linear slope, which can be measured in [rad/MHZ]. The steeper the slope, the higher is the gain in the feedback loop. At a detuning less than one cavity linewidth the phase difference peaks and then rolls off gently. At such large detunings the feedback system still has the correct sign.

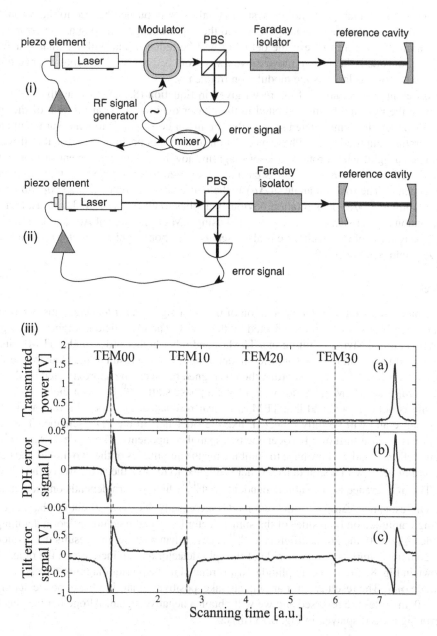

Figure 8.15: Frequency locking system. (i) Schematic layout for PDH locking and phase response of the cavity. (ii) layout for tilt-locking. (iii) Experimental error signals as demonstrated by D.A. Shadock et al. [Sha99]. (a) Transmitted power, (b) error signal generated via PDH locking, and (c) error signal generated via tilt locking technique.

However, it has a small gain and the system is only slowly pushed backed to the resonance. The regime of constant feedback sign is called the capture range. As long as the error signal stays within the capture range it the detuning will be reduced and will eventually arrive at resonance; outside it the system move move away from resonance. The capture range of this technique is large, as large as the modulation frequency.

A different interpretation of the result given in Equation (8.4.1) is to say that the vector describing the reflected beam is rotated in the phasor diagram in regard to that of the input beam. However, the demodulated signal after the mixer corresponds to a measurement along the quadrature amplitude axis. Phase modulation of the input beam is orthogonal to this axis. The reflected light, with a rotated phasor diagram, now has some component of modulation in this direction. Phase modulation is partially converted to amplitude modulation; an error signal appears. The rotation angle $\theta(\Delta\nu)$ is the half phase difference $\Delta\phi_{\text{out}}(\Delta\nu)$, the maximum rotation is $\pm\pi/2$. This rotation obviously applies to all types of modulations, and also to fluctuations. Error signals can be generated using AM or mixtures of AM and PM modulations. The rotation of the quadrature is also used as a self homodyning detector for quadrature squeezed light, see Section 9.3.

Tilt locking

An alternative technique for the generation of the error signal is *tilt locking*. This is based on the interference of spatial modes reflected by the cavity. The input field is slightly misaligned with respect to the cavity, resulting in a TEM$_{0,0}$ and other higher order modes TEM$_{i,j}$ inside the cavity, which are all resonant at different detunings $\Delta_{i,j}$ as a result of the Gouy phase shift, see Sections 5.3.3, 2.1.2. To first order, the misaligned beam can be approximated by the sum of the TEM$_{0,0}$ and TEM$_{0,1}$ modes which have a phase shift $\Phi_{0,1}^{0,0}$ between them. Assuming the cavity is not confocal and that TEM$_{0,0}$ is near resonance, the mode TEM$_{0,1}$ will be far off resonance and will be fully reflected, like the sidebands in the PDH scheme. The mode TEM$_{0,1}$ has a phase shift of π between the two spatial components. The mode TEM$_{0,0}$ will be partially reflected and we now have to combine the the amplitudes of the two modes and detect the resulting beam with a split detector that is aligned to resolve the two parts of the TEM$_{0,1}$ field. The interference is critically dependent on $\Phi_{0,1}^{0,0}$ which in turn depends on the detuning $\Delta_{0,0}$. On resonance, $\Delta_{0,0} = 0$, the two modes are exactly $\Phi_{0,1}^{0,0} = \pi/2$ out of phase and the resulting amplitude on both sides of the split detector has equal magnitude. For a detuning to one side, $\Delta_{0,0} < 0$, the phase difference $\Phi_{0,1}^{0,0}$ is larger than $\pi/2$ and this results in a reduction of the combined amplitude on one side of the split detector and an increase on the other side, as shown in Fig. 8.15(ii). The amplitude $\alpha_{0,0}$ is rotated in the phasor diagram in respect to the amplitude $\alpha_{0,1}$. The result is a, for arguments sake, positive signal. For the opposite detuning, $\Delta_{0,0} > 0$, the roles are reversed and we will obtain a negative signal. Altogether we generate the error signal as displayed in Fig. 8.15 (iii) c.

We see that both techniques are equivalent and produce similar error signals. Tilt locking requires less equipment but is more sensitive to mechanical instabilities. PDH locking is the more commonly employed technique. A third alternative, after Haensch and Couliard, is based on the response of the cavity to two input polarizations. Provided the cavity has a higher finesse for one polarization than the other the reflected light, when analysed in the appropriate polarization direction, will again produce an error signal with the required dispersion shape. Which particular technique should be chosen depends on the specific technical requirements.

The PID controller

The other component is the feedback amplifier. It processes the error signal. The gain should be as large as possible and the bandwidth should span as far as possible to control slow and fast frequency excursions. However, there are severe limitations set by the requirements for the phase of the feedback gain. As already discussed in Section 8.3, a reliable rule for stable operation is that the gain rolls off with high frequencies and the phase shift at the upper unity gain point should be less than π. A more precise definition can be obtained using a Nyquist diagram, see Fig. 8.10. As a consequence the delay times have to be kept short and any resonances, mechanical or electronic, have to be avoided. The gain is usually controlled by a PID amplifier, that means a combination of proportional amplifier (middle frequencies), integrator (low frequencies) and differentiator (high frequencies). All these components contribute to the electronic roundtrip phase and have to be carefully optimized.

Unfortunately, most electro-mechanical actuators have resonances in the acoustic range [kHz], see Fig.8.16, severely limiting the bandwidth of the amplifier which has to reach unity gain at frequencies well be below the resonance. One option is to split the control into several independent branches. For example, to use a piezo with large range (length changes of several optical wavelengths) at low frequencies to compensate for a drift of the frequencies and to use electro-optic components, such as a phase modulator inside the cavity, which has a smaller range but no resonance, for the control of the high frequency excursions. Extensive research has been carried out into these control systems, for example for the applications in frequency standards and precision interferometry.

8.4.1   How to mount a mirror

Where the control is done using with a PZT mounted mirror and the bandwidth of the feedback control system will be severely limited. The mirror, PZT and mirror mount effectively form a system with mechanical resonances $\Omega_{res,j}$ and since this system is inside the feedback loop we can expect stable operation only at frequencies below the lowest resonance $\Omega_{res,1}$. This frequency depends on the mass of the mirror, the mass of the mirror mount or counterweight, and the way the PZT and the mirror are connected. The first improvement is to use the smallest possible mirror substrate and PZT glued with a hard compound onto a massive backplate as the counterweight. This type of solution is found in many commercial systems, for example any laser system with a extended optical resonator.

The next step is to modify the mount by pre-loading it, that means we contain the mirror and PZT in a rigid cage and exert a pressure force onto the mirror. In practical terms the cage is pushing via a rubber O-ring onto the mirror. The result is that the effective spring constants are increased and that the resonance frequencies are increased. This trick has been used by many laboratories which specialize in laser frequency control and many different mechanical systems have been developed. One example was built and analysed by W. Bowen et al. [Bow03] which clearly demonstrates the effect. He built a system where the counterweight and the pre-loading tension can be systematically varied, mounted the mirror as part of

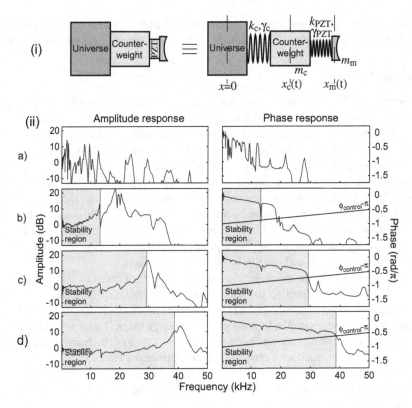

Figure 8.16: (i) Mirror mounted on a PZT modelled as a mass and spring system. The counterweight is the mass on which the PZT is pushing. This model includes the option of preloading. (ii) Amplitude and phase response of a PZT (a) with small counterweight, (b) with large counterweight, (c) with large counterweight and low tension pre-Loading, and (d) with large counterweight and high tension pre-loading. From W. Bowen [Bow03]

a cavity and recorded the cavity locking signal with an electronic network analyser which can record the amplitude and phase response. The results are shown in Fig. 8.16b. We can see that the resonances in the spectrum move toward higher frequencies. The shaded area indicates the useful locking range and this is extended by more and more preloading. The preloading force is limited to values that cause no damage to the mirror.

8.5   Injection locking

For a number of reasons high power lasers tend to have a much wider line width and not have a well defined output mode. The mechanical design is simple, allowing more mechanical vibrations, the control of the pump source is not very sophisticated, the electric current driving the pump source could be noisy or more than one pump source are used. On the other hand, very high quality lasers of low power but with excellent frequency and intensity noise quality

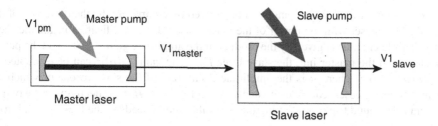

Figure 8.17: Schematic layout for injection locking.

and perfect spatial mode can be built. The trick is to combine the high output power of one laser with the high beam quality of the other. This can be achieved through the technique of *injection locking*. The basic idea is to seed the high power laser, referred to as the *slave laser*, with some of the output from the well controlled, high quality laser, the *master laser*, and force the high power laser to take on most of the mode qualities of the master laser.

The slave laser will change its output frequency and phase to that of the master, it will remain locked to this frequency. As a consequence the laser linewidth of the slave will be reduced to that of the master. If the slave laser was already operating in a single spatial mode no changes would occur, the spatial mode is determined by the geometry of the resonator of the slave laser. However, if it was operating in several modes one mode will get an advantage, through the injection of light, and the laser can change over this specific spatial mode. The output power of the slave laser is basically unaffected by the injection locking.

Injection locking is the optical equivalent of the locking of two oscillators, an effect that was known for a long time for mechanical systems. It was reported that in 1865 Christian Huygens, while confined to his bed by illness, observed that the pendula of two clocks in his room moved in synchrony if the clocks were hung close to each other, but became free running when hung apart. He eventually traced the coupling to mechanical vibrations transmitted through the wall from one clock to the other. This observation was later confirmed using electrical oscillators and has developed into a widespread technique in electronic signal processing and microwave technology. Thus it was well known and established before the laser was invented.

Optical injection locking will occur simply when the output beam of the master laser is matched into the resonator of the slave laser, both in the spatial waveform and by having the same polarisation state. In addition, the optical frequency ν_m of the master laser has to be within the locking range $\nu_s \pm \Delta_l$ of the slave laser. Inside this range locking will happen completely. Outside this range no locking will occur at all, the slave laser will operate as a free running laser. The change from one state to another is abrupt. In practice this means that the slave laser will track the master laser as long as their independent free running frequencies are within the locking range, $|\nu_s - \nu_m| = \Delta_{ms} < \Delta_l$. It is actually not easy to detect where the slave laser is within the locking range since the free running frequency cannot be observed. The only indication would be a phase difference between the injection locked output beam and the master beam. To maintain locking, it might be necessary to add a control system that suppresses frequency drifts of the slave to keep the detuning Δ_{ms} within the locking range. This can be achieved by a much simpler control system than would be necessary for a full frequency stabilization of the slave laser.

An expression for the locking range can be derived by comparing the the power P_s of the free running slave laser with the power of the master laser after amplification inside the slave resonator. This later value is given by the product of the amplitude gain $|G(\nu)|^2$ and the power P_m injected from the master into the slave laser P_m. The gain $|G(\nu)|^2$ can be estimated as $(2\kappa)^2/(2\pi(\nu - \nu_s))^2$ where κ is the amplitude decay rate of the slave resonator which, in many small coupling cases, can be estimated as $\kappa = (1 - R)/t$, R the reflectivity of the output coupler and t the round trip time. Setting the two values equal leads to the locking condition

$$|G(\nu)|^2 P_m \geq P_s \quad \text{or} \quad \Delta_l \leq \frac{2\kappa}{2\pi}\sqrt{P_m/P_s} \tag{8.5.1}$$

As a practical example, if the slave laser has a power of 10 [W], the laser resonator has a round trip length L of 1 [m], the cavity loss rate is 10% per round trip and the master laser has a power of 0.1 [W] the locking range will be $\Delta_l = 2(1 - R)c/2\pi L\sqrt{P_m/P_s} = 0.5$ [MHz].

The first observation of optical injection locking was carried out in 1966 by Stover and Steier who used two He-Ne lasers for this purpose [Sto66]. They controlled the frequencies of the two lasers and were able to confirm the predicted locking range. The technique is now widespread, it is used to control the the properties of the light emitted by high power lasers, to improve the mode quality and to build powerful tunable lasers by locking a low cost powerful diode laser to well controlled low power diode lasers.

On the other hand frequency locking can also be an undesirable process. It prevents the measurement of very small frequency differences. For example in laser gyroscopes, which are used in navigational applications, a beat signal between the two counter propagating modes in a ring laser is measured. The modes experience different cavity lengths, if the ring laser rotates in an inertial frame. They will beat and the frequency of the beat is a measure of the rate of rotation. These frequencies f_{beat} are very small, they are given by $2\pi f_{\text{beat}} = \Omega_{\text{rot}} A/p$ where A is the area and p the perimeter of the gyroscope. Ω_{rot} is the angular frequency of the rotation. For the earth's rotation the value, at the pole, is 2π in 24 hours or 75×10^{-6} [rad/sec]. For a 1 [m^2] ring gyroscope located at the pole the beat frequency is about 20 [Hz]. Locking of the two modes is very likely, but it obviously has to be avoided otherwise the beat signal is zero and it would appear that the earth stood still. Extraordinarily small scatter losses are sufficient to provide enough light to cause locking. As a consequence the demand for laser gyroscopes have been the motivation for most of the recent improvements in low loss optics and the advances in the technology of mirror manufacture. Quantum optics experiments have gained from these improvements.

It is interesting to ask what the noise properties of the injection locked laser system could be. Does the master laser dominate the system completely? Is it possible to build in this way a high power quantum noise limited laser? This question was addressed by several researchers in the regime of classical fluctuations [Far95]. More recently, complete laser models have been developed and, using the linearized quantum theory, it is possible to derive complete noise spectra for the injection locked systems. In these models the laser is literally described as an oscillator with two distinct internal modes, at the frequency of the master and the slave laser, and the combined system is solved within one laser model, similar to that described in Chapter 5 [Ral96].

The outcome of these models and of experiments [Har96] is that the characteristic of the laser intensity noise varies with the detection frequency. There are clearly distinguishable

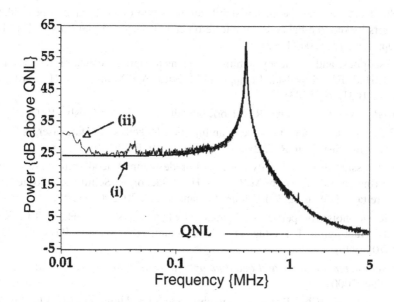

Figure 8.18: The normalized noise transfer spectrum of an injection locked laser: theory and experiment after C.C. Harb et al. [Har96]. (i) theory (ii) experiment.

regions of frequencies that show different properties. A practical example for the locking of two Nd:YAG lasers is shown in Fig. 8.18 One region, region II in Fig. 8.18, is centered roughly around the relaxation oscillation frequency of the slave laser. Here the slave laser operates effectively like an amplifier for the noise of the master laser. The power gain is given by $H = (P_m + P_s)/P_s$ and the noise is multiplied by the same factor. Note: this region acts as a linear control i.e. $V_{out} \approx 1 + V_{in} + (H - 1)$. Thus the variance of the output is the same as the variance of the master laser noise. Note that if the master laser is quantum noise limited this means output is considerably (H times) noisier than a QNL laser with the same power than the slave laser. At very high frequencies (region III) the output can reach the quantum limit, it maintains the properties of the slave laser. At very low frequencies (region I) the dominant source of noise is the pump source of the slave laser. Different strategies have to be followed to optimize the system at the different frequencies.

Bibliography

[All87]   *Time and frequency (time-domain) characterization, estimation, and prediction of precision clocks and oscillators*, D.W. Allan, ICEE Transact. on Ultrasonics, Ferro-electrics and frequency control, Vol. UFFC, No. 6, 647 (1987)

[Bow03]   Experiments towards a quantum information network with squeezed light and entanglement, PhD thesis, W. Bowen, Australian National University (2003)

[Day92]   Sub-Hertz relative frequency stabilisation of two diode laser pumped Nd:YAG lasers locked to a Fabry-Perot interferometer, T. Day, E.K. Gustafson, R.L. Byer, Opt. Lett. 17, 1204 (1992)

[Dre83]   Laser phase and frequency stabilisation using an optical resonator, R.W.P. Drever, J.L. Hall, F.V. Kowalski, J,Hough, G.M. Ford, A.J. Munleyand, H. Ward, Appl. Phys. B 31, 97 (1983)

[Dor92]   *Modern Control Systems*, R.C. Dorf, 6th edition, Addison Wesly (1992)

[Far95]   Frequency and intensity noise in an injection locked, solid state laser, A.D. Farinas, E.K. Gustafson, R.L. Byer, J. Opt. Soc. Am. B 12, 328 (1995)

[Har94]   Suppression of the intensity noise in a diode pumped neodymium YAG nonplanar ring laser, C.C. Harb, M.B. Gray, H-A. Bachor, R. Schilling, P. Rottengatter, I. Freitag, H. Welling, IEEE Jour.of Quant. Elect., 30, 2907 (1994)

[Har96]   Intensity-noise properties of injection locked lasers, C.C. Harb, T.C. Ralph, E.H. Huntington, I. Freitag, D.E. McClelland, H-A. Bachor, Phys. Rev. A 54, 4370 (1996)

[Hau00]   *Electromagnetic noise and quantum optical measurements*, H.A. Haus, Springer Verlag (2000)

[Hil87]   Response of a Fabry-Perot cavity to phase modulated light, D. Hils, J.L. Hall, Rev. Sci. Instrum., 1406 (1987)

[Innolight] http://www.innolight.de

[Kan85]   Monolithic, unidirectional single-mode Nd:YAG ring laser, T.J. Kane, R.L. Byer, Opt. Lett. 10, 65 (1985)

[Kan90]   Intensity noise in a diode pumped single frequency Nd:YAG laser and its control by electronic feedback, T.J. Kane, IEEE Photon. Technol. Lett. 2, 244 (1990)

[Koe96]   *Solid-state laser engineering*, W. Köchner, 4 th ed, Springer Verlag (1996)

[Lev85]   Squeezing of classical noise by non degenerate four-wave mixing in an optical fiber, M.D. Levenson, R.M. Shelby, S.H. Perlmutter, Opt. Lett. 10, 514 (1985)

[Mas94]   Sub-Poissonian light and photocurrent shot-noise suppression in a closed optoelectronic loop, A.V. Masalov, A.A. Putilin, M.V. Vasilyev, J. Mod. Opt. 41, 1941 (1994)

[Mer93]   Photon noise reduction by controlled deletion techniques, J. Mertz, A. Heidmann, J. Opt. Soc. Am. B 10, 745 (1993)

[Ral96]   Intensity noise of injection-locked lasers: quantum theory using linearized input-output method, T.C. Ralph, C.C. Harb, H-A. Bachor, Phys. Rev. A 54, 4359 (1996)

[Rob74]   *The theory of polarisation phenomena* B.A. Robson, Clarendon, Oxford (1974)

[Rob86]   Intensity stabilization of an Argon laser using electro-optic modulator: performance and limitation, N.A. Robertson, S. Hoggan, J.B. Mangan, J. Hough, Appl. Phys. B 39, 149 (1986)

[Sch03]   Stokes operator-squeezed continuous-variable polarization states, R. Schnabel, W.P. Bowen, N. Treps, T.C. Ralph, H-A. Bachor, P.K. Lam. Phys. Rev. A 67, 012316 (2003)

[Sha87] Theory of light detection in the presence of feedback, J.H. Shapiro, G. Saplakoglu, S.-T. Ho, P. Kumar, B.A.E. Saleh, M.C. Teich, J. Opt. Soc. Am. B 4, 1604 (1987)

[Sha99] Frequency locking a laser to am optical cavity by use of spatial mode interference, D.A. Shaddock, M.B. Gray, D.E. McClelland, Opt. Lett. 24, 1499 (1999)

[Sto66] Locking of laser oscillators by light injection, H.L. Stover, W.H. Staier, Appl. Phys. Lett. 8, 91 (1966)

[Tau95] Effects of intensity feedback on quantum noise, M.S. Taubman, H.M. Wiseman, D.E. McClelland, H-A. Bachor, J. Opt. Soc. Am. B 12, 1792 (1995)

[Tau96] A reliable source of squeezed light, and an accurate theoretical model, M.S. Taubman, T.C. Ralph, A.G. White, P.K. Lam, H.M. Wiseman, D.E. McClelland, H-A. Bachor, in *Lasers in Research and Engineering: Proceedings of Laser 95*, Springer-Verlag, 120 (1996)

[Tro91] Photon statistics in a negative-feedback optical system, A.S. Troshin, Opt. Spektrosk. 70, 389 (1991)

[Wal81] Reduced quantum fluctuations in resonance fluorescence, D.F. Walls, P. Zoller, Phys. Rev. Lett. 47, 709 (1981)

[Wis95] Feedback-enhanced squeezing in second-harmonic generation, H.M. Wiseman, M.S. Taubman, H-A. Bachor, Phys. Rev. A 51, 3227 (1995)

[Wis00] Squashed states of light: theory and applications to quantum spectroscopy, H.M. Wiseman, J. Opt. B 1, 459 (1999)

[Yue80] Optical communication with two-photon coherent states. i. quantum-state propagation and quantum-noise reduction, H.P. Yuen, J.H. Shapiro, IEEEtrans. Inf. Theory IT-26, 78 (1980)

9 Squeezing experiments

9.1 The concept of squeezing

So far we have worked with coherent states of light and the limits imposed by quantum noise which affect all quadratures equally. However, it is possible to generate light which has less noise in one selected quadrature than the QNL dictates. Light with this property is called *squeezed light*. Ultimately we want to be able to carry out optical measurements with this light and achieve a better signal to noise ratio than is possible with coherent light.

In this chapter, we start be describing nonlinear optical processes and how they affect the fluctuations. This is first done with in a classical approach – since it is possible to see some of the key results immediately. We use two examples examples which are intuitive and instructive. Next we introduce the concept of squeezed states of light rigorously. This will provide the mathematical tools to analyse the experiments. Finally we discuss the experiments which have been performed since 1985 to generate squeezed light in detail.

9.1.1 Tools for squeezing, two simple examples

The properties of quantum noise have already been introduced in detail in Chapter 4, now we wish to reduce the fluctuations of the light. We cannot suppress the fluctuations altogether, that would violate the uncertainty principle. But we can rearrange them between the quadratures. We can build experiments, or applications, which measure only one quadrature at a time. In this case only the noise in this one quadrature contributes to the final result. Consequently, it will be sufficient to reduce the fluctuations in this one quadrature only.

The remaining problem is: What are the tools that would reduce the noise in one quadrature? How can we generate states of light that have asymmetric quadrature noise distributions which are squeezed in one quadrature? We will see in this section that the trick is to start with a normal coherent state and to introduce correlations between the fluctuations of the two quadratures $\tilde{X}1$ and $\tilde{X}2$. To illustrate how correlations can reduce the quantum noise, the case of the Kerr effect is instructive [Lev90].

The Kerr effect

Consider a beam of light travelling through a medium which has an intensity dependent refractive index. The beam can be described as an optical field E_{in} with both amplitude and phase fluctuations:

$$\alpha(t) = \alpha_{\text{in}} + \delta X1_{\text{in}}(t) + \delta X2_{\text{in}}(t) \tag{9.1.1}$$

A Guide to Experiments in Quantum Optics, 2nd Edition. Hans-A. Bachor and Timothy C. Ralph
Copyright © 2004 Wiley-VCH Verlag GmbH & Co. KGaA
ISBN: 3-527-40393-0

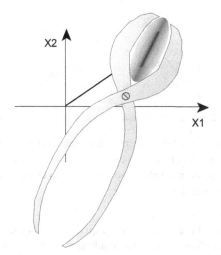

Figure 9.1: Tools for squeezing

This could be a QNL laser beam, in which case $\delta X1_{\text{in}}(t)$ and $\delta X2_{\text{in}}(t)$ represent the quadratures of the quantum noise. The nonlinear refractive index $n(\alpha)$ can be described by

$$n(\alpha) = n_0(1 + n_2\, \alpha^2) \tag{9.1.2}$$

As a consequence fluctuations in the intensity will modulate the refractive index. The refractive index, in turn, will modulate the phase of the transmitted light. For a medium of length L and light at the wavelength λ the output phase is given by :

$$\phi_{\text{out}} = \phi_{\text{in}} + \frac{2\pi nL}{\lambda} = \frac{2\pi L}{\lambda}\left(n_0 + n_2(\alpha_{\text{in}}(t) + \delta X1_{\text{in}}(t)\,)^2\right) \tag{9.1.3}$$

This is shown in Fig. 9.2. There is no change to the amplitude or intensity of the light

$$\alpha_{\text{out}} = \alpha_{\text{in}} = \alpha \qquad \text{and} \qquad \delta X1_{\text{out}}(t) = \delta X1_{\text{in}}(t)$$

At the output of the medium, the phase fluctuations are those of the input light plus a component which is linked to the input intensity fluctuations. Using the relationship between the quadratures $\alpha\delta\phi = \delta E2$, which is valid for small angles, this can be written in terms of the quadrature amplitudes

$$\begin{aligned}
\delta X2_{\text{out}}(t) &= \delta X2_{\text{in}}(t) + \frac{4\pi n_0 n_2 L}{\lambda}\alpha^2\delta X1_{\text{in}}(t)\\
&= \delta X2_{\text{in}}(t) + 2r_{\text{Kerr}}\delta X1_{\text{in}}(t)\\
\text{with} \quad r_{\text{Kerr}} &= \frac{2\pi n_0 n_2 L\alpha^2}{\lambda}
\end{aligned} \tag{9.1.4}$$

where r_{Kerr} is a parameter which includes the nonlinearity and the average amplitude α of the input field. Higher order terms in $\delta X1$ have been neglected since they are exceedingly

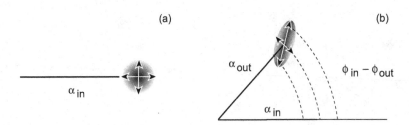

Figure 9.2: Effect of a Kerr medium on the electric field vector. (a) input field, (b) output field. Note that the field illustrated here has a large amplitude, thus the origin of the diagram is far away and not shown. Consequently any changes in the phase are orthogonal to changes in the amplitude. The change in phase is intensity dependent.

small, and only the terms linear in $\delta X1$ have been kept. We see from Equation (9.1.4), that the nonlinear medium couples the fluctuations in amplitude and phase quadratures. The amplitude α drives the nonlinear refractive index. This in turn controls the quadrature phase X2. At the output of the medium the fluctuations are no longer independent. The two quadratures are correlated.

What is the effect of this process on the statistics of the fluctuations? On the one hand, we have made the light noisier by adding fluctuations. On the other hand we now have correlated fluctuations. We can estimate the consequences by plotting the effect not just for one point but for a whole distribution of classical amplitudes. This is done in Fig. 9.3. The input and the output fields are represented by density plots, already discussed in Chapter 4 and shown in Fig. 2.9. The axes of the plot for Kerr effect, Fig. 9.3 a (i), are the input quadratures $X1_{in}$ and $X2_{in}$. The mean value is given by the point $(\alpha_{in}, 0)$. The fluctuating amplitude is represented by a group of dots, each dot gives one instantaneous value of $(\delta X1_{in}(t), \delta X2_{in}(t))$. The density of the dots indicates the probability of finding a specific value of the quadrature amplitude fluctuation, as introduced in Section 4.1. This simulation uses a random distribution, with 2-dim. Gaussian weighting, as described in Appendix A. The probability of measuring a certain value $X1(t)$ is given by a histogram of the projection of all dots onto one of the axes. Two such histograms, for both the $X1$ and $X2$ coordinates, are shown in Fig. 9.3 a (ii). Both of them are Gaussian curves with the same width.

Part (b) of Fig. 9.3 shows the distribution at the output of a nonlinear Kerr medium for a parameter $r_{Kerr} = 0.375$. Note that the total phase shift ϕ_{out} has been ignored. The axes are again the quadrature amplitudes $X1_{out}$ and $X2_{out}$. The distribution is no longer circularly symmetric. The area covered by the distribution is actually larger since we have added fluctuations, see Fig. 9.3 b (i). But note that the distribution is narrower in one special direction. The histogram of this distribution, taken for a projection axis $Y1$ rotated by the angle θ_s to the $X1$ axis, Fig. 9.3 b (ii), is narrower than the histogram for the input state, Fig. 9.3 a (ii. In this particular case the FWHM, or standard deviation, is reduced from 1 to 0.79. The histogram for the projection $Y2$, Fig. 9.3 b (ii) orthogonal to $Y1$, is wider, by a factor 1.23, than at the input. The width of the distribution in the squeezing direction is given by

$$VY1 = 1 - 2\, r_{Kerr}\sqrt{1 - r_{Kerr}^2} + 2\, r_{Kerr}^2 \qquad (9.1.5)$$

well below the QNL. This result is derived on page 275.

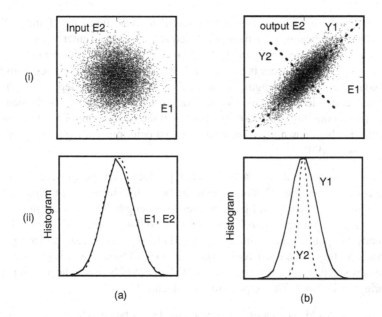

Figure 9.3: Simulation of the effect of a Kerr medium on the fluctuations of an electro-magnetic field. Left hand side (a) is the input field, right hand side (b) the output field. The top part (i) shows the probability distribution of the field, the bottom part (ii) shows the histograms of the samples above.

This concept can also be used for a quantum noise limited laser beam or coherent state. The dot diagram represents the measured probabilities of the observables $\hat{X}1$ and $\hat{X}2$. The variance of the rotated quadrature $\hat{Y}1$, rotated by θ_s with respect to $\hat{X}1$, is smaller than that for a coherent state. Just by introducing this correlation between the quadratures we have created light with noise reduced below the standard quantum limit in one specific quadrature. Note that in this example the uncertainty area has been increased. The output is no longer a minimum uncertainty state, but this is not of great importance for squeezing. We are only interested in measurements of the quadrature $\hat{Y}1$. The details of the size of the noise suppression and of the squeezing angle in a real experiment are described in Section 9.7.

The Kerr effect acts on all fluctuations, independent of their frequency. The model can be applied to each individual Fourier component of the fluctuations. For each detection frequency the process correlates the quadratures of the modes. The correlations are introduced separately for each Fourier component. In this way, the Kerr effect correlates the amplitude and the phase quadrature of the light, and this applies to modulations as well as the quantum noise.

Four wave mixing

A second example of a squeezing tool is *four wave mixing* (4WM). This is a well-known non-linear process that occurs in many resonant media, such as atoms. Imagine three laser beams that are interacting with the medium and are tuned close to resonance with one absorption line, as illustrated in Fig. 9.4(a). This process is best described in the frequency domain. Two of the laser beams, the pump beams, have the optical frequency ν_L. They enter from different angles. The third beam is at the frequency $\nu_L + \Omega$. The medium responds with the generation of a new beam at $\nu_L - \Omega$. This process is energetically allowed; the energy comes from the abundance of photons at ν_L. The strength, or probability, of this process depends on three factors: (i) a nonlinear coefficient $\chi^{(3)}$ which varies from material to material; (ii) the density of the sample; and (iii) a condition that describes how well the momentum of the photons is preserved. The latter is also known as phase matching and determines the geometry of the input and output beams [Lev90].

Conventionally, this process is used as a technique to produce a new beam which emerges at a different frequency or direction. This new beam can be used to investigate the properties of the nonlinear medium or as a source of light at the new wavelength. This process also works if we have only one single input beam which has a single modulation sideband at $\nu_L + \Omega$. In this case all frequency components are co-linear and 4WM produces a new sideband at $\nu_L - \Omega$, which is phase locked to the original sideband. The efficiency of the process is given by the relative size of the new sideband. It can be quantified by the ratio C between the amplitude of the new and the original sideband. This applies for any sideband.

Consider the effect of 4WM on an amplitude modulated laser beam. That means we have pairs of sidebands, each pair in phase with each other, i.e. their respective beat signals are in phase and constructively interfere. Each sideband is partially transferred to the other sideband respectively by 4WM, see Section 2.1.6. This will increase the sidebands, and thus the modulation by a factor $(1 + C)$. The medium provides gain for amplitude modulation, as shown in Fig. (9.4(c)). In contrast, phase modulation consists of two sidebands 180 degrees out of phase. Again, 4WM creates additional components at the opposite sideband frequency. The transfer from one sideband to the other will decrease the phase modulation by a factor $(1 - C)$, that means 4WM reduces phase modulation, see Fig. 9.4(d). In conclusion, the nonlinear process of 4WM provides a selective gain to one quadrature and reduces the other. 4WM is an example of a phase sensitive amplifier, see Section 6.2, which has different gain for each quadrature. This process applies both to modulation as well as noise sidebands. Any amplitude modulation noise is amplified, any phase modulation noise is reduced. Consequently, a beam which initially has equal amounts of both types of noise will turn into light with enhanced amplitude modulation and reduced phase modulation noise. This classical argument shows us that an input beam which is a coherent state will turn into a state with suppressed phase noise. This result is confirmed by a full quantum treatment [Slu87a].

It is useful to consider the 4WM process in terms of a phasor diagram, much like we have done for the Kerr effect. Again, we will consider one Fourier component of the fluctuations at a time. Any periodic excursion from the long term average value corresponds, in the phasor diagram, to an oscillation at the modulation frequency Ω. It can be a mixture of amplitude and phase modulation. This can be drawn, as shown in Fig. 9.4(c), as an additional vector, the sum of the vectors $\delta X1(\Omega)$ and $\delta X2(\Omega)$. The 4WM process will increase $\delta X1(\Omega)$ by a factor $(1 + C)$ and reduce $\delta X2(\Omega)$ by a factor $(1 - C)$, see Fig. 9.4(d). This is called *phase sensitive gain*. We can test the effect of 4WM on a truly noisy state, which is represented by a 2-dimensional random Gaussian distribution in time. The transformation we apply point by

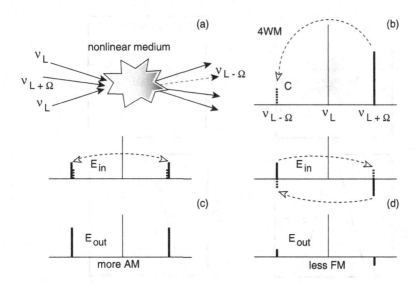

Figure 9.4: The effect of a 4WM medium: (a) The geometrical layout, a new beam is generated. (b) The effect of 4WM in the frequency domain is a transfer from one sideband to another. (c) Amplitude modulation is increased. (d) Phase modulation is reduced.

point is

$$\delta X1_{\text{out}}(t) = (1 + C)\, \delta X1_{\text{in}}(t) \quad \text{and} \quad \delta X2_{\text{out}}(t) = (1 - C)\, \delta X2_{\text{in}}(t) \quad (9.1.6)$$

Figure 9.5 shows an example of both input and output distributions Figs. 9.5(a), 9.5(b) respectively, based on the random sample from Fig. 9.3 and a value of $C = 0.3$. Clearly the output state has increased amplitude fluctuations and reduced phase fluctuations. The size of the uncertainty area is preserved. This phasor diagram represents a quantum noise limited input, clearly the output has phase fluctuations below the standard quantum limit. It is worth noting that the arguments used here are exactly the same as those applied in phase-conjugation. The only difference is that there the spatial phase contributions are reduced by the generation of a reflected wave with opposite phase while in 4WM we consider temporal fluctuations.

These two examples, Kerr and 4WM, are only illustrations of a concept. They show that correlations introduced by nonlinear processes can be used to generate a new type of light with smaller fluctuations, in one specific quadrature, as compared to the input light. The above classical examples can be used as a guide for a complete quantum model of this new type of light.

In Chapter 4, it was shown that the quadrature probability distribution for a quantum state can be described by the Wigner function. In particular, for a coherent state, the Wigner function is a two-dimensional Gaussian function with circular symmetry. The new state has a Wigner function which is compressed in one direction, expanded in the other. Certainly *squeezed state* is a fitting description. The first step for a complete description of this light is a formal and rigorous definition.

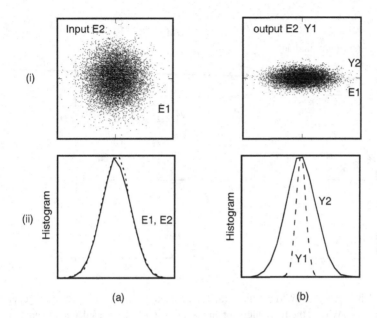

Figure 9.5: Simulation of the effect of a 4WM medium on the fluctuations of an electro-magnetic field. Part (a) shows the input state. The distribution is circularly symmetric. Part (b) shows the distribution after the 4WM medium. It is stretched. The top part (i) shows the probability distribution of the field, the bottom part (ii) shows the histograms of the samples above.

9.1.2   Properties of squeezed states

With squeezed states we mean states that have the following property: The distribution function of the quadratures is not symmetric, that is, the uncertainty area is no longer a circle and the width of the distribution function in one particular direction $Y1$ is narrower than the standard quantum limit. The direction $Y1$ can be at any angle with respect to the quadratures $X1$ and $X2$. Measurements which are designed to be sensitive only to $Y1$ and not the orthogonal quadrature $Y2$ contain noise less than the standard quantum limit. For this reason we call it *non-classical* light, meaning that it cannot be described by the normal theory of coherent light, given in Chapter 4, which prescribes a minimum uncertainty area that is independent of the quadratures. Quantum mechanics and non-linear optics have to be combined to avoid the symmetry and to go below the standard uncertainty limit.

The uncertainty area for a squeezed state is generally drawn as an ellipse. Why does it have this particular shape? The Wigner function for any coherent state is a Gaussian distribution in every direction. The contour line for this state is a circle. We found that this corresponds to fluctuations in the measurements of the variable such that $V(\Omega, 0) = V(\Omega, 90) = 1$ for all detection frequencies. It was already shown by W. Pauli [Pau30] that any minimum uncertainty state, even squeezed, will have 2-dimensional Gaussian distribution functions. The only requirement is that the widths of these functions satisfy the uncertainty principle $V(\Omega, \theta)\, V(\Omega, \theta + 90) = 1$ for any direction θ. One of the two variances can be less than 1.

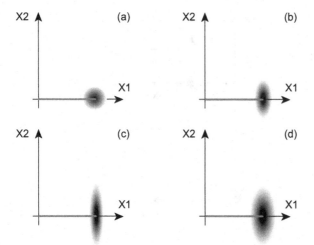

Figure 9.6: Phasor diagram comparing squeezed and other states: (a) Coherent state, (b) Minimum uncertainty squeezed state which is narrower than the coherent state in one direction, (c) Squeezed state with excess noise, (d) An asymmetric, noisy but not squeezed state – it is described by an ellipse, but no projection is narrower than the coherent state.

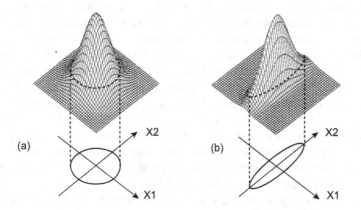

Figure 9.7: Wigner function of (a) coherent state and (b) squeezed state

The contour line for such a 2-dimensional Gaussian function is always an ellipse. An ellipse can be described by exactly two parameters: the direction of the narrowest axis $Y1$ and the ellipticity. Note that the area inside the contour can never be smaller than the area inside the uncertainty circle of a coherent state.

There are states with excess noise, that is, states with $V(\Omega, \theta)\, V(\Omega, \theta + 90) > 1$ for some or even all frequencies. Such states need not be described by 2-dimensional Gaussian with elliptical contours. However, we find that most realistic optical systems can be described by theoretical models which are linearized in the fluctuations, (see Section 4.5). These systems

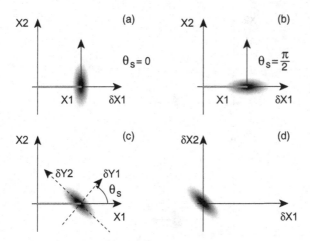

Figure 9.8: Different types of squeezed states. (a) amplitude, (b) phase squeezed state, (c) quadrature squeezed state, (d) vacuum quadrature squeezed state

can have nonlinear properties which are essential for the generation of squeezed light, but its description requires only first order terms in regard to the fluctuations. All such models result in Gaussian distribution functions – and the contours are ellipses. It would be of great interest to generate light with more exotic contours, with parts of the Wigner functions that have negative values and strong non-classical properties – but this has so far only been achieved in the single photon counting domain. Further details are given in Section 9.13. For the remainder we shall restrict ourselves to squeezed states which are described by elliptical contours.

Squeezed states require a two dimensional description. The quadrature amplitudes $Y1$ and $Y2$ provide the most useful coordinate system – the theoretical model is best derived using these parameters. Direct intensity measurements will only determine the intensity fluctuations. While this determines the width of the variance $V1(\Omega)$ of the amplitude, measured along the $X1$ axis, it does not measure the quadrature fluctuations in any other direction. We require experiments that can measure rotated quadratures, in particular, $V(\Omega, \theta_s)$ and $V(\Omega, \theta_s + 90)$. This is usually achieved by using a homodyne detector (8.1.4).

There are different types of squeezed states, as shown in Fig. 9.8. Each squeezed state ellipse has a narrow axis ($Y1$) and, at right angles, the axis with the widest distribution ($Y2$). If the narrow axis is aligned with the amplitude axis, i.e. ($Y1$) parallel to ($X1$), we call the light amplitude squeezed. If it is aligned with the quadrature phase axis, i.e. ($Y1$) parallel to ($X2$), it is called phase squeezed light. In general, ($Y1$) could be rotated at some arbitrary angle θ_s relative to the axis $X1$. This is called quadrature squeezed light.

What are the uses of these various types of squeezed light?

Amplitude squeezed light is useful in absorption or modulation measurements. It can be detected simply by direct detection. Due to the reduced amplitude noise the signal to noise ratio for amplitude modulation is improved. The fluctuations in the other quadrature, the phase quadrature, are above the standard quantum limit. There could even be excess noise in the phase quadrature. None of this would affect the direct intensity measurement. Amplitude squeezed light is obviously related to light with Sub-Poissonian counting statistics. It has a Fano-factor of less than unity. The detailed connection between squeezing and photon statistics has to be determined in each individual case. It depends on the actual photon probability distribution which could be very different to a Poissonian distribution. However, for the generally observed moderate degree of squeezing the link is simple: $F_\alpha = V1_\alpha$. This light is frequently called *bright squeezed* light. The generation of bright squeezed light using second harmonic generators [Siz90, Pas94, Ral95, Tsu95] or diode lasers is now a reliable experimental technique and is described in Section 9.6.

Quadrature phase squeezed light can be used to improve the signal to noise ratio of optical measurements. Care has to be taken that only the phase quadrature is measured. This can be achieved in interferometric arrangements where amplitude fluctuations are cancelled out and phase modulations are turned into modulation of the intensity at the output. In such an arrangement only the noise of the light in the phase quadrature will contribute to the final signal to noise ratio.

Finally, there is a peculiar state, centered at the origin of the phasor diagram, that has a squeezing ellipse. This state has an axis of reduced fluctuations. But for each fluctuation there is an equally likely fluctuation in the opposite direction. This state is given the confusing name *squeezed vacuum state*. We will see that it is closely related to the coherent vacuum state $|0\rangle$. Detecting a squeezed vacuum state directly would not result in any measurable intensity. To be precise, this state contains a certain small number of photon pairs, see Equation (9.2.5). The number of such pairs increases the more the state is squeezed, but this effect is far too small to be directly with photo-diode technology. Only with photon counting can such small intensities be seen. In any other regard this state has normal properties: it is a beam with a well defined mode, direction, size, frequency and polarization. However, once this beam is aligned with a homodyne detector and beats with a strong local oscillator beam we can see its properties: the quantum fluctuations vary with the detection phase. The fluctuations corresponding to the $(Y1)$ quadrature are smaller than those of the $(Y2)$ quadrature.

This squeezed vacuum state can have significant practical applications, as shown in Section 5.2. All interferometers, and other instruments based on beam splitters, contain input ports that are not used. That means there are locations in the experiment which could be used as the input for a real laser beam which would propagate right through the instrument, probably with some attenuation, and finally reach the detector. Such an input port has the properties of a real mode of light. In practice this port is not illuminated and it can be ignored in classical calculations. However, in a quantum model of such an instrument we cannot ignore this port. We have to represent it with a vacuum mode, the coherent state $|0\rangle$. These vacuum input modes are the origin of the quantum noise in the output, as was shown in Section 5.1. Consequently the performance of the instrument can be improved by illuminating the unused input port by a squeezed vacuum mode. The squeezing axis $(Y1)$ has to be aligned such that this quadrature is now the origin of the quantum noise. The result will be reduced quantum noise and a better signal to noise ratio. This was proposed by C. Caves [Cav81] for the case of interferometers and has been demonstrated in an impressive landmark experiment by M. Xiao et al. [Xia87] who achieved an improvement of the signal to noise ratio of 3 [dB]. The details and limitations of this technique will be discussed in the Chapter 10.

9.2   Quantum model of squeezed states

In Chapter 4 we described beams of light as coherent states and we now want to expand this definition. There is a much larger class of states, the squeezed states, which could exist. They have to be based on a wider definition and can be derived from the lowest number state, the vacuum state.

9.2.1   The formal definition of a squeezed state

Squeezed states are represented by

$$|\alpha, \xi\rangle \qquad \text{and} \qquad \xi = r_s \exp(i2\theta_s) \tag{9.2.1}$$

where α^2 is the intensity of the state, θ the orientation of the squeezing axis and r_s the degree of squeezing. The squeezed states can be generated from the vacuum state

$$|\alpha, \xi\rangle = \tilde{\mathbf{D}}(\alpha)\tilde{\mathbf{S}}(\xi)|0\rangle \tag{9.2.2}$$

where the squeezing operator $\tilde{\mathbf{S}}(\xi)$ is defined as

$$\tilde{\mathbf{S}}(\xi) = \exp\left(\frac{1}{2}\xi^*\tilde{\mathbf{a}}^2 - \frac{1}{2}\xi\tilde{\mathbf{a}}^{\dagger 2}\right) \tag{9.2.3}$$

The properties of the squeeze operator are:

$$\begin{aligned}
\tilde{\mathbf{S}}^{\dagger}(\xi) &= \tilde{\mathbf{S}}^{-1}(\xi) = \tilde{\mathbf{S}}(-\xi) \\
\tilde{\mathbf{S}}^{\dagger}(\xi)\tilde{\mathbf{a}}\tilde{\mathbf{S}}(\xi) &= \tilde{\mathbf{a}}\cosh(r_s) - \tilde{\mathbf{a}}^{\dagger}\exp(-2i\theta_s)\sinh(r_s) \\
\tilde{\mathbf{S}}^{\dagger}(\xi)\tilde{\mathbf{a}}^{\dagger}\tilde{\mathbf{S}}(\xi) &= \tilde{\mathbf{a}}^{\dagger}\cosh(r_s) - \tilde{\mathbf{a}}\exp(2i\theta_s)\sinh(r_s)
\end{aligned} \tag{9.2.4}$$

Note that the reverse order of $\tilde{\mathbf{D}}(\alpha)$ and $\tilde{\mathbf{S}}(\xi)$ in Equation (9.2.2) is possible. This results in the so called two photon correlated state which was introduced by H. Yuen even before the name squeezed state was coined [Yue76]. However, this distinction has little relevance for experiments. It makes no detectable difference for any realistic state. The formal description allows us to derive the noise properties of a squeezed state exactly. The squeezed states have the following expectation values for the annihilation and creation operators

$$\begin{aligned}
\langle \alpha, \xi | \tilde{\mathbf{a}} | \alpha, \xi \rangle &= \alpha \\
\langle \alpha, \xi | \tilde{\mathbf{a}}^2 | \alpha, \xi \rangle &= \alpha^2 - \cosh(r_s)\sinh(r_s)\exp(2i\theta_s) \\
\langle \alpha, \xi | \tilde{\mathbf{a}}^{\dagger}\tilde{\mathbf{a}} | \alpha, \xi \rangle &= |\alpha|^2 + \sinh^2(r_s)
\end{aligned} \tag{9.2.5}$$

The last equation means that the intensity of a squeezed state is slightly larger than that of the coherent state with the same value $|\alpha|$. However, this is a minute effect for any laser beam with detectable intensity. Using these properties we can derive the variances of the generalized quadrature $\tilde{\mathbf{X}}(\theta) = \exp(-i\theta)\tilde{\mathbf{a}} + \exp(i\theta)\tilde{\mathbf{a}}^{\dagger}$, the quadrature which would be measured at the rotation angle θ. This variance is given by

$$\text{Var}(\tilde{\mathbf{X}}(\theta)) = \cosh(2\,r_s) - \sinh(2\,r_s)\cos(2\,(\theta - \theta_s)) \tag{9.2.6}$$

This quite complex expression shows that the variance is a periodic function of the rotation angle, as one would expect from the concept of an ellipse being rotated. It has a minimum when $\theta = -\theta_s$ and a maximum in the orthogonal direction $\theta = -\theta_s + \pi/2$. This variance is plotted in Fig. 9.9 for one fixed squeezing angle $\theta_s = -\pi/3$ and for the range of squeezing parameters $r_s = 0.25; 0.5; 0.75$ and 1.00. Figure 9.9(a) is a linear plot of the variance while Fig. 9.9(b) shows the same data on a logarithmic scale, that means the vertical scale is the normalized variance in [dB] above and below the standard quantum limit. Part (c) shows the same data as (a) plotted in a polar diagram. Here the quantum limit corresponds to a circle with radius 1 (bottom of the plot) and the orientation of the suppressed quadrature Y1, the squeezing angle θ_s, is immediately visible.

It can be seen, that as the squeeze parameter is increased the minimum variance decreases and the maximum increases. This agrees with the concept of stretching the ellipse. What is not immediately obvious is the fact that the range of rotation angles for which the variance is below the standard quantum limit gets narrower with increasing squeezing. The points where the variance is equal to the standard quantum limit get closer to θ_s. This is particularly obvious from the logarithmic plot (Fig. 9.9(b)). This means that the rotation angle has to be very well specified in order to obtain the best noise suppression. The noise properties are most easily described by introducing the quadratures $\tilde{\mathbf{Y}}\mathbf{1}$ and $\tilde{\mathbf{Y}}\mathbf{2}$ which are rotated in regard to the standard quadrature amplitudes $\tilde{\mathbf{X}}\mathbf{1}$ and $\tilde{\mathbf{X}}\mathbf{2}$ by the angle θ_s

$$\tilde{\mathbf{Y}}\mathbf{1} + i\tilde{\mathbf{Y}}\mathbf{2} = (\tilde{\mathbf{X}}\mathbf{1} + i\tilde{\mathbf{X}}\mathbf{2}) \exp(-i\theta_s) \tag{9.2.7}$$

The variances for these two quadratures, that is the minimum and maximum of Equation (9.2.6), have particularly simple expressions

$$\langle \Delta\tilde{\mathbf{Y}}\mathbf{1}^2 \rangle \;=\; V_{\alpha,\xi}(\theta_s) = \exp(-2r_s)$$

$$\langle \Delta\tilde{\mathbf{Y}}\mathbf{2}^2 \rangle \;=\; V_{\alpha,\xi}\left(\theta_s + \frac{\pi}{2}\right) = \exp(2r_s) \tag{9.2.8}$$

Thus the minor and major axes of the contour ellipse are of the size $\exp(r_s)$ and $\exp(-r_s)$. Note that the contour has a value that corresponds to the square root of the variance. The details of the link between the variance and the Wigner function are discussed in Section 4.1

Coherent states are eigenstates of the annihilation operator. This is not true of squeezed states. The squeezed states are eigenstates of

$$\tilde{\mathbf{b}} = \cosh(r_s)\tilde{\mathbf{a}} + \sinh(r_s)\tilde{\mathbf{a}}^\dagger$$

and are thus not eigenstates of the annihilation operator. Notice that in the limit $r_s \to \infty$ the operator $\tilde{\mathbf{b}}$ approximates $\tilde{\mathbf{X}}\mathbf{1}$. Thus the squeezed states tend toward quadrature eigenstates.

At the photon detector we will have one single beam and we can arbitrarily set the amplitude α as real. The average number of photons in a squeezed state is evaluated in Eq. (9.2.5) and the variance in the photon number is given by

$$Vn_{\alpha,\xi} \;=\; |\alpha|^2[\,\cosh(2\,r_s) - \exp(-2i\theta_s)\sinh(2\,r_s)\,]^2 + 1/2\;\;\sinh^2(2\,r_s)$$

This complex expression is the variance measured by direct intensity detection. For quadrature squeezed states there is no longer a simple link between $Vn_{\alpha,\xi}$ with the variance $V_{\alpha,\xi}(\theta_s)$. The variance measured for the intensity behaves differently to the quadrature amplitude variance.

(a)

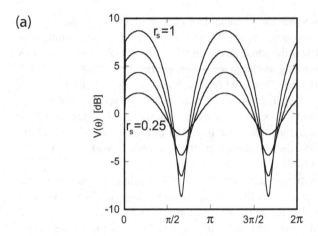

(b)

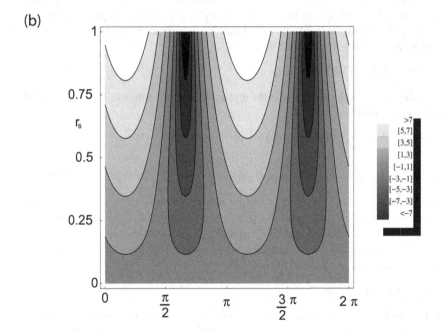

Figure 9.9: (a) A plot, on a logarithmic scale in [db] above QNL of the variance of a squeezed state with $\theta_s = \pi/3$ as a function of the squeezing angle and for different squeezing parameters $r_s = 0.25 \rightarrow$ 1.00. (b) contour plot of the same data.

There are two limiting cases which have simple properties: For bright squeezed light, $(\alpha \gg r_s)$ and $(\alpha \gg 1)$, the average photon number is close to the coherent state $|\alpha\rangle$. Thus intensity measurements of $|\alpha, \xi\rangle$ will provide the same average results as those of a coherent state and the intensity fluctuations will be proportional to V1. The noise in the quadratures will range from $\exp(-2r_s)$ to $\exp(2r_s)$ depending on the detection angle. In general the noise will depend on the detection frequency and we define the noise spectra

$$VY1_{\alpha,\xi}(\Omega) = \quad \mathcal{F}(V_{\alpha,\xi}(\theta_s)) \qquad \text{at frequency } \Omega$$
$$VY2_{\alpha,\xi}(\Omega) = \quad \mathcal{F}(V_{\alpha,\xi}(\theta_s + 90)) \qquad \text{at frequency } \Omega$$

Note that in many situations we are only interested in the best possible squeezing. This is expressed by the squeezing spectrum $S(\Omega)$. This is the value of $V_{\alpha,\xi}(\Omega, \theta)$ for the optimized squeezing angle θ_s, which can vary with the detection frequency. Thus, any rotation of the ellipse is simply ignored. In most theoretical papers the final result quoted is $S(\Omega)$. The other extreme case is a squeezed vacuum state $|0, \xi\rangle$. This state contains a small number of photons

$$n_{0,\xi} = \langle 0, \xi | \tilde{a}^\dagger \tilde{a} | 0, \xi \rangle = \sinh^2(r_s) \tag{9.2.9}$$

Any squeezed vacuum state cannot be perfectly empty of photons. It is not the minimum energy state, and therefore not a vacuum at all. However, for realistic degrees of squeezing ($r_s < 1$) the number of photons is so small compared to the coherent state used as the local oscillator in a homodyne detector, that it can be neglected. Direct detection of the energy of a squeezed vacuum state has not yet been done. As mentioned already, the squeezed vacuum state has many practical applications.

It is of great theoretical interest, but little practical consequence, to consider the case of a squeezed state with a few photons, typically $|\alpha|^2 < 10$. In these cases even more unusual effects can occur. Take for example photon statistics. A squeezed state has larger probabilities for even than for odd photon numbers. A strongly squeezed state contains mainly pairs of photons. For these cases the photon number distribution is not at all simple and for completeness it is stated here

$$\mathcal{P}(n) \quad = \quad \frac{1}{2^n n! \cosh(r_s)} (\tanh(r_s))^n \ \exp[-\alpha^2(1 + \tanh(r_s))]$$
$$\times \left| H_n \left(\frac{\alpha \exp(r_s)}{\sqrt{\sinh(2 r_s)}} \right) \right|^2 \tag{9.2.10}$$

where α is again real and H_n are Hermite polynomials. Please note that this formula is frequently misprinted. For illustration the photon probability functions for amplitude squeezed states ($\theta_s = 0$) with different average photon numbers ($\alpha^2 = 4, 36, 1000$) and a degree of squeezing ($r_s = 1$) are shown in Fig. 9.10. We see that for moderate squeezing, or for large photon numbers, the distributions are similar to a Poissonian with a narrower width. If the squeezing parameter r_s approaches the value of α the distribution develops extra oscillations. This is a sign that photon pairs are predominant in strongly squeezed states.

9.2.2   The generation of squeezed states

Conventional light, from either thermal sources or from lasers, is not squeezed. Laser light can be approximated by coherent states at sufficiently high detection frequencies. How can a coherent state be transformed into a squeezed state? In the introduction to this chapter, examples were given for this process: the Kerr effect and 4WM. The conclusion was that light which

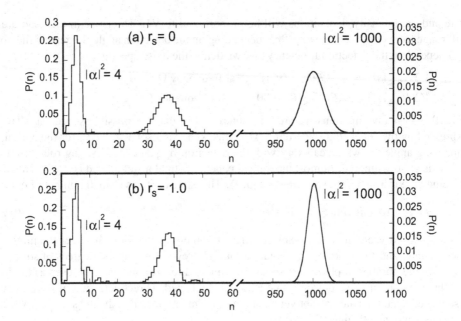

Figure 9.10: Photon probability functions for three different average photon numbers $\alpha^2 = 4, 36, 1000$ for a coherent state ($r_s = 0$) and (b) a squeezed state ($r_s = 1.0$).

contains correlated quadrature fluctuations can show a reduction in the variance of a specific quadrature. This result can be generalized. The feature that distinguishes any squeezed state from a coherent state are these correlations. While for coherent states all quadrature pairs $\tilde{X}(\theta), \tilde{X}(\theta + 90)$ are equal, the correlations will single out a particular set of quadratures $\tilde{Y}1, \tilde{Y}2$ and the fluctuations in $\tilde{Y}1$ will be below the QNL. We need a nonlinear process to achieve this correlation between the quadratures. Obviously there are several such interactions which can be used. Two classes of interactions are described by the generic Hamiltonian operators

$$\tilde{H} = i\hbar \; [\alpha \chi^{(2)} \tilde{a}^2 - \alpha \chi^{(2)} \tilde{a}^{\dagger 2}] \tag{9.2.11}$$

and

$$\tilde{H} = i\hbar \; [\alpha^2 \chi^{(3)} \tilde{a}^2 - \alpha^2 \chi^{(3)} \tilde{a}^{\dagger 2}] \tag{9.2.12}$$

where $\chi^{(2)}$ and $\chi^{(3)}$ are the second and third order nonlinear susceptibilities of the medium and α is the amplitude of the pump field which drives the nonlinear process, which is assumed to be a large classical field. The first Hamiltonian describes a degenerate parametric amplifier, the second 4WM or a Kerr medium. There are other more complex situations which also lead to squeezing. These will contain higher order terms or other combinations of $\tilde{a}$ and $\tilde{a}^{\dagger}$. In his landmark paper in 1983, D. Walls pointed out the general effects of these operators and coined the name *squeezed states* [Wal83].

The effect of these Hamiltonians on the fluctuations will be discussed in detail in Sections 9.6 and 9.5. In all practical cases, the linearization technique discussed in Section 4.5 can be used. These Hamiltonians are frequently over-simplifications, not a complete description of a realistic system where other effects will take place at the same time, but the Hamiltonians, Eqs. (9.2.11), (9.2.12), are essential for the generation of the correlations. In general, calculations based only on this part of the total Hamiltonian would overestimate the size of the correlations and the degree of squeezing. Additional processes will degrade the correlations and thus the noise suppression. A different type of Hamiltonian which generates the required correlation is of the form

$$\tilde{\mathbf{H}} = i\hbar\chi^{(2)}(\tilde{\mathbf{b}}^\dagger\tilde{\mathbf{a}}_1^\dagger\tilde{\mathbf{a}}_2 - \tilde{\mathbf{b}}\tilde{\mathbf{a}}_1\tilde{\mathbf{a}}_2^\dagger)$$

between the modes $\tilde{\mathbf{a}}_1, \tilde{\mathbf{a}}_2, \tilde{\mathbf{b}}$ and which converts, in a nonlinear process, the photons in the mode $\tilde{\mathbf{b}}$ into photons in the modes $\tilde{\mathbf{a}}_1, \tilde{\mathbf{a}}_2$ such that the energy is conserved. Examples are second harmonic generation from from pairs of photons in $\tilde{\mathbf{a}}$ to $\tilde{\mathbf{b}}$ and the optical parametric oscillator above threshold operating in the reverse direction. In this case the correlation is not created between the quadratures of one state but between the quadratures of two different states, for example the fundamental and the second harmonic beam. This process results in cross correlations between beams of different optical frequencies as well as auto correlations of the individual beams. In the following chapters experiments will be described which use a whole range of different processes and materials. In each case the noise spectrum is derived and the practical limitations and the relevant competing noise processes are discussed in detail.

9.2.3   Squeezing as correlations between noise sidebands

A very useful interpretation can be given for the properties of the squeezed states, using the engineering point of view described in Section 4.5.5. The quantum noise of a coherent state at a specific detection frequency Ω can be interpreted as the contributions to the beat signal by the sidebands of the complex amplitude. There are two contributing side-bands and these have frequency independent fluctuations (white noise). The two side-bands $\delta E_{+\Omega}, \delta E_{-\Omega}$, at the frequency $+\Omega$ and $-\Omega$, are not correlated. In contrast, squeezed states have correlated sidebands. The observed noise suppression is a consequence of both sidebands. For this reason most theoretical books call this *two mode squeezing*. All squeezing experiments are based on nonlinear effects and work due to such a correlation.

For a coherent state the noise which is generated by these two side-bands adds in quadrature. The RMS value of this noise corresponds to the quantum noise. The fluctuations at the side bands can have a random phase to each other. They are described by complex numbers. Graphically they can be represented by a vector located on the frequency axis at $\nu_L + \Omega$ and $\nu_L - \Omega$ with a random orientation, shown in Fig. 9.11(a). Only the RMS size is fixed, which is is indicated by the circles at $\nu_L + \Omega$ and $\nu_L - \Omega$. The amplitude of the field is described by

$$\alpha(t) = \alpha_0 \exp(2\pi i\nu_L) + \delta E_{+\Omega}(t)\exp(2\pi i\nu_L + \Omega) + \delta E_{-\Omega}(t)\exp(2\pi i\nu_L - \Omega)$$

Squeezed states have correlated sidebands, the observed noise suppression is a consequence of both sidebands. For this reason theoretical books refer to *two mode squeezing*. Consider the following extreme cases: If the sidebands are completely correlated the wave would be modulated. A complete positive correlation $\delta E_{+\Omega}(t) = \delta E_{-\Omega}(t)$ means the left and the right side bands fluctuate in phase with each other. This corresponds to an amplitude modulated beam. Since the fluctuations at one frequency are random in time this will be a noisy amplitude. The beam has intensity noise. It corresponds to the modulated light introduced in Section 4.5.5. The corresponding quadrature diagram is a line in the $X1$ directions shown in Fig. 9.11(c). Similarly, a complete anti-correlation $\delta E_{+\Omega}(t) = -\delta E_{-\Omega}(t)$ leads to a vector which oscillates in the $X2$ direction, see Fig. 9.11(e). The two components are completely out of phase with each other. This is closely related to a phase modulated light beam which is equivalent to a beam with a whole series of sidebands, see Equation (2.1.25). The oscillating term $\delta E_{+\Omega}(t)$ corresponds exactly to the first side bands. The higher order side bands (at $\pm 3\Omega$, $\pm 5\Omega$, ...) can be neglected since we deal with extremely small modulations. This is light with random phase modulation but with no amplitude noise and direct detection will show no extra noise at all.

More generally we can have complete correlation with a fixed phase angle θ. That means any contribution of a fluctuation $\delta E_{+\Omega}(t)$ is accompanied by a fluctuation $\delta E_{-\Omega}(t) = \exp(i\theta)\delta E_{+\Omega}(t)$. The result is again a straight line in the phase diagram, see Fig. 9.11(f). The light is randomly modulated in a quadrature $Y1$. When detected with a homodyne detector with local oscillator phase $\theta/2$, the full noise is detected, at local oscillator phase $(\theta/2 + \pi/2)$ no noise is detected.

Now let us turn our attention to partially correlated side bands. That means that $\delta E_{+\Omega}(t)$ and $\delta E_{-\Omega}(t)$ contain both fluctuations which are completely correlated and some other components which are completely random and independent. If we add the contributions in a quadrature phase diagram the correlated part will again be represented by a line. The random, uncorrelated part corresponds to a circle, with a diameter smaller than the standard quantum limit. The total result is a convolution of the two. The vector is a sum of the instantaneous components of both. If the partial correlation is positive we obtain an extended, elliptically shaped area along the $X1$ direction, Fig. 9.11(b). This light is tending towards amplitude modulation. Its quadrature phase is squeezed. Similarly, a partial anti-correlation leads to an uncertainty area along the $X2$ direction, Fig. 9.11(d). This light is tending towards phase modulation. Thus the light is quadrature amplitude squeezed. If the correlation is of a fixed phase angle θ_s the uncertainty area is drawn out along the $Y1$ axis, Fig. 9.11(f). This light is quadrature squeezed.

The question is: Can this semiclassical interpretation of squeezed light be used for any quantitative calculations? The answer is yes. This model will produce a result as reliably as the full quantum model. The degree of correlation between the quadratures and the correlation angle are equivalent to the squeezing spectrum $V(\Omega, \theta)$ and the squeezing angle θ_s. This is because the Wigner function for Gaussian states is always positive and can be interpreted as a classical probability distribution for the quadrature measurement.

The only assumption in this interpretation which is not classical is the fixed size of the noise side bands which correspond to the QNL, as discussed in Section 4.5.5. This is based entirely on the quantum nature of light and is mathematically equivalent to the inclusion of the vacuum state in the full quantum calculations. As stated in Section 4.5.5, the sidebands $|\delta E_{+\Omega}(t)|$ and $|\delta E_{-\Omega}(t)|$ can be associated with the operators $\tilde{\delta}a$ and $\tilde{\delta}a^\dagger$. Once the variances of the fluctuations $|\delta E_{+\Omega}(t)|$ and $|\delta E_{-\Omega}(t)|$ have been set everything else can be calculated

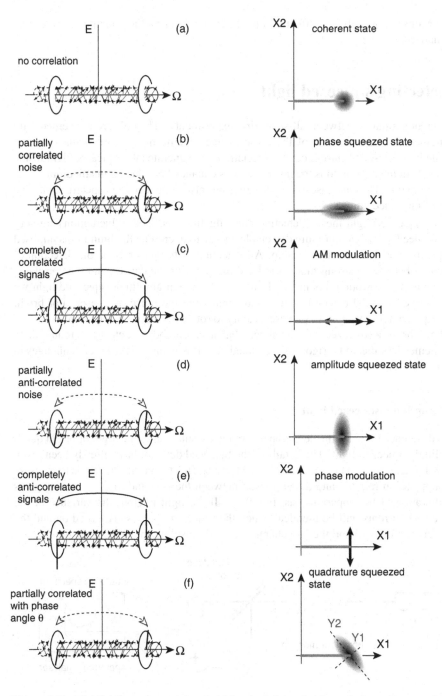

Figure 9.11: The link between correlated sidebands, left hand side, and phase diagrams for squeezed states, right hand side.

classically. More importantly, the effect of a nonlinear medium on the quantum noise can be
clearly visualized.

9.3   Detecting squeezed light

There is a great similarity between all squeezing experiments. They all are concerned with
the measurement of the reduction of the noise in one specific quadrature of the light. A
nonlinear medium is used to introduce the correlation that generates the squeezed light inside
the experiment. In many situations the nonlinearity is enhanced by placing the medium inside
an optical resonator. This increases the squeezing but also influences the spectrum and the
orientation of the squeezing.

Detecting squeezed light means detecting noise fluctuations. These fluctuations are very
small, and we need to have a minimum detectable energy to overcome the limitations imposed
by electronic noise in our detection system. All detection techniques rely on the detection of
the beat signals between a strong mode, the local oscillator or the carrier, and the correlated
noise sidebands at $\pm\Omega$ around this mode. In the case of bright amplitude squeezed light we
can use the same balanced detector as we used to measure the noise of a laser. For bright
quadrature squeezed bright light we can use a cavity to rotate the angle of detection. For weak
squeezed light, or for a squeezed vacuum state, a balanced homodyne detector is required. In
all these schemes loss has to be rigorously avoided, and this is one of the chief challenges in
squeezing experiments.

Detecting amplitude squeezed light

There is no difference between the detection technique for sub-Poissonian light or bright quadrature amplitude squeezed light. The details of the balanced detector have already been given
in Section 8.1 where it is used to measure classical noise. This technique allows us to measure
the noise suppression reliably and a comparison between the $(+)$ and the $(-)$ current gives
a direct calibration of the apparatus, see Fig. 9.12. If the light is QNL the variance of the
$(+)$ and the $(-)$ currents will be identical, when the squeezing process is turned on and the
difference between the $(+)$ and the $(-)$ channels directly displays the noise reduction.

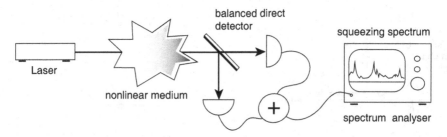

Figure 9.12: Detection scheme for bright squeezed light: the balanced detector.

Detecting quadrature squeezed light

For quadrature squeezed light we use the homodyne detection scheme, discussed in Section 8.1.4. The experiment looks like a Mach Zehnder interferometer with the nonlinear medium in one of the arms. A typical squeezing experiment is shown in Fig. 9.13. The light emerges from the squeezing medium with the correlated fluctuations in one quadrature. A strong local oscillator is generated and sent around the squeezing experiment. Its phase is controlled and the two beams are recombined. The alignment procedure is identical to that of an interferometer, in particular the spatial modes of the two interfering beams have to be matched. They have to overlap and contain identical wave fronts and essentially form one large interference fringe. Both output beams are detected, the difference $(-)$ of the two photo-currents is formed. The optical power of the local oscillator is chosen to be far higher than that of the squeezed light. An analysis of this apparatus, see Appendix E, shows that the noise level of the photo-current is proportional to the optical power of the local oscillator and contains only the fluctuations of one quadrature of the squeezed light. Any local oscillator noise is suppressed.

The detection frequency is determined by the spectrum analyser that operates with zero span, that means it detects only one single frequency. The quadrature is chosen by scanning the phase of the local oscillator. In this way all quadratures can be investigated, one at a time. The result is an oscillating trace, showing the variances for all quadratures θ, as derived in Equation (9.2.6) and plotted in Fig. 9.9. The apparatus has to be calibrated with an independent Poissonian light source. This arrangement is the essential scheme for the detection of squeezed vacuum states. The whole experiment is one large interferometer and thus has to be protected from vibrations. A change in the local oscillator phase during the measurement will smear out the trace, and this tends to reduce the best noise suppression. This is particularly noticeable when the squeezing is large.

Using a cavity to measure quadrature squeezing

For bright squeezed light there is an alternative of using a cavity to analyse the degree and the quadrature of the squeezing. In the conventional homodyne detector the rotation of the quadratures is achieved by controlling a phase shift between the sidebands and the local oscillator. We could use the squeezed light itself as the local oscillator, if we had a device which distinguishes between the two noise sidebands and the centre component of the output beam and gives one a phase shift in respect to the other. This can be achieved by reflecting the light from a cavity that has a linewidth narrower than the detection frequency, see Fig. 9.14. Once the cavity is detuned a phase shift $\theta(\Delta\nu)$ is introduced. The details of this technique have already been described in Section 8.4, where we found that the effect of a detuned cavity is to rotate the quadratures in the reflected light. The rotation is almost linear with the detuning particularly if the detuning is less than one cavity linewidth. With a phase shift of θ the noise sidebands that are measured correspond to the fluctuations of the light in the quadrature $X(\theta)$. By varying the detuning $\Delta\nu$ the angle θ can be selected.

In practice, the cavity has to be locked to a fixed detuning $\Delta\nu$ during each measurement. This can be achieved by a servo control system and by using the Pound-Drever-Hall technique described in Section 8.4. The phase modulation for the locking should be at frequencies far away from the detection frequency for the squeezing, see Fig. 9.14, to avoid contamination of

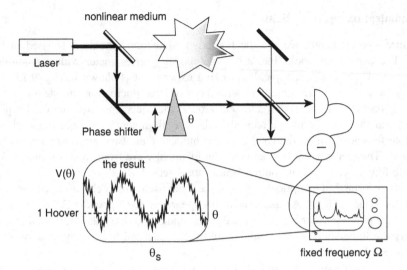

Figure 9.13: A typical squeezing experiment. The fluctuations of the quadratures are selected by tuning the phase of the local oscillator. One single detection frequency is selected. $V_{\alpha,\xi}(\Omega)$ is displayed as a function of the detection angle θ.

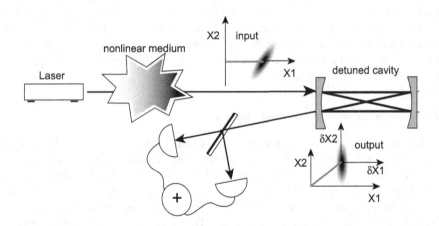

Figure 9.14: The detection of quadrature squeezed light using a phase shifting cavity.

the signal. The actual detuning, and thus θ, has to be determined separately. Note that the noise level for the QNL, which is proportional to the optical power on the detector, also varies with the detuning. We have to record the DC photo-current for each detuning and calibrate each measurement individually. Typical examples of the power attenuation and the phase shift of a detuned cavity are given in Fig. 5.12 and we notice that it is quite possible to achieve rotations of $\pm\pi/2$. This detection technique was first used in the fibre Kerr experiment by M. Levenson et al. [She86, Lev85] and has been described in detail by P. Galiato et al. [Gal91].

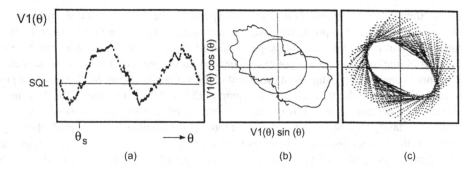

Figure 9.15: Comparison of (a) measured variances, (b) the polar plot of the variances, known as a lemniscate and (c) the reconstruction of the contour of the Wigner function.

9.3.1   Reconstructing the squeezing ellipse

The link between the measured variance $V(\theta)$ and the squeezing ellipse is complex. The ellipse is the contour of the quasi probability distribution or Wigner function. It indicates the point where the quasi probability is half the maximum value. The variance can be derived from histograms obtained after all points in the distribution have been projected on to the direction θ. These histograms should not be confused with cross sections of the Wigner function. A cross section depends on the values along one single line – all other parts of the Wigner function do not influence the cross section. In principle there could be many different distribution functions which have the same histogram. For example one could imagine a Wigner function with a valley in the centre. This would result in double peaked cross sections with a minimum at the centre, but the histograms which include the projections of all points in the 2-dimensional plane, not only along a line still has a maximum. It can generally be shown that the entire 2-dimensional function can be reconstructed from a large number of histograms, which is the purpose of quantum state tomography, as discussed in Section 9.13.

However, for realistic systems these fine distinctions are not important. We can reliably assume that the distribution function is a 2-dimensional Gaussian function (see Section 4.1). Consequently the measured variances allow us to reconstruct the function with the contour, i.e. the squeezing ellipse. Actually, the measurement of the minimum and the maximum variances would give us all the information required to uniquely determine the parameters r_s and θ_s. However, where experimental data is involved it is not advisable to rely on a few data points only. It is useful to plot a likely contour based on all the measured values of the variance, without any assumption about the shape of the Wigner function.

This can be done in two steps: Firstly, the measured data is converted into a polar plot. All the variances $V(\theta)$ are plotted as a function of θ. For a squeezed state this results in a peculiar figure, not unlike a bow tie, see Fig. 9.15(b). The mathematical shape, for the ideal case of a minimum uncertainty squeezed state is a lemniscate. It has pointed minima at the squeezing angle θ_s. The dip inside the circle represents a coherent state. From θ_s the line rapidly expands outwards until it reaches a broad maximum at $(\theta_s + 90)$. Secondly, one takes the square root of all the values, since the point with value 0.5 in a normalized Gaussian curve has the distance of one standard deviation, or RMS, from the maximum. We cannot transfer each variance into exactly one point on the contour since the relationship is not directly from point to point. However, we can indicate a limit for the extent of the Wigner function by drawing two tangents, perpendicular to the line with direction θ. The contour of the distribution function should not extend beyond these tangents. From a measurement at θ we cannot conclude anything about the extent of the Wigner function, but from drawing many of these tangents a reliable 2-dimensional map of the contour is obtained. The diagram with all the tangents drawn, Fig. 9.15(c), leaves a space in between all the lines which corresponds to the uncertainty area, the area inside the contour. This technique is very useful when applied to data with considerable noise. It mixes together many measurements and provides a good indication of the quality of the data. All that is required is a complete scan of $V(\theta)$ for a range of θ of 180 degrees. For illustration the measured variance, the lemniscate and the reconstructed contour lines are shown in Fig. 9.15 for an experimental case, namely data obtained from the squeezing experiment with Ba atoms reported in Section 9.8.

9.3.2 Summary of different representations of squeezed states

We have seen that there are a number of graphical representations for a coherent state, each concentrating on one particular aspect. These are compiled in Fig. 4.7. Similarly, there are a number of graphical representations for squeezed states, which are summarized in Fig. 9.16. A coherent state can be represented by just one parameter, $|\alpha\rangle$. We need more parameters to define a squeezed state. Any minimum uncertainty squeezed state could be represented by $|\alpha, \xi\rangle$ and could be described by three parameters (α, r_s, θ). A squeezed state with excess noise is mixed and requires a fourth parameter, which describes the ratio of the minimum variance to the maximum variance. The properties of the pure squeezed states can be summarized as follows:

$$
\begin{aligned}
VY1 &= \exp(-2r_s) \quad \text{and} \quad VY2 = \exp(2r_s) \\
n_{\alpha,\xi} &= |\alpha|^2 + \sinh^2(r_s) \\
Vn_{\alpha,\xi} &= (\alpha \cosh(r_s) - \alpha^* r_s e^{2i\theta_s} \sinh(r_s))^2 + 2\cosh^2(r_s)\sinh(r_s)
\end{aligned}
$$

Both r_s and θ_s can be functions of the detection frequency. The best noise suppression, for given frequency Ω, is quoted as the *squeezing spectrum*

$$
S(\Omega) = \exp(-2r_s(\Omega)) \quad \text{with optimised} \quad \theta_s(\Omega)
$$

9.3.3 Propagation of squeezed light

A major concern in all experiments is the sensitivity of the squeezed light to losses. The light will have to propagate through various optical components; it needs to be detected and has to overlap with other beams. Each such loss, mismatch or inefficiency will reduce the degree of squeezing. This is very different from conventional optics experiments, where losses can be tolerated and can always be compensated by an increase in input intensity. The quantitative effect of losses is easy to predict since all passive losses behave in the same way and can be simulated by beam splitters. Only nonlinear processes are excluded from this model.

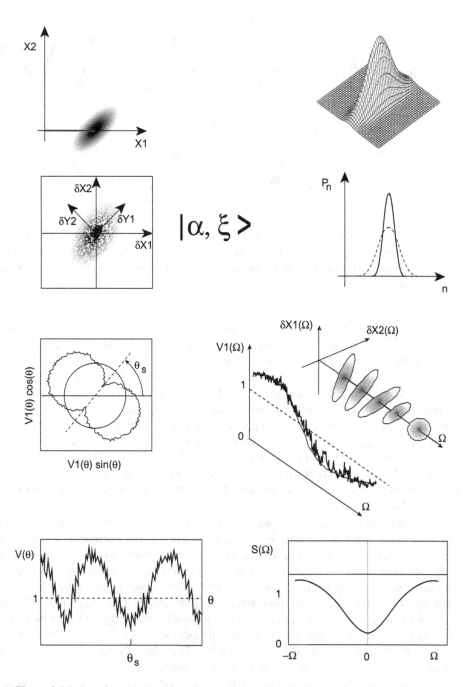

Figure 9.16: Summary of the various graphical representations of a squeezed state.

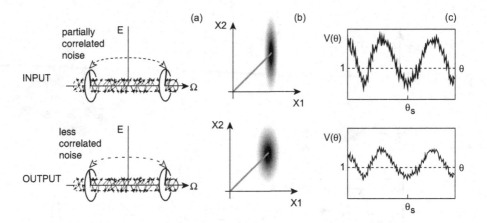

Figure 9.17: The effect of a beamsplitter on squeezed light, input at the top and output at the bottom. Three representations are used: (a) correlated sidebands, (b) phasor diagrams, (c) quadrature variances. The top row are representative of the light at the input, the bottom row at the output of the beamsplitter.

In terms of the sideband picture the effect of the beam splitter is easy to describe. The squeezed light corresponds to light with partially correlated sidebands $\delta E_{+\Omega}$ and $\delta E_{-\Omega}$. For a beamsplitter with reflection ϵ, transmission $\eta = (1 - \epsilon)$, the correlated part of the noise sidebands is reduced by $\sqrt{\eta}$, in the same way as modulation sidebands are reduced, see Section 2.1, while simultaneously the uncorrelated part is increased to keep the magnitude of the quantum noise sidebands constant. This is equivalent to including the quantum noise sidebands of the vacuum state from the empty port of the beamsplitter. The combined sidebands $\delta E_{+\Omega,\text{out}}$, $\delta E_{-\Omega,\text{out}}$ at the output are less correlated than those at the input, the degree of squeezing is reduced, Fig. 9.17a. In terms of the phasor diagrams the beamsplitter replaces the squeezed Wigner function with one that is less squeezed Fig. 9.17b, that means the contour ellipse is not as narrow. The corresponding variance $V(\theta)$ at the output is to the QNL, see Fig. 9.17c.

Loss transforms the contour from an elongated ellipse towards a circle, as shown in Fig. 9.18. This might not be obvious, if one considers that the fluctuations of the input of the beamsplitter simply add. Just adding the two Wigner functions would suggest a rather more complicated result, as indicated in Fig. 9.18b, for the case of adding two squeezed states with different squeezing angle θ_s. However, all features of the Wigner function have to be taken into account, including the phase terms – and the answer is actually simple: The result is another ellipse with the new orientation θ_s and smaller ellipticity. It has to be an ellipse since we know that any solution of the linearized model will result in a 2-dimensional Gaussian quasi probability distributions, see Section 9.2. Thus the only parameter that can vary is the variance $VY1$ of the squeezed state after the beamsplitter.

If the input state is a squeezed state with the general parameters (α, r_s, θ_s), the output state, after the beamsplitter, is $(\alpha', r_s', \theta_s')$. We know that the intensity is reduced such that $|\alpha'|^2 = \eta |\alpha|^2$. The beam splitter has a second input port for vacuum fluctuations and is

(a)

(b)

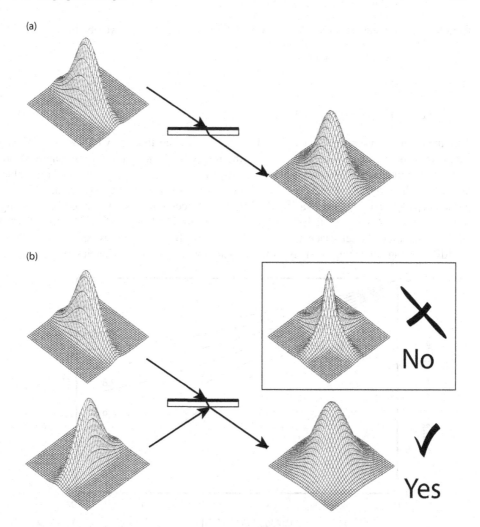

Figure 9.18: Wigner functions cannot be added directly, the phase terms must be taken into account. Two examples: (a) attenuation of a squeezed state by a beamsplitter results in a state with reduced squeezing. (b) the combination of two squeezed states with different θ_s values results in a normal squeezed state, not a Wigner function with four contributions.

described in the standard way

$$\delta\tilde{\mathbf{X}}(\theta)_{\text{out}} = \sqrt{\eta}\,\delta\tilde{\mathbf{X}}(\theta)_{\text{in}} + \sqrt{1-\eta}\,\delta\tilde{\mathbf{X}}_{\text{vac}} \tag{9.3.1}$$

The calculation can be carried out independently for each angle θ. As a result, the minimum will remain at θ_s since $|0\rangle$ has no angular dependence. Thus the squeezing angle is unchanged, $\theta'_s = \theta_s$. We have to evaluate the variances and this is a linear operation which leads to the sum of two contributions. The second term simply results in $\sqrt{1-\eta}$ since any contribution from the vacuum state results in $V_{\text{vac}} = \langle |\delta \tilde{\mathbf{X}}(\theta)|^2 \rangle = 1$. As a result we obtain

$$VY1_{\text{out}} = \eta VY1_{\text{in}} + (1 - \eta)$$

or

$$(1 - VY1_{\text{out}}) = \eta(1 - VY1_{\text{in}}) \tag{9.3.2}$$

This means the noise suppression $(1 - VY1_{\text{out}})$ is reduced linearly with loss η. The noise suppression is affected in the same way as the intensity. Note that the noise suppression is commonly quoted on a logarithmic scale, in [dB] for the normalized variance and this value is not reduced linearly. The effect of η is shown in Fig. 9.19 where the output variance $VY1_{\text{out}}$ is plotted in [dB] versus the input variance $VY1_{\text{in}}$. The effect of losses is more severe on strongly squeezed light. A loss of 50%, ($\eta = 0.5$) transforms $VY1_{\text{in}} = -3$ [dB] into -1.3 [dB] while for the much better squeezing of $VY1_{\text{in}} = -10$ [dB] the output is very similar, namely -2 [dB]. Strong squeezing is easily lost in a system with limited efficiencies.

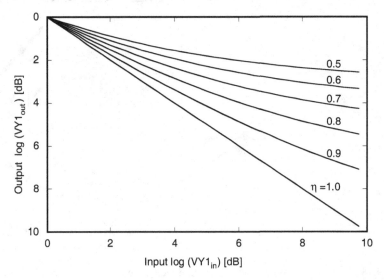

Figure 9.19: The quantitative effect of different efficiencies η on the minimum variance of the squeezed state.

The minimum variance of a squeezed state is $VY1_{\alpha,\xi} = \exp(-2r_s)$. One could formally try to link the squeezing parameter r'_s to r_s. The link is $\exp(-2r'_s) = \exp(-2r_s)\eta + (\eta - 1)$. This is not a simple relationship and the squeezing parameter is not practical for this type of calculation. Since the conversion of the squeezing due to losses is constantly required in the evaluation of squeezing data the squeezing parameter is rarely used by experimentalists. The most common parameter cited is $(1 - VY1)$ either on a linear or logarithmic scale and this scales easily, as stated in Equation (9.3.2).

Note that a beamsplitter, or any other phase independent loss, does not change the spectrum of the squeezing. Equation (9.3.2) holds equally well for all detection frequencies. Both

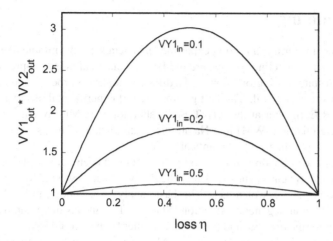

Figure 9.20: The uncertainty product for a squeezed state after loss. Three different degrees of squeezing are shown.

the shape of the spectrum and the squeezing angle are not affected by losses, only the degree of squeezing is reduced.

The squeezed light produced by some devices, for example a below threshold OPO, can be minimum uncertainty state, a pure state with $VY1\ VY2 = 1$. Does it remain a minimum uncertainty state after losses? The beamsplitter will mix in a fraction of the vacuum state, itself a minimum uncertainty state, and the combined output will be a mixed state, above the minimum uncertainty value. How much above? We can directly calculate

$$VY1_{out}\ VY2_{out} = (\eta VY1_{in} + (1-\eta))\ (\eta VY2_{in} + (1-\eta))$$

but do not obtain a simple answer. The result is easiest interpreted from a diagram. Figure 9.20 shows the values for the product of the variances as a function of η for three degrees of squeezing: $VY1_{in} = 0.5, 0.2$ and 0.1. We see that the product rises quickly and for $\eta \approx 0.5$, reaches a maximum which depends on the initial degree of squeezing. For large attenuation all these states approach again a minimum uncertainty, the coherent state.

If the light is too intense for the detector we should not simply attenuate the light but reflect it off a cavity. If a cavity with smaller linewidth than the detection frequency is chosen the cavity will fully reflect the noise sidebands while the carrier component is partially transmitted. In this way the optical power can be attenuated without losing the squeezing. An interesting extension would be to incorporate an active gain medium into the cavity [Ral95]. This now becomes a noiseless optical power amplifier, it amplifies the energy but leaves the noise sidebands outside the cavity linewidth unaffected. Alternatively we could detune the cavity and thereby rotate the squeezing ellipse, as discussed in the previous section. Amplitude quadrature squeezing can be turned into phase quadrature squeezing and vice versa. We have all the components available to modify the squeezing parameters, the only thing we cannot do is to increase the degree of squeezing.

9.4   Four wave mixing

We are now discussing the pioneering squeezing experiments in detail. By the middle of 1984 several research groups were competing to measure the first nonclassical noise suppression; to detect the first squeezed state. Each group favoured a different nonlinear process, and eventually all these attempts were successful. The first group to report experimental observations of squeezing was led by R.E. Slusher at the AT&T Bell Laboratories in Murray Hill. Their experiment used four wave mixing (4WM) in a Na atomic beam [Slu85]. This is a resonant medium. The nonlinearity was enhanced by an optical cavity.

The first proposal of squeezing with a 4WM mixing medium had been published as early as 1974 by H.P. Yuen and J.H. Shapiro [Yue79]. The difficulty was to achieve sufficient 4WM gain while avoiding all other sources of noise. The basic analysis of squeezing by 4WM confirmed that we required a non-degenerate situation [Slu87a]. That means the pump beam, which provided the energy for the process, and the signal and idler beams should have separate frequencies. This led to the classical interpretation of 4WM as correlating two sidebands, each detuned by $\pm\Omega$ from the pump frequency.

One important noise source that could swamp any squeezing was spontaneous emission. Light from the pump beam, which was absorbed by the atoms, was re- emitted either at the same frequency or at the atomic resonance frequency. Such light produced fluctuations in the photo-current which raise the noise level above the standard quantum noise level (SQL). It was necessary to detune the pump wave from the atomic resonance by many atomic linewidths. However, this reduced the nonlinearity and a compromise between nonlinearity and spontaneous emission noise had to be achieved. One technique to reduce the spontaneous emission, at a given detuning, was to reduce the Doppler broadening. Consequently most experiments were performed with atomic beams which are essentially Doppler free as long as all optical beams are aligned at right angles to the propagation axis of the atoms. The atomic beams had to be optimized for high densities. In Slusher's experiment an atomic absorption of $\exp(-0.5)$ on line centre was measured.

The experiment consisted of a beam of Na atoms which was illuminated, pumped by an optical pump beam, and detuned by about 2 [GHz] away from the hyperfine components of the Na D1 transition at 590 [nm]. Figure 9.21 shows the experimental layout and Fig. 9.22 the frequencies of all the optical beams in the experiment. A second beam, the probe, was aligned through the same atoms at a small angle to the pump beam. This probe beam was aligned into a heterodyne detection system together with a local oscillator beam which was split off from the pump beam. Two photo-diodes detected the beams and generated a beat signal at the difference frequency between probe and pump, in this case at 595 [MHz]. In the absence of the atoms, and with the probe beam blocked the noise level recorded was the standard quantum limit (1 Hoover). It can be interpreted as the noise due to the vacuum state entering the beam splitter of the heterodyne detector.

With the atoms present and the probe beam turned on, the 4WM medium generated a second sideband with exactly the opposite detuning to that of the probe. Now two beams were present and were detected together. In this way the probe beam was used to measure the classical 4WM gain and to align the detection system. The 4WM process can be interpreted as phase sensitive amplification, it increases amplitude modulation and reduces phase modulation sidebands. This gain correlates the two sidebands, as discussed in Section 9.2. Even with the

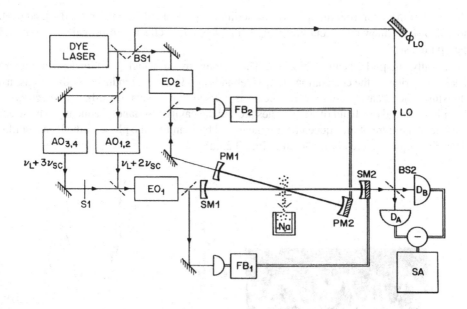

Figure 9.21: Layout of the atom 4WM experiment, after R. Slusher et al. [Slu85]

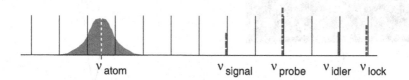

Figure 9.22: The different frequencies involved in the 4WM experiment. The cavity modes are shown as thin solid lines. They are shifted by the dispersion of the atoms.

probe beam switched off this correlation exists. The ever present quantum noise sidebands, which enter the medium on the axis that was defined by the probe beam, will be correlated, resulting in a squeezed vacuum state.

For a single transit of the Na beam this effect was far too small to be detectable. The 4WM process, and the resulting correlation, could be built up by a cavity which is tuned to both sidebands simultaneously. Each sideband was on resonance with a different cavity mode. In this experiment they were not even adjacent modes, they were separated by nine free spectral ranges, see Fig. 9.22. Again, the probe beam was used for the alignment of the cavity. A locking system kept the cavity on resonance. This used an optical beam, which probed the cavity at yet another optical frequency, avoiding the pump and the probe frequency. One practical problem was the fact that the atoms produced a strongly dispersive refractive index which shifts all cavity resonances. The cavity modes were asymmetric to the probe frequency, but by using AO modulators the locking beam could be shifted until both the probe and the newly generated beam were in resonance. The pump intensity was raised until it approached

the level of the saturation intensity of the sodium vapour at a detuning of 2 [GHz], equivalent
to 400 atomic linewidths. and the pump beam was retro-reflected to provide an even higher
pump intensity.

Finally, the probe beam is blocked. The vacuum mode, which is collinear with the probe
beam, experiences the correlation at the sidebands at $+\Omega$ and $-\Omega$ induced by the atoms inside
the single sided cavity. A squeezed beam is created that travels towards the detectors. The
light is squeezed in a band of frequencies around the cavity resonance frequency. The strength
and the quadrature of the squeezing are measured by scanning the phase of the local oscillator.
This *first* squeezing result is shown in Fig. 9.23(a).

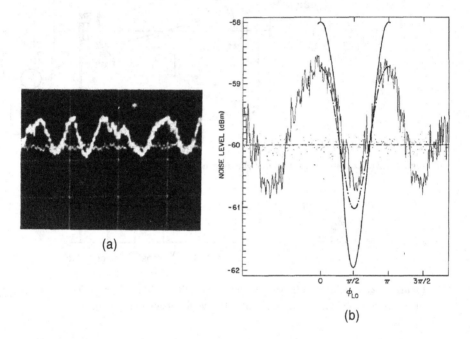

Figure 9.23: The first squeezing results: (a) photograph of the first trace [Slu85]. (b) quantitative comparison between experimental results and theoretical predictions after months of further
work. After R. Slusher et al. [Slu87a]

Figure (9.23(b)) shows the result after continuous improvements for 15 months after the
first observation [Slu85]. The dotted line in Fig. 9.23(b) gives the QNL, obtained with the cavity blocked. The measured noise increases 1.3 [dB] above and decreases -0.7 [dB] below the
SQL. The electronic noise (amplifier noise) is -10 [dB] below the SQL and contributes about
10% to the noise level. The detection frequency is 594.6 [MHz], detected with a resolution
bandwidth of 300 [kHz] and a video bandwidth of 100 [Hz].

The theoretical model predicts ± 2 [dB] (solid curve) for an ideal measurement. The efficiency of the photo-detectors is $\eta_{\text{det}} = 0.8$, the efficiency of the interferometer is $\eta_{vis} = 0.81$
resulting in a total efficiency of 0.65. Taking this and the electronic noise into account the
theoretically predicted noise is given as the dashed-dotted line in Fig. 9.23(b). It shows good

agreement with the enhanced noise and moderate agreement with the squeezed noise quadrature, namely -0.7 [dB] observed versus -1.05 [dB] predicted.

One practical limitation is the mechanical stability of the interferometer required for the homodyne detection. Any mechanical vibration can wash out the noise suppression. This effect, described as phase-jitter, is cited as the reason for the disagreement with the prediction. It can explain this particular set of results. However, at higher intensities the disagreement gets even larger and other processes have to be considered. Nonlinear coupling between the light used for the locking of the cavity to the laser frequency and the mode of the squeezed light is considered by the authors as a possible source of excess noise.

This experiment clearly demonstrates the existence of a squeezed state of light, in this case a squeezed vacuum state. All predictions about the properties of the squeezed light and the detection systems were confirmed qualitatively. The observed noise reduction is modest. Despite the predictions of squeezing of up to -10 [dB], 4 Wave Mixing has never been really successful as a squeezing process. The technical difficulties with the atomic beam were frustrating and this type of experiment has not been repeated by any other group. It remains a milestone in the early exploration of squeezed light.

9.5   Optical parametric processes

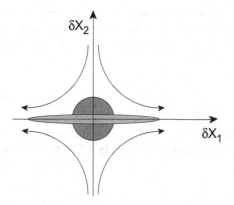

Figure 9.24: A parametric medium inside a cavity. The various input, output and intra-cavity modes are labelled. This system has a threshold and can operate below threshold as a phase sensitive amplifier of the vacuum modes and above threshold as an oscillator. The subscript p refers to the fundamental frequency and the subscript d refers to the subharmonic frequency.

Some of the most successful results in the early period of squeezing came from the group led by H.J. Kimble at the University of Texas [Wu87]. He used an optical parametric amplifier to generate squeezing. We have already seen in Chapter 6 that the OPA is ideally suited since it is phase sensitive and amplifies one quadrature while it attenuates the other as shown in Fig.9.25. The trick to enhance the process is to build a cavity around the OPA as shown in Fig.9.24. The experiments with the nonlinear OPO are carried out with a cavity containing a χ^2 material, usually a crystal such as $LiNbO_3$. A single mode laser illuminates the cavity.

Following the derivations in Section 6.3.3, we know that a parametric oscillator below threshold will produce quite different gain for the two quadratures, similar to 4WM, and thereby suppresses the noise selectively. Figure 9.25 symbolically shows how the quadratures are affected and the conversion of a vacuum input state (circle) into a squeezed vacuum (ellipse).

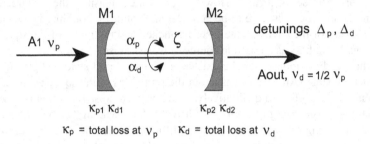

Figure 9.25: Symbolic representation of parametric gain in the phase diagram

From Equation (6.3.19) we can derive a noise spectrum and find that the effect is greatest at low frequencies ($\Omega \ll$ cavity linewidth). For such an oscillator the crucial experimental parameter is given by the normalized pump amplitude $d = \alpha_{\text{in}}/\alpha_{\text{in}}^c$, where α_{in}^c is the input amplitude required to reach threshold. We wish to operate with $d < 1$ and can optimize the performance by approaching $d \to 1$ by changing the input intensity. In terms of the cavity parameters, losses and nonlinear gain, we find $d = \chi/\kappa_a$ and consequently we can derive the variance from Eq. (6.3.11)

$$V2(\Omega) = 1 - \frac{4\,d}{(1+d)^2 + (2\pi\Omega)^2} \tag{9.5.1}$$

which is identical to the squeezing spectrum, $S(\Omega) = V2(\Omega)$, since the noise suppression always occurs in the phase quadrature. An example is shown in Fig.9.26. At very low frequencies, $\Omega \to 0$, we get the optimum value

$$V2(\Omega) = \frac{(1-d)^2}{(1+d)^2} \tag{9.5.2}$$

and at the same time the noise enhancement in the orthogonal quadrature

$$V1(\Omega) = \frac{(1+d)^2}{(1-d)^2} \tag{9.5.3}$$

We see that the fluctuations in the $X2$ quadrature tend to zero. On the other hand the fluctuations in the $X1$ quadrature diverge. The effect for the two quadratures are shown in Fig. 9.27. The system has a critical point at the threshold. Perfect squeezing is possible at threshold, at least in theory but in practice the operation gets unstable and in most situations values no larger than $d = 0.8$ are chosen. Interestingly the noise suppression improves only slowly for intensities larger than half the threshold value ($d^2 = 0.5$). The efficiency η of the

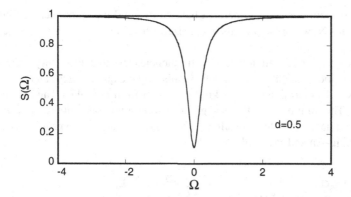

Figure 9.26: The squeezing spectrum for an OPO below threshold for a fixed value of $d = 0.5$. The modulation frequencies are scaled in units of the cavity bandwidth.

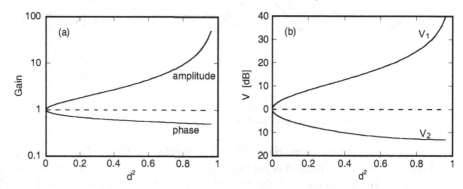

Figure 9.27: Phase dependent gain and resulting variances $V1$ and $V2$ for the two quadratures of a OPO below threshold as a function of the normalized input power d^2. This calculation is for very low detection frequencies, less than the linewidth of the cavity, $\Omega = 0$.

apparatus has to be taken into account and the complete equation for the noise suppression is

$$V2(\Omega) = 1 - \frac{\eta \, 4d}{(1+d)^2 + (2\pi\Omega)^2} \tag{9.5.4}$$

The first realization of this experiment was carried out by Wu et al. in 1986. The layout of their experiment is shown schematically in Fig. 9.28. The light source is a CW Nd:YAG ring laser, in this case lamp pumped. The light is frequency doubled using an intra cavity $Ba_2NaNb_5O_{15}$ crystal. The laser cavity has two output beams, at 1064 [nm] and 532 [nm] with orthogonal polarizations. The green light is sent to the parametric oscillator, which is a MgO doped $LiNbO_3$ crystal inside a linear cavity. When operated above threshold as a degenerate OPO the cavity sends out a beam at exactly the laser frequency. This beam is recombined in a Mach Zehnder interferometer with the infrared output from the laser cavity. This arrangement forms the homodyne detector for the output of the parametric oscillator. The laser provides the local oscillator. When operated below threshold the output ceases, but the apparatus is still aligned. Now the OPO emits a squeezed vacuum state which is detected by the homodyne detector.

All measurements were carried out at detection frequencies less than the cavity linewidth. The phase angle θ of the local oscillator is scanned. In this way the quadratures can be probed. The vacuum noise level is determined by blocking the beam from the OPO. The results are shown in Fig. 9.29(a). The minimum of the trace gives the optimum noise suppression. It was difficult to stabilize the phase of the local oscillator, thus mainly scans of the projection angle θ were published by Ling-An and Wu et al. [Wu87].

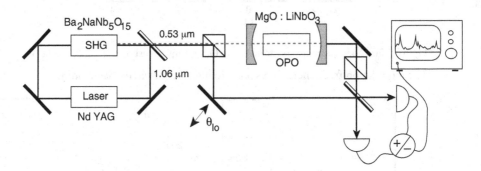

Figure 9.28: The schematic layout of the experiment by Wu et al. [Wu87] generating a squeezed vacuum state with a sub threshold OPO.

We can investigate the properties of the OPA in different ways. The noise suppression will improve when the threshold is approached. This has been tested and results are given in Fig. 9.29a. The agreement with the theoretical predictions, Equation (9.5.4) is excellent. The best noise suppression observed is $V_1 = 0.33$. Once the detector efficiencies and the losses are taken into account the best inferred squeezing corresponds to a factor of 10, or -10 [dB] for the light emitted by the cavity. Finally, this experiment is also a most impressive demonstration of the concept of a minimum uncertainty state. In Fig. 9.29b both the minimum and the maximum variances for each input amplitude are plotted. According to the theory, the product of the two variances should be unity, since the parametric amplification preserves the minimum uncertainty area. The data points should lie on a hyperbola and this is indeed the case. Note that this result can only to be expected for data describing directly the OPO, with perfect efficiency, since any loss would have turned these minimum uncertainty states into mixed states. The data for Fig. 9.30 are inferred, after removing the effect of the detector loss.

This is a landmark experiment. Some of the most striking features of squeezed light have been demonstrated in a single experiment: noise suppression, noise enhancement in the opposite quadrature, and preservation of the minimum uncertainty. And the light generated, a squeezed vacuum beam, is certainly not classical in any regard. The work of Linang Wu et al. is a remarkable achievement and remained unsurpassed for many years. Only recently have better results been reported and a noise suppression of -6 [dB] has been observed [Pol92, Bre95].

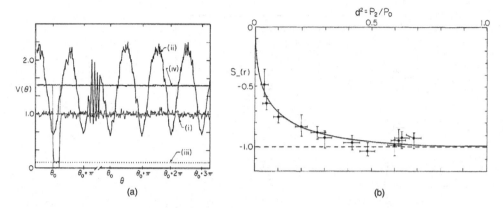

Figure 9.29: Experimental results (a): Dependence of noise power on the local oscillator phase θ. With the output of the OPO blocked the vacuum field entering the detector produces trace (i). With the OPO input present, trace (ii) exhibits phase-sensitive deviations both below and above the SQL level. Trace (iii) is the amplifier noise. (b) Noise suppression for different normalized intensities d^2. Comparison of the inferred noise suppression $V2(d^2)=1+S_-(r)$ after correction for the quantum efficiencies derived from measurements such as shown in (a) with optimized phase angle with those given by theory (solid curve). Detection frequency Ω is fixed. Perfect squeezing corresponds to $V2 = 0$ and is indicated by the dashed line. After Wu et al. [Wu87].

The best squeezing with an OPO was reported by P.K. Lam et al. at the ANU in Canberra with a noise suppression of -7.1 [dB], as shown in Fig. 9.31 [Lam99]. The main advance had been in the reduction of losses inside the OPO which in this case is of monolithic design. The limitations are now the efficiencies of the detector and technical fluctuations in the phase of the local oscillator θ which wash out the very sharp minima. The same machine, with some modifications required to inject a beam into the OPO, produces -5 [dB] of bright squeezed light locked for hours. Similar work at Caltech, Konstanz and Shanxi University have established the OPO/OPA as the most reliable source for squeezed light.

9.6 Second harmonic generation

The process of frequency doubling or second harmonic generation (SHG) has been used to generate amplitude squeezed light. It is presently the most reliable and efficient process to generate light with high intensity and good noise suppression. A qualitative description why SHG leads to amplitude squeezed light is obtained from the photon model. A coherent input beam can be thought of as a stream of photons with Poissonian distribution. The time separation between photons is random. This beam illuminates a nonlinear crystal, which converts a fraction of the light into a new beam with twice the optical frequency. Two beams emerge, the remaining fundamental (f) beam and a second harmonic (SH) beam. The crystal converts pairs of photons from the incoming fundamental beam into single photons of the second harmonic beam. If the probability for the conversion were constant, the statistics of the two emerging beams would be identical to that of the input beam.

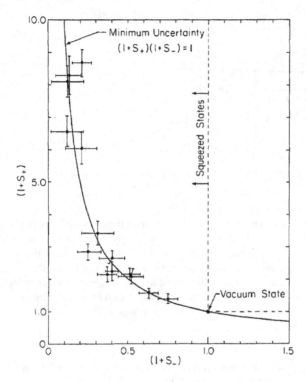

Figure 9.30: Variances V1 and V2 determined from measurements. The solid curve is the hyperbola V1 V2 = 1, which defines the class of minimum uncertainty states. Squeezed states are those states for which either V1 or V2 < 1 and that lie in the region bounded by the hyperbola and the dashed lines.

However, the conversion probability is proportional to $\alpha^2 \chi^{(2)}$, that is the instantaneous intensity of the fundamental beam multiplied by the second order nonlinearity. This means the conversion is more likely when the pump intensity is high, when the photons in the pump are closely spaced. Closely spaced pairs of photons are most likely to be converted into a single SH photon. As a consequence the statistics of the SH photons is closely linked to the statistics of pairs in the fundamental beam. The SH photons come in a more regular stream than the pump photons. If the pump beam has a Poissonian photon statistics the SH beam has a sub-Poissonian distribution. Similarly, the remaining fundamental light from which the pairs have been deleted has a more regular stream of photons; it also has a sub-Poissonian photon statistics. Similar arguments apply to nonlinear absorption which could also produce squeezing. This simple interpretation is suitable only for a single pass through the nonlinear SHG material. In CW experiments the nonlinearity is not large enough and the crystal is placed inside a cavity to build up a mode with high intensity at the fundamental frequency. The first successful observations were carried out by S. Pereira et al. [Per88] who detected a noise suppression of -0.6 [dB] in the fundamental beam.

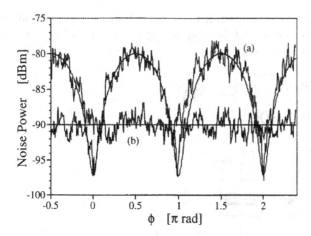

Figure 9.31: Experimental data: (a) The typical scan through the variances indicating squeezing and anti-squeezing. For comparison a theoretical curve with a noise suppression of -7.1 [dB] is shown. (b) shows the monolith used in this experiment. After P.K. Lam et al. [Lam99]

These early experiments were plagued by additional technical noise sources inside the buildup cavity. A breakthrough occurred when it became possible to polish and coat the nonlinear crystal directly. In such a monolithic design the internal losses can be kept to a minimum, the fundamental light is built up to increase the nonlinearity and the SH field is reflected at one end to have all of the SH field escape on one side. Care has to be taken that the fundamental and second harmonic waves stay in phase, which is achieved by selecting the correct phase matching temperature. The coatings are designed such that phase shifts upon reflection preserve the optimum phase relation for a large number of round trips.

Figure 9.32: The monolithic design for second harmonic generation (SHG) using CW lasers. The fundamental field (FH) is enhanced by the cavity, the second harmonic field is reflected at one end and forms one single output beam. After R. Paschotta et al. [Pas94]

Such a monolith was first designed and operated by the group in München [Siz90]. A. Sizmann et al. measured a noise suppression of -1.0 [dB] in the fundamental light in the presence of significant electronic noise from which they inferred a noise suppression of at least -2.2 [dB]. This was one of the earliest applications of the balanced detector system (see Section 8.1. Unfortunately the system was not stable, squeezing was observed only for milliseconds at a time. Systematic improvements were carried out by the group in Konstanz who

achieved very reliable squeezing of considerable magnitude. They used a monolithic cavity.
R. Paschotta et al. [Pas94] measured −0.6 [dB] squeezing (−2.0 [dB] inferred) on the SH
beam. This system showed great reliability.

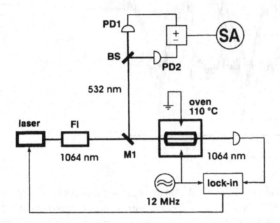

Figure 9.33: Typical layout of a SHG squeezing experiment. The monolithic resonator is
pumped through an isolator (FI) by a single mode ND:YAG laser at 1064 [nm]. The reflected
second harmonic light is detected by a balanced detector, the sum and difference currents from
the two detectors (D1 and D2) are measured with a spectrum analyser (SA). Parameters as used
by R. Paschotta et al. [Pas94] are: Material MgO:LiNbO$_3$, cavity length 7.5 [mm], radius of
curvature 10 [mm], input reflectivities are $R_{FH} = 99.9\%$, $R_{SH} < 1\%$, output reflectivities
were close to perfect for both fields.

A quantitative analysis of the fluctuations can be obtained from a description of the cavity
using quantum operators. The input mode $\hat{A}_{in}$ at frequency ν_L enters the cavity with the decay

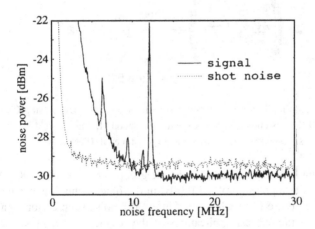

Figure 9.34: Squeezing spectrum obtained by R. Paschotta et al. [Pas94]

constants κ_{f1} and losses γ. It builds up an intra-cavity mode $\hat{a}_f$. A mode $\hat{a}_{SH}$ at frequency $2\,\nu_L$ is generated and leaves the cavity. The remaining fundamental mode forms the output mode $\hat{\mathbf{A}}_f$. The decay of the internal mode due to losses is described by $(\kappa_{f1} + \kappa_{f2})$ which includes losses due to scattering. A corresponding amount of vacuum noise is coupled in. An equivalent equation describes the generation of the second harmonic field in the presence of losses (κ_{shl}) and output coupling (κ_{sh1} and κ_{sh2}). This process is described by

$$\frac{d\hat{a}_f}{dt} = -(\kappa_f + i\Delta_f)\hat{a}_f + \zeta\hat{a}_f^\dagger\hat{a}_{sh} + \sqrt{2\kappa_{f1}}\,\delta\hat{\mathbf{A}}_{f1} + \sqrt{2\kappa_{f2}}\,\delta\hat{\mathbf{A}}_{f2} + \sqrt{2k_{fl}}\,\delta\hat{\mathbf{A}}_{fl}$$

$$\frac{d\hat{a}_{sh}}{dt} = -(\kappa_{sh} + i\Delta_{sh})\hat{a}_{sh} - \frac{1}{2}\zeta|\alpha_f|^2 + \sqrt{2\kappa_{sh1}}\,\delta\hat{\mathbf{A}}_{sh1} + \sqrt{2\kappa_{sh2}}\,\delta\hat{\mathbf{A}}_{sh2}$$

$$+ \sqrt{2\kappa_{shl}}\,\delta\hat{\mathbf{A}}_{shl} \qquad (9.6.1)$$

where $\delta\hat{\mathbf{A}}_{sh1} = \delta\hat{\mathbf{A}}_{sh2} = \delta\hat{\mathbf{A}}_{shl} = \delta\hat{\mathbf{A}}_{vac}$. This equation was used by M. Collett and later R. Paschotta [Col85, Pas94]. They concluded that the squeezing was best in the limit of low detection frequencies ($\Omega \to 0$) and that the variance of the SH beam could ideally reach the value of $V1_{SH} = 0.11$ or -9.5 [dB]. The squeezing extended up to frequencies of the order of the linewidth of the cavity. This model assumed that the laser was driven by a coherent state. However, in the experiment the laser noise swamped the noise suppression at very low frequencies. The best noise suppression observed, see Fig. 9.34 was only $V1_{SH} = 0.8$, or -1.0 [dB], at a frequency of 16 [MHz]. The full spectrum can be calculated using the formalism of linearization. It was shown that the spectrum of the second harmonic light is given by a noise transfer function [Ral95]

$$V1_{SH}(\Omega) = 1 - \frac{8\kappa_{nl}^2 - 8\kappa_{nl}\kappa_{f1}(V1_{las} - 1)}{(3\kappa_{nl} + \kappa_f)^2 + (2\pi\Omega)^2} \qquad \text{where} \quad \kappa_{nl} = \frac{\zeta^2}{2\kappa_{sh}}|\alpha|^2 \qquad (9.6.2)$$

At low frequencies the large excess noise of the laser $V1_{las}$, derived in Chapter 6, dominates the measurement. An optimum detection frequency exists. The noise that masks the squeezing is not a technical imperfection, it is intrinsic to the laser. It is possible to reduce the laser noise by placing a mode cleaner cavity between the laser and the squeezing cavity. This cavity will reflect the high frequency noise for frequencies larger than its linewidth. The system can now be modelled by two sequential transfer functions, one for each cavity. This theoretical model was tested by A.G. White et al. [Whi96b]. The experimental layout shown in Fig. 9.35 is similar to that by R. Paschotta (see Fig. 9.33). The light source is a diode laser pumped monolithic Nd:YAG laser. The nonlinear material is MgO doped LiNbO$_3$. The crystal is phase matched by tuning the temperature. The operating temperature is at about 110°C and has to be optimized to within ± 2 [mK]. The SH light is separated by a dichroic mirror and detected by a balanced detector. The detectors receive optical powers of up to 15 [mW] each. The efficiency of the photo-detectors (EG&G FND100) is low ($\eta_{det} = 0.55$). In a repeat of the experiment a mode cleaner cavity, locked to the laser frequency, is placed between laser and the crystal.

The experimental noise spectra are shown in Fig. 9.36 for the frequency range 2–34 [MHz]. The quantum noise limit was determined by measuring the difference of the currents from the two detectors. The electronic noise, typically -6 [dB], has been subtracted. The experimental trace (a) shows the noise suppression for the monolithic cavity. The optimum of -0.6 [dB]

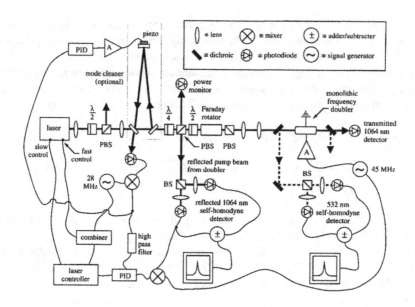

Figure 9.35: Schematic layout of the improved SHG experiment with mode cleaner. After A.G. White et al. [Whi96b].

is measured at 13 [MHz]. A theoretical prediction, trace (b), based on Equation (9.6.2) and using measured values for the nonlinearity, losses, intensities and cavity reflectivities agrees very well. In a second experiment the mode cleaner cavity is inserted. As shown in trace (c), the high frequency noise from the laser (2–15 [MHz]) is substantially rejected. The best observed squeezing is increased to -1.6 [dB] at 10 [MHz] (3 [dB] inferred). Trace (d) shows the theoretical prediction obtained by including the transfer function for the mode cleaner cavity, that means by replacing $V1_{las}$ in Equation (9.6.2) with the output variance from the cavity. The good agreement shows that all effects are included in the combined transfer function.

This experiment is very reliable: consistent noise reduction can be obtained for many hours. Later experiments with better photo-detectors produced a noise suppression of -2.1 [dB] (-2.6 [dB] inferred). How can these results be improved? Can we get closer to the limit of $V1_{SH} = 1/9$? A systematic study shows that the main obstacles are the limited ratio of nonlinearity to absorption in available nonlinear materials and competing nonlinearities.

When the ratio of nonlinearity to absorption is high, it enables experiments to be built which achieve high nonlinear conversion at reasonable powers and with high escape efficiencies. That means most of the squeezed light leaves the cavity, rather than being absorbed. Until recently $LiNbO_3$ has been the preferred material since it showed the lowest internal losses. An alternative material is $KNbO_3$, which has a higher nonlinearity. It tended to suffer from large internal losses. However, work by Tsuchida [Tsu95] has shown that appropriately cut crystals can be used to generate large conversion efficiencies and degrees of squeezing. In his experiment a crystal of 5 [mm] length was placed inside an external ring cavity and pumped by an Ti:Sapphire laser operating at 862 [nm]. The Finesse of the cavity was 34 with a free

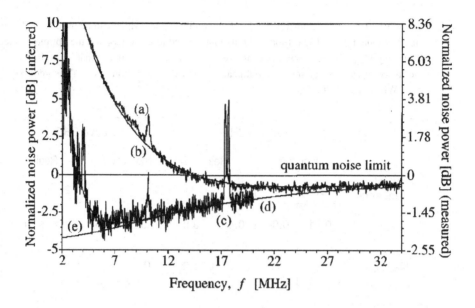

Figure 9.36: Comparison between theory and experiment for the squeezing generated by second harmonic generation as shown in Fig. 9.35. The linearized model describes the complete system in a modular approach through the use of transfer functions. (a) and (b) no cavity, (c) and (d) with mode cleaner cavity.

(e) model with no laser noise after A.G. White et al. [Whi96b]

spectral range of 480 [MHz]. The noise reduction observed is -2.4 [dB] with an inferred noise reduction of -5.2 [dB]. Clearly this is the best result obtained to date with SHG.

As was pointed out earlier, the equations that describe the OPO and SHG processes are almost identical, the only difference being that in the former case the driving energy is provided by the high frequency $(2\nu_L)$ field; in the latter by the low frequency (ν_L) field. If driving fields are present at both frequencies then the system exhibits a continuum of behaviours, with SHG and OPO as the two limiting cases. So far we have assumed that these are the only processes that can occur inside the cavity. However, it is possible for other processes to occur that compete with the nominal process – this has been termed competing χ^2 nonlinearities. Consider, for example, singly resonant SHG. When there is sufficient power in the second harmonic field, at $(2\nu_L)$, it can drive a non-degenerate optical parametric oscillation, which generates two fields at frequencies $\nu \pm \Delta$, the signal and the idler. For this to happen the signal and idler must be simultaneously resonant with the fundamental, leading to a Triply Resonant Optical Parametric Oscillation (TROPO). This process has a threshold. At threshold the power of the transmitted fundamental beam is clamped – i.e. it remains constant independent of the pump power. The squeezing diminishes both due to excess noise imposed by the OPO and the extra vacuum noise introduced by the additional modes [Whi97]. The opposite is true in the quadruply resonant case, where the second harmonic mode is also resonant, the competition leads to improved squeezing in previously inaccessible power regimes [Mar96].

Table 9.1: This table compares the different configurations which can be used for SHG, in the singly resonant case. The estimated limits in noise reduction $V1_{min}$ and the optimum detection frequency $\Omega_{optimum}$ are shown, both for realistic, presently available materials (Real) and for the hypothetical case of infinite nonlinearities and no absorption losses (Theoretical). After A.G. White et al. [Whi96a].

	Passive SHG				Active SHG			
	Theoretical		Real		Theoretical		Real	
optical frequency	2ν	ν	2ν	ν	2ν	ν	2ν	ν
$V1_{min}$	0.11	0.66	0.40	0.8	0.5	0.5	no	no
$\Omega_{optimum}$	0	0	γ	1.2γ	0	0		

Large noise reductions can be expected if the system is doubly resonant, that means both the fundamental and the second harmonic modes inside the cavity are simultaneously on resonance. This provides optical feedback for the second harmonic wave, the energy can oscillate forwards and backwards between the two modes. Such a system can produce very strong correlations. The fundamental theoretical limit for this system is perfect squeezing. A more realistic systems has been evaluated and noise suppression of -6 [dB] seems feasible. To maintain double resonance is technically very difficult, particularly with a monolith cavity since both the cavity length and the laser have to be locked simultaneously. The best results have been achieved by P. Kürz et al. [Kue93] using a $LiNbO_3$ crystal that had been polished as a monolithic resonator. The length of the crystal was controlled by a transverse field, the laser frequency was controlled independently. He observed a noise reduction of -3 [dB] and was able to operate his system for periods of about 10 [sec]. This remains one of the most difficult squeezing experiments performed.

One appealing, and technically simple, alternative to the external cavity frequency doubler would be to incorporate the crystal into the laser cavity. Such internal, or active, systems show very high conversion efficiencies for SHG. Early theoretical studies predicted large squeezing in these systems, but a careful analysis by White et al. [Whi96b], including the intensity and dephasing noise of the laser, shows that in practice no squeezing can be expected. This prediction applies to a wide a range of lasers and thus active doubling is not worth considering for experiments. An overview of all the options for using SHG as a process for the generation of squeezed light is given in Table 9.1. SHG is already a very reliable technique for generating bright squeezed light.

9.7 Kerr effect

A third process for squeezing that had been proposed very early is the Kerr effect. It was favoured by the group of M.D. Levenson & R.M. Shelby at IBM Almaden Research Laboratories. It led to some of the earliest published squeezing results and was later extended by H.A. Haus & K. Bergman at MIT to the pulsed regime where it achieves very impressive, -3.5 [dB] noise reduction. The responses of the Kerr medium was briefly discussed in Section 9.1. Here we analyse it in detail.

9.7.1 The response of the Kerr medium

The classical description of light propagating through a nonlinear Kerr medium has already been given in Equations (9.1.1, 9.1.4). The fluctuations of the quadrature phase of the output field can be described as by quadrature operators, either in the time or the frequency domain

$$\tilde{X1}_{out} = \tilde{X1}_{in}$$
$$\tilde{X2}_{out} = \tilde{X2}_{in} + 2r_{Kerr}\,\tilde{X1}_{in} \quad \text{with} \quad r_{Kerr} = \frac{2\pi n_0 n_2 L \alpha^2}{\lambda} \tag{9.7.1}$$

This output field is detected using a homodyne detector by beating the output signal against a local oscillator. The detected quadrature is a linear superposition of $\delta X1$ and $\delta X2$ given by

$$\tilde{X}(\theta) = \tilde{X1}\,\cos(\theta) + \tilde{X2}\,\sin(\theta) \tag{9.7.2}$$

The variance is a measure of the noise in the quadrature and is calculated as shown in Section 4.1.1. It depends on the Kerr parameter r_{Kerr} and is normalized to the variance of a coherent state. Note that the amplitude remains unchanged and that for $\theta = 0$, the amplitude quadrature variance is always equal to the QNL. This particular projection is independent of the nonlinearity.

$$V(\theta, r_{Kerr}) = 1 + 2r_{Kerr}\sin(2\theta) + 4r_{Kerr}^2\sin^2(\theta) \tag{9.7.3}$$

where $V(\theta = 0, r_{Kerr}) = 1$ is the QNL and for $V(\theta, r_{Kerr}) < 1$ we have achieved squeezing. The angle θ_s for the best squeezing, the minimum of $V(\theta, r_{Kerr})$, can be found by evaluating $d/d\theta(V(\theta, r_{Kerr})) = 0$ which leads to

$$\cot(2\theta_s) = -r_{Kerr} \quad \text{or} \quad \theta_s = \frac{1}{2}\arctan\left(\frac{-1}{r_{Kerr}}\right) \tag{9.7.4}$$

In the limit of a small nonlinear parameter ($r_{Kerr} \to 0$) the best squeezing is observed for $\theta_s = -0.25\pi$. For large squeezing parameters the angle approaches asymptotically the value -0.5π. Figure 9.37 shows an example of the normalized variance $V(\theta, r_{Kerr})$ for three values of the nonlinearity, namely $r_{Kerr} = 0.25, 0.5\ 0.75$ and 1.0. It can be seen that the noise suppression is improving with larger nonlinearity and that the angle of best squeezing is rotated. The noise suppression increases steadily with the non-linearity and the optimum squeezing is given by

$$V(\theta_s) = 1 - 2\,r_{Kerr}\sqrt{1 + r_{Kerr}^2} + 2\,r_{Kerr}^2 \tag{9.7.5}$$

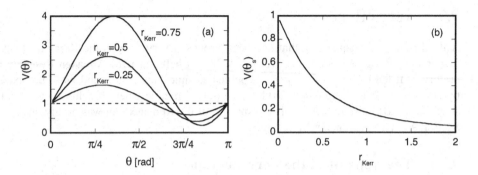

Figure 9.37: (a) $V(\theta)$ for three values of r_{Kerr}. (b) dependence of optimum value $V(\theta)$ on r_{Kerr}.

The calculation did not assume any specific detection frequency Ω. This is correct for a situation where the non-linearity does not depend on Ω. In practice, this applies only to the case of a travelling wave through a non-resonant Kerr medium. The two different cases of (i) a broadband Kerr medium inside a cavity and (ii) of a resonant Kerr medium, such as atoms close to an absorption line, will require a description of $n(\alpha^2)$ as a function of Ω and have to be evaluated for each detection frequency individually.

It is useful to consider the Kerr effect in view of quantum noise sidebands. At the core of the above calculation is the correlation between the quadratures stated in Equation (9.7.1). This correlation enforces a link between the noise side bands. Without such a correlation δE_Ω and $\delta E_{-\Omega}$ would be completely random. There would be independent contributions from δE_Ω and $\delta E_{-\Omega}$ to both $\delta X1$ and $\delta X2$. The correlation requires that for each component of δE_Ω there is also an additional component $-2\, r_{Kerr}\delta E_\Omega$ at the other side band, and vice versa. The side bands at the output are larger than at the input, as can be seen in Fig. 9.4. In detection the correlation leads to a reduction of the noise in one quadrature.

Optimizing the Kerr effect

The Kerr effect has been used in both CW and pulsed systems. It was actually one of the first successful demonstrations of squeezing. The main practical problem is to achieve a large Kerr parameter r_{Kerr} and this requires the largest possible values for the nonlinear coefficient $\chi^{(3)}$, interaction length L and pump intensity I. There are four strategies for obtaining a large value for r_{Kerr}:

i) Optical fibre. The coefficient for the nonlinearity is small, but the length of the medium and the intensity can be increased by using several hundred meters of single mode fibre. Values for r_s in the range 0.1–0.5 have been obtained for the case of a CW squeezing and in the range 0.5–2.0 have been achieved using pulses (see Section 9.9).

ii) Non resonant medium inside a cavity. The same material (quartz) as used in the optical fibre can be used in a high finesse resonator. While the physical length is short (typically 10 mm) the effective length and the high intra- cavity intensity in a high finesse cavity could compensate this shortcoming. No results have yet been published.

iii) Resonant media. The nonlinear coefficient $\chi^{(3)}$ can be enhanced several orders of magnitudes by operating close to an atomic resonance. Every absorption line shows a strong dispersive effect, that means a strong linear refractive index. The spectral line shape of the refractive index is affected by the driving field. The transition can be saturated and power broadened which in turn provides a nonlinear refractive index. In the experiments this effect is again enhanced by a cavity around the medium.

iv) The use of cascaded $\chi^{(2)}$ processes. It is possible to create a process which has the properties of the Kerr effect by combining two $\chi^{(2)}$ processes. It was shown [Li93] that the simultaneous occurrence of OPO and SHG is equivalent to one single process. Provided the system is degenerate, the excitation of the pump and down-converted mode will be coupled in a way that resembles a Kerr effect for the pump mode. This was demonstrated in single pass experiments [Hag94] and for the CW case with a nonlinear crystal in a cavity [Whi97].

These strategies suggest that large Kerr parameters should be achievable. However, when we analyse the experiments in detail we will find that the reality of squeezing experiments is not that simple. All four approaches have severe limitations since, in all experiments, there are extra sources of noise and competing nonlinear effects which will degrade the squeezing. In the actual experiments special tricks are used to suppress the extraneous noise and a maximum of the nonlinearity is traded off against a minimum of degradation due to competing effects.

9.7.2   Fibre Kerr Squeezing

Fiber-optic quantum noise reduction was eventually developed to reliable and strong photon-number noise reduction through three following generations of experiments [Siz99]. After the pioneering experiments with continuous-wave laser light described here, the next generations used ultrashort pulses to achieve a larger nonlinear phase shift with less Brillouin scattering noise in shorter fibers. These experiments are described in the sections on pulsed squeezing in Kerr media (9.9.2) and soliton squeezing (9.9.4).

With short pulses, however, the distortion of pulse shape and phase becomes an issue in coherent detection schemes. The elegant solution is to employ solitons which propagate free of chirp. For solitons, the local oscillator phase can easily be optimized for the entire pulse in the phase-sensitive detection of quadrature-amplitude squeezing. No phase-matched local oscillator was needed in the two latest generations of experiments in fiber-optic Kerr squeezing. Using the methods of spectral filtering and asymmetric nonlinear interferometers with picosecond or sub-picosecond solitons, directly detectable photon-number squeezing was produced. These experiments will be described in Section 9.9.4 on soliton squeezing. A comprehensive summary of fiber-optic Kerr squeezing experiments is given in the tables of ref. [Siz99].

One of the pioneering experiments was carried out by M.D. Levenson, R.M. Shelby and their collaborators at IBM in 1985 [Lev85]. The idea was very straightforward: use a quartz fibre with a Kerr coefficient to turn a coherent state, generated by a Krypton laser, into quadrature squeezed light. The experimental arrangement is shown in Fig. 9.38.

The detection system used the scheme of a detuned cavity which provides attenuation for the light without destroying the squeezing and provides a variable projection of the uncer-

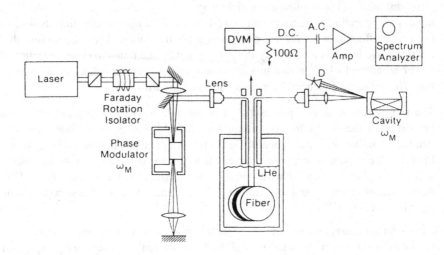

Figure 9.38: The schematic layout of the Kerr squeezing experiment in an optical fibre at IBM [Mil87].

tainty area as a function of detuning. The details of this scheme are described in Section 9.3. Phase dependent noise was readily observed. However, the initial squeezing results were rather disappointing. The apparatus appeared to be dominated by classical noise, frequently significantly larger than the QNL. The reason is that optical fibres show Brillouin Scattering which is driven by the thermal excitations of the solid state material. This drives small fluctuations in the refractive index. Inside the fibre, which has a well defined refractive index guiding geometry, acoustic waves can be set up which have a complex spectrum of resonances. A consequence of these effects is a broadband spectrum of phase noise, called guided acoustic wave Brillouin Scattering (GAWBS), see Section 5.4.5. This adds to the noise in the squeezing quadrature and can mask all squeezing. The GAWBS noise can be partially suppressed by cooling the fibre, thus the experiments by Levenson et al were performed at 4° [K] using liquid Helium cooled fibres [Mil87]. The GAWBS effect also varies with the materials and the geometry used in the fibre. Improved materials have more recently reduced the effects of GAWBS [Ber92]. It is possible to suppress many of these phase fluctuations by cancelling the classical phase noise due to GAWBS in an interferometric arrangement. Here, the local oscillator and the squeezed beam both propagate through the fibre in opposite directions, both beams experience the same classical linear phase fluctuations which then are cancelled in the detection scheme.

Finally, there is stimulated Brillouin scattering in a fibre. This is a nonlinear phenomenon with a threshold, above which a significant fraction of the light in the fibre is back-scattered. The transmitted light is extremely noisy, thereby limiting the intensity that can be used in a fibre experiment. This effect can be partially overcome by using not a single mode but a multi mode input, where each mode is squeezed individually and has its own stimulated Brillouin threshold. The experiment has to be arranged such that all these squeezing features add up in

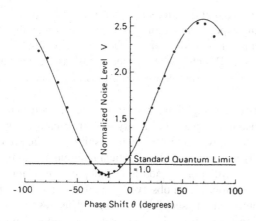

Figure 9.39: Early squeezing results from the fibre Kerr squeezing experiment at IBM [Mil87]

phase. The results of the experiment are shown in Fig. 9.39. The data clearly show squeezing at a squeezing angle of 28 degrees. The best noise suppression observed is $V1(\theta_s) = 0.85$ (-0.7 [dB]), which corresponds to an inferred squeezing of $V1(\theta_s) = 0.7$ for the light emitted by the fibre.

What are the alternatives? For the non resonant medium in a cavity the main obstacle is the problem associated with achieving a sufficiently high Finesse and the effects of losses in the cavity. The cavity will have to be locked actively to the laser frequency, using the techniques described in Section 8.4. Squeezing can be achieved only at sidebands around each cavity resonance. That means the squeezing spectrum will show maxima at multiples of the free spectral range of the cavity. Any dispersion effects have to be carefully avoided. Otherwise the FSR will change with the optical frequency and the sidebands at plus and minus one FSR cannot be brought into resonance simultaneously.

How high a Finesse is required? In the fibre the interaction length is typically 100 [m], the input intensity is 100 [mW] and the waist size is 3 [μm]. In a resonator with the same input power, with a waist size of 10 [μm] and a material length of 10 [mm] a Finesse of about $100,000$ would be required to achieve the same parameter r_{Kerr} for the interaction. For a single port cavity this requires a total loss of about 10^{-5} in optical power, including all absorption and scatter losses. Such low values require extremely pure and homogeneous materials which are not yet available.

9.7.3 Atomic Kerr squeezing

The use of a resonant medium relies on the ability to tune the input light, which is to be squeezed, close to resonance with an absorption line of the medium. The nonlinear refractive index is enhanced. However, for these small detunings the spontaneous emission becomes an extraneous source of noise, as with 4WM, Section 9.4. The light absorbed and re-emitted by the atoms produces intensity fluctuations which are detected alongside the fluctuations due to the quantum noise of the input beam. This produces uncorrelated noise at the detection frequency and swamps the carefully arranged correlations between the sidebands which we want to observe. Therefore the squeezing experiment requires detuning larger than the atomic linewidth absorption. For thermal atoms this would mean the Doppler linewidth. To avoid this the experiments by A. Lambrecht et al. [Lam95] were carried out in laser cooled atoms. A calculation for the Kerr effect generated by a sample of cold atoms inside an atomic trap predicts significant noise suppression.

In practice several complications arise: The high intensities inside the cavity lead to instabilities. The cavity with a nonlinear medium displays bistability and irregular switching. In addition, real atoms have many states due to hyperfine and Zeeman splitting – the prediction of a simple 2- level-model can frequently not be achieved and optical pumping has to be taken into account. After overcoming all these difficulties, good noise suppression of more than -1.8 [dB] below the standard quantum limit has been observed [Lam95]. These are difficult experiments and despite an intensive optimization, such higher atomic densities in magneto-optical traps or even the proposal of using a Bose Einstein condensate as the nonlinear sample and much smaller detunings, the atoms are not a promising Kerr medium. The best results were probably achieved by the group led by P. Grangier [Roc97] who optimized the medium for the demonstration of the QND effect, as described in Chapter 11.

Atomic polarization self-rotation

A related mechanism for squeezing is the rotation of the polarization of the light induced by an atomic vapour that occurs at high intensities. Interaction of the near resonant atoms with elliptically polarized light can cause the initially isotropic $\chi^{(3)}$ nonlinear medium to display circular birefringence: the two orthogonal circularly polarized components of the light will propagate at different velocities. The result is a rotation of the linear polarization of a light beam. This in turn corresponds to a shear in phase space and the distortion of the initial symmetric Wigner function into an ellipse [Rie03]. The effect was demonstrated in Rubidium and the experiment can be very simple: an atomic vapour cell, heated to achieve the correct atomic density and shielded from stray magnetic fields, is illuminated by carefully controlled polarized light. The output beam is separated with a polarizing beamsplitter into a squeezed vacuum beam and a local oscillator beam which are combined, after rotation of the polarization of the local oscillator, in a standard homodyne detector. While a simple theory predicts squeezing of up to -6 to -8 [dB] [Mat02] the experiment achieved so far a squeezing of -0.85 [dB] [Rie03]. The limitations are largely due to self-focussing of the vapour, resonance fluorescence and phase mismatch. It remains to be seen if this tantalisingly simple experiment can achieve its full potential.

9.8 Atom-cavity coupling

Atoms inside a cavity can form a nonlinear medium. The atoms themselves show a nonlinear response once the intensity is sufficiently high and the detuning is small enough such that the driving field reaches saturation. Unfortunately, for a sample of atoms, such as those in an atomic beam, these nonlinearities are weak. However, placing the atoms inside a cavity with modes tuned close to the atomic resonance makes a coupled system with a strong nonlinearity.

An extensive theoretical investigation of the various cases was carried out by M. Reid [Rei88] and she and others predicted squeezing for an number of different cases. This led to two distinct sets of experiments, both with different physical explanations and both successful in generating bright quadrature squeezed light.

The first case to be explored experimentally was that of atoms in a short cavity where the lifetime of the cavity and the spontaneous lifetime of the atoms were matched. The experiments by L.A. Orozco & M.G. Raizen used a Doppler-free Na beam [Oro87]. First the atoms were optically pumped into sublevels which can be described as a two level system. These atoms then entered a cavity which was tuned close to resonance with the transition. It is reasonable to describe the atom/cavity system as similar to a Kerr medium in that the excitation of atoms influence the optical path length of the cavity. The strength of this influence depends on the input field and thus a coupling between the light intensity and the optical cavity length exists. This in turn affects the intensity inside the cavity and feedback sets in. At high atom densities the whole system can be driven into a bistable regime. Quite clearly the quadratures of the field inside are now linked and a correlation between the quadrature phase and the quadrature amplitude is created. A detailed description of the processes is significantly more complicated than the ordinary Kerr effect since all details of the atomic response have to be taken into account. The result is a bright squeezed beam of light with noise suppression of a quadrature between $X1$ and $X2$.

It should be noted that the cavity and the atoms no longer have independent energy levels. It is necessary to consider the atom cavity system as a whole. The coupling can be made sufficiently strong that we have to consider the energy values of the combined system, which are shifted from both the free atom and the empty cavity energy values. Further details of this coupling are discussed by L. Orozco et al. [Oro87]. The combined system can be described by the following independent parameters: atomic detuning Δv_{atom}, cavity detuning Δv_{cav}, atomic cooperativity C, intra-cavity field α and the ratio of the cavity losses to the atomic rate decay μ. A systematic search for the optimum squeezing within this five dimensional parameter space found that the best condition was a cavity that is close to bistable operation.

In the experiment the cavity is illuminated by a single mode of light from a tunable dye laser. The cavity is short (about 1.00 [mm]) with a Finesse of 600–1200. It has a peak transmission of about 10^{-3}. The laser frequency and the two squeezing sidebands are all within a single cavity resonance. Typical detection frequencies are 100–600 MHz. The predicted squeezing spectrum for this condition is shown in Fig. 9.40. It gives a wide range of detection frequencies. The best squeezing is about $VY1 = 0.2$ (-0.7 [dB]). The squeezing angle varies slightly. The experimental results and detection parameters are given in Fig. 9.41 which show a measured noise suppression of -1.1 [dB]. This was the first experiment to use atom cavity coupling and to exploit the ideas of using a bistability point for squeezing. It should be pointed out that for most other parameters, where either the cavity or the atoms independently dominate the dynamics of the system, the squeezing disappears.

An independent search for the optimum parameters was carried out by the group in Canberra [Hop92]. In their experiment Ba was used which allowed the simplification that the atoms have a two level structure and thus optical pre-pumping is no longer necessary and the theoretical model is more compact. This investigation found a second, completely separate case for the generation of squeezed light, namely the high transmission regime of an atom-cavity system which is tuned close to bistable operation. The layout of the experiment

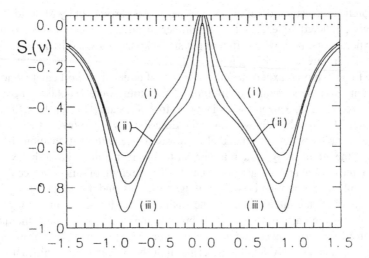

Figure 9.40: This spectrum of the squeezing reflects the complex response of the coupling between atoms and cavity. An operating point at the lower transmission branch of the bistability curve is chosen. The horizontal axis is the detection frequency. A broad minimum for the squeezing $S(\Omega)$ exists around 400 [MHz]. After M. Orosco et al. [Oro87].

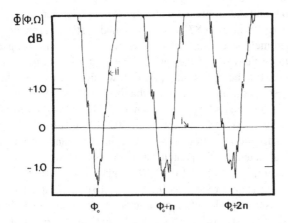

Figure 9.41: Variance measured for different quadratures. The local oscillator is scanned through more than 2π. Parameters are the same as used in Figure 9.32. Detection frequency $\Omega = 280$ [MHz]. Resolution bandwidth 300 [kHz], video bandwidth 159 [Hz]. Trace (i) experimentally determined quantum noise level. Trace (ii) phase sensitive fluctuations. Best observed squeezing -1.1 [dB]. After L. Orosco et al. [Oro87].

resembles that of the Orozco experiment. A high temperature, high density Ba atomic beam generates an optically thick sample of atoms inside a single sided optical cavity. The input

power is 10 [mW], the peak transmission is 0.05. The squeezed light is analysed with a homodyne detector using a local oscillator beam of about 6 [mW]. The intensity of the squeezed
light is sufficient to cause interference with the local oscillator, the power at each of the two
detectors varies when the local oscillator phase is scanned.

In the experiment both the projection angle θ and cavity detuning are scanned simultaneously, but at different rates. θ is varied rapidly, allowing us to see the best noise suppression
and also the opposite, noisy quadrature for the different detunings. The detuning is scanned
slowly from below to above the cavity resonance. The detection frequency and atomic detuning are kept fixed. The data from the time scan were converted into a squeezing ellipse,
using the procedure described in Section 9.3.1, and shown in Fig. 9.15(b). This was one of
the first reported full reconstructions of a squeezing ellipse. Superimposed in this figure is
a squeezing ellipse with the parameters $r_s = 0.2$ and $\theta_s = 34$ degrees. The experimental
results fit very well to the theoretical simulations [Hop92], the size of the noise suppression,
the squeezing angle and the dependence of these parameters on experimental conditions such
as atomic detuning, detection frequency and intensity have all been verified, but despite this
effort no simple physical interpretation has emerged. We are certain that the results cannot be
modelled successfully as a standard Kerr interaction and again the whole complex interaction
of the atoms with the cavity has to be considered. While these experiments are an interesting
example of the complexity of atom-cavity coupling they have not produced strong reliable
noise suppression.

9.9 Pulsed squeezing

Some of the best squeezing results have been achieved using the very high intensities and
nonlinearities that can be achieved with optical pulses.

9.9.1 Quantum noise of optical pulses

The concepts of quantum noise and squeezing can be extended to pulsed light sources. Consider a series of optical pulses, each pulse with a length t_{pulse} repeated at time intervals
$t_{\text{rep}} \gg t_{\text{pulse}}$, as shown in Fig. 9.42(a). The spectrum of the photo-current will contain frequency peaks at the repetition frequency Ω_{rep} and its higher harmonics. The width of each
peak will depend on the length of the pulse train measured, or the integration time of the
detection. The distribution of power amongst, or the envelope of, the frequency peaks is set
by the Fourier transform of the pulse shape of the individual pulses. This is schematically
shown in Fig. 9.42(b). For the short pulses which can be generated with mode-locked lasers,
(Ω_{rep} typically 100 [MHz], $t_{\text{pulse}}0.2$–100 [ps]), the envelope is very wide but only a few of the
fundamental peaks fit within the detection bandwidth. In between these peaks the spectrum
contains a quantum noise floor. This is equivalent to the quantum noise spectrum of a CW
laser. Thus we can this *quasi CW*. One can interpret this noise as being the sum of many pairs
of quantum noise sidebands, each beating with one of the fundamental peaks in the optical
spectrum.

Any part of the recorded spectrum at frequency $\Omega < \Omega_{\text{rep}}$ represents the Fourier component of the variation in the energy of many pulses. For example, the pulsed source could

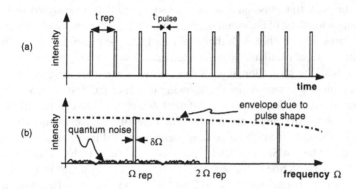

Figure 9.42: Optical pulses in the time domain (a) and the frequency domain (b). Quantum noise is present at frequencies between the fundamental components associated with the pulses.

be intensity modulated at a frequency Ω_{mod}. This would create spectral sidebands to all fundamental peaks of the optical spectrum, that means signals at $n\Omega_{rep} \pm \Omega_{mod}$. The quantum noise is the noise floor in such a sideband. In these experiments noise represents the lack of reproducibility of the pulses. An ideal classical light source would produce a sequence of perfectly reproducible pulses, and thus no noise outside the fundamental components would be detected. Coherent light cannot have exactly identical pulses, the energy will vary from pulse to pulse. We are comparing the variation of one group of pulses with that of the next group – and observing the noise floor. This quantum noise is independent of the detection frequency, as long as it is less than the repetition frequency, and a QNL pulsed laser will have a flat, white quantum noise spectrum between the fundamental peaks.

The detection scheme for pulsed light is almost identical to that for CW laser light. The same photo-diodes and spectrum analysers are used. A single detector can reveal intensity fluctuations. A balanced detector will be able to distinguish between classical and quantum intensity noise and a balanced homodyne detector can detect noise of specific quadratures. However, three additional requirements have to be satisfied in the detection of pulses:

(i) The detector should not respond to fundamental frequency components and their harmonics, otherwise saturation of the detector and the electronic amplifiers will occur even at very low optical powers. This can be achieved by using sophisticated filter networks with notch filters suppressing any response at Ω_{rep}. Such detectors require extensive development. The only exception is the detection at very low frequencies $\Omega \ll \Omega_{rep}$, where the detector sensitivity can simply be rolled off with lowpass filters.

(ii) The pulses have to be carefully aligned in the time domain. For example, the pulses used as the signal and local oscillator beams in the homodyne detector have to arrive simultaneously, with a time difference of much less than the pulse length. This requires an accurate adjustment of all distances in the experiment. Any cavity would have to have roundtrip times which are an integer of t_{rep}.

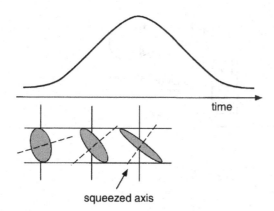

squeezed axis

Figure 9.43: The development of the squeezing during a light pulse, indicated by the change of the squeezing ellipse. From H. Haus et al. [Hau95].

(iii) The pulses contain very high intensities. While this is an advantage in that the nonlinear response is significantly enhanced, it can introduce at the same time other nonlinear processes, such as self focusing, which limit the beam quality and thus the detectability of the light.

The fundamental advantages of pulsed experiments are the high intensities and consequently the very large nonlinear effects that can be observed. The intensity during a pulse make the build-up cavities, which are necessary in CW experiments, obsolete. Single pass SHG or OPO operation is possible. Similarly, the Kerr effect in single pass is sufficiently strong to introduce a correlation between noise sidebands and therefore induce squeezing. However, there are some effects that limit the noise suppression. The squeezing occurs only during the pulse if we assume that the response of the material is instantaneous. When calculating the degree and the quadrature of the measurable squeezing it is necessary to take the time development of the squeezing into account and to average over the temporal development of the noise properties, and the squeezing, during the duration of a pulse. Figure 9.43 shows an example where the degree of squeezing and the orientation vary during the pulse. The measurement averages the noise contributions, which are weighted by the intensity at the detector at any one time. Although the squeezing at the peak intensity gets the highest weighting, the observed total squeezing will be considerably less than predicted for the peak intensity alone. In addition the squeezing angle is again a weighted average and squeezing in different quadratures can partially cancel each other.

9.9.2   Pulsed squeezing experiments with Kerr media

A very successful series of squeezing experiments with optical pulses was carried out by the MIT group [Ber91], [Hau95]. Here the optical medium is a single mode fibre, the nonlinear process is the Kerr effect. The light source is a mode-locked Nd:YLF laser generating 100 [ps] pulse with a repetition frequency of 100 [MHz], $\Omega_{\text{rep}} = 10$ [ns]. The wavelength is 1.3 [μm],

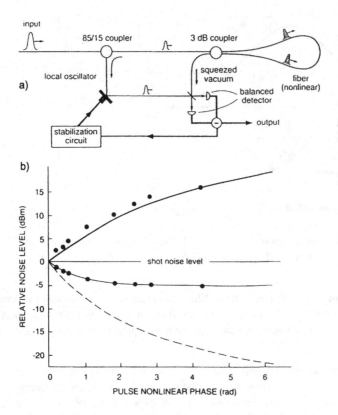

Figure 9.44: The pulsed Kerr experiment after H. Haus et al. [Hau95]. (a) the schematic layout, (b) noise suppression results from the pulsed Kerr experiment.

which is the zero dispersion wavelength of the fibre and assures that the pulses maintain their pulse shape during the experiment. The experiment is shown schematically in Fig. 9.44a.

The experimental arrangement is equivalent to a nonlinear Sagnac interferometer. The pulses from the laser enter the interferometer via a 3 [dB] fibre coupler, equivalent to a 50/50 beam splitter. There they are separated into two pulses of identical intensity which move in opposite directions through the fibre. After a full traverse the pulses arrive simultaneously back at the coupler. They interfere and form a pulse which is moving back towards the laser. A pulse of squeezed vacuum light leaves through the other port of the interferometer. This pulse is detected with a homodyne detector. A fraction of the reflected light is used as a local oscillator, generated by the 85/15 coupler. The arrival time of the local oscillator has to be synchronized with the pulse of squeezed light. The phase of the local oscillator is set and stabilised with a piezo controller.

The effect is seen in different experiments in two frequency bands: at low frequencies (80–100 [kHz]) and at high frequencies ($\geq$ 100 [MHz]). The Sagnac interferometer is essential for such measurements. All fluctuations induced by the fibre, such as phase shifts and intensity variations due to mechanical vibrations, are experienced by both counter-propagating

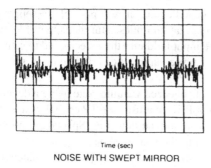

Time (sec)

NOISE WITH SWEPT MIRROR

Figure 9.45: The temporal trace of the noise from the pulsed experiment. Detection frequency is fixed at $\Omega = 55$ [kHz]. After K. Bergman et al. [Ber91].

pulses. They are cancelled when the pulses interfere after one roundtrip and do not appear on the output. This also applies to phase noise induced by the fibre, such as GAWBS. Equally, noise in the laser light frequency or intensity are compensated by the interferometer. In this particular example the suppression was so good that all noise with the exception of large relaxation oscillations were cancelled. What remains is the correlation between the quadratures, induced by the Kerr effect. The squeezing in this case is so strong that they did not even need a spectrum analyser, it was directly detectable with an oscilloscope in the time traces of the laser intensity. Figure 9.45 shows alternating noise traces with the squeezing process turned on and off. The difference between these traces gives directly the amount of squeezing. A noise suppression of -3.5 [dB] was achieved at frequencies as low as 55 [kHz].

9.9.3   Pulsed SHG and OPO experiments

The high peak intensities of short laser pulses allow the realization of direct single pass travelling wave second harmonic generation (TW SHG) and the associated squeezing. Consider that in a single pass through a crystal of length $\mathcal{L}$ a beam of input power P(0) generates a beam of power $P_{\text{SHG}}(\mathcal{L}) = \Gamma P(0)$. We call Γ the single pass conversion efficiency. At the same time the beam will experience an effective nonlinear phase shift ϕ_{eff} between the SH light generated at position $z = \mathcal{L}$ and SH light generated at position $z = 0$ and propagating to $z = \mathcal{L}$. Best SHG is achieved with complete phase matching, or $\phi_{\text{eff}} = 0$. It was shown by the group of P. Kumar [Li93] that in this case the predicted noise suppression at low frequencies is

$$S(\Omega \to 0) = 1 - \Gamma \tag{9.9.1}$$

This indicates the possibility of almost perfect squeezing provided that most of the light is converted into the second harmonic. Even with a moderate amount of phase mismatch the noise suppression should be very strong, detailed calculations predict squeezing as large as 10 [dB] [Li93]. However, the corresponding experiments were so far not successful, a noise suppression of only 0.3 [dB] has been observed [You96]. This was achieved with a Na doped KTP crystal. A SHG efficiency of $\Gamma = 0.15$ was obtained at a peak intensity of about 250 [MW/cm^2]. The main technical challenge has been in isolating the squeezed from the fundamental light. By using a type II phase matched crystal it was possible to separate the SHG beam from the fundamental beam using polarization optics. But the achievable extinction ratio was too small. Excessive leakage of the input field into the detector led to saturation of the detection system and limited the observed noise suppression. This type of squeezing requires further technical developments.

Far more successful has been the generation of squeezed vacuum state using a degenerate optical parametric amplifier (DOPA). The group of P. Kumar used a KTP crystal in single pass operation. The fluctuations are measured with a homodyne detector. Special attention was given to the matching of the local oscillator beam to the squeezed beam in both the spectral and temporal domains. Noise suppression of 5.8 [dB] was observed [Kim94], matching the best results from the corresponding CW cavity experiment.

9.9.4   Soliton squeezing

One of the requirements for quadrature squeezing in the pulsed experiments is to maintain the temporal shape of the pulse throughout the medium. This is commonly achieved by working at the zero-dispersion wavelength of an optical fibre. However, the material shows self phase modulation which results in a chirp of the frequencies. Consequently a non-soliton pulse does not allow the local oscillator phase to be optimized for the pulse as a whole. This limited early pulsed fiber squeezing experiments where Gaussian zero-dispersion pulses where used. A clever alternative was proposed by P. Drummond and his group [Dru87] and in parallel the group of H. Haus [Hau89]. A special pulse form can be used to form *temporal solitons* which preserve their shape by a balance between dispersion and nonlinear refractive effects and propagate free of chirp. Solitons have almost particle-like properties, they are robust against a variety of perturbations and preserve their shape, energy and momentum even after numerous interactions with each other [Zab65]. The pulse shape is a hyperbolic secant. The pulse width t_{FWHM} is dependent on the peak power P_{peak} because shorter pulses experience stronger dispersive broadening and consequently require a higher peak power to balance dispersion. A fundamental soliton is obtained for a wide range of pulse energies and wide varieties of initial pulse shapes such as Gaussian or square pulses. After some initial self-reshaping the exact balance of pulse width and peak power is eventually attained with a smooth soliton shape.

It is convenient to introduce the pulse parameter $N^2 = L_{\mathrm{D}}/L_{\mathrm{NL}} \propto P_{\mathrm{peak}} t_{\mathrm{FWHM}}^2$ which indicates whether dispersive pulse broadening ($N < 1$) or nonlinear pulse compression ($N > 1$) dominates the initial pulse evolution. The N-parameter is the ratio of the length scale for dispersive pulse broadening, $L_{\mathrm{D}} \propto (t_{\mathrm{FWHM}})^2$, and the length per one radian nonlinear phase shift in the pulse peak, $L_{\mathrm{NL}} \propto (P_{\mathrm{peak}})^{-1}$. These length scales are given by the dispersive and nonlinear coefficients of the fibre. A fundamental soliton is also called a $N = 1$ soliton because it exactly balances dispersion with nonlinear effects.

The quantum mechanical description, using a nonlinear Schrödinger equation, was able to describe the soliton's noise properties and unique quantum soliton effects were found [Lai89, Hau89]. The first such effect was amplitude quadrature squeezing, predicted by Carter et al. [Car87]. It was observed at IBM by M. Rosenbluh & R.M. Shelby [Ros91], who used a colour centre laser to produce soliton pulses with a pulse length of 200 [fs] at 1500 [nm] propagating through a polarization preserving single mode fibre. The Kerr effect in the fibre generated the squeezing. Noise suppression of -1.67 [dB] was measured with a homodyne detector.

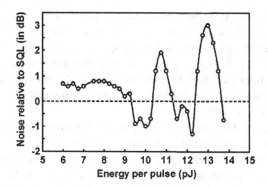

Figure 9.46: Squeezing calculations for soliton propagation and spectral filtering. In (a) the output energy of the filtered soliton is shown plotted against the input energy, in multiples of the fundamental soliton energy. In (b) the noise levels, normalized to the QNL, for filtered solitons are shown, for both the heuristic model (solid line) and the quantum field theory model (rectangular markers). After S. Friberg et al. [Fri96].

A second quantum effect is the "entanglement" of pairs of solitons. Their quadratures get linked, which can be exploited for quantum non-demolition experiments, see Chapter 11 and quantum information experiments, Chapter 13.

9.9.5   Spectral filtering

A third quantum effect is soliton photon-number squeezing by spectral filtering which was unexpectedly found by S. Friberg et al. [Fri96] in 1995. The experimental method was fairly simple: a pulse launched into a fibre was spectrally band-pass filtered after emerging from the fibre end and was then directly detected – an unusual idea since filtering deliberately introduces losses to achieve squeezing.

The effect can be heuristically explained by considering the propagation of solitons in the fibre. As they propagate, self-phase modulation and group velocity dispersion interact to cause a periodic broadening and shrinking of the soliton spectrum. For a fixed length of fibre the soliton energy, or photon number, determines the periodicity. When the soliton, after leaving the fibre, is attenuated by a fixed bandwidth spectral filter the losses depend on the soliton's pulse energy. The combination of propagation and filtering results in a nonlinear input-output relationship, i.e. nonlinear absorption. At specific lengths, or pulse energies, the intensity noise is suppressed below the QNL – photon number squeezing is seen. This simple model is qualitatively confirmed by full quantum field calculations, and one example is given in Fig. 9.46.

Predictions for noise reduction by spectral filtering show that up to -6.5 [dB] of photon-number noise reduction can be expected when an optimized bandpass filter removes the outlying sidebands of a fundamental ($N = 1$) soliton and transmits 82 % of the pulse energy [Mec97]. With more energetic pulses, e.g. second-order ($N = 2$) solitons, up to -8.4 [dB] of photon-number squeezing were predicted [Sch00].

The underlying squeezing mechanism can be best understood in terms of the multimode quantum structure of solitons. Spectrally resolved quantum noise measurements showed that anticorrelations in photon number fluctuations emerge within the spectrum of the soliton as it propagates down the fiber [Wer96, Spa98]. If these anticorrelated spectral domains are selectively transmitted through a spectral filter and then detected simultaneously and coherently, their fluctuations cancel each other in the observed photon flux. This anticorrelation produces a photocurrent noise reduction far below shotnoise.

The pioneering experiment was carried out with 2.7 [ps] pulses with a repetition rate of 100 [MHz] centred at 1455 [nm] with a bandwidth of 1.25 [nm]. After propagating through 1.5 [km] of fibre the pulses emerge with a pulse width of 1.3 [ps] and a spectral bandwidth of 1.65 [nm]. They were filtered with a variable bandpass filter to remove the high and low cut-off frequencies, resulting in the observation of -2.3 [dB] squeezing with $N = 1.2$ solitons [Fri96]. The method was further investigated in Erlangen by A. Sizmann & G. Leuchs et al. with sub-picosecond solitons and a variety of filter functions, fibre lengths and pulse energies below and above the fundamental soliton energy [Spa98]. When optimized in the experiment, up to $-3.8(\pm0.2)$ [dB] of photocurrent noise reduction below shotnoise were achieved in direct detection. If corrected for linear losses, a reduction in the photon-number uncertainty by -6.4 [dB] can be inferred [Spa98]. This technique is very elegant and has significant practical advantages: It requires no coherent noise measurements and thus is insensitive to phase (GAWBS) noise. It is broadband and all that is required is a pulsed soliton source and suitable fibre. By using an in-line filter this device can be made exceedingly lossless and robust.

9.9.6 Nonlinear interferometers

The most promising and most successful method of photon-number noise reduction with solitons so far are highly asymmetric nonlinear interferometers, Fig. 9.47a. The continuous-wave [Kit86] and soliton-pulse analysis [Wer98, Sch98a, Lev99a] show that the beam splitting ratio in the interferometer must be highly asymmetric in order to produce strong photon-number noise reduction, in contrast to earlier experiments with symmetric nonlinear fibre interferometers that produced a squeezed vacuum. More than 10 [dB] of noise reduction was predicted.

After the first experimental demonstration in Erlangen by S. Schmitt et al. using 130 [fs] solitons, the interferometer parameters were optimized [Sch98a] and the observation of less than -5 [dB] in the photocurrent noise was reported, Fig. 9.47c. This and parallel work reported by K. Bergman [Kry98] are the strongest noise reductions observed for solitons to date. When corrected for linear losses, the inferred photon-number squeezing was -7.3 [dB] [Sch98b]. A simple model shows a step-like input-output energy transfer function. Around each minimum the transmission is nonlinear, the output energy is stabilized at a fixed value and the photon-number uncertainty is reduced below the shotnoise limit.

However, the squeezing mechanism can be better understood in terms of number-phase correlations induced by the Kerr nonlinearity [Lev99b], similar to the continuous wave case studied by M. Kitagawa and Y. Yamamoto [Kit86]. A Kerr-squeezed state from the high-power path of the interferometer is slightly rotated by the addition of the weak field from the low-power path of the interferometer, such that the amplitude quadrature of the total output

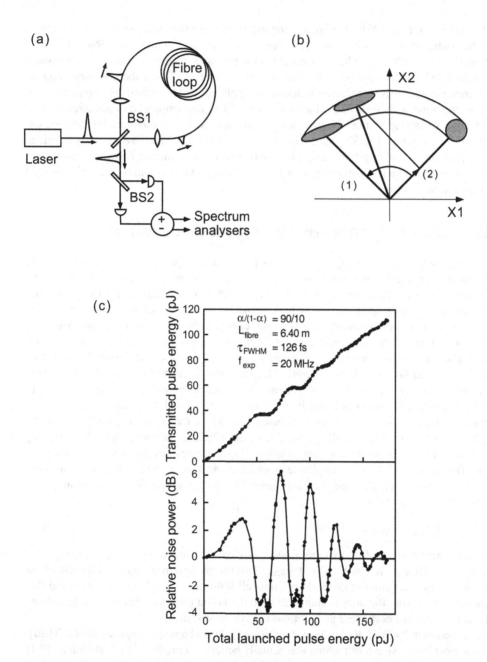

Figure 9.47: Intensity squeezing of solitons with the nonlinear interferometer. (a) experimental arrangement, (b) the concept of projecting the squeezing ellipse after the Kerr effect, (c) experimental results with solitons of different orders N. The top trace shows the transmission and the bottom trace the measured noise suppression. From A. Sizmann & G. Leuchs [Siz99].

field is squeezed, Fig. 9.47b. This squeezing mechanism implies that noise reduction grows with the nonlinear phase shift in one arm of the nonlinear interferometer [Lev99b]. This is in contrast to spectral filtering, where noise reduction originates from multimode photon-number correlations, which does not grow with propagation distance. Today the nonlinear loop interferometer is seriously considered in technical applications for stabilizing communication systems operating with Terabaud information rates. It is also employed in basic research as an effective squeezing source for generating entangled beams in quantum information experiments, see Chapter 13. To summarize, optical solitons in fibres have recently provided the perspectives of building fibre-integrated sources of noise reduction and gave new insight into multimode quantum correlations with a Terahertz bandwidth through spectrally resolved quantum measurements.

9.10   Amplitude squeezed light from diode lasers

The simplest intensity squeezing experiment is to produce amplitude squeezed light directly with a laser of high efficiency which is driven by a sub-Poissonian current. The laser can be driven inside a high impedance circuit with a very quiet current ($V i_{dr} \ll 1$). The conversion efficiency of the diode laser can be as high as $\eta_{las} = 0.5 - 0.75$. An important distinction is the fact that a diode laser has an internal cavity, to build up the radiation. The laser also involves both spontaneous and stimulated emission. Consequently, it has laser dynamics and spontaneous emission noise. The earliest models for diode lasers were developed by Y. Yamamoto and his coworkers [Mac89]. More recently linearized laser models have been developed [Ral96]. It is possible to apply them to the diode laser system (see Chapter 6) and to calculate directly the transfer function between the fluctuations of the pump source (i_{dr}) and the light. As Equation (6.1.17) shows, the noise of the laser intensity is completely dictated by the pump noise. For an electrically pumped laser the efficiency of pumping is close to unity and, in diode lasers, the output coupling is large. Thus the efficiency η_{las} is basically given by κ_m/κ. The variances $V1$ of the amplitude quadrature and VN of the intensity are equivalent and it is common practice to use the Fano factor for this situation. With these assumptions Equation (6.1.17) turns into

$$(1 - V1_{las}) = \eta_{las}(1 - Vi_{dr}) \tag{9.10.1}$$

This is the same result as for any other passive device. The maximum noise reduction for a perfectly quiet drive current is $V1_{las} = 1 - \eta_{las}$. Two further important assumptions have been made in this model, namely that the laser emits all light into a single output mode and that all the current entering the laser is contributing to the lasing process. Bias currents have been neglected. In practice both these assumptions are not easy to justify.

An experiment by W.H. Richardson et al. [Ric91] showed a noise suppression of 6.5 [dB]. This is a most impressive result which was actually better than predicted by Equation (9.10.1) and the overall efficiency of the laser used. No definite explanation has been given to date for this result. However, it is possible that the diode laser had significant internal parallel currents which did not participate in the lasing process. Thus the actual laser efficiency η_{las} might have been better than measured directly. Alternatively, some unknown process might have correlated the drive current with the light. Unfortunately, later experiments were not able to reproduce these early measurements, they showed less noise reduction and tended to confirm the prediction of Equation (9.10.1). In particular Inoue observed noise reduction up to −4.5 [dB] in diode lasers cooled to 20 [K] [Ino92].

It has been observed that diode lasers will excite more than one laser mode. These might not contribute much to the intensity of the output beam but they can influence the noise behaviour dramatically. A particularly systematic study was carried out by the group in France [Mar95]. They showed that a free running diode laser was emitting light into over 50 individual modes. The total intensity was found to be sub-Poissonian. However, by measuring only some of the modes, rejecting the others with a spectrograph, the noise increased drastically. The noise of the single central mode, which carried the vast majority of the power, was up to 40 [dB] above the Poissonian level. This suggests that the various modes have anti-correlated noise and the noise term cancel each other when all modes are detected simultaneously. This behaviour can be predicted by a simple multi mode laser model which includes the competition between the laser modes [Mar95]. Only some of the modes will be operating well above threshold, others will remain at threshold and only contribute fluctuations. The competition incorporates a limitation of the total gain based on the available number of excited atoms. The division between the modes is not stochastic but balances the fluctuations, similar to a conservation of the total number of available photons. The result is anti-correlation between the modes. More detailed theoretical models have been published to take mode competition properly into account [Esc96], but further study is still required.

It is possible to enforce single mode operation, and simultaneous noise suppression, by either injection locking or optical feedback from an external cavity. When these techniques are optimized the laser system will operate on a truly single mode output and generate amplitude squeezed light. M.J. Freeman et al. [Fre93] used injection locking and H. Wang et al. [Wan93] used external cavities to achieve single mode operation. They achieved a noise suppression of 1.8 [dB] with lasers at room temperature. More recently, D.C. Kilper et al. observed a noise reduction by 4.5 [dB] in an injection locked laser at 15 [K]. On the whole, these results fit well to Equation (9.10.1). An overview of all these results is given in Fig. 9.53.

However, it appears that not all diode lasers show reproducibly the same degree of noise suppression. Some commercial systems do not perform as well as earlier prototype versions of the same laser design. Large variations in the noise performance have been found even within one commercial series of lasers with otherwise identical specifications. The detailed reasons for this variation has still to be investigated. Recently Y. Yamamoto's group has succeeded for the first time in designing and manufacturing diode lasers which operate in single mode due to their geometry and which reliably operate at −2.8 [dB] below the Poissonian limit [Lah99].

Finally, one experimental problem is that the maximum efficiency of the lasers occurs at the highest drive currents, well above threshold, where the output powers are large. For a measurement of the noise suppression all the light has to be detected. Thus the beam cannot be attenuated or otherwise the noise suppression is lost. The limit of detectable power is set by the saturation properties of the detector, presently only a maximum of 50 [mW] can typically be detected with fast detectors capable of measuring fluctuations at RF frequencies (≥ 10 [MHz]). This means that some of the best noise suppression cannot actually be detected.

In summary, we can expect that diode lasers will soon be reliable, compact and low cost sources of non-classical light.

9.11   Twin photon beams

Closely related to amplitude squeezed light is the generation of *twin photon beams*. In this case two optical beams are generated by one above threshold optical parametric oscillator (OPO), see Sections 9.5 and 6.3.3.

The single input beam is converted, in a nonlinear crystal inside a resonant cavity pumped above threshold, into two output beams which have identical statistical properties. In a simple picture one can imagine that each input photon generates two identical photons, one for each output beam, but unlike the parametric downconverter we have here not correlated pairs of photons but two single mode beams with correlated photon flux. By using type II materials, that means the two modes have orthogonal polarizations, see Section 9.5, the beams can easily be separated and detected individually. The OPO in most cases is non-degenerate, that means the optical frequencies of the beams are different but this has no consequences since they are not recombined optically.

Theoretical calculations [Fab89, Fab90] show that each beam individually has intensity fluctuations which depend on the dynamics of the OPO, as discussed in Section 6.3.3. In particular, close to threshold each beam has very large fluctuations, $V1(\Omega) \gg 1$, at four times the threshold the beams are at the QNL and far above the threshold each beam should be squeezed, $V1(\Omega) = 0.5$. This has not yet been observed.

However, if we detect the difference between photo currents these fluctuations cancel because of their strong correlations, and we can observe a noise level well below the QNL. The QNL in this case corresponds to the noise measured on the difference between the intensities of "classical twin beams", obtained by splitting any light beam on a 50 : 50 beamsplitter. It is equal to the shot noise level for the combined photocurrent. This experiment was first carried out with CW lasers by the Paris group, led by E. Giacobino & C. Fabre [Hei87] who recorded a noise suppression of 27 % ($V1 = 0.73$). Next came an experiment with mode locked, Q-switched lasers by O. Aytür & P. Kumar [Ayt90]. C.D. Nabors & R.M. Shelby [Nab90] used a doubly resonant OPO to generate two beams of different colours and reported a noise suppression of -3.1 [dB]. Systematic improvements of their apparatus by the group in Paris led to observations of noise suppression by -8.5 [db], after correction for the dark currents, which is a record that stood for a long time. It was beaten by the Taiyuan group [Gao98] with a noise reduction of -9.2 [dB] and finally improved again by the Paris group to -9.5 [dB] [Lau03]. Since these experiments involve only intensity measurements, and no local oscillators, they avoid the complications of mode matching and beam overlap and the data are limited purely by the quantum efficiency of the detectors and the escape efficiency. Note also that twin photon beams are obtained at the two outputs of a 50/50 beamsplitter which has at its two input ports a squeezed vacuum state and a local oscillator of appropriate phase. This is actually what is done in any homodyne measurement of squeezing.

It is possible to generate one beam which has sub-Poissonian statistics by measuring one beam and using this information to control, via electro-optics feedforward, the intensity of the other. This was demonstrated by J. Mertz et al. [Mer91] who achieved a noise suppression of -4 [dB] of one beam compared to the QNL of this individual beam. This new output beam can be directly used for all the applications of sub-Poissonian light, it forms a direct alternative to the quiet diode lasers mentioned previously. Twin photon beams can also be directly used to beat the QNL in the spectroscopy of very weak absorption lines by -1.9 [dB] [Sch97], and the measurement of very weak absorptions by -7.0 [dB] [Gao98].

Finally, the Paris group showed that twin beams offer the opportunity to introduce post-selection processes, commonly used in the photon counting regime, to the regime of continuous variables [Lau03]. If the fluctuations of the two photocurrents at a fixed detection frequency Ω_{det} are recorded directly and stored, as it was described in the section on state tomography (9.13), we can now produce a sequence of recordings each containing short intervals of the fluctuations. Thanks to the quantum correlations we can use one beam to select intervals where $V1(\Omega_{det})$ is within a narrow, well defined band and select from the other beam the corresponding intervals. The selected light describes a beam with unusual properties. For example, we could select information that describes a beam with strong sub-Poissonian distribution, which was demonstrated experimentally and produced an intensity noise reduction of -4.4 [dB] below the QNL. Alternatively the selection could be done with an electro-optical modulator. The resulting beam is now no longer continuous but intermittent, depending on the selection process which rejects the majority of the time intervals but is has the advantage of representing a strongly non-classical quantum state.

9.12 Polarization squeezing

The polarization of a beam of light is another property which is limited by quantum noise. In order to detect the polarization we require a combination of two detectors. Each is producing quantum noise and consequently one expects a QNL for the polarization. The polarization is well described by the four Stokes parameters S_0, S_1, S_2, S_3, see Section 3.5.3, which describe the light in terms of three degrees of freedom for linear and circular components and one normalization factor. The state of polarization can be described by a 3-dimensional diagram, the Poincaré sphere, defined by the radius $S = (S_1^2 + S_2^2 + S_3^2)^{\frac{1}{2}}$. Each point on the sphere is one polarization, and the radius of the sphere is given by the intensity of the beam. An analysis of the quantum properties of polarization shows that they obey the cyclical commutation relation

$$[\tilde{\mathbf{S}}_1, \tilde{\mathbf{S}}_2] = 2i\tilde{\mathbf{S}}_3 , \quad [\tilde{\mathbf{S}}_2, \tilde{\mathbf{S}}_3] = 2i\tilde{\mathbf{S}}_1, \quad [\tilde{\mathbf{S}}_3, \tilde{\mathbf{S}}_1] = 2i\tilde{\mathbf{S}}_2 \qquad (9.12.1)$$

This means that the polarization has an uncertainty volume, the polarization is only defined within a small sphere around each point on the Poincaré sphere, as shown in Fig. 9.48 (i)

The possibility of achieving polarization below the QNL was investigated extensively by the group around A.S. Chirkin in Moscow. In particular N. Korolkova and A.P. Alodjants investigated the possibility of using amplitude squeezed light to influence the polarization [Kor96, Alo98]. It was shown that two squeezed beams can be combined on a beamsplitter and by selecting the correct polarization and phase between the two beam the output beam will have one or more of the Stokes variables measurable below the QNL [Kor02]. The uncertainty volume is compressed and can take the shape of a cigar, with the two Stokes parameters $\tilde{\mathbf{S}}_1 \tilde{\mathbf{S}}_3$ suppressed at the same time, or alternatively like a pancake, with suppression of only one parameter, $\tilde{\mathbf{S}}_2$.

The first experimental demonstrations were made by the Bell group [Gra87] and the Aarhus group [Sor98]. Recently the ANU group achieved the goal where several Stokes parameters were squeezed simultaneously, and below the -3 [dB] limit. They generated two independent squeezed beams [Sch03] with two seeded OPAs driven by one Nd:YAG laser

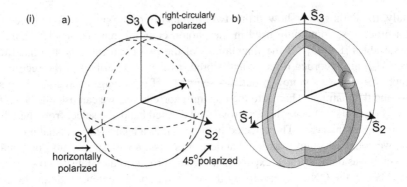

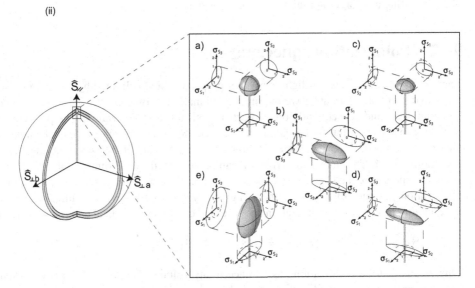

Figure 9.48: (i) Diagram of (a) the classical and (b) quantum Stokes vectors mapped on a Poincaré sphere. The ball at the end of the vector visualises the quantum noise of the polarization states. (ii) Measured quantum polarization noise at 8.5 [MHz] from different combinations of input beams. (a) single coherent beam, (b) coherent beam and squeezed beam, (c) bright squeezed beams, (d) two phase squeezed beams, and (e) two amplitude squeezed beams after W.P. Bowen et al. [Sch03].

and frequency doubler. The squeezed beams were combined on one beamsplitter. The apparatus is shown in Fig. 9.49 as an example of the actual complexity of a current quantum optics experiments. By generating two bright squeezed beams with about 1 [mW] power and about −4 [dB] of squeezing each they were able to achieve both the cigar and pancake states, with suppression of the Stokes parameters simultaneously by better than −3 [dB] at detection

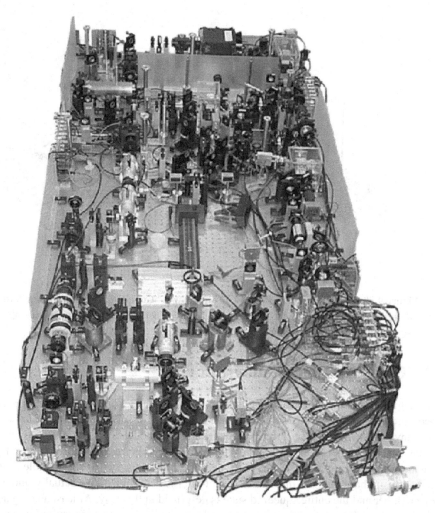

Figure 9.49: The double squeezer experiment, with 2 OPAs, shows the complexity of a current quantum optics experiment. The multitude of components, all individually aligned and opti-mized, is required to produce the pump, seed and local oscillator beams, to produce the two squeezed beams, to combine them on a beamsplitter and to detect the properties of the output state. The main challenges are to reduce the loss of all components, to optimize the mode-match between all the beams and to maintain the frequency locking between all the cavities, such as lasers, mode-cleaners and OPOas. For this purpose the experiment contains 4 temperature control systems and 7 optical locking systems.

frequencies around 8 [MHz]. Figure 9.48(ii) shows the experimental results in the form of reconstructed uncertainty areas on the Poincaré sphere. This type of light can now be used to couple the quantum properties of the light to the quantum properties of atoms, as discussed in Section 10.3.

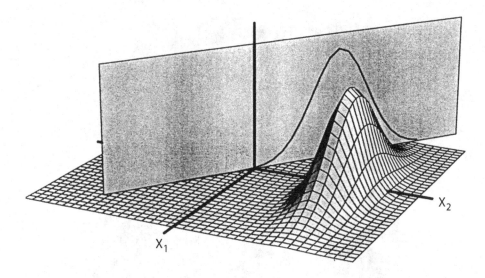

X_1

X_2

Figure 9.50: The concept of quantum state tomography: Complete noise traces are recorded for different projection angles, one such example is given here. These data can be recombined to reconstruct the Wigner function of the quantum state, after U. Leonhardt [Leo97].

9.13   Quantum state tomography

So far we have concentrated only on the measurements of the variances of quantum noise. It has been shown that this is fully sufficient for a description of squeezed states, or to be more general, any states which can be described by linearizable models. We saw in Section 4.5 that these states have 2-dimensional Gaussian Wigner functions. For a quadrature squeezed state, with unknown squeezing angle, we have to measure the variance $V(\theta)$ for a whole range of angles and from these data the minimum value $V(\theta_s)$ is determined. We actually quote only this one number and the entire squeezed state is described in this way. At least as long as it is a minimum uncertainty state, otherwise a second number, $V(\theta_s + 90)$ is required to describe the state completely.

On the other hand techniques have been developed to measure the Wigner function. This is of fundamental interest since some QPDs exhibit unusual features and it would be nice to show diagrams of completely reconstructed QPDs. The key idea to these techniques is to measure higher order moments of the noise statistics. Rather than measuring only the variance for a certain projection angle θ the full noise trace is recorded, see Fig. 9.51. This was first done, for pulsed light, by Raymer's group at Oregon [Smi93] by recording a histogram of the energy of many pulses. This corresponds to a recording of the full noise characteristics of the pulses. Many such histograms were recorded for different angles θ and the underlying Wigner function can be reconstructed using tomographic techniques [Smi93].

For CW experiments the process is similar. The full time history of the fluctuations is recorded. This can be achieved by directly measuring the noise fluctuations of the photo-current and demodulating the signal with a fixed single detection frequency. Alternatively

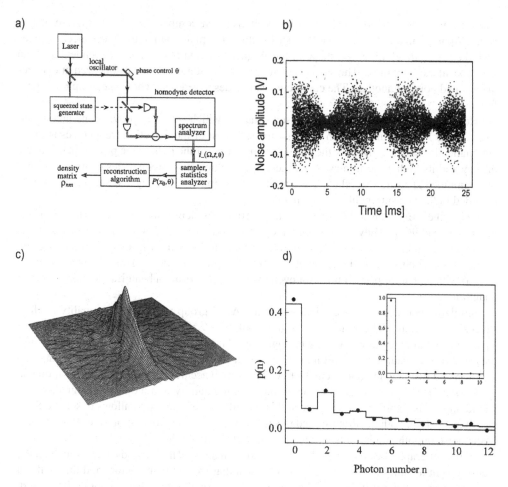

Figure 9.51: Quantum tomography experiment. Part (a)shows the schematic layout. Results for a squeezed vacuum state are give: (b) the phase dependent noise, (c) the squeezed 2-dimensional Wigner function and (d) the reconstructed photon probability distribution. From S. Schiller et al. [Sch96b].

an electronic spectrum analyser is used, as shown in Fig. 9.51a. By leaving the video band-width sufficiently large the full noise trace is recorded. For a squeezed state these traces vary dramatically with the projection angle, as shown Fig. 9.51b. The many traces are then transformed into histograms of the fluctuations for this specific projection angle and detection frequency Ω. The restriction to specific values of Ω is particularly important for those squeezing sources which have a distinct squeezing spectrum $S(\Omega)$. Here it is important to detect only those sidebands which are strongly squeezed. For each detection frequency a function $W(\Omega)$ can be reconstructed using the tomographic technique [Leo93, Leo96]. Note that all the measured Wigner functions will be centred around zero, they show the fluctuations, which are independent of the mean value of the intensity. The quadrature of the squeezing is only determined as a rotation angle, which is measured either with a homodyne detector or, in the case of a bright squeezed source with large intensity, in comparison with the phase of this beam. Apart from a rotation of the Wigner function, the phase and amplitude squeezed states look the same. In addition, the shape of the Wigner function is frequency dependent and can vary considerably with Ω. The same light source can simultaneously be squeezed at one frequency and noisy at another. The complete beam is described by a whole spectrum of Wigner functions.

How does this compare with the pulsed situation which seems to provide one unique answer for the Wigner function? In this case the selection of the detection frequencies is made automatically; they are given by the Fourier components of the pulses, the pulse repetition frequency and its higher harmonics. Only these specific detection frequencies contribute. Only one combined answer, a weighted average of several single frequency Wigner functions, was recorded in the measurements by Raymer et al..

Using their excellent OPO, the Konstanz group has demonstrated the capability of tomography [Sch96b]. They investigated a squeezed vacuum, measured the noise projections and converted them into Wigner functions and photon distributions. One example is given in Fig. 9.51. Their results, the complete *squeezed* Wigner function (c) and the photon statistics (d) with enhanced probabilities for even photon numbers are a beautiful demonstration of quantum optics.

Similar experiments were carried out by the ANU group to analyse the quality of their squeezing apparatus. The front cover of this book Fig. 9.52 shows the Wigner function of a squeezed vacuum beam, recorded by tomography [And03]. We can clearly see the noise suppression in one quadrature. Careful tests show that the Wigner function is indeed a Gaussian function and that any deviations can be traced back to inaccuracies in the data acquisition and processing. This technique can be used to measure the quality of a modulation signal, which would appear as a shift in phase space at this modulation frequency. It allows to see the SNR simultaneously in both quadratures and to see the quality of the modulation as well as the size of the noise in both quadratures within one measurement [Wu98].

Will tomography be able to produce more exotic results? This depends not so much on the measurement techniques but on the types of states that we can manufacture and this will not be easy. As pointed out in Section 4.5 we would require a system that cannot be linearized. There are basically two types of states that could be considered: either a state with a very small photon number, which only be detected by photon counting techniques and here strongly nonclassical Wigner functions have been recorded, as discussed in Chapter 13. Alternatively we would require a CW system which is extremely nonlinear, for example close to an instability point, and where the coefficients for the nonlinearities (e.g. $\chi^{(2)}$ or $\chi^{(3)}$) change within the uncertainty limit of the state. These are extremely difficult to achieve since we do not have such nonlinearities. It remains a challenge for the future.

9.14   Summary of squeezing results

Squeezing moved from the stage of a cute theoretical idea to the first experimental demonstration in a remarkably short time. From 1980 to 1985 the details of possible squeezing were explored theoretically, then within one year several squeezing experiments were performed more or less simultaneously.

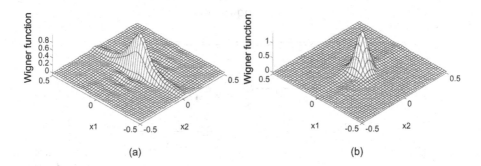

Figure 9.52: Tomographic reconstruction of the quasi-probability distribution of (a) a squeezed vacuum state in comparison with (b) a normal vacuum state. After U. Andersen et al. [And03].

Since the first demonstration of squeezed states in 1985 we have seen a steady improvement in the size and the reliability of the noise suppression. During the first few years several unexpected new noise sources were discovered hampering the progress. These had to be avoided, in most cases through improvements of the materials used. Examples are: the new fibres used by K. Bergman et al. have significantly less GAWBS noise than the earlier fibres used by M.D. Levenson et al.; the early noise problems encountered by H.J. Kimble et al. with external cavity second harmonic generators were avoided in the more recent monolithic devices; the limitation in squeezing with the OPO was set by noise created inside the crystal, this can now be avoided by lower loss materials. New processes were found, such as the nonlinear interferometer for pulsed squeezing. Very quiet monolithic diode pumped Nd:YAG lasers have replaced the much noisier lamp pumped lasers used in the earlier experiments. And even the detectors have been improved with a quantum efficiency of more than 0.9 now available for the visible (Si) as well as the near infrared spectrum (InGaAs). All this technology had matured by about 1998.

More than 40 different experiments have been reported until 2003, too many to list separately. Instead, Fig. 9.53 shows the historical development of the experiments. The observed noise suppression is plotted for several groups of experiments producing squeezed light. As can be seen, the OPO generating a squeezed vacuum state was the first to show large squeezing and it still remains the best process. At this point in time a suppression to -6 [dB] is reliably achievable. One could expect a suppression to -10 [dB] once all the components and materials are optimized. This is a realistic estimate of the level of suppression that will be achieved within the next few years. The main advance is reliability. In 1984 these were "hero" experiments which worked at 2 AM. "Daylight squeezing" during normal businesses hours is now the norm. Commercialization is a possibility, limited only by the demand but not the availability of the device.

The CW experiments with atomic systems, both 4WM and strong cavity coupling, and with Kerr media were beautiful demonstrations of concepts and test of the theory. They were essential for our understanding but are less likely to play a role in applications.

The singly resonant SHG is rapidly improving and can already be regarded as a reliable source of bright amplitude squeezed light. A noise suppression to -5 [dB] is a realistic ex-

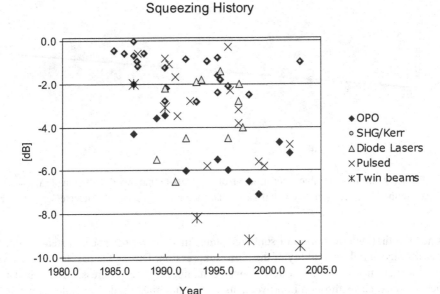

Figure 9.53: Observed noise suppression in different squeezing experiments 1985–2003. Boxes are CW vacuum squeezing using a sub-threshold OPO, diamonds are CW bright squeezed light (SHG, atoms, Kerr), triangles are diode laser noise suppression, crosses are pulsed squeezing experiments (OPO, SHG, solitons etc.), circles are twin-photon beam experiments.

pectation. These systems can be made very robust, small and reliable and with the use of periodically poled materials we might see even simpler and better devices.

For the pulsed experiments the results look equally impressive. Here the Kerr effect, soliton squeezing and the nonlinear interferometer all have the potential to achieve a noise suppression to -10 [dB]. The potential of new fibres with engineered bandgap structures has not been explored.

The only exception to this positive trend have been the diode lasers. Some outstanding results have been achieved very early on, but they could not be confirmed. The best of the present results are supporting the direct theoretical prediction and further understanding and refinements of the manufacturing process could make -3 [dB] of noise suppression available in commercial, low cost lasers.

Squeezed vacuum states can be transformed into various types of squeezed states, the squeezing axis can be rotated, coherent amplitude can be added through reflection off a cavity. We can control the squeezing independently of the coherent amplitude. In addition we can produce polarization squeezed light and spatially squeezed light. Fibre based squeezing and diode lasers have the potential to be readily integrated into practical systems.

Table 9.2: Loopholes in the quantum description

Limit	Loophole
QNL	Squeezing
$V1 = V2 = 1$	$V1 < 1,\ V2 > 1$
Attenuation	Noiseless beamsplitter
$V1_{out} = \eta V1_{in} + (1 - \eta)$	$V1_{out} = \eta V1_{in}$
Amplification	Noiseless amplifier
$V_{out} = G\,V_{in} + 1$	$V_{out} = GV_{in}$
Measurement	Quantum non-demolition
$V_{out} > V_{in}$	$V_{out} = V_{in}$
Communication	Teleportation
$V1_{out} > V1_{in}$	$V1_{out} = V1_{in}$
$V2_{out} > V2_{in}$	$V2_{out} = V2_{in}$

9.14.1 Loopholes in the quantum description

Squeezing is only only one way we can improve on the quantum noise limit associated with the coherent state of light produced by a laser. There are further loopholes in the quantum description of light that can be explored, and these will be described in the following chapters. Table 9.2 provides an overview of the quantum limits of the standard solution, the coherent state, also referred to as classical light. The description of these limits in terms of the variances and the loopholes we can exploit.

Further reading

Quantum Squeezing, P. Drummond, Z. Ficek, Springer (2003)
Measuring the quantum state of light, U. Leonhardt, Cambridge University Press (1997)
Quantum Optics in Phase Space, W.P. Schleich, E. Mayr, D. Krähmer, Wiley-VCH (1997)

Bibliography

[And03] Four modes of optical parametric operation of squeezed state generation, U.L. Andersen, B.C. Buchler, P.K. Lam, J.W. Wu, J.R. Gao, P.K. Lam, H-A. Bachor Eur. Phys. J. D, 1 (2003)

[Alo98] Polarization quantum states of light in nonlinear distributed feedback systems, A.P. Alodjants, S.M. Arakelian, A.S. Chirkin, Appl. Phys. B. 53, (1998)

[Ayt90] Pulsed twin beams of light, O. Aytür, P. Kumar, Phys. Rev. Lett. 65, 1551 (1990)

[Ber91] Squeezing in fibres with optical pulses, K. Bergman, H.A. Haus, Opt. Lett. 16, 663 (1991)

[Ber92]   Analysis and measurement of GAWBS spectrum in a nonlinear fiber ring, K. Bergman, H.A. Haus, M. Shirasaki, Appl. Phys. B 55, 242 (1992)

[Ber94]   Squeezing in a fiber interferometer with a gigahertz pump, K. Bergman, H.A. Haus, E.P. Ippen, M. Shirasaki, Opt. Lett. 19, 290, (1994)

[Bow02]   Polarization squeezing of continuous variable stokes parameters, W.P. Bowen, R. Schnabel, H-A. Bachor, P.K. Lam, Phys. Rev. Lett. 88, 093601(2002)

[Bre95]   Squeezed vacuum from a monolithic optical parametric oscillator, G. Breitenbach, T. Mueller, S.F. Pereira, J.-Ph. Poizat, S. Schiller, J. Mlynek, J. Opt. Soc. Am. B 12, 2304 (1995)

[Bre97]   Measurement of the quantum states of squeezed light, G. Breitenbach, S. Schiller, J. Mlynek, Nature 387, 471 (1997)

[Car87]   Squeezing of quantum solitons, S.J. Carter, P.D. Drummond, M.D. Reid, R.M. Shelby, Phys. Rev. Lett. 58, 1841 (1987)

[Cav81]   Quantum-mechanical noise in an interferometer, C.M. Caves, Phys. Rev. D 23, 1693 (1981)

[Col84]   Squeezing of intracavity and travelling-wave light fields produced in parametric amplification, M.J. Collett, C.W. Gardiner, Phys. Rev. A 30, 1386 (1984)

[Col85]   Squeezing spectra for nonlinear optical systems, M.J. Collett, D.F. Walls, Phys. Rev. A 32, 2887 (1985)

[Col87]   Output properties of parametric amplifiers in cavities, M.J. Collett, R. Loudon, J. Opt. Soc. Am. B 4, 1525 (1987)

[Dru87]   Quantum field theory of squeezing in solitons, P.D. Drummond, S.J. Carter, J. Opt. Soc. Am. B 4, 1565 (1987)

[Esc96]   Master-equation theory of multimode semiconductor lasers, A. Eschmann, C.W. Gardiner, Phys. Rev. A 54, 760 (1996)

[Fab89]   Noise characteristics of a non-degenerate optical parametric oscillator – application to qunatum noise reduction, C. Fabre, E. Giacobino, A. Heidmann, S. Reynaud, J. Physique 50, 1209 (1989)

[Fab90]   Squeezing in detuned degenerate optical parametric oscillators, C. Fabre, E. Giacobino, A. Heidman, L. Lugiato, S. Reynaud, M. Vadachino, Wang Kaiye, Quantum Opt. 2, 159-187 (1990)

[Fab92]   Squeezed states of light, C. Fabre, Physics Reports, 219 (1992)

[Fre93]   Wavelength-tunable amplitude-squeezed light from a room-temperature quantum well laser, M.J. Freeman, H. Wang, D.G. Steel, R. Craig, D.R. Scrifes, Opt. Lett. 18, 2141 (1993)

[Fri92]   Quantum-non demolition measurements of the photon number of an optical soliton, S.R. Friberg, S. Machida, Y. Yamamoto, Phys. Rev. Lett. 69, 3165 (1992)

[Fri96]   Observation of optical soliton photon-number squeezing, S.R. Friberg, S. Machida, M.J. Werner, A. Levanon, T. Mukai, Phys. Rev. Lett. 77, 3775 (1996)

[Gal91]   System control by variation of the squeezing phase, P. Galatola, L.A. Lugiato, M.G. Porreca, P. Tombesi, G. Leuchs, Opt. Commun. 85, 95 (1991)

[Gao98] Generation and application of twin beams from an optical parametric oscillator including an α-cut KTP crystal, J. Gao, F. Cui, C. Xue, C. Xie, K. Peng, Opt. Lett. 23, 870 (1998).

[Gat95] Quantum images and critical fluctuations in the optical parametric oscillator below threshold, A. Gatti, L.A. Lugiato, Phys. Rev. A 52, 1675 (1995)

[Gra87] Squeezed-light-enhanced polarization interferometer, P. Grangier, R.E. Slusher, B. Yurke, A. LaPorta, Phys. Rev. Lett. 59, 2153 (1987)

[Hag94] Phase-controlled transistor action by cascading of second order nonlinearities in KTP, D.G. Hagan, Z. Wang, G. Stegemann, E.W. VanStryland, M. Sheik-Bahae, G. Assanto, Opt. Lett. 19, 1305 (1994)

[Hau89] Quantum nondemolition measurement of optical solitons, H.A. Haus, K. Watanabe, Y. Yamamoto, J. Opt. Soc. Am. B6, 1138 (1989)

[Hau95] From classical to quantum noise, H.A. Haus, J. Opt. Soc. Am. B12, 2019 (1995)

[Hei87] Observation of quantum noise reduction on twin laser beams, A. Heidmann, R.J. Horowicz, S. Reynaud, E. Giacobino, C. Fabre, Phys. Rev. Lett. 59, 2555 (1987)

[Ho91] Single beam squeezed-state generation in sodium vapour and its self focussing limitations, S.T. Ho, N.C. Wong, J.H. Shapiro, Opt. Lett. 16, 846 (1991)

[Hop92] The atom-cavity system as a generator of quadrature squeezed states, D.M. Hope, H-A. Bachor, D.E. McClelland, A. Stevenson, Appl. Phys. B 55, 210 (1992)

[Ino92] Quantum correlation between longitudinal-mode intensities in a multimode squeezed semiconductor laser, S. Inoue, H. Ohzu, S. Machida, Y. Yamamoto, Phys. Rev. A 46, 2757 (1992)

[Kim92] Squeezed states of light: an (incomplete) survey of experimental progress and prospects, H.J. Kimble, Physics Reports 219, 227 (1992)

[Kim94] Quadrature-squeezed light detection using a self-generated matched local oscillator, C. Kim, P. Kumar, Phys. Rev. Lett. 73, 1605 (1994)

[Kit86] Number-phase minimum-uncertaintystate with reduced number uncertaintyin a Kerr nonlinear interferometer M. Kitagawa, Y. Yamamoto, Phys. Rev. A 34, 3974 (1986)

[Koe96] *Solid-state laser engineering*, W. Köchner, 4 th ed, Springer Verlag (1996)

[Kor96] Formation and conversion of the polarization-squeezed light, N.V. Korolkova, A.S. Chirkin, J. Mod. Opt. 43, No5, 869 (1996)

[Kor02] Polarization squeezing and continuous-variable polarization entanglement, N. Korolkova, G. Leuchs, R. Loudon, T.C. Ralph, Ch. Silberhorn, Phys. Rev. A 65, 052306 (2002)

[Kry98] Amplitude-squeezed solitons from an asymmteric fiber interferometer, D. Krylov, K. Bergman, Opt. Lett. 23, 1390 (1998)

[Kue93] Bright squeezed light by second-harmonic generation in a monolithic resonator, P. Kürz, R. Paschotta, K. Fiedler, J. Mlynek, Europhys. Lett. 24, 449 (1993)

[Lah99] Transverse-junction-stripe GaAs-AlGaAs lasers for squeezed light generation, S. Lahti, K. Tanaka, T. Morita, S. Inoue, H. Kan, Y. Yamamoto, AIEEE J. Quantum Electron. 35, 387 (1999)

[Lam95]   Cold atoms: A new medium for quantum optics, A. Lambrecht, J.M. Courty,
          S. Reynaud, E. Giacobino, Appl. Phys. B 60, 129-134 (1995)

[Lam99]   Optimization and transfer of vacuum squeezing from an optical parametric oscil-
          lator P.K. Lam, T.C. Ralph, B.C. Buchler, D.E. McClelland, H-A. Bachor, J. Gao,
          J. Opt. B. Qu. Semmiclass. Opt. 1, 469 (1999)

[Lai89]   Quantum theory in solitons in optical fibres. I. Time-dependent Hartree approxima-
          tion, Y. Lai, H.A. Haus, Phys. Rev. A 40, 844 (1989)

[Lau03]   Conditional preparation of quantum states in the continuous variable regime : Gen-
          eration of sub-Poissonian states from twin beams, J. Laurat, T. Coudreau, N. Treps,
          A. Maître, C. Fabre, accepted Phys. Rev. Lett. (2003)

[Leo93]   Realistic optical homodyne measurements and quasiprobability distributions,
          U. Leonhardt, H. Paul, Phys. Rev. A 48, 4598 (1993)

[Leo96]   Sampling of photon statistics and density matrix using homodyne detection,
          U. Leonhardt, M. Muroe, T. Kiss, Th. Richter, M.G. Raymer, Opt. Comm. 127,
          144 (1996)

[Leo97]   *Measuring the quantum state of light*, U. Leonhardt, Cambridge University Press
          (1997)

[Lev85]   Squeezing of classical noise by non degenerate four-wave mixing in an optical fiber,
          M.D. Levenson, R.M. Shelby, S.H. Perlmutter, Opt. Lett. 10, 514 (1985)

[Lev89]   Stochastic noise in TEMoo laser beam position, M.D. Levenson, W.H. Richardson,
          S.H. Perlmutter, Opt. Lett. 14, 779 (1989)

[Lev90]   *Introduction to Nonlinear Optics*, 2nd edition, M.D. Levenson, S.S. Kano, Aca-
          demic Press (1988)

[Lev99a] Perturbation theory of quantum solitons: continuum evolution and optimum
          squeezing by spectral filtering D. Levandovsky, M.V. Vasilyev, P. Kumar Opt. Lett.
          24, 43 (1999)

[Lev99b] Soliton squeezing in a highly transmissive nonlinear optical loop mirror, D. Levan-
          dovsky, M.V. Vasilyev, P. Kumar Opt. Lett. 24, 89 (1999)

[Li93]   Squeezing in travelling-wave second harmonic generation, Ruo-Ding Li, P. Kumar,
          Opt. Lett. 18, 1961 (1993)

[Lou89]   Graphical representation of squeezed states-variances, R. Loudon, Opt. Commun.
          70, 109 (1989)

[Mac89]   Observation of amplitude squeezing from semiconductor lasers by balanced direct
          detectors with a delay line, S. Machida, Y. Yamamoto, Opt. Lett. 14, 1045 (1989)

[Mae96]   Bright squeezing by singly resonant second-harmonic generation: effect of funda-
          mental depletion and feedback, J. Maeda, K. Kikuchi, Opt. Lett. 11, 821 (1996)

[Mar95]   Squeezing and inter-mode-correlations in laser diodes, F. Marin, A. Bramati, E. Gi-
          acobino, T.-C. Zhang, J-Ph. Poizat, J-F. Roche, P. Grangier, Phys. Rev. Lett. 75,
          4606 (1995)

[Mar96]   Nonlinear Dynamics and quantum noise for competing χ^2 nonlinearities, M. Marte,
          J. Opt. Soc. Am. B 12, 2296 (1995)

[Mat02]   Vacuum squeezing in atomic media via self-rotation, A.B. Matsko, I. Novikova,
          G.R. Welch, D. Budker, D.F. Kimball, S.M. Rochester Phys. Rev. A 66, 043815
          (2002)

[Mec97] Linearized quantum-fluctuation theory of spectrally filtered optical solitons, A. Mecozzi, P. Kumar, Opt. Lett. 22, 1232 (1997)

[Mer91] Generation of sub-Poissonian light using active control with twin beams, J. Mertz, A. Heidmann, C. Fabre, Phys. Rev. A 44, 3229 (1991)

[Mil87] Optical-fiber media for squeezed-state generation, G.J. Milburn, M.D. Levenson, R.M. Shelby, S.H. Perlmutter, R.G. DeVoe, D.F. Walls, J. Opt. Soc. Am. B 4, 1476 (1987)

[Nab90] Two-color squeezing and sub-shot noise signal recovery in doubly resonant optical parametric oscillators, C.D. Nabors, R.M. Shelby, Phys. Rev. A 42, 556 (1990)

[Oro87] Squeezed-state generation in optical bistability, L.A. Orozco, M.G. Raizen, M. Xiao, R.J. Brecha, H.J. Kimble, Opt. Soc. Am. B 4, 1490 (1987)

[Ou96] Observation of nonlinear phase shift in CW harmonic generation, Z.Y. Ou, Opt. Comm. 124, 430 (1996),

[Pas94] Bright squeezed light from a singly resonant frequency doubler, R. Paschotta, M. Collett, P. Kürz, K. Fiedler, H-A. Bachor, J. Mlynek, Phys. Rev. Lett. 72, 3807 (1994)

[Pau30] *General Principles of Quantum Mechanics*, W. Pauli, Springer (1980)

[Per88] Generation of squeezed light by intercavity frequency doubling, S.F. Pereira, M. Xiao, H.J. Kimble, J.L. Hall, Phys. Rev. A 38, 4931 (1988)

[Pol92] Atomic spectroscopy with squeezed light for sensitivity beyond the vacuum-state limit, E.S. Polzik, J. Carri, H.J. Kimble, Appl. Phys. B 55, 279 (1992)

[Ral95] Squeezed light from second-harmonic generation: experiment versus theory, T.C. Ralph, M.S. Taubman, A.G. White, D.E. McClelland, H-A. Bachor, Opt. Lett. 20, 1316 (1995)

[Ral96] Intensity noise of injection locked lasers:quantum theory using a linearised input/output method, T.C. Ralph, C.C. Harb, H-A. Bachor, Phys. Rev. A 54, 4370 (1996)

[Rar92] Quantum Correlated Twin Beams, J.G. Rarity, P.R. Tapster, J.A. Levenson, J.C. Garreau, I. Abraham, J. Mertz, T. Debuisschert, A. Heidmann, C. Fabre, E. Giacobino, Appl. Phys. B Photophysics & laser chemistry 55, 250 (1992)

[Rei88] Quantum theory of optical bistability without adiabatic elimination, M. Reid, Phys. Rev. A 37, 4792 (1988)

[Ric91] Quantum correlation between the junction-voltage fluctuation and the photon-number fluctuation in a semiconductor laser, W.H. Richardson, Y. Yamamoto, Phys. Rev. Lett. 66, 1963 (1991)

[Rie03] Experimental vacuum squeezing in Rubidium vapor vis self-rotation, J. Ries, B. Brezger, A.I. Lvovsky, quant-ph-0303109 (2003)

[Roc97] Quantum nondemolition experiments using cold trapped atoms, J-F. Roch, K. Vigneron, Ph. Grelu, A. Sinatra, J-Ph. Poizat, P. Grangier, Phys. Rev. Lett. 78, 634 (1997)

[Ros91] Squeezed optical solitons, M. Rosenbluh, R.M. Shelby, Phys. Rev. Lett. 66, 153 (1991)

[Sch96a] Generation of continuous-wave bright squeezed light, S. Schiller, G. Breitenbach, S.F. Pereira, R. Paschotta, A.G. White, J. Mlynek, SPIE Vol. 2378, 91 (1995)

[Sch96b] Quantum statistics of the squeezed vacuum state by measurement of the density matrix in the number state representation, S. Schiller, G. Breitenbach, S.F. Pereira, T. Müller, J. Mlynek, Phys. Rev. Lett. 77, 2933 (1996)

[Sch97] Sub shot noise high sensitivity spectroscopy with OPO twin beams, C. Schwob, P.H. Souto Ribeiro, A. Maître, C. Fabre, Opt. Lett. 22, 1893 (1997)

[Sch98a] Photon-number squeezed solitons from an asymmetric fiber-optic Sagnac interferometer S. Schmitt, J. Ficker, M. Wolff, F. König, A. Sizmann, G. Leuchs, Phys. Rev. Lett. 81, 2446 (1998)

[Sch98b] Investigation of strongly photon-number squeezed solitons from an asymmetric fiber-Sagnac interferometer S. Schmitt, J. Ficker, A. Sizmann, G. Leuchs, Annual Report Lehrstuhl fuer Optik Universität Erlangen, p. 63 (1998) Photon-number noise reductionfrom a nonlinear fiber loop mirror S. Schmitt, F. König, B. Mikulla, S. Spälter, A. Sizmann, G. Leuchs, IQEC OSA Technical Digest series 7, 195 (1998)

[Sch98c] Generation of strongly squeezed continuous-wavelight at 1064 nm, K. Schneider, M. Lang, J. Mlynek, S. Schiller, Opt. Express 2, 64 (1998)

[Sch00] Quantum noise of damped N-solitons, E. Schmidt, L. Knöll, D-G. Welsch, Opt. Commun. 179, 603 (2000)

[Sch03] Stokes operator-squeezed continuous-variable polarization states, R. Schnable, W.P. Bowen, N. Treps, T.C. Ralph, H-A. Bachor, P.K. Lam. Phys. Rev. A 67, 012316 (2003)

[She86] Generation of squeezed states of light with a fiber-optics ring interferometer, R.M. Shelby, M.D. Levenson, D.F. Walls, A. Aspect, G.J. Milburn, Phys. Rev. A 33, 4008 (1986)

[Siz90] Observation of amplitude squeezing of the up-converted mode in second harmonic generation, A. Sizmann, R.J. Horowicz, E. Wagner, G. Leuchs, Opt. Commun. 80, 138 (1990)

[Siz99] The optical Kerr effect and quantum optics in fibers A. Sizmann and G. Leuchs, in: Progress in Optics **XXXIX**, E. Wolf (Ed.), p. 373 (Elsevier, Amsterdam 1999)

[Slu85] Observation of squeezed states generated by four-wave mixing in an optical cavity, R.E. Slusher, L.W. Hollberg, B. Yurke, J.C. Mertz, J.F. Valley, Phys. Rev. Lett. 55, 2409 (1985)

[Slu87a] Squeezed-light generation by four-wave mixing near an atomic resonance, R.E. Slusher, B. Yurke, P. Grangier, A. LaPorta, D.F. Walls, M. Reid, J. Opt. Soc. Am. B 4, 1453 (1987)

[Slu87b] Pulsed squeezed light, R.E. Slusher, P. Grangier, A. La Porta, B. Yurke, M.J. Potasek, Phys. Rev. Lett. 59, 2566 (1987)

[Smi93] Measurement of the Wigner distribution and the density matrix of a light mode using optical homodyne tomography: Application to squeezed states and the vacuum, D.T. Smithey, M. Beck, M.G. Raymer, A. Faridani, Phys. Rev. Lett. 70, 1244 (1993)

[Sok01] Quantum holographic teleportation, I.V. Sokolov, M.I. Kolobov, A. Gatti, L. Lugiato, Opt. Commun. 193, 175 (2001)

[Sor98] Quantum noise of an atomic spin polarization measurement, J.L. Soerernsen, J. Hald, E.S. Polzik, Phys. Rev. Lett. 80, 3487 (1998)

[Spa97] Photon number squeezing of spectrally filtered sub-picosecond optical solitons, S. Spälter, M. Burk, U. Strössner, M. Böhm, A. Sizmann, G. Leuchs, Europhys. Lett. 38, 335 (1997)

[Spa98] Propagation of quantum properties of subpicosecond solitons in a fiber, S. Spälter, M. Burk, U. Strößner, A. Sizmann, G. Leuchs, Opt. Expr. 2, 77 (1998)

[Tsu95] Generation of amplitude-squeezed light at 431 nm from a singly resonant frequency doubler, H. Tsuchida, Opt. Lett. 20, 2240 (1995)

[Wal83] Squeezed states of light, D.F. Walls, Nature 306,141 (1983)

[Wam93] *Quantum Optics*, D.F. Walls, G. Milburn, Springer (1993) and (1997)

[Wan93] Squeezed light from injection-locked quantum well lasers, Hailin Wang, M.J. Freeman, D.C. Steel, Phys. Rev. Lett. 71, 3951 (1993)

[Wer96] Quantum statistics of fundamental and higher-order coherent quantum solitons in Raman active waveguides, M.J. Werner, Phys. Rev. A. 54, R2567 (1996)

[Wer98] Quantum soliton generation using an interferometer, M.J. Werner, Phys. Rev. Lett. 81, 4132 (1998)

[Whi96a] Experimental test of modular noise propagation theory for quantum optics, A.G. White, M.S. Taubman, T.C. Ralph, P.K. Lam, D.E. McClelland, H-A. Bachor, Phys. Rev. A 54, 3400 (1996)

[Whi96b] Active versus passive squeezing by second harmonic generation, A.G. White, T.C. Ralph, H-A. Bachor, J. Opt. Soc. Am. B13, 1337 (1996)

[Whi96c] Cascaded second order nonlinearity in an optical cavity, A.G. White, J. Mlynek, S. Schiller, Europhys. Lett. 35, 425 (1996)

[Whi97] PhD Andrew White, Australian National University (1997)

[Wu87] Squeezed states of light from an optical parametric oscillator, Ling-An Wu, Min Xiao, H.J. Kimble, J. Opt. Soc. Am. B4, 1465 (1987)

[Wu98] Optical tomography of an information carrying laser beam, Jinwei Wu, P.K. Lam, M. Gray, H-A. Bachor, Optics Express 3, 154 (1998)

[Xia87] Precision Measurement beyond the shot-noise limit, Min Xiao, Ling-An Wu, H.J. Kimble, Phys. Rev. Lett. 59, 278 (1987)

[You96] Observation of sub-Poissonian light travelling wave second harmonic generation, Sun Hyun Youn, Sang-kyung Choi, P. Kumar, Ruo-Ding Li, Opt. Lett. 21, 1597 (1996)

[Yue76] Two photon coherent state of the radiation field, H.P. Yuen, Phys. Rev. A 13, 2226 (1976)

[Yue79] Generation and detection of two-photon coherent states in degenerate four-wave mixing, H.P. Yuen, J.H. Shapiro, Opt. Lett. 4, 334 (1979)

[Zab65] Interaction of "solitons" in a collisionless plasma and the recurrence of initial states, N.J. Zabusky and M.D. Kruskal, Phys. Rev. Lett. 15, 240 (1965)

10 Applications of squeezed light

Squeezed states allow us to, in principle, circumvent some of the quantum limits that apply to coherent states of light. Quantum measurements provide opportunities that classical systems do not have. Will this have any practical implications? Three areas stand out as the most likely applications: *Optical communication* could benefit if information can transmitted with non-classical light. *Optical sensors* could benefit from squeezed light and a whole range of *spatial effects and imaging* have quantum equivalents. Finally interferometric detectors for *gravitational waves* are the most likely example for the use of squeezed light. For all these cases we will discuss the standard limits which apply to coherent states and ways to overcoming them.

10.1 Optical communication

Optical communication has been one of the motivations for the development of quantum optics technology. In the last two decades there has been a tremendous advance in optical communication technology: low loss fibres, high quantum efficiency photo-diodes, diode lasers and laser amplifiers have all improved the system performance. It can be foreseen that we will soon reach the situation where the quantum properties of light become an important limit for the performance of optical communication systems. At that stage squeezed light might play a role in overcoming these limits and in enhancing the performance. In this chapter we want to estimate the size of such possible improvements.

In optical systems the information is encoded on a mode of light which is propagated via an optical fibre and decoded at the receiver. The attraction of optical systems is the wide modulation bandwidth and the ability to wavelength multiplex. In principle, a communication system contains the components shown in Fig. 10.1. Different encoding and detection schemes are in use, with the simplest, and most commonly used one direct incoherent modulation. In this case the transmitter is a diode laser, the information is encoded by an amplitude modulator, and the receiver is a simple photo-detector. More advanced systems use coherent schemes such as phase shift keying (PSK). This requires a phase modulator in the transmitter and a homodyne, or heterodyne, detector with a local oscillator beam that is phase locked to the input mode. These systems can operate within a few decibels of the QNL.

The question of how nonclassical light can improve the performance of the system was the motivation for the early work by Takahashi [Tak65] and H. Yuen and Shapiro [Yue78]. The techniques for the encoding, propagating and decoding of information on a quantum system were reviewed by Y. Yamamoto and H.A. Haus [Yam86] and by Caves and Drummond [Cav94]. Non classical light can be used in several ways to improve components of the system: i) The laser can produce sub Poissonian or quadrature squeezed light, and thus possibly allow

A Guide to Experiments in Quantum Optics, 2nd Edition. Hans-A. Bachor and Timothy C. Ralph
Copyright © 2004 Wiley-VCH Verlag GmbH & Co. KGaA
ISBN: 3-527-40393-0

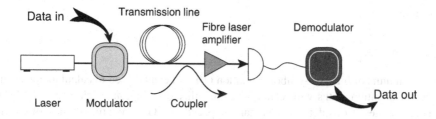

Figure 10.1: Components of an optical communication system.

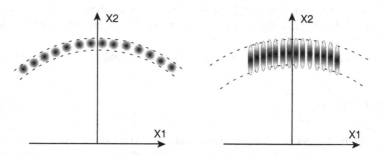

Figure 10.2: Illustration of the number of distinguishable states for incoherent detection. The states within a ring have the same intensity, and are thus counted as the same information. (a) coherent states (b) squeezed states.

a better signal to noise ratio. This is particularly attractive and practical when diode lasers are used. ii) The phase insensitive laser amplifier can be replaced by a phase sensitive parametric amplifier. iii) The beamsplitter, or coupler, can be equipped with a squeezed vacuum input which reduces the noise introduced at this point or alternatively a QND device can be used. In all cases the detection system remains unchanged.

The overall improvement can be illustrated by considering the amount of information that can be encoded on a mode of light with given maximum optical power. For an incoherent system, where only intensity information is processed, this information content can be estimated by evaluating the number of states with different intensities which can be distinguished. This is illustrated in Fig. 10.2 where all possible coherent states are shown. The maximum available power is represented by the large circle. Within this power constraint each different state is represented by an uncertainty area, symmetric circles in the case of coherent states. An incoherent system recognises different intensities, the phase information is ignored. That means states within a particular ring, with a width given by the uncertainty area, can be distinguished. Each ring represents one different piece of information, like one letter in the alphabet. With sub-Poissonian light the rings can be made narrower, allowing for more information to be encoded as illustrated in Fig. 10.2.

A quantitative figure of merit for a communication system is the channel capacity, a concept that describes the maximum amount of information that can be transmitted based on statistical arguments. The Shannon capacity [Sha48] of a communication channel with Gaussian noise of power (variance) N and Gaussian distributed signal power S operating at the bandwidth limit is

$$C = \frac{1}{2} \log_2 \left[1 + \frac{S}{N} \right]$$ (10.1.1)

where C is in units of bits per symbol. Equation (10.1.1) can be used to calculate the channel capacities of quantum states with Gaussian probability distributions such as coherent states and squeezed states. Consider first a signal composed of a Gaussian distribution of coherent state amplitudes all with the same quadrature angle. The signal power V_s is given by the variance of the distribution. The noise is given by the intrinsic quantum noise of the coherent states, $V_n = 1$. Because the quadrature angle of the signal is known, homodyne detection can in principle detect the the signal without further penalty thus the measured signal to noise is $S/N = V_s/V_n = V_s$.

In general the average photon number per bandwidth per second of a light beam is given by

$$\bar{n} = \frac{1}{4}(V^+ + V^-) - \frac{1}{2}$$ (10.1.2)

where V^+ (V^-) are the variances of the maximum (minimum) quadrature projections of the noise ellipse of the state. These projections are orthogonal quadratures, such as amplitude and phase, and obey the uncertainty principle $V^+V^- \geq 1$. In the above example one quadrature is made up of signal plus quantum noise such that $V^+ = V_s + 1$ whilst the orthogonal quadrature is just quantum noise so $V^- = 1$. Hence $\bar{n} = 1/4 V_s$ and so the channel capacity of a coherent state with single quadrature encoding and homodyne detection is

$$C_c = \log_2 \left[\sqrt{1 + 4\bar{n}} \right]$$ (10.1.3)

Establishing in an experiment that a particular optical mode has this capacity would involve: (i) measuring the quadrature amplitude variances of the beam, V^+ and V^-, (ii) calibrating Alice's signal variance and (iii) measuring Bob's signal to noise. If these measurement agreed with the theoretical conditions above then Shannons theorem tells us that an encoding scheme exists which could realize the channel capacity of Eq. (10.1.3). An example of such an encoding is given in [Cer01].

For photon numbers $\bar{n} > 2$ improved channel capacity can be obtained by encoding symmetrically on both quadratures and detecting both quadratures simultaneously using heterodyne detection or dual homodyne detection. Because of the non-commutation of orthogonal quadratures there is a penalty for their simultaneous detection which reduces the signal to noise of each quadrature to $S/N = 1/2V_s$. Also because there is signal on both quadratures the average photon number of the beam is now $\bar{n} = 1/2V_s$. On the other hand the total channel capacity will now be the sum of the two independent channels carried by the two quadratures. Thus the channel capacity for a coherent state with dual quadrature encoding and heterodyne detection is

$$\begin{aligned} C_{ch} &= \frac{1}{2} \log_2 \left[1 + \frac{S^+}{N} \right] + \frac{1}{2} \log_2 \left[1 + \frac{S^-}{N} \right] \\ &= \log_2 \left[1 + \bar{n} \right] \end{aligned}$$ (10.1.4)

which exceeds that of the homodyne technique (Eq. 10.1.3) for $\bar{n} > 2$.

The above channel capacities are the best achievable if we restrict ourselves to a semi-classical treatment of light. However the channel capacity of the homodyne technique can be improved by the use of non-classical, squeezed light. With squeezed light the noise variance of the encoded quadrature can be reduced such that $V_{ne} < 1$, whilst the noise of the unencoded quadrature is increased such that $V_{nu} \geq 1/V_{ne}$. As a result the signal to noise is improved to $S/N = V_s/V_{ne}$ whilst the photon number is now given by Eq. 10.1.2 but with $V^+ = V_s + V_{ne}$ and $V^- = 1/V_{ne}$ where a pure (i.e. minimum uncertainty) squeezed state has been assumed. Maximizing the signal to noise for fixed $\bar{n}$ leads to $S/N = 4(\bar{n} + \bar{n}^2)$ for a squeezed quadrature variance of $V_{ne,opt} = 1/(1 + 2\bar{n})$. Hence the channel capacity for a squeezed beam with homodyne detection is

$$C_{sh} = \log_2 \left[1 + 2\bar{n}\right] \tag{10.1.5}$$

which exceeds both coherent homodyne and heterodyne for all values of $\bar{n}$.

Although squeezing gives an improvement in channel capacity this improvement is rapidly washed out by non-idealities in the system such as propagation loss and other inefficiencies. It still seems some way off before squeezing might be considered a viable option for increasing channel capacity.

In principle, a final improvement in channel capacity can be obtained by allowing non-Gaussian states. The absolute maximum channel capacity for a single mode is given by the Holevo bound and can be realized by encoding in a maximum entropy ensemble of Fock states and using photon number detection. This ultimate channel capacity is

$$C_{Fock} = (1 + \bar{n}) \log_2 \left[(1 + \bar{n})\right] - \bar{n} \log_2 \left[\bar{n}\right] \tag{10.1.6}$$

which is the maximal channel capacity at all values of $\bar{n}$.

10.2 Spatial squeezing and quantum imaging

All our concern was so far devoted to the temporal fluctuations of a radiation field measured on a large detector that records the total intensity of the light beam. The following question arises: if one uses instead a spatially resolving photo detector (e.g. diode array, CCD detector,...), what will be the spatial distribution of the quantum fluctuations correlations in the transverse plane? Can one have local squeezing in small parts of the beam, or correlations between different parts? Could there be a spatial equivalent to squeezing? The study of such questions forms a growing domain in quantum optics, called *quantum imaging*, which was pioneered by M.D. Levenson and the St. Petersburg group and which has been extensively studied from the mid 90's. For an overview see: [Kol99, Lug02].

All the experiments described above are concerned with a single transverse mode of the field. In 1989, M. Levenson commented that strictly speaking a single spatial mode is not possible, that a real laser beam would always have some fluctuating contributions from other, higher order spatial modes, which do not affect the intensity of the laser, but bring fluctuations to its transverse variation. In a discussion aptly named 'Why a laser beam can never go straight' [Lev89] he described the consequences of the uncertainty principle on the spatial mode distribution. Using a diode laser he showed that the presence and the noise of such higher order spatial modes can be detected and are in agreement with simple models.

One can show that in a highly squeezed single mode beam the local quantum fluctuations, which can be measured on a very small photodetector, are at the QNL [Lug93], the local fluctuations at different points of the transverse plane being correlated in such a way that they cancel when all these contributions are summed up on a large area detector. It is only when the many transverse modes are simultaneously squeezed that one can get a local noise reduction below the QNL. Spatial quantum effects and multi (transverse) mode quantum optics are thus intimately related. M. Kolobov et al. [Kol89] showed through their theory that travelling wave frequency-degenerate parametric amplification in a thin crystal pumped by a plane wave was able to produce such a multimode squeezed light, when the parametric gain is large, because it creates quantum correlations between all the tilted plane wave modes at the subharmonic frequency with wave vectors k and k' such that $k + k' = k_{\mathrm{pump}}$, in the same way as in the time domain the correlation between symmetrical sideband modes creates broad band squeezing. L. Lugiato et al. [Lug93] theoretically showed that the same property holds if one places the parametric crystal in a resonant cavity with plane mirrors, provided the system is operated just below the oscillation threshold. In both situations strong quantum correlations should appear between the symmetrical pixels in the far field, that means correlations between the k vectors. These spatial quantum properties disappear when one uses curved mirrors instead of plane mirrors [Lug95], (which has the practical advantage of considerably reducing the OPO threshold), because such cavities are usually resonant for a single transverse mode at a given length. Only transverse degenerate cavities are likely to produce spatial intensity quantum effects. L. Lugiato and P. Grangier [Lug97] have shown that frequency degenerate OPOs inside a confocal cavity produce below threshold squeezed vacuum states whatever the size and shape of the detector sensitive area, provided that it is symmetrical with respect to the cavity axis.

Almost all the temporal quantum effects detailed in the chapter on squeezing can be extended into the spatial domain. For example, one can produce spatially resolved quantum correlated beams using parametric amplification [Nav02] and noiseless parametric amplification of images can be performed [Kol95]. Parametric down conversion is not the only way of producing such spatial quantum effects : Second Harmonic Generation can also produce interesting spatial quantum effects [Sco03]; Spatial solitons propagating either in $\chi^{(2)}$ or $\chi^{(3)}$ media have also interesting quantum spatial properties [Tre00]. One can show for example that more squeezing on a spatial soliton can be measured if one takes only the central part of it [Tre00]. Such a spatial filtering is an exact analogy of the spectral filtering which has been shown to improve the squeezing in temporal solitons, see Section 9.9.4.

Possible applications are numerous, such as the detection of weak phase and amplitude images [Kol93], noiseless amplification and teleportation of images [Sok01], two-photon microscopy [Fei97], microlithography [Bot00], detection of small changes in an image [Fab00]. Multimode squeezed light can also improve the ultimate resolution in imaging [Kol00] using the so-called super-resolution or reconstruction techniques.

There has been so far only a few experiments that showed such spatial effects. A pioneering experiment on the noiseless amplification of a double slit image by frequency degenerate travelling wave degenerate amplification was performed by P. Kumar [Cho99]. Recently a nontrivial spatial distribution of correlations inside twin beams produced by a confocal OPO has been reported [Mar03].

A different way to produce multimode nonclassical light was chosen in the experiments at the ANU, where the idea is to synthesize such a beam from single mode squeezed beams. It aims at measuring with maximum accuracy the transverse displacement d of a light beam, using a split detector . The difference between the photocurrents measured on the two segments produces a very sensitive signal proportional to the transverse displacement from the centre. The quantum noise from the two segments cannot be subtracted and sets the quantum noise limit d_{QNL} for the measurement of the displacement. This value is equivalent to the size Δ of the beam at the detector divided by number of photons n detected during one measurement,

$$d_{\text{QNL}} = \Delta/\sqrt{n} \tag{10.2.1}$$

In order to go beyond this limit it was shown by the Paris group that a normal intensity squeezed beam is not useful, instead one must have a mixture of two transverse modes: a normal squeezed TEM_{00} vacuum mode and a bright coherent "split mode", which has the same Gaussian intensity distribution as the TEM_{00} mode, but with a π phase shift between the two halves. This solution is in true analogy to the homodyne detector, Section 8.1.4, where two beams, one bright one vacuum, are combined on a beamsplitter and one of the four beams recorded by the two detectors has a π phase shift.

Experimentally such a multimode state is "synthesized" from normal TEM_{00} modes using split phaseplate with a π retardation left to right and combining the beams either with a beamsplitter [Tre02] or a ring cavity [Tre03] which transmits the symmetric mode and reflects the asymmetric mode with minimum losses, see Fig. 10.3(a). In an first experiment a noise reduction of -1.7 [dB] below the QNL in the position measurements was achieved, allowing us to measure beam displacements of a less than 1 [nm]. The measurement was extended to 2 dimensions, which requires the mixing of three spatial modes, one bright coherent and two squeezed vacuum beams. A noise reduction of -3.5 [dB] in the horizontal and -2.0 [dB] in the vertical direction was obtained.

This "quantum laser pointer" can measure spatial modulations of the beam with a detection limit a factor two smaller than a conventional laser, as is shown in Fig. 10.3(b). This is a truly sub QNL multimode spatial state and we can now search for applications.

10.3   Optical sensors

Sensors using squeezed light

Any sensor that is already QNL can, in principle, benefit from the use of squeezed light and the SNR of the measurements can be further increased. A number of experiments have already demonstrated such improvements. For the case of the polarisation sensors, discussed above, the group at ATT demonstrated an improvement by 2.8 [dB] in the SNR of detecting a polarisation generated by a EO modulator [Gra87]. Their experiment consisted of a polarimeter, operating with two input beams, one direct CW laser beam the other a squeezed vacuum beam generated by an OPO. For details see Section 10.4.4.

The same concept can be applied to the detection of absorption. In this case amplitude squeezed light is used, as generated with quiet diode lasers, see page 160, or from a SHG experiment, Section 9.6. The experiments are identical to conventional absorption sensors or

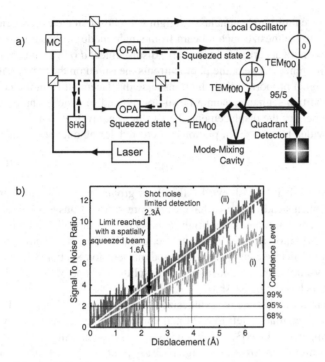

Figure 10.3: The quantum laser pointer: (a) the experimental layout which shows the mixing of three beams, two vacuum beams and one bright coherent beam with different spatial phase distributions using a ring cavity, (b) the measurement of a spatial oscillation with increasing amplitude (horizontal scale). The recorded value (vertical scale) is noisy due to the short recording intervals. The noise floor for the squeezed beams (II) is below the noise floor for the conventional beam (I) and the value for smallest detectable oscillation is reduced by a factor of two. After N. Treps et al. [Tre03]

absorption spectroscopy with the only change being that the input beam is squeezed. The absorption has to be modulated at a frequency Ω_{em} within the bandwidth of the squeezing spectrum. Care has to be taken that the overall efficiency of the apparatus and the detectors is very high, otherwise the degree of squeezing and the improvement in SNR rapidly disappears. One of the first demonstrations was made by E. Polzik et al. [Pol92] who demonstrated an improvement of 2 [dB] for the SNR for saturated absorption spectroscopy of a Cs vapour. They used an OPO as a source for squeezed light and the major challenge was to make this light source tunable. Another fine demonstration was carried out by C.D. Nabors and R.M. Shelby [Nab90] who showed that a modulation can be sensed with improved SNR using the bright squeezed light from an above threshold doubly resonant monolithic OPO.

One of the difficulties is to modulate the absorption at a sufficiently high frequency. The experiments by the group in Shanxi University [Hai97] use an elegant trick: twin photon beams generated by an above threshold OPO. One beam is used as a reference. The other is modulated, at Ω_{mod} and split into another pair of beams, probe 1 and probe 2 and only

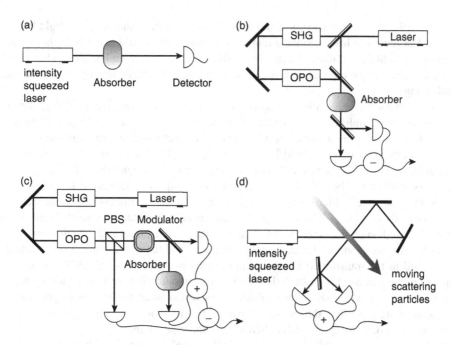

Figure 10.4: Different schemes for using squeezed light for improved SNR in optical sensors. (a) absorption using amplitude squeezed light, (b) Absorption using tunable OPO, after E. Polzik et al. [POL92], (c) measuring DC absorption using twin photon beams, after Hai et al.[HAI97], (d) Detecting scattered light laser Doppler velocimetry, after Li et al. [LI97].

probe 1 is sent through the sample, see Fig. 10.4(c). By first forming the sum the two probe photo-currents and then the difference between the total probe and the reference a current is generated that contains an AC signal at Ω_{mod} which is proportional to the strength of the absorption. In this way a slowly varying absorption can be observed with high sensitivity.

A similar approach can be used to detect the intensity of scattered light, see Fig. 10.4(d). It was shown by Li et al. [Li97] that the light scattered by fast moving smoke particles can be detected using an optical heterodyne detectors. The Doppler shift provided the necessary frequency shift and the spectrum was used to detect the density and the velocity distribution of the sample. This technique is known as Doppler velocimetry. Using amplitude squeezed light from a diode laser Li et al. was able to demonstrate an increase in the SNR of 1 [dB] compared to an experiment with classical light.

Making squeezed light robust

One of the principle problems for the use of squeezed light in sensors is its fragility. Any attenuation, or loss of the light will reduce the quantum noise suppression, see Section 9.3. Thus it is not realistic to consider such practical systems as the combination of sensor and optical fibres to transmit the optical beam to a remote sensor. One possible solution is to amplify both the signal, generated by the sensor, and the noise, of the squeezed light to levels well above the SQL. Such a beam could have the same SNR as the squeezed light but would be robust in regard to losses, since attenuation of this beam would bring both the signal and noise closer to SQL, without affecting the SNR. Such a noisy beam has clear advantages for the communication of information; it can be attenuated or conventionally amplified without disadvantage.

The generation of such a beam requires a noiseless phase sensitive amplifier. Any other phase insensitive amplifier (PIA) would reduce the SNR, in the case of large gain by 3 [dB], see Section 6.2. Possible techniques are the complete detection and re-emission of the light using high efficiency detectors and LEDs [Goo93, Roc93]. The disadvantage here is that no phase reference or coherence exists between the input and the output beam. Optical parametric amplifiers (OPA) could be used [Lev93], but they are complex in design and have limited gain.

As an alternative, which does not require any non-linear process whilst its output still retains optical coherence, P.K. Lam et al. used electro-optic feed forward [Lam97]. This scheme is based on partial detection of the light with a beamsplitter and detector, Fig. 10.5(a). The light reflected from the beam splitter is detected and the photo-current is amplified and fed forward to the transmitted beam via an electro-optic modulator (EOM). The electronic gain and phase can be chosen such that the intensity modulation signals carried by the input light are amplified, whilst the vacuum fluctuations which enter through the empty port of the beam splitter are cancelled, thus making the amplification noiseless. Since not all of the input light is destroyed, the output is still coherent with the input beam.

In the setup used by P.K. Lam et al., a half-waveplate controlled the transmittivity, ε_1, of the polarizing beamsplitter. On the in-loop beam, a high efficiency ($\varepsilon_2 = 0.92 \pm 0.02$) photo-detector was used. The photo-current is then passed through multiple stages of RF amplification and filtering to ensure sufficient gain. An amplitude modulator is formed by using the EOM in conjunction with a polarizer and finally the fluctuation spectrum of the output is measured using detector D_{out} and a spectrum analyser.

This system can be modelled with the linearized operator approach as used for the electro-optic noise-eater, see Section 8.3. We obtain the following transfer functions for the spectrum of the output beam normalized to the QNL

$$
\begin{aligned}
V_{out}(\Omega) &= \varepsilon_3 \left| \sqrt{\varepsilon_1} + \lambda\sqrt{(1-\varepsilon_1)\varepsilon_2} \right|^2 V_{in}(\Omega) \\
&+ \varepsilon_3 \left| \sqrt{(1-\varepsilon_1)} - \lambda\sqrt{\varepsilon_1\varepsilon_2} \right|^2 V_1 \\
&+ \varepsilon_3 \left| \lambda\sqrt{(1-\varepsilon_2)} \right|^2 V_2 \\
&+ (1-\varepsilon_3)\, V_3,
\end{aligned}
\tag{10.3.1}
$$

where the electronic gain $\lambda(\omega)$ is in general a complex number. $V_{in}(\Omega)$ is the amplitude noise spectrum of the input field. The vacuum noise spectra due to the beamsplitter V_1, the in-loop detector efficiency V_2 and the out-of-loop losses V_3, are shown separately to emphasize their origins. All vacuum inputs are quantum noise limited, i.e.., $V_1 = V_2 = V_3 = 1$. Due to the opposite signs accompanying the feedback parameter λ in Eq. (10.3.1), it is possible to amplify the input noise (first term), while cancelling the vacuum noise from the feed-forward beamsplitter (second term). The third and fourth terms of Eq. (10.3.1) represent unavoidable experimental losses. In particular if we choose $\lambda = \sqrt{1-\varepsilon_1}/\sqrt{\varepsilon_1\varepsilon_2}$, the vacuum fluctuations

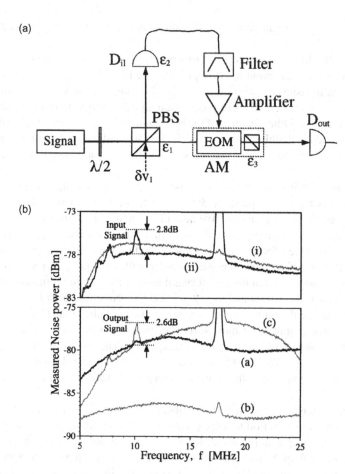

Figure 10.5: (a) Schematic of the feed-forward experiment. D_{il}: in-loop detector; D_{out}: out-of-loop detector; PBS: Polarizing beam splitter; $\lambda/2$: half-waveplate; AM: Amplitude Modulator. (b) Top: Noise spectra of the squeezed input beam. Bottom: Noise spectra of the output beam. (A) Direct detection of the input light after lossy (86 %) transmission. (B) Output noise spectrum without feed-forward (effectively 98.6 % loss). (C) With optimum feed-forward gain the signal and noise are amplified with little loss of SNR. After P.K. Lam et al. [Lam97]

from the beamsplitter, V_1, are exactly cancelled. Then, under the optimum condition of unit efficiency detection and negligible out-of-loop losses ($\varepsilon_2 = \varepsilon_3 = 1$), we find

$$V_{out}(\Omega) = \frac{1}{\varepsilon_1} V_{in}(\Omega). \tag{10.3.2}$$

That is, the fluctuations are noiselessly amplified by the inverse of the beamsplitter transmittivity. Hence our system ideally can retain the SNR for a signal gain of $G = 1/\varepsilon_1$. This system was used to demonstrate low noise amplification of amplitude squeezed light. The second harmonic output from a singly resonant frequency doubling crystal, Section 9.6, provided the input beam.

The top half of Fig. 10.5(b) shows the input noise spectra. Trace (i) shows the QNL, and trace (ii) is the noise spectrum of the input light. Regions where (ii) is below (i) are amplitude squeezed. The maximum measured squeezing of 1.6 [dB] is observed in the region of 8–10 [MHz] on a 26 [mW] beam. A small input modulation signal is introduced at 10 [MHz] which has $\text{SNR}_{\text{in}} = 1.10 \pm 0.03$. Other features of the spectra include the residual 17.5 [MHz] locking signals of the frequency doubling system, see Section 9.6 and the low frequency roll-off of the photo-detector, introduced to avoid saturation due to the relaxation oscillation of the laser.

The bottom half of Fig. 10.5(b) shows the noise spectra obtained from the single output detector. The input beam is made to experience 86% downstream loss. As trace (A) shows the SNR is greatly reduced by the attenuation. We now perform signal amplification by first introducing a beamsplitter of reflectivity 90%. This further attenuates the output beam to $\varepsilon_{\text{tot}} = \varepsilon_1 \varepsilon_3 = 0.014$. With no feed-forward gain, as trace (B) shows, the modulation signal is completely lost and the trace itself is quantum noise limited to within 0.1 [dB] over most of the spectrum. Finally, by choosing the optimum feed-forward gain, trace (C) shows the amplified input signal with $\text{SNR}_{\text{out}} = 0.82 \pm 0.03$ and G = 9.3 ± 0.2[dB]. Note that both (B) and (C) are of the same intensity, hence the output signal is now significantly above the QNL. This is the reason why the amplified output is far more robust to losses than the input. The results are best described by the transfer coefficient $T_s = \text{SNR}_{\text{out}}/\text{SNR}_{\text{in}}$ The spectra in Fig. 10.5(b) corresponds to a value of $T_s = 0.75 \pm 0.02$, in good agreement with the theoretical result of $T_s = 0.77$. This is to be compared with the performance of an ideal PIA, with similarly squeezed light, of $T_s \approx 0.4$.

An unavoidable consequence of this signal amplification technique is the reduction in intensity of the beam. However, injection locking can be used to amplify the intensity of the output beam without affecting the signal or noise, see Section 8.5. When the signal frequency is much larger than the linewidth of the laser cavity, the output spectrum of the injection locked field is the same as the input spectrum ($V_{\text{out}} = V_{\text{in}}$). Because the injection locked output is of a higher optical intensity, we can thus regain the optical intensity lost by the feed-forward amplification scheme. This was demonstrated by using the output of the feed-forward set up to injection lock a Nd:YAG nonplanar ring oscillator. A transfer coefficient of $T_s = 0.88 \pm 0.05$ with signal gain of $G = 9.3 \pm 0.2$ [dB] was obtained for a range of output powers from 14 [mW] to 350 [mW]. Hence with this set-up it is possible to independently vary the amount of signal and carrier amplification with minimal loss of signal to noise ratio. The sensor can be integrated into an optical network.

The future: nonclassical spectroscopy

All these experiments are aiming at an improvement in the SNR – they show similar effects as conventional sensors but with better sensitivity. However, it might be possible to consider nonclassical effects which are only occur with squeezed light. There have been a number of theoretical proposals showing that an atom imbedded in a squeezed vacuum should have different absorption or emission spectra. An extensive review of the properties of atom interacting with squeezed light was prepared by B.J. Dalton et al. [Dal99] which outlines the theoretical methods as well a numerous special cases.

In practice however, this idea has one fundamental problem: while in theory it is possible to squeeze the entire vacuum field, in practice only very few of the many vacuum modes acting on the atom could be squeezed. Thus the only feasible experiments will be those where the atom interacts preferentially with a single mode – that means an atom coupled closely to a cavity. This type of experiments, also known as cavity QED [Geo96, Pol97], are presently in progress and one can expect results, such as new spectral shapes, modified multiphoton absorption rates, which are entirely different to those obtained with classical light.

Form the point of view of quantum information processing the interaction of light and atoms will play a pivotal role. While will see in Chapter 13 that nonclassical light and entangled laser beams are exceptionally useful for sending quantum information we need a storage medium. And this will have to be done with atoms. A number of beautiful experiments, in particular by the group of E. Polzik and K. Moelmer in Aarhus, have already demonstrated ways of transferring quantum states from light to atoms through transferring polarisation information to the spin of atoms [Hal99, Hal01]. The spin system can be developed into the medium for storage and possibly processing. This can be the beginning of a wide field of work concerned with the concept of exchange of entanglement between atoms and light.

10.4   Gravitational wave detection

One of the most likely and promising applications of squeezed light will be the improvement of the sensitivity of interferometers, in particular those designed to measure gravitational waves (GW). These instruments are already optimized and are designed to operate at the quantum noise limit and with minimum losses. The extra benefit of squeezed light could have an important scientific impact.

10.4.1   The origin and properties of GW

One of the most fascinating consequences of the general theory of relativity is the coupling between mass and the geometry of space leading to the prediction that changes in the mass distribution can produce a perturbation of the geometry which propagates through space like a wave. These ripples in spacetime, or gravitational waves, propagate at the speed of light and carry with them information about the dynamics of those cosmological objects that created them. Collapsing or exploding stars will emit bursts of waves and rotating systems will constantly emit waves. The detection of gravitational waves (GW) has the potential for an alternative type of astronomy, complementing the conventional optical, infrared, radio and x-ray astronomy.

The gravitational wave induces a change of the geometry in a plane transverse to the direction of propagation. It causes an expansion in one direction and simultaneously a contraction in the orthogonal direction. This change will oscillate in time. A length, measured in one direction of the plane, will alternate between expansion and contraction. The frequency of this oscillation, f_{GW}, is linked to the frequency of the movement of the mass in the object that emitted the wave. GW waves are analogues to quadrupole radiation, they follow the higher order moments of the movement of the mass. We should be able to detect the GW by continuously monitoring the length in two orthogonal directions. However, the extremely small size of the effect makes the gravitational waves so very elusive.

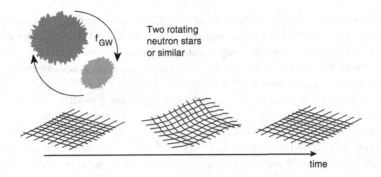

Figure 10.6: A typical source for gravitational waves and its effect on space and time.

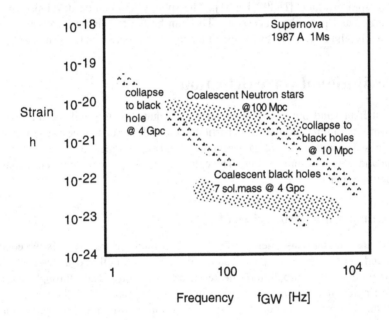

Figure 10.7: Predictions of the strain induced by various sources of gravitational waves.

The size of the effect of a GW is given by the relative change in distance, or the relative strain h induced. A value of $h = 10^{-20}$ means a change in a length measurement, for example from one point on this page to another point somewhere on the surface of the earth, by one part in 10^{20}. A list of recent predictions of cosmological sources of GW is shown in Fig. 10.7 which has been taken from the design study for the American LIGO project [LIGO]. It incorporates most of the astrophysical predictions available in 1994. Various types of objects differ in frequency and amplitude. The likely frequencies for GW range from 10^{-6} to 10^3 [Hz]. Some objects, such as spinning pulsars, have fixed frequencies. Others wil change their dynamics rapidly, over days to seconds, and move through an entire spectral evolution.

Each object has its own typical signature. The details depend critically on the spatial distribution, the symmetry and size of the masses involved. More details can be calculated using the astrophysical models that are presently being developed and over the next few years we can expect to get detailed results from these models. These types of simulations are presented on conferences such as the regular Marcel Grossman meetings [MGM]. It has occasionally been suggested that the argument above contains a logical trap. One could argue that not only the distance between the mirrors changes as a consequence of the GW but that simultaneously also the time it takes for the light to travel from the beamsplitter to the mirror is modified. After all, general relativity affects both distance and time. Would these two effects cancel each other? No; when the calculations are carried out one has to make a choice which set of equations should be used, either a set with normalized time coordinate or, alternatively a set with normalized space coordinates. The results of both calculations are identical, but they can not be used simultaneously. A detailed description of these calculations can be found in [Sau95].

The predicted strain amplitude ranges from 10^{-25} to 10^{-20}. Over what length should we measure the change in length? At any one time the largest effect will be at two points with a separation that corresponds to half the wavelength λ_{GW} of the gravitational wave. At any one place the biggest effect can be seen between two time intervals separated by $\frac{1}{2}t_{GW} = \frac{1}{2f_{GW}}$. For example, let us assume a GW at $f_{GW} = 500$ [Hz], that means $t_{GW} = 2$ [ms] and $\lambda_{GW} = 6 \times 10^5$ [m] $= 600$ [km]. In this case the largest amplitude is a length change of $\Delta l = \frac{1}{2}h\lambda_{GW} = 3 \times 10^{-16}$ [m]. This is ridiculously small, less than the typical size of a nucleus of an atom. However, we should recognise that we will measure the movement of surfaces, not individual atoms. Surfaces are typically made of more than 10^{20} atoms, the measurement averages over all these atoms and the idea of such a small shift of the surface becomes meaningful again.

The displacement of a mirror surface could possibly be measured with this accuracy. The idea is to use a mirror which has no internal motion, that means it is a free mass without internal oscillations and reflects light back to the source. If we use a beamsplitter as the source we can measure in two orthogonal directions, and compare the displacement of the mirrors. The biggest effect is achieved by positioning the mirror at a distance $\frac{1}{4}\lambda_{GW}$ from the beamsplitter. In this case the distance to one mirror will be at maximum expansion when the other will has the maximum contraction. This is shown in Fig. 10.8. Note that this instrument has an optimum response only for GW that approach from a direction orthogonal to the plane of the interferometer. Two optical waves leave through the beamsplitter to the output port which are displaced by 3×10^{-16} [m]. Using light with a wavelength of 1000 [nm] we will get a maximum phase shift of 5×10^{-9} [rad]. This is clearly an incredibly small value. To measure it will require the use of all tricks known in interferometry.

10.4.2   Quantum properties of the ideal interferometer

Before we get drawn into the intricacies of a specific configuration we shall determine the quantum noise properties of a simple, idealized instrument, the Michelson interferometer. What is the fundamental limit in sensitivity of an interferometer? This question was analysed by V.B. Braginsky [Bra68] and eloquently answered by C. Caves and R. Loudon in 1980 [Cav80, Lou81]. Their results actually stimulated the interest in squeezed states of light.

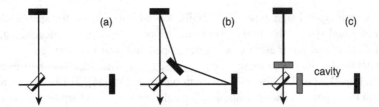

Figure 10.8: Three configurations for the detection of gravitational waves (a) Michelson interferometer and (b) Sagnac interferometer, (c) Michelson interferometer with internal cavities.

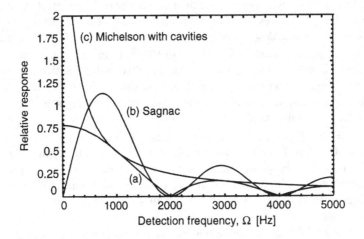

Figure 10.9: Comparison of the spectra of the response for a small displacement of the mirror for (a) Michelson interferometer, (b) Sagnac interferometer and (c) Michelson interferometer with internal cavities.

C. Caves analysed the ideal, loss less interferometer. The interferometer performs best when the difference in the length of the arms is chosen such that the light interferes destructively at one output port, it forms a *dark fringe*. The intensity at this port is detected with a single detector. The intensity noise depends on the quadrature fluctuations of the two input fields, the mode of light from the laser represented by $\tilde{A}_{\text{las}}$ and the mode $\tilde{A}_u$ entering the unused port. The sidebands are represented by their quadratures $\delta\tilde{X}1_{\text{las}}$ and $\delta\tilde{X}2_u$. It also depends on the momentum and position state of the mirror, described by the expectation values of the momentum operator $\tilde{q}$ and position operator $\tilde{p}$. The total variance of the intensity of the output light, at detection frequency Ω can be summarized as [Wal93]

$$\begin{aligned}
\langle\delta\tilde{X}1_{\text{out}}^2\rangle &= A\langle\delta\tilde{X}2_u\delta\tilde{X}2_u^\dagger\rangle + B\langle\delta\tilde{X}1_{\text{las}}\delta\tilde{X}1_{\text{las}}^\dagger\rangle \\
&\quad + C\langle\delta\tilde{q}\delta\tilde{q}^\dagger\rangle + D\langle\delta\tilde{p}\delta\tilde{p}^\dagger\rangle \\
&\quad + E\langle\delta\tilde{X}2_u\delta\tilde{X}1_{\text{las}}^\dagger\rangle + F\langle\delta\tilde{q}\delta\tilde{p}^\dagger\rangle
\end{aligned} \qquad (10.4.1)$$

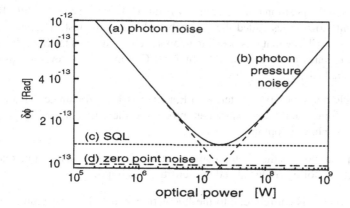

Figure 10.10: The noise contributions in an ideal interferometer as a function of optical input power. (a) photon noise (dashed line), (b) photon pressure noise (dashed line) The solid line gives the total noise. The standard quantum limit (c) is given by the horizontal dotted line. (d) The zero point quantum fluctuations of the mass are even lower (dash dotted line). The parameters for this diagram are: $m = 1$ [kg], $\tau = 1$ [s], $\lambda = 1$ [μm], $f_{GW} = 100$ [Hz].

The first term describes the effect of the intensity quantum noise, or photon noise, a consequence of the phase quadrature fluctuations of the unused port, see Section 5.2. The factor A depends only on the optical power P_{in} of the input light and the detection interval τ. The second term describes the effect of radiation pressure fluctuations induced by the amplitude quadrature fluctuations of the laser mode, see Section 5.2.2. The factor B depends on the mass M of the mirror and on P_{in} and τ. The third and fourth terms are the fluctuations of the mirror position and momentum and the mechanical vibration modes inside the mirror have been ignored. This calculation is correct for a free mass or for a mirror suspended by a pendulum with an eigen-frequency much less than the detection frequency.

In this regime we find the following noise spectra: The photon noise is frequency independent. The variance of the radiation pressure noise scales with Ω^{-4}. The mirror noise scales with Ω^{-2}. In principle higher frequency signals have lower noise and are easier to detect.

The coefficients A and B in Equation (10.4.1) are intensity dependent. With increasing intensity the effect of photon noise, or shot noise, is reduced, while that of radiation pressure noise is increased. There is an optimum intensity at which the best performance can be achieved. Figure (10.10) shows such a noise balance. The optimum value is called the *standard quantum limit of the interferometer*, or SQL, trace (c) in Fig. 10.10 and is given by

$$\langle\, \delta\tilde{\mathbf{X}}1_{out}^2 \,\rangle = \frac{1}{L}\sqrt{\frac{2\hbar\tau}{M}} \qquad\qquad (10.4.2)$$

This result is independent of the quantum mechanical noise of the free mass M. It represents a balance of two noise effects, the noise in the measurement of the light and the noise induced by the light on to the mass and is a typical example of the ideas of Heisenberg applied to this optical measurement. This noise limit can be interpreted as the consequence of the back-action of the measurement, see Chapter 11. For heavy masses the SQL is much larger than the zero point fluctuations in the position of a free mass which is shown as trace (d) in Fig. 10.10. This fundamental limit, sometimes called the Heisenberg limit of the test mass, cannot be achieved completely. We will see that it is possible to approach it, and pass the SQL, by using correlated nonclassical states of light [Kim01]. In 1980 C. Caves drew several important conclusions from his analysis:

i) For any reasonable mass the first two terms in Equation (10.4.1) dominate. That means the interaction with the light, the measurement process, is the origin of the largest quantum fluctuations. We have to optimize this process.

ii) At reasonable optical powers the photon counting noise dominates. With an increase in the optical power the noise is reduced and the sensitivity is improved.

iii) The noise at the output is entirely due to the fluctuation in the phase quadrature of the unused port of the interferometer. Illuminating the empty port with a squeezed vacuum state which is phase locked to the laser can reduce the noise. For a squeezed state with minimum uncertainty $\langle \delta \tilde{\mathbf{X}}2_u^2 \rangle = V2 = \exp(-2r_s)$ the improvement is equivalent to an increase in the optical power of a factor $\exp(r_s)$.

Configurations of interferometers

An alternative to the Michelson interferometer is to measure the interference of two optical waves that have travelled from the beamsplitter to both mirrors and back, but in the opposite sequence. One beam travels first to mirror one and then to mirror two, the other in the opposite order. The crucial point for achieving the maximum effect is the travel time. It should be $\frac{1}{2}t_{GW}$. This can be achieved with an optical arrangement that is basically similar to a Sagnac interferometer, see Fig. 10.8(b). One can imagine a number of other configurations with identical travel, or storage, time and with beams travelling in orthogonal paths in opposite order. They all will produce the same result. Both systems have the optimum sensitivity at frequency $f_{opt} = 1/t_{RT}$. At lower frequencies, $f_{GW} < f_{opt}$, the response of the Sagnac interferometer drops, proportional to $1/f$, simply because the slow changes in the length will affect both counterpropagating beams similarly. The Sagnac is insensitive to DC differences in the arm length while here the Michelson interferometer has its full sensitivity. At larger frequencies, $f_{GW} > f_{opt}$, the responses are similar and at $f_{GW} = 2/t_{RT}$ both instruments do not register the GW at all, since we are comparing the position of the mirror at time intervals separated by a full period of the gravitational wave. For even higher frequencies detection is possible again but only for frequencies between the response minima at multiples of $2/t_{RT}$. For a broadband detector we would choose the round trip time to be the highest GW frequency we wish to detect. These sensitivity spectra are shown in Fig. 10.9.

The idea of the round trip time can be generalized. With the Michelson interferometer we do not have to send the two beams the full distance $\frac{1}{4}\lambda_{GW}$. We could send it forwards and backwards from the location of the beamsplitter to the far mirror several times until it has covered a total travel distance of $\frac{1}{2}\lambda_{GW}$, or equivalently has spent the time $\frac{1}{2}t_{GW}$ in this arm of the interferometer before it interferes with the other beam. Any practical GW detector will use this trick since λ_{GW} is far too large for earth bound systems, it is typically several hundreds of kilometers. For this purpose, some designs include so called delay lines, two mirrors of radius of curvature $\mathcal{R}_0$ placed at a separation of $\mathcal{R}_0(1 + 1/j\mathcal{C})$. This configuration has the remarkable property that an incoming beam is reflected j times between the pair of mirrors and emerges with the same beam size and wave front curvature it had at the input. The beam is well confined by this geometry. The number of reflections that one chooses depends on the amount of technical noise that is added at each reflection. In practice more than 50 reflections are not useful. Thus a delay line GW interferometer is typically 1 - 4 [km] long.

This arrangement is suitable for both the Michelson and Sagnac interferometers. The major difference between the two is that in the Sagnac all the components are sampled by both beams, which means that the two beams will have very similar wave fronts when they emerge and the visibility of the interference will be high. Secondly any length changes which occur on time scales slower than the round trip time will be sampled by both beams, the interferometer is immune to technical noise at frequencies $f \ll 1/t_{RT}$. Both these points are advantages for the Sagnac configuration. On the other hand the Sagnac interferometer requires an exactly balanced beamsplitter, since one beam experiences two transmissions, the other two reflections. A detailed comparison is given by Sun et al. [Sun96].

An alternative way of obtaining the necessary storage time for the light is to place long cavities with storage time t_{st} into the arms of the interferometer. The length changes will now be integrated. The effect is that for $f_{GW} \leq 1/t_{st}$ the response of the instrument with cavities is similar to that of a simple Michelson interferometer. At larger frequencies, $f_{GW} > t_{st}$,the sensitivity of the instrument decreases due to the integration property of the cavity. The effects induced by the GW inside the cavity cannot be detected in the reflected light. A comparison of the spectrum of the signal response for all three types of instruments is shown in Fig. 10.9.

Recycling

The interferometer that will finally be used to detect gravity waves will have a more complex structure than the Michelson or Sagnac we considered here for simplicity. The instrument will use some form of *recycling*. This can be *power recycling*, which remedies one glaring defect of the interferometer, namely that all the light which is not lost by absorption is reflected back towards the laser. Whilst we actually use optical isolators to dump it effectively, this is obviously a considerable waste. Drever and Schilling [Dre83] recognized that by introducing a mirror M_{PR} between the input laser and the interferometer the light reflected by the instrument can be used again. M_{PR} essentially forms a cavity with the whole interferometer acting as one compound mirror. This is an elegant and efficient way to enhance the laser power. In practice all we have to achieve is to keep the mirror locked to a resonance position.

Alternatively, the sensitivity can be enhanced by placing a mirror M_{SR} at the output and reflecting the sidebands back into the interferometer. This leads to a buildup of the sidebands which contain the information, with the consequence of increased sensitivity. The position of M_{SR} can be chosen such that the laser frequency, the carrier, is resonantly built up or, alternatively, such that the cavity formed by M_{SR} is detuned and only one of the signal sidebands is built up. The latter is known as *tuned signal recycling* and allows us to tune the entire instrument to a specific signal frequency f_{GW}. Both signal and power recycling can be combined, dual recycling. The major penalty is loss in bandwidth and increased complexity of the instrument and the control system required to keep all components locked simultaneously. The ideas of recycling were pioneered by B. Meers [Mer88, Str91] and extended by Mizuno et al. [Miz93, Hei96]. Recycling allows an improvement of the instrument beyond the limits set by the simple Michelson interferometer. It allows the detection of weaker GW at specific frequencies. Alternatively, they would allow more compact, and cheaper, instruments with enhanced sensitivity at specific frequency. Many of these ideas will be implemented in the German/British GEO project, which attempts to achieve the sensitivity of a large instrument of 4[km] with a reduced baseline of 600[m].

10.4.3   The sensitivity of real instruments

The effect of the GW has to be measured in the presence of noise, both in the position of the mirrors as well as noise of the light. The quality of the signal is described by the signal to noise ratio (SNR), as defined in Section 2.1. Consider first the signal; the GW induces a length change $\Delta l(t)$ which is transformed by the interferometer into a phase change $\Delta\phi_s(t)$ between the two interfering output waves. The link between $\Delta l(t)$ and $\Delta\phi_s(t)$ depends on the type of instrument used. For a simple Michelson interferometer we obtain

$$\Delta\phi_s(t) = \frac{4\pi\Delta l(t)}{\lambda} \tag{10.4.3}$$

If we consider one Fourier component f_{GW} only, we can simulate the effect of the GW as a sinusoidal modulation of the phase term and get the phase difference between the two optical waves as

$$\Delta\phi_s(t) = \phi_0 + \phi_s \sin(2\pi f_{\mathrm{GW}} t + \chi_s) \tag{10.4.4}$$

where ϕ_0 is the DC phase offset and χ_s is the signal phase offset at time $t = 0$. The optical power at the output port is:

$$P_{\mathrm{out}} = P_{\mathrm{in}}(1 + \mathcal{V} \, \cos(\Delta\phi_s(t))) \tag{10.4.5}$$

where P_{in} is the optical power incident on the interferometer and $\mathcal{V}$ is the fringe visibility. A perfect instrument would have a visibility of one, but in most cases scattering and wave-front distortions will result in $\mathcal{V} \leq 1$. Combining Equation (10.4.4) and Equation (10.4.5) leads to an expansion in term of Bessel functions J_n of the first kind (see Appendix F). For small phase modulations, $\Delta\phi \ll 1$, harmonics above the fundamental modulation frequencies are small and the higher order terms can be neglected. Using a detector with responsivity $\rho(\lambda)$ [Ampere/Watt] this optical power can be converted into a photo-current. The DC current is

$$i_{\mathrm{DC}} = \frac{1}{2}\rho P_{\mathrm{in}}\big(1 + \mathcal{V}J_0(\phi_s)\cos(\phi_0)\big) \tag{10.4.6}$$

while the variance of the AC components is

$$V_i(f_{\mathrm{GW}}) = \frac{1}{2}\big(\rho P_{\mathrm{in}}\mathcal{V}J_1(\phi_s)\sin(\phi_0)\big)^2 \tag{10.4.7}$$

As expected (see Section 5.2) the received signal is strongest for $\phi_0 = \pi/2$, midway between a bright and a dark fringe. No signal is received at the turning points ($\phi_0 = 0, \pi, 2\pi \ldots$). However, the best SNR is obtained at a dark fringe. Realistic interferometers contain imperfections, or noise sources, which will reduce the performance of the idealized system discussed by Caves. These additional noise sources will add to the quantum noise, and thus reduce the sensitivity. In order to optimize the performance of the instrument the operating conditions will be different from those of the QNL interferometer. The important noise sources are seismic noise $\langle \Delta i_{sm} \rangle$, thermal noise in the suspension $\langle \Delta i_{sp} \rangle$, thermal noise in the test mass $\langle \Delta i_{tm} \rangle$, and electronic noise $\langle \Delta i_{el} \rangle$. All these noise sources are independent of each other and add in quadrature. The total noise budget present in the instrument is

$$
\begin{aligned}
\langle \Delta i_{\text{noise}}^2 \rangle &= \langle \Delta i_{sm}^2 \rangle + \langle \Delta i_{sp}^2 \rangle + \langle \Delta i_{tm}^2 \rangle + \langle \Delta i_{el}^2 \rangle \\
&= \langle \Delta i_{qn}^2 \rangle (1 + \varepsilon(f_{\text{GW}}))
\end{aligned}
\tag{10.4.8}
$$

The combined additional noise term $\epsilon(f_{\text{GW}})$, scaled in units of the quantum noise, has a strong and complex dependence on the detection frequency. Inserting the absolute value for the quantum noise, see Chapter 4, we obtain the signal to noise ratio as a function of the offset angle ϕ_0, the optical power P_{in} and the detection bandwidth RBW. After expanding the Bessel functions for small signals we obtain

$$
\text{SNR} = \frac{\rho P_{\text{in}} \phi_s^2}{8\, e\, \text{RBW}} \frac{\mathcal{V}^2 \sin^2(\phi_0)}{(1 + 2\varepsilon) + \mathcal{V} \cos(\phi_0)}
\tag{10.4.9}
$$

with an optimum value, for the quantum noise limited interferometer ($\varepsilon = 0$) and perfect visibility ($\mathcal{V} = 1$) of

$$
\text{SNR}_{\max} = \frac{\rho P_{\text{in}} \phi_s^2}{4\, e\, \text{RBW}}
\tag{10.4.10}
$$

The value of $\Delta \phi$ that can be measured with a signal to noise ratio of 1 corresponds to the quantum noise limited value $\Delta \phi_{\text{QNL}}$ which is derived in Section 5.2. The sensitivity of the GW detector is quoted as the relative strain Δh which produces a signal to noise ratio of 1 measured with a bandwidth of 1 [Hz]. A plot of Δh as a function of the detection frequency f_{GW} provides an overview of the frequency response of the instrument. This spectrum is influenced by both the frequency dependence of the various noise contributions and the response function of the interferometer. A practical example of the already achieved sensitivity of a 40m prototype is shown in Fig. 10.11 [LIGO, Abr96]. This is used to predict the sensitivity of the large baseline interferometers under construction in the LIGO project.

These instruments will be limited at low frequencies, ($f_{\text{GW}} < 10$ [Hz]), by seismic noise. Suspension and test mass thermal noise will dominate at frequencies 10 [Hz] $< f_{\text{GW}} <$ 100 [Hz]. The instrument will be QNL for frequencies above 100 [Hz], where the sensitivity worsens due to the effect of the cavities in the interferometer arms.

Modulation techniques

It was shown in Section 5.2, that in the absence of noise, the largest signal can be obtained with $\phi_0 = \pi/2$. This is also the best operating condition for a classical interferometer where the relative intensity noise is constant and independent of ϕ_0. In contrast, for a QNL interferometer, where the noise is proportional to the square root of the intensity, the best performance is achieved very close to $\phi_0 = 0$ or $\phi_0 = \pi$, that means at a dark fringe.

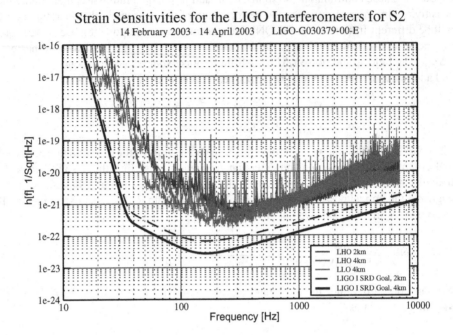

Figure 10.11: The measured performance of several simultaneously operating interferometers in 2003. The experimental curve shows a broad noise background and additional spikes which are due to mechanical resonances, violin modes, and mirror modes. Also shown are predicted levels for suspension (low frequencies), internal thermal (medium frequencies) and photon counting noise (higher frequencies). For updates check the website: http://www.ligo.caltech.edu/

In practical instruments, there will always be some additional noise, $\varepsilon > 0$ in Equation (10.4.8), and thus the optimum operating condition will be further offset from the exact dark fringe, $\phi_0 = 0$. The signal could be detected directly and laser noise would be no problem. All laser noise, in frequency and in intensity, is common to both waves in the interferometer and the ideal interferometer, with matched arms, has common mode suppression to reject this noise. In practice, however, the common mode suppression is rather limited. Thus intensity noise has to be kept within stringent limits set by many imperfections of the real instrument. Similarly, the length of the two interferometer arms will not be perfectly balanced, in particular when cavities are used inside the interferometer and not only the physical length but also the storage time has to be matched. In this case the instrument is sensitive to laser frequency noise. Typical specifications cannot be derived from first principles. For a large instrument they typically are: $RIN < 10^{-6}$ and $\frac{d\nu}{\nu} < 10^{-16}$ [LIGO], values which can be achieved with the latest control technology.

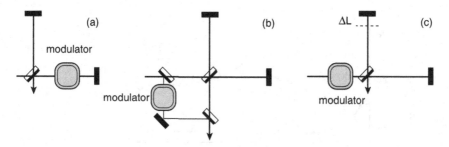

Figure 10.12: Three types of modulation schemes (a) internal modulation, (b) external modulation and (c) frontal modulation.

Measurements at very low frequencies, such as 10 [Hz], are prone to a large number of difficulties, the most severe being the laser noise. Consequently all interferometer designs use modulation techniques. By phase modulating the light it is possible to transfer all information to frequencies Ω_{mod} where the laser is QNL. As a result the photo-current will have a strong Fourier component at $2\Omega_{\mathrm{mod}}$. All information is contained in the sideband of this spectral component, called the carrier. Demodulation using an electronic mixer and an electronic local oscillator signal, derived from the modulator driving signal, allows very sensitive heterodyne detection, as shown in Fig. 10.13. There are three alternatives for modulation: internal modulation requires a modulator inside one of the interferometer arms. Frontal, or inline, modulation involves a phase modulation of the light before it enters the interferometer. The phase modulated light propagates through both arms. The arms have to have different lengths to convert common mode into differential mode phase modulation. External modulation uses an optical local oscillator with phase modulation, which is mixed on a beamsplitter with the output beam from the interferometer. This latter scheme is analogous to an optical heterodyne detector as discussed in Section 8.1.4.

To demonstrate the improvement that can be achieved with modulation techniques Fig. 10.13 shows a comparison of experimental results obtained with the same interferometer using either (a) direct detection or (c) internal modulation [Ste94]. A signal is generated at 621 [kHz] (arrow) and is almost completely buried in the laser intensity noise, which consists mainly of the RRO oscillation centered at about 450[kHz]. With internal modulation at $\Omega_{\mathrm{mod}} = 25$ [MHz] the information is shifted to a frequency where this laser (a Nd:YAG miniature ring oscillator) is QNL. It is retrieved by demodulation and trace (b) shows that the signal to noise ratio is dramatically improved, by about 30 [dB], and the signal is easily detected. However, we have to pay a slight penalty, a reduction in the sensitivity. The signal to noise ratio with internal modulation, assuming perfect optics ($\mathcal{V} = 1$) and no electronic noise, is given by [Ste94]

$$\mathrm{SNR_{int}} = \frac{\rho P_{\mathrm{in}} \phi_s^2}{8\, e\, \mathrm{RBW}} \frac{J_1^2(\Phi_m)}{1 - (1 - \phi_s^2/4) J_0(\Phi_m)} \tag{10.4.11}$$

where Φ_m is the modulation depth in [rad]. By selecting the optimum modulation depth and

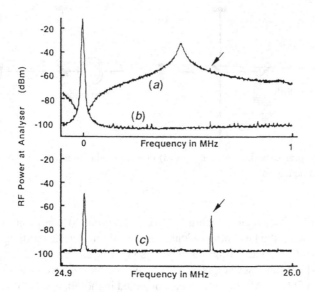

Figure 10.13: Direct comparison of the signal strength measured with an interferometer using: trace (a) show direct detection and (c) internal modulation. Trace (b) is the detector noise. The signal is generated at 621 [kHz]. In the case (c) a modulation at 25 [MHz] is used. An improvement of SNR by about 30 [dB] is achieved. After A. Stevenson et al. [Ste94].

phase a maximum value of

$$\text{SNR}_{\text{int,max}} = \frac{8\rho P_{\text{in}}\phi_s^2}{12\, e\, \text{RBW}} = \frac{2}{3}\text{SNR}_{\text{max}} \tag{10.4.12}$$

can be achieved. This means the instrument with internal modulation has an optimum performance only 2/3 as good as with direct detection, see Equation (10.4.12). This is a consequence of loss of information; modulation and demodulation are carried out by the same waveform and in the process information is transferred to other harmonic frequencies. This effect would be particularly strong for modulation functions other than a *sine* function. In addition, information from other harmonic frequencies is brought in, via demodulation. Uncorrelated noise is added in the process. It was suggested [Nie91] and demonstrated [Mio92, Gra93] that the use of inverse pairs of modulation, demodulation functions would assure that all Fourier components are fully recovered and no noise is added. But the technical complications in generating and processing matching functions are prohibitive. A similar analysis can be made for an instrument with external modulation. For the optimized instrument, with $\Phi_m = 1.84$ [rad], $\mathcal{V} = 1$, $\epsilon = 0$, the limit is given by $\text{SNR}_{\text{ext,max}} = 0.67\,\text{SNR}_{\text{max}}$, a result which is very similar to the case of internal modulation [Ste94, Gra96].

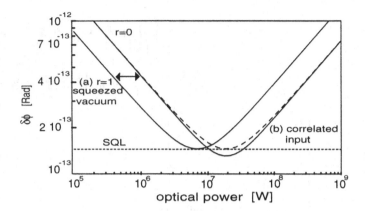

Figure 10.14: The effect of squeezed light on the total noise budget: (a) squeezing the phase quadrature of the light at the vacuum input is equivalent to a higher laser power; (b) squeezing both quadrature phase and amplitude reduces the standard quantum limit.

10.4.4 Interferometry with squeezed light

All instruments have basically the same quantum noise limit. As can be seen from Fig. 10.11 the quantum noise is the most important effect in a GW detector for higher frequencies. It should be possible to improve the sensitivity using squeezed states of light. In most instruments the laser intensity will be sufficiently low that photon counting noise, the first term in Equation (10.4.1), dominates and improvements can be achieved by using a squeezed vacuum state as the input to the second port of the beamsplitter. A reduction of the noise term $\langle \delta \tilde{\mathbf{X}} 2_u \delta \tilde{\mathbf{X}} 2_u^\dagger \rangle$ would lower the total noise. As shown in Fig. 10.14 this is equivalent to increasing the laser power, or shifting the noise curve along the horizontal axis, in Fig. 10.14. This experiment requires that the squeezed light is phase-locked to the laser input. It has to be generated by a beam from the same laser and carefully aligned with the axis of the empty port. The effect is rather surprising, blocking off the invisible squeezed vacuum beam will increase the noise at the detector.

This effect was demonstrated in a beautiful experiment by M. Xiao and J. Kimble [Xia87]. Their aim was to show that squeezed light, which they had generated with an OPO, see Section 9.5, could improve a practical measurement by at least 3 [dB]. Their apparatus is shown in Fig. 10.15. A Nd:YAG laser with internal frequency doubler generated a beam at 1064 [nm] which illuminated a Mach-Zehnder interferometer. The beam at 532 [nm] is used to drive a degenerate OPO. Inside the interferometer are two electro-optic phase modulators. One controls the DC phase shift and thereby sets the operating point, namely the dark fringe on the output. The other is used to introduce the signal, an AC phase shift at 80 [MHz]. The vacuum port of the interferometer is blocked. The amplitude of the signal ϕ_s is chosen to provide a SNR of 3 [dB], as shown in Fig. 10.15(b). Now squeezed light is injected into the interferometer and the noise floor drops, Fig. 10.15(c). Note that the displayed signal trace also drops since it is the quadrature sum of signal and noise. The signal to noise ratio is improved by about

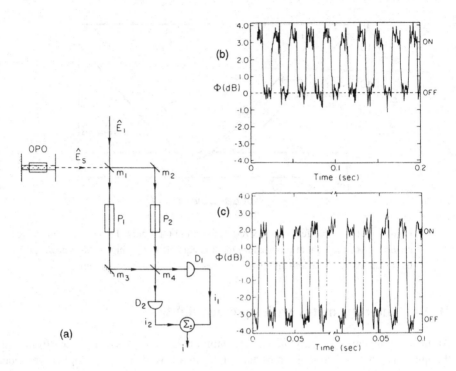

Figure 10.15: A demonstration of SNR improvement using squeezed light: (a) Schematic layout of the experiment. Experimental results using (b) vacuum input (c) squeezed input. After M. Xiao et al. [Xia87].

3 [dB]. These landmark experiments showed beyond any doubt that squeezing can be a powerful tool. Almost simultaneously Slusher's group at ATT-Bell Laboratories obtained similar results using a polarimeter and pulsed squeezed light [Gra87]. They achieved an improvement of 2.8 [dB].

The light used for the empty port does not actually have to be a squeezed vacuum state. The requirement is only that the power of the beam at this port is much weaker (≤ 10 percent) than that in the other port, otherwise the uncorrelated quantum noise of the laser would appear in the output signal. The second condition is that the correct quadrature (X2) is squeezed. The group at Stanford [Ino95] realized that this can be achieved with two phase locked lasers, one for each input port. A high power diode laser illuminates one input, a low power, intensity squeezed diode laser illuminates the other. The squeezing quadrature is selected by choosing the absolute phase difference between the two waves. If a beam, which is squeezed in the X1 quadrature, is shifted by $\frac{\pi}{2}$ the reduced fluctuations appear in the X2 quadrature as seen by the interferometer. The initially surprising result is that intensity squeezed light can be used to improve the sensitivity of an interferometer.

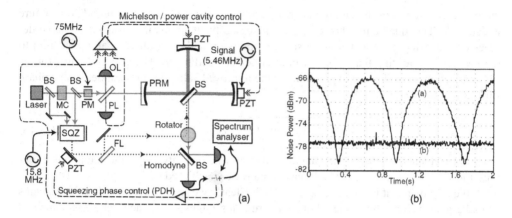

Figure 10.16: Experimental demonstration of an improvement in the SNR of a power recycled Michelson interferometer using squeezed light. (a) Schematic layout. The control servos for the interferometer are shown by the dashed lines. BS=beamsplitter; CL=power cavity locking detector; FL=flip mirror; OL=offset locking detector; PM=phase modulator; SQZ=squeezed state generation scheme. (b) Noise spectrum showing the improvement of the SNR with, and without, power recycling (PRM). Electronic dark noise lies at -93.5 [dB]. RBW = 100 [kHz], VBW = 30 [Hz]. After K. McKenzie et al. [McK02].

A realistic demonstration

A very impressive experimental demonstration of the capability of squeezed light to improve the SNR was given by the Canberra group [McK02]. They built a tabletop Michelson interferometer which included power recycling and had most of the features of a real GW detector. Due to smaller scale all frequencies were scaled up into the [MHz] regime. In this experiment squeezed light from an OPA (-3.5 [dB]) was coupled through optical isolators into a Michelson with locked power recycling cavity, as shown in Fig. 10.16a. The Michelson was actively locked to a dark fringe. A PZT driving a mechanical resonance of the mirror at 5.46 [MHz] simulated the GW wave. When squeezing was introduced a noise suppression to -1.8 ± 0.2 [dB] below the QNL was observed, Fig. 10.16b. The recycling mirror provided a signal enhancement by a factor of ≈ 9, and the corresponding noise level is suppressed to -2.3 ± 0.2 [dB] below QNL. The SNR was enhanced accordingly. This experiment clearly demonstrates that it is practical to combine squeezing and GW detection technology. However, one major technical problem is the generation of squeezed light at sufficiently low frequencies, typically 100 [Hz]. This will require extraordinary good suppression of the laser noise, using mode cleaners and possibly the active cancellation of technical noise by cancellation between two squeezed noise sources as demonstrated by W. Bowen et al. [Bow02].

In real GW experiments noise suppression will be obscured by the imperfections of the instrument. The limited quantum efficiency η of the detector, and other losses, reduce the effect of squeezed light. The fringe visibility will be less than unity, which mixes in components of the orthogonal quadrature. Gea-Banacloche & G. Leuchs [Gea87] considered the effect of an imbalance between the two arms of the interferometer and found that not only the phase quadrature of the empty port but also a small fraction of the amplitude quadrature is detected. This quadrature, however, has increased fluctuations, in particular for large degrees of squeezing. The actual noise suppression is limited – and there is an upper limit to the degree of squeezing that is useful in a particular instrument. The effect of squeezing is summarized in the following expression for the improved value for a lowest detectable phase shift $\Delta\phi_{sq}$ [Gea87]

$$\Delta\phi_{sq} = \sqrt{\sqrt{2(1-\mathcal{V})} + 1 - \eta + \eta V2_u} \ \ \Delta\phi_{\text{QNL}} \qquad (10.4.13)$$

The improvements depend on the quantum efficiency η, visibility $\mathcal{V}$ and the actual noise suppression $V2_u$. For parameters which have already been achieved ($V2_u = 0.4, \eta = 0.95, \mathcal{V} = 0.99$) the improvement is $\Delta\phi_{sq}/\Delta\phi_{\text{QNL}}$ is 0.76. Clearly we require very low loss instruments to make good use of squeezing. Future instruments might soon reach parameters such as ($V2_u = 0.1, \eta = 0.97, \mathcal{V} = 1 - 10^{-5}$) [Kim01] resulting in $\Delta\phi_{sq}/\Delta\phi_{\text{QNL}} = 0.36$. This is equivalent to an increase of the laser power by a factor three.

Squeezing might appear to be a complicated and difficult technique to obtain an improvement which could be achieved by simply increasing the laser power. However, higher intensities have their own negative consequences, such as wave front distortions due to thermal effects. An upper limit in intensity is likely to be set by nonlinear effects that cause distortions in the beam. Here squeezing can provide an extra gain in sensitivity not available otherwise. For the ideal case ($V2_u = 0, \eta = 1, \mathcal{V} = 1$) Equation (10.4.13) would predict unlimited improvements. This is not correct, since it ignores noise such as radiation pressure fluctuations. Also it will be easier to approach the SQL first by eliminating other noise sources and increasing the power. The present generation of GW detectors, e.g. LIGO I (2002–2005) operate at intensities where the photon counting noise and thermal noise dominates, see Fig. 10.11. The next generation of instruments, e.g. LIGO II (2006-2008) will have improvements of all sources of noise and an increase in the intensity that will allow it to operate at the SQL. This will mean circulating powers of about 1 [MW/cm^2] which is stretching the limits of the components inside the interferometer. Presently there is considerable design work for improving the performance to below the SQL for LIGO III (2010 and beyond) since it is unlikely that the intensity can be increased much higher without causing other nonlinear noise sources.

Beyond the SQL

To go below the SQL we have to consider all terms of Equation (10.4.1) since Equation (10.4.13) considers only the 1st term. By using squeezed light of the correct quadrature the fifth term can reduce the noise in the output. We also have to recognise that the optimum squeezing quadrature depends on the intensity, at low intensities $I < I_{\text{SQL}}$ the light should be phase squeezed ($\theta_s = 90$), at high intensities where radiation pressure dominates $I > I_{\text{SQL}}$ the light should be amplitude squeezed ($\theta_s = 0$) and at the SQL the squeezing angle should be between ($\theta_s = 45$). Furthermore, we find that for GW detectors which contain resonant storage cavities the squeezing quadrature, or θ_s, depends on the detection frequency. Consequently, we will require squeezing with detection frequency dependent squeezing angle. This is an additional requirement which can be satisfied by using a standard squeezing source ($\theta_s = 90$),

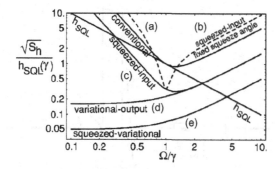

Figure 10.17: Prediction of the improvements in sensitivity of a GW detector with storage cavities for different detection frequencies Ω. The instrument is chosen to be limited by the SQL for frequency $\Omega = 1$. (a) with conventional interferometer, which uses coherent light and no squeezing. (b) with interferometer with squeezed input with fixed squeezing angle θ_s, (c) with interferometer with squeezed input with frequency dependent θ_s, (d) interferometer with squeezing at the output., (e) the interferometer uses squeezing at the input and the output. After H.J. Kimble et al. [Kim01].

such as an OPA, and two cavities to rotate the squeezing quadrature. This is shown in the study by H.J. Kimble et al. [Kim01] and in Fig. 10.18a. For a GW detector these mode cleaners have to have the full length, e.g. for LIGO 4 [km]. When operated at the SQL and with $V_u = 0.1$ this design would allow an improvement $\Delta\phi_{sq}/\Delta\phi_{SQL} = 0.3$. There are alternative configurations which can be used to improve an interferometer below the SQL. The study by H.J. Kimble et al. [Kim01] evaluates a scheme where the squeezed, of suitable frequency dependent quadrature is used in a homodyne arrangement to suppress in the detection. Again full length mode cleaners have to be used, as shown in Fig. 10.18b. The gain in sensitivity is similar.

So far we have assumed that the two inputs, laser and vacuum state, have independent fluctuations. What would happen if both inputs were strongly correlated? The case of two number states has been investigated by M. Holland et al. [Hol93], who showed that an improvement in sensitivity can be achieved that is as good as the use of vacuum squeezed light. In theory the sensitivity can be improved such that $\Delta\phi$ no longer scales with $1/\sqrt{I}$ but with $1/I$. However, their scheme is not practical since it requires number states and is extremely sensitive to losses.

An alternatives, which is are currently being investigated is to create the correlation between the photon counting noise and the radiation pressure noise inside the actual interferometer. In recycled interferometers, such as LIGO II, the dynamics of the motion of the mirror could be much more complex than the dynamics of a free mass and as consequence and it could be possible to reshape the noise curve and to push the noise limit below the SQL, without the use of squeezed light [Buo02]. Yet another option is to consider to incorporate a second interferometer which is sensitive to only the radiation pressure and not the photon counting noise. This second instrument could provide a reference signal and using feedback control at the quantum noise level the effects of the radiation pressure could be actively suppressed. This

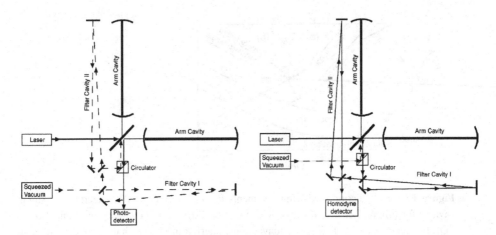

Figure 10.18: Schematic diagram of improved interferometers. (a) interferometer with squeezed input. The two full length mode cleaner cavities provide the required rotation of the squeezing angle. (b) an alternative where the squeezed light is used at the output to improve the SNR through a homodyne detection with a squeezed beam. Again two full length mode cleaner cavities are used. After H.J. Kimble et al. [Kim01].

would lower the total noise budget in the high intensity limit and also allow operation below the SQL [Cou03]. Both proposals show that the SQL is not going to be the ultimate limit for the sensitivity of interferometers.

In summary, interferometers, and in particular optical GW detectors which are already optimized for low losses, can clearly benefit from squeezed light. The extra complexity is well within the nature and budget of these instruments and it is likely that we will find this type of applications implemented in the near future.

Bibliography

[Abr96]   Improved sensitivity in a gravitational wave interferometer and implications for LIGO, A. Abramovici, W. Althouse, J. Camp, D. Durance, J.A. Giaime, A. Gillespie, S. Kawamura, A. Kuhnert, T. Lyons, F.J. Raab, R.L. Savage Jr, D. Shoemaker, L. Sievers, R. Spero, R. Vogt, R. Weiss, S. Whitcomb, M. Zucker, Phys. Lett. A 218, 157 (1996)

[Bot00]   Quantum interferometric optical lithography: Exploiting entanglement to beat the diffraction limit, A.N. Boto, P. Kok, D.S. Abrams, S.L. Braunstein, C.P. Williams, J.P. Dowling, Phys. Rev. Lett. 85, 2733 (2000)

[Bow02]   Recovery of continuous wave squeezing at low frequencies, W.P. Bowen, R. Schnabel, N. Treps, H-A. Bachor, P.K. Lam, J. Opt. B Q. Semiclss. Opt 4, 421 (2002)

[Bra68]   V.B. Braginsky, Sov. Phys. JETP 26 (831 (1968) NASA Technical translation TT F-672 (U.S. Technical Information Service, 1972)

[Buo02]   Signal recycled laser-interferometer -wave detectors as optical springs, A. Buonanno, Y. Chen, Phys. Rev. D 65, 042001 (2002)

[Cav80]   Quantum-mechanical noise in an interferometer, C.M. Caves, Phys. Rev. D 23, 1693 (1981)

[Cav94]   Quantum limits on bosonic communication rates, C.M. Caves and P.D. Drummond, Rev. Mod. Phys. 66, 481 (1994)

[Cer01]   Cloning and Cryptography with Quantum Continuous Variables, N.J. Cerf, S. Iblisdir and G.V. Assche, Eur. Phys. J. D 18, 211 (2002).

[Cho99]   Noiseless optical amplification of images, S.K. Choi, M. Vasilyev, P. Kumar, Phys. Rev. Lett. 83, 1938 (1999)

[Cou03]   Qunatum locking of mirrors in interferometers, J-M. Courty, A. Heidmann, M. Pinard, Phys. Rev. Lett. 90, 083601 (2003)

[Dal99]   Topical review: atoms in squeezed light fields, B.J.dalton, Z. Ficek, S. Swain, Journal of Modern Optics 46, 379 (1999)

[Dre83]   R.W.P. Drever, in *Gravitational Waves* eds. N. Deruelle and T. Piran, 321, North Holland, Amsterdam, (1983)

[Fab00]   Quantum limits in the measurement of very small displacements in optical images, C. Fabre, J-B. Fouet, A. Ma"tre, Opt. Lett. 25, 76 (2000)

[Fei97]   Entanglement-induced two-photon transparency, Hong-Bing Fei, B.M. Jost, S. Popescu, B.E.A. Saleh, M.C. Teich, Phys. Rev. Lett. 78, 1679 (1997)

[Gea87]   Applying squeezed states to non ideal interferometers, J. Gea-Banacloche, G. Leuchs, J. Opt. Soc. Am. B 4, 1667 (1987)

[Geo96]   Frequency metrology by use of quantum interference, N. Georgiades, E.S. Polzik, H.J. Kimble, A.S. Parkins, Opt. Lett. 21, 334 (1979)

[Goo93]   Experimental realization of a semiconductor photon number amplifier and a quantum optical tap, E. Goobar, A. Karlsson, G. Björk, Phys. Rev. Lett. 71, 2002 (1993).

[Gra87]   Squeezed-light enhanced polarisation interferometer, P. Grangier, R.E. Slusher, B. Yurke, A. La Porta, Phys. Rev. Lett. 59, 2153 (1987)

[Gra93]   Harmonic demodulation of non stationary shot-noise, M.B. Gray, A.J. Stevenson, H-A. Bachor, D.E. McClelland, Opt. Lett. 18, 759 (1993)

[Gra96]   External phase-modulation interferometry, M.B. Gray, A.J. Stevenson, C.C. Harb, H-A. Bachor, D.E. McClelland, Appl. Opt. 35, 1623 (1996)

[Hai97]   Optical measurement of weak absorption beyond shot-noise limit, Wang Hai, Xie Changde, Pan Qing, Xue Chenyang, Zhang Yun, Peng Kunchi in *Laser Spectrocopy XIII* Proc. 13 th Int. Conf. Laser Spectroscopy, Hangzhou (1997)

[Hal99]   Spin squeezed atoms: a macroscopic entangled ensemble created by light, J. Hald, J.L. Sörensen, C. Schori, E.S. Polzik, Phys. Rev. Lett. 83, 1319 (1999)

[Hal01]   Mapping a quantum state of light onto atoms, J. Hald, E.S. Polzik, J. Opt. B 3, S83 (2001)

[Hei96]   An experimental demonstration of resonant sideband extraction for laser interferometric gravitational wave detectors, G. Heinzel, J. Mizuno, R. Schilling, W. Winkler, A. Rüdiger, K. Danzmann, Phys. Lett. A 217, 305 (1996)

[Hol93]   Interferometric detection of optical phase shifts at the Heisenberg limit, M.J. Holland, K. Burnett, Phys. Rev. Lett. 71, 1355 (1993)

[Ino95]   Subshot noise interferometry with amplitude squeezed light from a semiconductor laser, S. Inoue, G. Bjork, Y. Yamamoto, Proc. SPIE 2378, 99 (1995)

[Kah89]   1 Gb/s PSK homodyne transmission system using phaselocked semiconductor lasers, J.M. Kahn, IEEE Photon. Technol. Lett., Oct (1989)

[Kim01]   Conversion of conventional gravitational-wave interferometers into quantum nondemoloition interferometers by modifying their input and/or output optics, H.J. Kimble, Y. Levin, A.B. Matsko, K.S. Thorne, S.P. Vyatchanin, Phys. Rev. D. 65, 022002 (2001)

[Kol89]   Squeezed states of light and noise-free optical images, M.I. Kolobov, I.V. Sokolov, Phys. Lett. A 140, 101 (1989)

[Kol93]   Sub-shot-noise microscopy: imaging of faint phase objects with squeezed light, M.I. Kolobov, P. Kumar, Opt. Lett. 18, 849 (1993)

[Kol95]   Noiseless amplification of optical images, M.I. Kolobov, L. Lugiato, Phys. Rev. A 52, 4930 (1995)

[Kol99]   The spatial behaviour of nonclassical light, M.I. Kolobov, Rev. Mod. Phys. 71, 1539 (1999)

[Kol00]   Quantum limits on optical resolution, M. Kolobov, C. Fabre, Phys. Rev. Lett. 85, 3789 (2000)

[Lam97]   Noiseless signal amplification using positive electro-optic feedforward, P.K. Lam, T.C. Ralph, E.H. Huntington, H-A. Bachor, Phys. Rev. Lett. 79, 1471(1997)

[Lev89]   Stochastic noise in TM_{oo} laser beam position, M.D. Levenson, W.H. Richhardson, S.H. Perlmutter, Opt. Lett. 14, 779 (1989)

[Lev93]   Quantum optical cloning amplifier, J.A. Levenson, I. Abram, T. Rivera, P. Fayolle, J.C. Garreau, P. Grangier, Phys. Rev. Lett. 70, 267 (1993)

[Li97]   Sub-shot-noise laser Doppler anemometry with amplitude-squeezed light, Y.Q. Li, P. Lynam, M. Xiao, P.J. Edwards, Phys. Rev. Lett. 78, 3105 (1997)

[LIGO]   for details on the detetction of gravitational waves, the present LIGO interferometer design and future LIGOII and LIGOIII upgrades see: http://www.ligo.caltech.edu/

[Lou81]   Quantum limit on the Michelson interferometer used for gravitational-wave detection, R. Loudon, Phys. Rev. Lett. 47, 815 (1981)

[Lug93]   Spatial structure of a squeezed vacuum, L.A. Lugiato, A. Gatti, Phys. Rev. Lett. 70, 3868 (1993)

[Lug95]   Qunatum spatial correlations in the optical parametric oscillatorwith spherical mirrors, L.A. Lugiato, I,Marzoli, Phys. Rev. A, 52, 4886 (1995)

[Lug97]   Improving quantum-noise reduction with spatially multimode squeezed light, L.A. Lugiato, P. Grangier, J. Opt. Soc. Am. B,14, 225 (1997)

[Lug02]   Quantum Imaging, L.A. Lugiato, A. Gatti, E. Brambilla, J. Opt. B 4, S176 (2002)

[Mar97]   Spatial quantum signatures in parametric down-conversion, I. Marzoli, A. Gatti, L.A. Lugiato, Phys. Rev. Lett. 78, 2092 (1997)

[Mar03]   Spatial distribution of quantum intensity correlations in a confocal type II optical parametric oscillators, M. Martinelli, N. Treps, S. Ducci, S. Gigan, A. Ma"tre, C. Fabre, Phys. Rev. A 67, 023808 (2003).

[Mer88] Recycling in laser-interferometric gravitational wave detectors, B.J. Meers, Phys. Rev. D 38, 2317 (1988)

[McK02] Experimental demonstration of a squeezing-enhanced power-recycled Michelson interferometer for gravitational wave detection, K. McKenzie, D.A. Shaddock, D.E. McClelland, B.C. Buchler, P.K. Lam, Phys. Rev. Lett. 88, 231102 (2002)

[MGM] The proceedings of the Marcel Grossman meetings are published in book format every three years.

[Mio92] Observation of an effect due to non stationary shot-noise, N. Mio, K. Tsubono, Phys. Lett. A 164, 255 (1992)

[Miz93] Resonant sideband extraction: a new configuration for interferometric gravitational wave detectors, J. Mizuno, K.A. Strain, P.G. Nelson, J.M. Chen, R. Schilling, A. Ruediger, W. Winkler, K. Danzmann, Phys. Lett. A 175, 273 (1993)

[Nab90] Two-color squeezing and sub-shot-noise signal recovery in doubly resonant optical parametric oscillators, C.D. Nabors, R.M. Shelby, Phys. Rev. A 42, 556 (1990)

[Nav02] Spatial entanglement of twin quantum images, P. Navez, E. Brambilla, A. Gatti, L. Lugiato, Phys. Rev. A 65, 13813 (2002)

[Nie91] Nonstationary shot noise and its effect on the sensitivity of interferometers, T.M. Niebauer, R. Schilling, K. Danzmann, A. Rüdiger, W. Winkler, Phys. Rev. A 43, 5022 (1991)

[Pac93] Quantum limits in interferometric detection of gravitational radiation, A.F. Pace, M.J. Collett, D.F. Walls, Phys. Rev. A 47, 3173 (1993)

[Pet63] Gravitational radiation from point masses in a Keplerian orbit, P.C. Peters, J. Mathews, Phys. Rev. 131, 435 (1963)

[Pol92] Atomic spectroscopy with squeezed light for sensitivity beyond the vacuum-state limit, E.S. Polzik, J. Carri, H.J. Kimble, Appl. Phys. B 55, 279, (1992)

[Pol97] Spin polarised atoms: beyond the standard quantum limit, E.S. Polzik, J. Sorensen, J. Hald, A. Kuzmich, K. Molmer, Proc. 5th Int. Conference on Squeezed States, Hungary (1997)

[QC94] *Special issue on Quantum Communication*, J. Mod. Opt. 12, December (1994)

[Roc93] Sub-shot-noise manipulation of light using semiconductor emitters and receivers, J.-F. Roch, J.-Ph. Poizat, P. Grangier, Phys. Rev. Lett. 71, 2006 (1993).

[Sau95] *Fundamentals of interferometric gravitational wave detectors*, P.R. Saulson, World Scientific (1994)

[Sco03] All-optical image processing with cavity type-II Second Harmonic Generation, P. Scotto, P. Colet, M. San Miguel to appear in Opt. Lett.

[Sha48] C.E. Shannon, Bell System Tech. J. 27, 623 (1948)

[Sok01] Quantum holographic teleportation, I.V. Sokolov, M.I. Kolobov, A. Gatti, L. Lugiato, Opt. Commun. 193, 175 ((2001)

[Ste93] Quantum-noise-limited interferometric phase measurements, A.J. Stevenson, M.B. Gray, H-A. Bachor, D.E. McClelland, App. Opt. 32, 3481 (1993)

[Ste94] Interferometers with internal and external phase modulation: experimental and analytical comparison, A. Stevenson, M. Gray, C.C. Harb, D.E. McClelland, H-A. Bachor, Aust. J. Phys, 48, 971 (1995)

[Str91]   Experimental demonstration of dual recycling for interferometric gravitational-wave detectors, K.A. Strain, B.J. Meers, Phys. Rev. Lett. 66, 1391 (1991)

[Sun96]   Sagnac interferometer for gravitational-wave detection, Ke-Xun Sun, M.M. Fejer, E. Gustafson, R.L. Byer, Phys. Rev. Lett. 76, 3053 (1996)

[Tak65]   Information theory of quantum-mechanical channels H. Takahashi, in *Advances in Communication Systems* A.V. Balakrishnan Ed., Orlando FL, Academic Press, 227 (1965)

[Tre00]   Transverse distribution of quantum fluctuations and correlations in spatial solitons, N. Treps, C. Fabre, Phys. Rev A 62, 033816 (2000)

[Tre02]   Crossing the quantum limit for high sensitivity measurements in optical images using non classical light, N. Treps, U. Andersen, B. Buchler, P.K. Lam, A. Maitre, H-A. Bachor, C. Fabre, Phys. Rev. Lett. 88, 203601 (2002)

[Tre03]   The quantum laser pointer, N. Treps, W. Bowen, N. Grosse, C. Fabre, H-A. Bachor,P.K. Lam, Science 308, 940 (2003)

[Wal93]   *Quantum Optics*, D.F. Walls, G. Milburn, Springer (1993)

[Xia87]   Precision measurement beyond the shot-noise limit, M. Xiao, L.A. Wu, H.J. Kimble, Phys. Rev. Lett. 59, 278 (1987)

[Yam86]   Preparation, measurement and information capacity of optical quantum states, Y. Yamamoto, H.A. Haus, Rev. Mod. Phys. 58, 1001 (1986)

[Yue78]   Optical communication with two-photo coherent states. Part I Quantum state propagation and quantum-noise reduction, H.P. Yuen, J.H. Shapiro, IEEE Trans. Inform. Theory Vol. IT-24, 657 (1978)

11 QND

One exciting question in quantum optics is whether we could measure and record a state of light and maintain it at the same time. While it is clear that we could not achieve this for all the properties of a laser beam or photon state simultaneously, it looks possible as a goal if we restrict ourselves to one quadrature, or one property, at a time. Such an experiment would be called a Quantum Non-Demolition, or QND, measurement, and we will see that through the use of squeezed states we can come close to this ideal. In this chapter we will discuss techniques that allow optical measurements with a quality better than possible with coherent light and conventional optics.

11.1 The concept of QND measurements

The properties of a beam of CW light are described by its two quadratures $X1$ and $X2$. They are conjugate variables, linked via the commutation relation and that means, more descriptively, they are linked through the uncertainty principle. Quantum mechanics of coherent states predicts that a measurement of one of these properties will perturb the quantity which is measured. There is a back action which adds noise to the system.

A very simple example is the measurement of the modulation of the $X1$ amplitude quadrature of a QNL beam of light using a beam splitter. We have seen that the remaining light, which was not used for the measurement, will still be at the QNL, but the signal to noise ratio (SNR) of the modulation will be reduced. Some part of the light, and with it some of the SNR, has been demolished in the experiment. Is there any chance of avoiding this back-action? Can we invent any technique to measure a property of the light without adding noise to this particular quadrature?

Historically, QND techniques were first discussed in the context of the problem of detecting the motion of a mechanical oscillator driven by a very small force. The limitations imposed by the quantum mechanical properties of the oscillator had to be avoided. V.B. Braginsky [Bra75, Bra80] developed these ideas first for mechanical systems and then generalized them to all harmonic oscillators. Different authors, including C.M. Caves, K.S. Thorne and W.G. Unruh developed the theoretical formalism [Cav80, Unr79]. More recently, a number of experiments have demonstrated that, in principle, an optical QND measurement is feasible. The basic idea is to introduce a link between the quadratures, through a nonlinear process, and to arrange the experiment such that all the back-action noise is transferred into the quadrature which is not interrogated. In this way the back-action is not avoided completely but is directed towards a quadrature which is not important for that particular measurement. As long as we

A Guide to Experiments in Quantum Optics, 2nd Edition. Hans-A. Bachor and Timothy C. Ralph
Copyright © 2004 Wiley-VCH Verlag GmbH & Co. KGaA
ISBN: 3-527-40393-0

Standard light detection

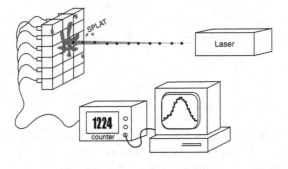

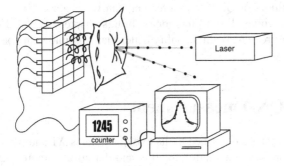

Quantum non-demolition measurement

Figure 11.1: CARTOON for QND

QND device

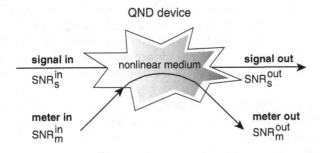

Figure 11.2: Schematic layout of a QND device

ask only questions about one quadrature we can, in principle, avoid the noise effects on the measurement.

An idealized QND experiment can be described by the interaction between two beams: One is the signal beam with quadratures $X1_s$ and $X2_s$. Let us assume we wish to measure amplitude modulation, that means the properties of the $X1_s$ quadrature. The second beam is a probe, or meter beam with the quadratures $X1_m$ and $X2_m$. It will somehow interact with the signal beam. We are going to detect one quadrature of the meter beam, say $X2_m$, and learn from this about $X1_s$. All the back-action induced by this measurement should influence only $X2_s$ and it should leave $X1_s$ unchanged. The meter is distinguished from the signal through its frequency, polarization or direction of propagation. We can describe the QND measurement by a set of equations which reflect the coupling of the quadratures

$$
\begin{aligned}
\tilde{X1}_s^{\text{out}} &= \tilde{X1}_s^{\text{in}} \\
\tilde{X2}_s^{\text{out}} &= \tilde{X2}_s^{\text{in}} + G\tilde{X1}_m^{\text{in}} \\
\tilde{X1}_m^{\text{out}} &= \tilde{X1}_m^{\text{in}} \\
\tilde{X2}_m^{\text{out}} &= \tilde{X2}_m^{\text{in}} + G\tilde{X1}_s^{\text{in}}
\end{aligned}
\tag{11.1.1}
$$

The first equation describes the fact that there is no back-action on the quadrature we wish to measure. The second shows the back-action on the orthogonal quadrature of the signal beam. The last equation shows that the information contained in $X1_s^{\text{in}}$ is transferred to $X2_m^{\text{out}}$, one of the quadratures of the meter beam.

In the ideal QND experiment, as proposed by V.B. Braginsky [Bra92], the output beam has the same properties as the input beam. In particular, no change in the optical power or frequency has occurred. The quadrature under investigation is not changed at all and the orthogonal quadrature is perturbed only as much as absolutely necessary. Such a measurement is almost impossible to carry out, but we can approach the ideal case. The question is: What type of interaction should be used? The best interactions are very similar to those used to create squeezed light.

Again, the Kerr effect comes to mind. In a Kerr medium the amplitude quadrature of the signal influences the refractive index of the medium. This in turn influences the phase quadrature of the meter beam propagating through the sample and a measurement of the the the meter will reveal the modulation of the signal amplitude. The amplitude quadratures of both signal and meter will not be affected, since the medium is transparent. However, the phase quadrature of the signal will experience the refractive index change induced by the meter beam. This is the back-action of the measurement. What we have to avoid is any self action of one quadrature to the other quadrature of the same beam. In this case the parametric gain induced in Equation (11.1.1) is simply linked to the Kerr coefficient of the medium and the intensities of the beams. The perfect QND measurement corresponds to an infinite gain G, which is clearly not attainable in practice.

A true QND experiment can be performed on one signal beam several times in a row, without introducing noise or affecting the quality of the quadrature under investigation. The key factor of QND measurements is repeatability [Cav80]. In practice, other sources of noise accompany any system with large parametric gain. The reasons are basically the same as with squeezing experiments: all noise sources have to be suppressed, but this gets harder and harder with increasing intensity and nonlinearity. Thus it is useful to characterize the degree of performance for a given QND measurement and to quantify how close a given experiment comes to the ideal QND device.

11.2   Classification of QND measurements

The characterization of a QND device can be carried out by comparing the four variances of the signal and meter beams before and after the interaction. Alternatively, one can perform correlation measurements between the fluctuations of the two beams, as proposed by N. Imoto [Imo89]. The standard for any comparison should always be the ideal, loss-less beamsplitter, a device that introduces only the absolutely minimal amount of noise. Please note that the perfect beam splitter is not a QND device – it is used as a reference. Many practical devices could actually be much worse than the ideal beamsplitter. An excellent summary of the characterization of QND measurements is given by J-Ph. Poizat et al. [Poi94].

One proposal to quantify the quality of a QND measurement was made by M. Holland et al. [Hol90]. The performance of a QND measurement device can be described by correlation coefficients between the input and the output modes. A coefficient C_m links the signal input to the meter output and it characterizes the fact that a measurement has been carried out.

$$
\begin{aligned}
C_m &= C(\delta \tilde{\mathbf{X}} 1_s^{\text{in}}, \delta \tilde{\mathbf{X}} 2_m^{\text{out}}) \\
&= \frac{\langle \delta \tilde{\mathbf{X}} 1_s^{\text{in}} \delta \tilde{\mathbf{X}} 2_m^{\text{out}} \rangle}{\sqrt{V 1_s^{\text{in}} V 2_m^{\text{out}}}}
\end{aligned}
\tag{11.2.1}
$$

The values for C_m are $-1 < C_m < 1$. A large correlation coefficient indicates a strong transfer from signal to meter. For a perfect measurement $C_m = 1$. A second coefficient C_s describes the ability of the device to avoid back-action. It compares the input and output properties of the device

$$
\begin{aligned}
C_s &= C(\delta \tilde{\mathbf{X}} 1_s^{\text{in}}, \delta \tilde{\mathbf{X}} 1_s^{\text{out}}) \\
&= \frac{\langle \delta \tilde{\mathbf{X}} 1_s^{\text{in}} \delta \tilde{\mathbf{X}} 1_s^{\text{out}} \rangle}{\sqrt{V 1_s^{\text{in}} V 1_s^{\text{out}}}}
\end{aligned}
\tag{11.2.2}
$$

The values for C_s are $-1 < C_s < 1$. A system that introduces no back-action has $C_s = 1$. These values cannot easily be obtained in an experiment, since they require knowledge of the input fluctuations $\delta \tilde{\mathbf{X}} 1_s^{\text{in}}$, which in turn would require a QND device for the unperturbed measurements.

Any practical characterization will be based on readily available experimental data. This was first described by N. Imoto and S. Saito at NTT Japan [Imo89] and later P. Grangier and his group showed that the measurement of signal to noise ratios is both practical and can be used to quantify the performance of a QND experiment [Poi94]. This is best done by transfer coefficients T_m and T_s which describe the transfer of the signal to noise ratio (SNR) from the input signal to the output of the meter and the signal respectively

$$
T_m = \frac{\text{SNR}_m^{\text{out}}}{\text{SNR}_s^{\text{in}}}
$$

$$
T_s = \frac{\text{SNR}_s^{\text{out}}}{\text{SNR}_s^{\text{in}}}
\tag{11.2.3}
$$

It can be shown that for a perfect beamsplitter with reflectivity ϵ one obtains the values

$$
T_{s,BS} = 1 - \epsilon \quad \text{and} \quad T_{m,BS} = \epsilon
\tag{11.2.4}
$$

and thus the result $T_{m,BS} + T_{s,BS} = 1$. For the ideal QND device we have $T_s = 1, T_m = 1$ and $T_m + T_s = 2$. In this way we have a criterion that a QND device should satisfy

$$T_m + T_s > 1. \tag{11.2.5}$$

The inverse value $\frac{1}{T_s}$ is known as the noise figure. Any classical device will have a noise figure > 1. To complement this criterion we need a second, independent quantity which measures the quantum correlations between the two outputs. This is the conditional variance, which is defined as

$$V_{s|m} = \langle (\delta \tilde{\mathbf{X}} \mathbf{1}_s^{\text{out}})^2 \rangle (1 - C_m^2) \tag{11.2.6}$$

For a perfect measurement this should vanish, $V_{s|m} = 0$. A determination of $V_{s|m}$ requires a measurement and comparison of the fluctuations of the photo-current at the output. The conditional variance is already normalized to the quantum noise level. For the ideal beamsplitter we obtain $V_{s|m,BS} = 1$. The achievement of $V_{s|m} < 1$ can be considered as conditional squeezing, it is non-classical in the same sense as squeezed light. The value of $V_{s|m}$ is identical to the best noise suppression that could be obtained in a feedback control system (see Section 8.3) using the meter beam as the input for the control loop. The experimental determination of the conditional variance involves the addition and subtraction of fluctuations of Fourier components of the photo-currents. To achieve the best interference between the two terms an optimum attenuation should be used as discussed in Section 7.5. This can be found empirically during the experiment and should agree with the correlation factor in Eq. (11.2.2) [Poi94].

The classification of a QND device requires two independent criteria and is best represented by a two dimensional diagram with the axis $V_{s|m}$ and $T = T_s + T_m$ as shown in Fig. 11.3a. The reference point is the ideal beamsplitter with $V_{s|m} = 1$ and $T = 1$. The extreme cases are the perfect twin beams with $V_{s|m} = 0$ and $T = 1$, the quantum duplicator with $V_{s|m} = 1$ and $T = 2$ and the perfect QND experiment with $V_{s|m} = 0$ and $T = 2$. The parameter space can be divided into four quadrants, each with a certain distinct meaning. The classical quadrant, top left, describes experiments with no quantum correlations which can be performed with normal linear components. Any imperfections in the system or in the measurement, such as losses and non-ideal quantum efficiencies, will move the result away from the optimum point. A value $T > 1$ indicates noiseless amplification as it can be achieved with a phase sensitive amplifier, described in Section 6.2. On the other hand, $V_{s|m} < 1$ indicates *quantum state preparation*. The combination of the two, represented by the bottom right quadrant in Fig. 11.3, indicates a true QND device.

Finally, the best and most important test for a QND device is that it can be operated in sequence, that means several QND measurements can be applied to the one mode of light, each producing the same result and each adding as little noise as possible. A complete demonstration of a QND measurement would show that the state is not more perturbed than absolutely necessary in accordance with the quantum description of the process.

This classification is not without dispute. It is probably not complete since it allows devices to be classified as QND which one would otherwise not regard as such. One example is to use a combination of photo-detector, amplifier/splitter and light source to detect a beam of light, generate a new beam and measure the intensity fluctuations electronically in between. In this experiment all the light is detected and, assuming the very good quantum efficiencies for both detection and generation, the intensity fluctuations, including the quantum noise, can be measured without back effect [Roc93, Goo93], shown as point (15) in Fig. 11.3. This good result is not surprising since in this case we measure current fluctuations, not quadrature fluctuations of the light. However, only information of the square of one quadrature is available, the information about the other quadrature is completely destroyed; the only property that is preserved is the intensity. It has been argued that such a destructive device should not be called QND device, and that a third parameter is required to classify the experiment. This additional parameter would measure how close the overall effect on all quadratures is to the minimum destructiveness required by quantum rules. Unfortunately, no generally accepted definition of such a third parameter has been given so far and the review by P. Grangier et al. [Gra98] still summarizes the results very well.

11.3   Experimental results

The fundamental difficulties in QND measurements are very similar to those of squeezing experiments. We require strong nonlinear effects which couple the quadratures of the signal and meter modes, no losses or noise and very efficient detectors. The measurements of both $V_{s|m}$ and T can be achieved with the same instruments as used for squeezing measurements. The only addition is the optimization of the electronic gain in order to measure the combination of electronic voltages which corresponds to the parameters $V_{s|m}$ and T. A summary of the presently achievable results is given in Fig. 11.3. Early experiments, such as the work by M.D. Levenson et al. [Lev86, Bac88], point (1), used the nonlinear properties of optical fibres to introduce correlation between the quadratures. They obtained some small degree of quantum correlation but insufficient transfer.

A whole group of experiments used optical parametric down-converters.   La Porta et

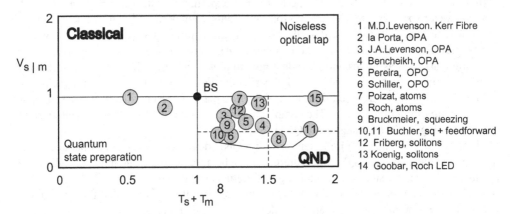

Figure 11.3: Characterization of quantum measurement devices by Poizat et al. [Poi94]. The two parameters $V_{s|m}$ and $T = T_m + T_s$ classify all possible devices. The conditions for a true QND device are satisfied with $V_{s|m} < 1$ and $T > 1$. The best experimental results to date are shown by circles, referenced in the text. This diagram is similar to the review by P. Grangier et al. [Gra98].

al. [Lap89] carried out an experiment with a KTP crystal as the OPO. This was driven by pulses at 532 [nm] from a frequency doubled Nd:YAG laser. A phase locked ring cavity was used to recirculate these green pump pulses through the crystal. Two infrared beams at 1064 [nm], which are distinguished by their polarization, served as the signal and meter. The results, point (2) in Fig. 11.3, show that the conditional variance is smaller than in a classical device but the transfer values are not sufficient. This experiment demonstrates quantum state preparation. More recent experiments using better materials and with smaller losses satisfy all the requirements of a QND device. J.A. Levenson [Lev93] and K. Bensheikh [Ben95], points (3) and (4) respectively, used a KTP crystal pumped by pulses from a frequency doubled Nd:YLF laser a single pass travelling wave OPA. The signal and meter beams are at 1054 [nm] and the meter beam is generated inside the OPA.

Improved results were achieved by S. Pereira [Per94] with a CW experiment. A KTP crystal, inside a resonant cavity, is optically pumped. Signal and meter beams, represented by orthogonally polarized beams at 1080 [nm], are measured in direct detection. Their results, point (5), show a clear QND performance. S. Schiller and his group [Sch96] utilized the properties of a $LiNbO_3$ OPO that operated as the generator for a strongly squeezed vacuum state. This device did not only squeeze, but introduced quantum correlations on to a second beam injected into the same cavity. This is clearly a QND result, shown as point (6) in Fig. 11.3.

A completely different experimental approach was chosen by P. Grangier and his group [Gra91]. They used the Kerr effect of a sodium atomic beam to correlate the quadratures. The signal beam at 589 [nm] was coupled into a resonant cavity, with the atoms inside. A second beam, the probe at 819 [nm], tuned to a second cascaded atomic resonance, picks up a modulation of the phase introduced by the intensity fluctuations of the signal beam. Losses and competing noise sources that appear when the light is tuned close to the atomic resonance limit this particular scheme. While these first results produced only a conditional variance of 1 the improved experiments by J.P. Poizat et al. [Poi93], point (7), clearly showed sufficient transfer. More recent experiments from this group used laser cooled atoms [Roc97] in order to take advantage of the reduced atomic linewidth and higher densities, see Fig. 11.4(a). This is one of the most difficult and most advanced quantum optics experiment based on atoms and produced some of the best QND results, as shown by point (8) in Fig. 11.3.

An alternative approach for QND is to improve the properties of a normal beam splitter by filling the unused port with squeezed light. This was proposed by J. Shapiro [Sha80] and was first realized by R. Bruckmeier et al. [Bru97] who used an CW OPO to generate the squeezed. We note that the beamsplitter will attenuate the light, but will leave the information on the beam largely unperturbed. The results were already impressive, with good T and V, point (9).

This concept was extended by B.C. Buchler et al. [Buc99] who showed that the combination of the squeezed light on the beamsplitter with an electro-optic feedforward control, as described in Section 10.3, will enhance the QND measurement. The layout is shown in Fig. 11.4(b). Analogous to the noiseless amplifier, the feedforward will improve the signal to noise ratio, that means T, while retaining the correlation $V_{s|m}$. This was demonstrated in a recent experiment [Buc01] where the values for a QND measurement with a beamsplitter and squeezed input beam was improved from the original values of $T = 1.12$, $V_{s|m} = 0.46$ to $T = 1.81$, $V_{s|m} = 0.55$ through the use of feedforward and optimized gain. This is represented by the points (10) and (11) in Fig. 11.3.

Finally, it should be mentioned that the QND measurement of optical solitons, which was proposed by H.A. Haus et al. [Hau89] was demonstrated by S. Friberg and his group [Fri92], point (12) in Fig. 11.3. The underlying effects are identical to those responsible for the generation of squeezed solitons as described in Section 9.9.4. This technology was extended by F. König, A. Sizmann and the group in Erlangen who used spectral filtering [Koe02]. In a series of experiments they showed the QND measurement of the soliton photon number and achieved, with optimized gain, the values $T = 1.37$, $V_{s|m} = 0.92$, point (13) in Fig. 11.3, and the a best value of $V_{s|m} = 0.73$.

The crucial tests for repeatability were recently carried out in the groups led by J.A. Levenson [Ben97] and S. Schiller [Sch96]. Both could show that while the second measurement is not quite as good as the first both consecutive measurements can be classified as QND measurements. These experiments open the way for the design of optical taps, devices that can be used to distribute, or eavesdrop, on information sent via optical fibres. QND promises the ability to tap into information transferred optically without reducing the SNR to a large degree. In practice, this is unlikely to be practical due to the experiments' sensitivity to losses. However, A. Levenson has already demonstrated that two cascaded optical taps which produce a better transfer of information than a perfect beamsplitter.

11.4   Single photon QND

We have seen in Chapter 3 that the coupling between light and atoms inside a cavity can be used as a direct evidence for the existence of photons. Very clear demonstrations exist for the coupling of microwave radiation to atoms in highly excited Rydberg states. In such a micromaser the radiation is contained inside an almost lossless superconducting microcavity and the advantages are the very long lifetime of the atoms, several milliseconds, the ability to avoid thermal noise by cooling the cavity and the ability to prepare the monitor the atoms, which are injected into the cavity and are analysed, via field ionization when they leave the cavity. Such an arrangement is shown in Fig. 11.5.

The optical state of interest is contained inside the cavity, it interacts closely and is modified by the atoms and we can gain information of the light by analysing the quantum state of the atoms coming from the cavity. The optical field cannot be detected directly, instead the information is indirectly gained from the properties of the atoms. The atoms and light are in a combined state, sometimes referred to as a *mesoscopic superposition state*, and a measurement of the atoms, leaving the cavity, provides information about light inside the cavity [Bru96], with minimum back-action. The atoms outside are destroyed, but the properties of the photons inside the cavity can be inferred. The measurement of the atoms is equivalent to a QND measurement of the light in the cavity. Like all single photon experiments the data acquisition relies on post-selection, that means using only those events where exactly one atom is present.

In the first experiments, see Section 3.7, it was possible to see the quantization of the microwave field by observing the Rabi oscillations of the atoms. By carefully selecting the velocity of the atoms, the magnitude of the Rabi frequency provides a measure of the field strength inside, and the apparatus is sufficiently sensitive to distinguish individual photons.

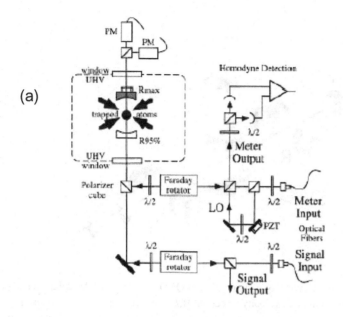

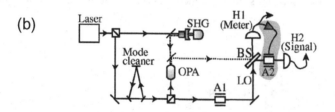

Figure 11.4: Schematic layout of two QND experiments. (a) experiment using the cross correlation in laser cooled atoms, after J.F. Roch et al. [Roc97] and (b) beamsplitter with squeezed input and feedforward, after B.C. Buchler et al. [Buc01].

Next we can use the interaction inside the cavity to create special states of light and it has been shown, both in the experiment in München [Var00] and Paris [Ber02], under the respective guidance of H. Walther and S. Haroche, that results closely resembling one or two photon Fock states can be generated. An example is given in Fig. 11.6 and shows a remarkable agreement between theory and experiment. In similar experiments in Paris it is now possible to carry out full state tomography and to reconstruct the full Wigner function of such a one photon Fock state [Ber02].

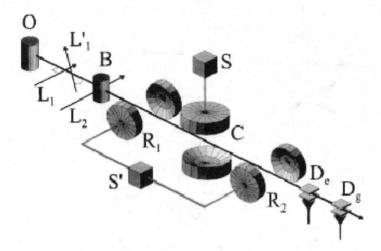

Figure 11.5: Schematic layout of a single photon QND experiment. The components are: O oven, L_1 optical pumping and velocity selection, B,L_2 optical excitation to Rydberg state, R1 state preparation by $\pi/2$ pulses from microwave source S', C cooled, low loss microwave cavity, R2 second $\pi/2$ microwave pulse for state preparation, and D field ionization. After M. Brune et al. [Bru96].

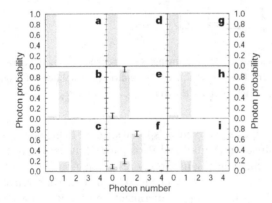

Figure 11.6: Results from the micromaser experiment showing the photon number probabilities for microwave photon number, or Fock, states produced inside a cavity. The results are for three conditions optimized to generate the states $|0\rangle$, $|1\rangle$ and $|2\rangle$. (a)–(c) theoretical simulation, (d)–(f) experimental results, (g)–(i) predictions for future optimized experiment. After B.T.H. Varcoe et al. [Var00].

Bibliography

[Bac88]   Quantum nondemolition measurements in an optical fiber ring resonator, H-A. Bachor, M.D. Levenson, D.F. Walls, S.H. Perlmutter, R.M. Shelby, Phys. Rev. A38, 180 (1988)

[Ben95]   Quantum nondemolition demonstration via repeated backaction evading measurements, K. Bencheikh, J.A. Levenson, Ph. Grangier, O. Lopez, Phys. Rev. Lett. 75, 3422 (1995)

[Ben97]   Cascaded amplifying quantum optical taps: a robust, noiseless optical bus, K. Bencheikh, C. Simonneau, J.A. Levenson, Phys. Rev. Lett. 78, 34 (1997)

[Ber02]   Direct measurement of the Wigner function of a one-photon fock state in a cavity, P. Bertet, A. Auffeves, P. Maioli, S. Osnahgi, T. Meunier, M. Brune, J.M. Raimond, S. Haroche, Phys. Rev. Lett. 89, 200402 (2002)

[Bra75]   Quantum-mechanical limitations in macroscopic experiments and modern experimental techniques, V.B. Braginsky, Yu. I. Vorontsov, Sov. Phys.-Usp. 17, 644 (1975)

[Bra80]   Quantum nondemolition measurements, V.B. Braginsky, Yu. I. Vorontsov, K.S. Thorne, Science 209, 547 (1980)

[Bra92]   V.B. Braginsky, F. Ya. Khalili, in:*Quantum Measurement*, Ed. K.S. Thorne, Cambridge University Press (1992)

[Bru96]   Observing the progressive decoherence of the "meter" in a quantum measurement, M. Brune, E. Hagley, J. Dreyer, X. Maitre, A. Maali, C. Wunderlich, J.M. Raimond, S. Haroche, Phys. Rev. Lett. 77, 4887 (1996)

[Bru97]   Realization of a paradigm for quantum measurements: the squeezed light beam splitter, R. Bruckmeier, H. Hansen, S. Schiller, J. Mlynek, Phys. Rev. Lett. 79, 43 (1997)

[Buc99]   Enhancement of quantum nondemolition measurements with an electro-optic feed-forward amplifier, B. Buchler, P.K. Lam, T.C. Ralph, Phys. Rev. A 60, 4943 (1999)

[Buc01]   Squeezing more from a quantum nondemolition measurement, B.C. Buchler, P.K. Lam, H-A. Bachor, U.L. Andersen,T.C. Ralph, Phys. Rev. A 65, 011803 (2001)

[Cav80]   On the measurement of a weak classical force coupled to a quantum-mechanical oscillator. I. Issues of principle C.M. Caves, K.S. Thorne, R.W. Drever, V.D. Sandberg, M. Zimmermann, Rev. Mod. Phys. 52, 341 (1980)

[Fri92]   Quantum-nondemolition measurements of the photon number of an optical soliton, S.R. Friberg, S. Machida, Y. Yamamoto, Phys. Rev. Lett. 69, 3165 (1992)

[Gra91]   Observation of backaction-evading measurement of an optical intensity in a three-level atomic nonlinear system, Ph. Grangier, J.F. Roch, G. Roger, Phys. Rev. Lett. 66, 1418 (1991)

[Gra98]   Quantum-non-demolition measurements in optics, P. Grangier, J.A. Levenson, J-Ph. Poizat, Nature, 396, 537 (1998)

[Goo93]   Experimental realization of a semiconductor photon number amplifier and a quantum optical tap, E. Goobar, A. Karlsson, G. Bjoerk, Phys. Rev. Lett. 71, 2002 (1993)

[Hau89]   Quantum-nondemolition measurement of optical solitons, H.A. Haus, K. Watanabe, Y. Yamamoto, J. Opt. Soc. Am. B 6, 1138 (1989)

[Hol90]   Nonideal quantum nondemolition measurements, M.J. Holland, M.J. Collett, D.F. Walls, M.D. Levenson, Phys. Rev. A 42, 2995 (1990)

[Imo89]   Quantum nondemolition measurement of photon number in a lossy optical Kerr medium, N. Imoto, S. Saito, Phys. Rev. A 39, 675 (1989)

[Koe02]   Soliton backaction-evading measurement using spectral filtering, F. König, B. Buchler, T. Rechtenwald, G. Leuchs, A. Sizmann, Phys. Rev. A 66, 043810 (2002)

[Lap89]   Back-action evading measurements of an optical field using parametric down conversion, A. La Porta, R.E. Slusher, B. Yurke, Phys. Rev. Lett. 62, 28 (1989)

[Lev86]   Quantum nondemolition detection of optical quadrature amplitudes, M.D. Levenson, R.M. Shelby, M. Reid, D.F. Walls, Phys. Rev. Lett. 57, 2473 (1986)

[Lev93]   Quantum optical cloning amplifier, J.A. Levenson, I. Abram, T. Rivera, P. Fayolle, Phys. Rev. Lett. 70, 267 (1993)

[Per94]   Backaction evading measurements for quantum nondemolition detection and quantum optical tapping, S.F. Pereira, Z.Y. Ou, H.J. Kimble, Phys. Rev. Lett. 72, 214 (1994)

[Poi93]   Experimental realization of a quantum optical tap, J.P. Poizat, P. Grangier, Phys. Rev. Lett. 70, 271 (1993)

[Poi94]   Characterization of quantum non-demolition measurements in optics, J-Ph. Poizat, J.F. Roch, P. Grangier, Ann. Phys. Fr. 19, 265 (1994)

[Roc93]   Sub-shot-noise manipulation of light using semiconductor emitters and receivers, J.F. Roch, J-Ph. Poizat, P. Grangier, Phys. Rev. Lett. 71, 2006 (1993)

[Roc97]   Quantum nondemolition measurements using cold trapped atoms, J.F. Roch, K. Vigneron, Ph. Grelu, A. Sinatra, J-Ph. Poizat, Ph. Grangier, Phys. Rev. Lett. 78, 634 (1997)

[Sch96]   Quantum nondemolition measurements and generation of individually squeezed beams by a degenerate optical parametric amplifier, S. Schiller, R. Bruckmeier, M. Schalke, K. Schneider, J. Mlynek, Europhys. Lett. 36, 361 (1996)

[Sha80]   Optical waveguide tap with infinitesimal insertion loss, J.H. Shapiro, Opt. Lett. 5, 351 (1980)

[Spa97]   Photon number squeezing of spectrally filtered sub-picosecond optical solitons, S. Spälter, M. Burk, U. Strössner, M. Böhm, A. Sizmann, G. Leuchs, Europhys. Lett. 38, 335 (1997)

[Siz98]   The optical Kerr effect and quantum optics is fibres, A. Sizmann, G. Leuchs in *Progress in Optics XXXIX* Ed. E. Wolf, Elsevier, 373 (1998)

[Unr79]   Quantum nondemolition and gravity-wave detection, W.G. Unruh, Phys. Rev. D 19, 2888 (1979)

[Var00]   Preparing pure photon number states of the radiation field, B.T.H. Varcoe, S. Brattke, M. Weidinger, H. Walther, Nature 403, 743 (2000)

12 Fundamental tests of quantum mechanics

Quantum mechanics is undoubtedly the most successful theory yet devised in terms of its ability to accurately predict physical phenomena. However, the physical picture that emerges from quantum mechanics is so foreign to the one experienced in the macroscopic world that many have rebelled against it and sought a more "sensible" physical picture hidden beneath. For many years it was not technologically possible to test the fundamental tenets of quantum mechanics directly. The advent of the laser and hence the birth of quantum optics changed that situation and now many of the Gedanken experiments proposed in the early days of the development of quantum theory have been demonstrated experimentally. In this chapter we discuss experiments aimed at testing three of these concepts: complementarity, or more particularly wave-particle duality, indistinguishability, and non-locality.

12.1 Wave-Particle duality

Quantum systems can display both particle and wave-like properties. This was first realized by A. Einstein in his description of the photo-electric effect [Ein05] but later extended by L.deBroglie to massive particles [DeB23]. However, any experiment which clearly delineates the particle aspects of a system will completely fail to see the wave aspect and vice versa. This is an example of the concept of complementarity first described by N. Bohr in 1928 [Boh28]. The canonical example of wave-particle duality is the interferometer, whether optical or matter, in which wave interference disappears if precise path information is obtained. In some situations wave particle duality is upheld by a straightforward application of the uncertainty principle [Fey63], while in others the connection is more subtle [Scu91], [Wis95].

In Chapter 3 we described experiments in which it could be postdicted that only one photon had passed through the interferometric apparatus at a time. We saw that provided we had no information about the path taken by the photon then perfect interference fringes could be observed. On the other hand if the path taken was determined, say by placing a detector in one arm, then no interference will occur.

This latter case is trivial if the photon is detected but notice that the absence of a detection in one arm also gives *which-path* information, i.e. that the photon took the other path. In this case the photon arrives at the output, but no interference takes place. This effect can be used to detect the presence of an object without depositing any energy into it. Consider a balanced interferometer set up such that photons entering say, input port "1" always exit by output port "1" due to destructive interference. Now suppose an absorber is placed in one path of the interferometer such that no interference will occur. It is now possible for a photon to exit through output port "2". If the observer finds a photon at output port "2" they can be sure

A Guide to Experiments in Quantum Optics, 2nd Edition. Hans-A. Bachor and Timothy C. Ralph
Copyright © 2004 Wiley-VCH Verlag GmbH & Co. KGaA
ISBN: 3-527-40393-0

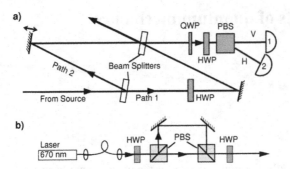

Figure 12.1: Experimental set-up of Schwindt et al. [Sch99] designed to quantify wave-particle duality.

that an absorber was present, even though there was no energy transfer to the absorber. This effect is referred to as an interaction free measurement [Eli93] and has been demonstrated experimentally [Kwi95a].

These are extreme cases of all or nothing path information: what if we obtain partial information about the path taken? It turns out that a strict bound can be placed on the quality of the interference versus the which-path knowledge obtained [Eng96]. The quality of the interference is quantified in the usual way by the fringe visibility

$$V = \frac{n_{\max} - n_{\min}}{n_{\max} + n_{\min}} \tag{12.1.1}$$

where $n_{\max}$ ($n_{\min}$) is the number of particles counted at the fringe maximum (minimum). Which-path information is quantified by the likelihood P_L. This is the probability that the experimenter will correctly guess the path taken by the particle. The path knowledge, K, can be defined as

$$K = 2P_L - 1 \tag{12.1.2}$$

which has the property of being zero when no path information is obtained and one when the information is perfect. Given these definitions, the following relation is obeyed

$$V^2 + K^2 \leq 1 \tag{12.1.3}$$

The inequality is saturated for pure input states and optimal readout of which-path information.

A few experiments have been carried out to probe wave-particle duality in this intermediate domain, both with massive particles [Sum87] and in optics [Mar92]. Perhaps the most sophisticated experiment with massive particles were the atom-interferometer experiments carried out by the group of G. Rempe [Dur98]. Here increasing which-path information stored in the internal states of the atoms was seen to progressively erase the atomic interference effects.

A quantitative test of wave-particle duality was carried out in optics by P.D.D. Schwindt, P.G. Kwiat and B-G. Englert [Sch99]. In this experiment a highly attenuated 670 [nm] laser

was used to illuminate a Mach-Zehnder interferometer and single photon counting was performed at one of the output ports. Inside the Mach-Zehnder a half-wave plate in one of the
arms was used to encode partial which-path information on the photons. The set-up is shown
in Fig. 12.1. The attenuation was sufficient that the maximum count rates at the output were
$\leq 50,000[s^{-1}]$. Given a transit time through the interferometer of 1[ns], the average number
of photons in the interferometer at any time was $\leq 10^{-4}$ and hence, given the Poissonian photon statistics of a coherent field, the probability of more than one photon travelling through
the interferometer at one time was negligible. By *post-selecting* only those events which
triggered the detector an excellent approximation to a single photon source is achieved, see
Section 13.2. The single photon quantum state inside the interferometer (see section 5.2), but
before the wave plate, is given by

$$\frac{1}{\sqrt{2}}\left(|1\rangle_{1V}|0\rangle_{2V} + e^{i\phi}|0\rangle_{1V}|1\rangle_{2V}\right)|0\rangle_{1H}|0\rangle_{2H} \tag{12.1.4}$$

where the ket $|n\rangle_{jK}$ indicates photon occupation n of the jth spatial and Kth polarization
mode. The input modes to the interferometer are assumed to be vertically polarized. Here ϕ is
the phase difference between the two arms of the Mach-Zehnder. This phase can be adjusted
experimentally by varying the path length of one arm by small amounts using a piezo-electric
transducer mounted on one of the mirrors. After passing through the half-wave plate the state
becomes

$$\frac{1}{\sqrt{2}}\left(\sqrt{1-\epsilon^2}|1\rangle_{1V}|0\rangle_{2V}|0\rangle_{1H}|0\rangle_{2H} + \epsilon|0\rangle_{1V}|0\rangle_{2V}|1\rangle_{1H}|0\rangle_{2H}\right)$$
$$+\frac{1}{\sqrt{2}}e^{i\phi}|0\rangle_{1V}|1\rangle_{2V}|0\rangle_{1H}|0\rangle_{2H} \tag{12.1.5}$$

There is now an amplitude $(1/\sqrt{2})\,\epsilon$ to find a photon in one of the horizontal modes, but this
can only occur if the photon took path (1) through the interferometer. After recombining the
interferometric paths at the output beamsplitter we have the output state

$$\frac{1}{2}\left((\sqrt{1-\epsilon^2} + e^{i\phi})|1\rangle_{1V}|0\rangle_{2V} + (\sqrt{1-\epsilon^2} - e^{i\phi})|0\rangle_{1V}|1\rangle_{2V}\right)|0\rangle_{1H}|0\rangle_{2H}$$
$$+\frac{1}{2}\epsilon\,|0\rangle_{1V}|0\rangle_{2V}\left(|1\rangle_{1H}|0\rangle_{2H} - |0\rangle_{1H}|1\rangle_{2H}\right) \tag{12.1.6}$$

It is clear that when the wave plate is oriented such that it produces no rotation of the polarization, $\epsilon = 0$, then perfect visibility is realized. That means: if we look at one spatial output,
say $|0\rangle_1|1\rangle_2$, then we see perfect extinction of counts when the phase shift is zero, $\phi = 0$, and
hence we obtain $\mathcal{V} = 1$. On the other hand, no which-path information is obtained as the photons always emerge vertically polarized. Qualitatively one can see from Equation (12.1.6) that
as polarization rotation is introduced the interferometric visibility is decreased whilst which-
path information is increased. When the polarization of a photon taking path (1) is completely
flipped, $\epsilon = 1$, the visibility is reduced to zero (the number of counts at an output is unchanged
by varying the path length) but knowledge of the path taken is perfect as a vertically polarized
photon at the output definitely took path (2) whilst a horizontally polarized photon definitely
took path (1).

   The visibility of the probability distribution was measured by counting the total number
of photons emerging from one of the outputs when the path length difference was adjusted to

produce maximum counts, giving $n_{\max}$, and when the path length was adjusted to produce a minimum, giving $n_{\min}$.

To find the which-path knowledge the output polarization analyser, comprising a half-wave plate, quarter wave plate, polarizing beamsplitter and two detectors was employed, see Fig. 12.1. This combination allows measurements to be made in an arbitrary polarization basis. By blocking one of the paths through the interferometer and looking at the relative count rates on the two detectors for some particular polarization basis and repeating this process with the other path blocked, it is possible to evaluate the likelihood for that particular basis. In particular, consider the case where more counts are registered on detector 1 than detector 2 when the path (1) through the interferometer is taken and vice versa when the path (2) is taken. If we now send a single photon through the interferometer with both paths open and we find that detector 1 "clicks", then we would guess that the photon took the first path. The likelihood that this guess is correct is given by [Woo79]

$$L(\lambda) = \frac{n_{1,1}(\lambda) + n_{2,2}(\lambda)}{n_{1,1}(\lambda) + n_{2,2}(\lambda) + n_{1,2}(\lambda) + n_{2,1}(\lambda)} \tag{12.1.7}$$

where $n_{i,j}(\lambda)$ is the count rate on detector i when the photons are made to take the jth path and the polarization analysis is in the λ basis. More generally we can write

$$L(\lambda) = \frac{\max[n_{1,1}(\lambda), n_{1,2}(\lambda)] + \max[n_{2,2}(\lambda), n_{2,1}(\lambda)]}{n_{1,1}(\lambda) + n_{2,2}(\lambda) + n_{1,2}(\lambda) + n_{2,1}(\lambda)} \tag{12.1.8}$$

The results obtained in the experiment by Schwindt et al. [Sch99] are shown in Fig. 12.2 along with theoretical curves. The trade-off between visibility and knowledge is clearly seen. The results from using two measurement bases are shown: horizontal/vertical and the optimized basis. The optimal basis is $(\phi_1 + \phi_2)/2 + 45^0$ and $(\phi_1 + \phi_2)/2 - 45^0$ where ϕ_j is the polarization angle of the photons if they traverse the jth path. When the optimal measurement basis is used the sum $V^2 + K^2$ is within a few percent of 1 over the whole parameter range. This is a powerful experimental demonstration of the fundamental quantum property of complementarity.

12.2   Indistinguishability

Another fundamental difference between classical and quantum physics is the indistinguishability of quantum particles. In classical mechanics one can always label different particles in an ensemble by their position and momentum. However, in quantum theory the uncertainty principle makes this not possible in general. Consider the situation depicted in Fig. 12.3. A collision occurs within a region smaller than the position uncertainty of the particles. It is thus not possible, even in principle, to distinguish between the two trajectories shown, and hence not possible to "keep track" of a particular particle.

The analogous situation for photons was described in Section 5.1. If two photons which cannot be distinguished by their spatial or spectral modes, collide on a beamsplitter then it is not possible to distinguish between events in which both were reflected or both were transmitted. Because photons are bosons, the amplitudes for these two events have opposite signs and, if transmission and reflection are equally likely, will exactly cancel.

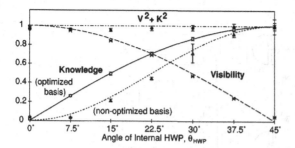

Figure 12.2: Experimental data and theoretical curves from the experiment by Schwindt et al. [Sch99]. The results are for a vertically polarized input as a function of the orientation of the half-wave plate in path (1). The crosses are the measured visibilities. The dotted line and the triangles correspond to K measurements in the horizontal/vertical basis, and the solid line and diamonds correspond to measurements in the optimal basis.

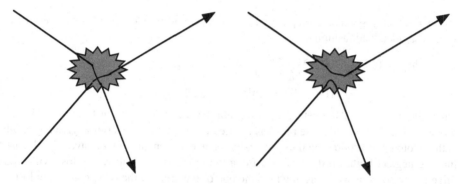

Figure 12.3: Particle collision in which uncertainty region, represented by shading, makes it impossible to distinguish alternate trajectories.

The first experimental demonstration of this effect was performed by C.K. Hong, Z.Y. Ou and L. Mandel in 1987 [Hon87]. A parametric down converter, see Section 6.3.4, constructed from a non-linear crystal of potassium dihydrogen phosphate (KDP) and pumped by the 351 [nm] argon-ion laser line, was used to produce pairs of visible photons. The photons were sent down paths of almost equal length and then mixed on a 50:50 beamsplitter as shown in Fig. 12.4. The difference in the path lengths followed by the two photons could be varied from zero to $\pm$ a few tens of [μm] by shifting the position of the beamsplitter. The beams were then passed through pin-holes and interference filters with a pass-band of order 5×10^{12} [Hz] before striking single photon counters. The photon counters were connected to a coincidence counter which is set to register only events which occurred within a 7.5 [ns] time window.

The pin-holes serve to erase spatial information about the paths taken by the photons and are oriented to isolate two well-defined spatial modes which overlap well at the beamsplitter. Similarly the interference filters serve to erase frequency information which might distinguish the photons.

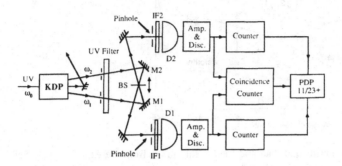

Figure 12.4: Set-up of the Hong-Ou-Mandel experiment [Hon87].

In frequency space the output of a non-degenerate down-converter can be written in a somewhat idealized fashion as

$$|\phi\rangle = |0\rangle \ + \ \int d\Omega \frac{2\chi\kappa_a}{\kappa_a^2 + (2\pi\Omega)^2}(|0\rangle_{\Omega,A}|1\rangle_{-\Omega,A}|1\rangle_{\Omega,B}|0\rangle_{\Omega,B}$$
$$+ \ |1\rangle_{\Omega,A}|0\rangle_{-\Omega,A}|0\rangle_{\Omega,B}|1\rangle_{\Omega,B}) \ + \ \cdots \tag{12.2.1}$$

where the kets $|n\rangle_{\Omega,J}$ indicate the photon number occupation, n, at frequency Ω, in spatial mode J. The line width of the photons emitted by the down-converter is given by κ whilst χ is the photon pair production rate. Typically, in an experiment we will have $\chi \ll 1$, justifying that we neglect higher order terms in Equation (12.2.1). In reality, any loss will produce a mixed state comprising terms in which the one or the other of the two photons are lost as well as a term where both are present. However, the coincidence detectors will only record events in which both photons make it through the experimental set-up and are successfully detected. As a result the coincidence detectors will select only the pure, two photon part of the state, see Section 13.2, and the analysis need only consider the partial state

$$|\phi\rangle_{PS} = \int d\Omega \frac{2\chi\kappa_a}{\kappa_a^2 + (2\pi\Omega)^2}(|0\rangle_{\Omega,A}|1\rangle_{-\Omega,A}|1\rangle_{\Omega,B}|0\rangle_{\Omega,B}+|1\rangle_{\Omega,A}|0\rangle_{-\Omega,A}|0\rangle_{\Omega,B}|1\rangle_{\Omega,B})$$

Writing a single fourier component of this equation in operator form gives

$$|\phi\rangle_{PS}(\Omega) = \frac{2\chi\kappa_a}{\kappa_a^2 + (2\pi\Omega)^2} \left(e^{i2\pi\Omega\tau}\tilde{\mathbf{A}}^\dagger(\Omega)\tilde{\mathbf{B}}^\dagger(-\Omega) + e^{-i2\pi\Omega\tau}\tilde{\mathbf{A}}^\dagger(-\Omega)\tilde{\mathbf{B}}^\dagger(\Omega) \right) \ |0\rangle$$

A time delay, τ, has been imposed on the $\tilde{\mathbf{A}}$ mode which appears as the $e^{\pm i2\pi\Omega\tau}$ phase terms in Fourier space. Transmission through the beamsplitter transforms the modes in the usual way, see Section 5.1, giving

$$|\phi'\rangle_{PS}(\Omega) \ = \ \frac{2\chi\kappa_a}{\kappa_a^2 + (2\pi\Omega)^2}[\cos(2\pi\Omega\tau)(\tilde{\mathbf{A}}^\dagger(\Omega)\tilde{\mathbf{A}}^\dagger(-\Omega) - \tilde{\mathbf{B}}^\dagger(\Omega)\tilde{\mathbf{B}}^\dagger(-\Omega))$$
$$+ \ i\sin(2\pi\Omega\tau)(\tilde{\mathbf{A}}^\dagger(\Omega)\tilde{\mathbf{B}}^\dagger(-\Omega) - \tilde{\mathbf{B}}^\dagger(\Omega)\tilde{\mathbf{A}}^\dagger(-\Omega))]|0\rangle \tag{12.2.2}$$

Finally the beams pass through the interference filters. In the experiment these had a Gaussian lineshape and a linewidth, σ, much narrower than the natural linewidth of the down converted photons, $\sigma \ll \kappa$. As a result this linewidth completely dominates and we can write the state post-selected by the coincidence detection as

$$|\phi''\rangle_{PS}(\Omega)$$
$$= \frac{\sqrt{2\pi}\chi'}{\sqrt{\sigma}} \exp\left[\frac{-(2\pi\Omega)^2}{2\sigma}\right] i \sin(2\pi\Omega\tau) \left(\tilde{\mathbf{A}}^\dagger(\Omega)\tilde{\mathbf{B}}^\dagger(-\Omega) - \tilde{\mathbf{B}}^\dagger(\Omega)\tilde{\mathbf{A}}^\dagger(-\Omega)\right)|0\rangle$$

Taking the Hermitian conjugate of this equation and integrating over all frequencies gives the expected coincidence rate for the experiment

$$C = \frac{\sqrt{2\pi}\chi'}{\sqrt{\sigma}} \int d\Omega \exp\left[\frac{-(2\pi\Omega)^2}{\sigma}\right] \sin^2(2\pi\Omega\tau)$$
$$= \frac{\chi'}{2}(1 - e^{-\sigma\tau^2}) \tag{12.2.3}$$

As expected the coincidence count rate vanishes for zero time delay ($\tau = 0$), but asymptotes to half the total count rate for large time delays, the result expected for distinguishable particles. The width of the dip reflects the range of times for which the finite width of the photon wave packets lead to partial indistinguishability.

The experimental results obtained by C.K. Hong, Z.Y. Ou and L. Mandel are shown in Fig. 12.5. Plotted are the number of coincidences obtained in a 10 [min] interval as a function of the path length difference. The rate of accidental coincidences has been measured and subtracted from the data. A strong dip in the coincidence rate is observed at a beamsplitter position corresponding to equal path lengths, experimentally demonstrating the destructive interference of the indistinguishable "collision" events of double transmission and double reflection. This is known as the HOM dip. A theoretical trace based on Equation (12.2.3) is shown for comparison.

Since this ground-breaking experiment much work has been carried out into phenomena associated with photon indistinguishability, first by L. Mandel's group (see [Man99]) and later by many others. Notable achievements include the demonstration by the group of Y. Shih et al. who showed that photon indistinguishability, and hence two-photon interference could be produced even in situations where the photons never actually overlapped at a beamsplitter [Pit96] and by J.G. Rarity and P.R. Tapster who showed that a Hong-Ou-Mandel dip could be observed between independently produced photons [Rar97]. The main limit to the visibility in Hong-Ou-Mandel interferometers is the quality of the spatial mode-matching at the beamsplitter. By conducting the experiment in a single mode optical fibre and employing a fibre beamsplitter, near perfect mode matching can be achieved with correspondingly high visibilities; for example 99.4% in the 2003 experiment of T.B. Pittman et al. [Pit03].

In 2001 E. Knill, R. Laflamme and G.J. Milburn [Kni01] made the remarkable discovery that indistinguishability of photons allows massive non-linear interactions between single photons to be conditionally induced through photon number measurement. This allows the possibility of performing quantum computation tasks using photons as qubits as will be discussed in Section 13.1.

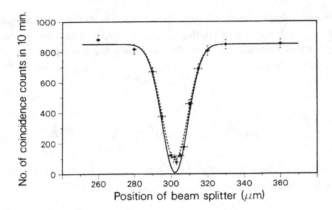

Figure 12.5: Measured number of coincidences in the Hong-Ou-Mandel experiment as a function of the path length difference. From [Hon87].

12.3   Nonlocality

A unique feature of quantum mechanics is its non-local character. Consider two photons in the entangled state

$$|H\rangle_A|H\rangle_B - |V\rangle_A|V\rangle_B \qquad\qquad (12.3.1)$$

where $|K\rangle_J$ indicates that a single photon occupies the Kth polarization mode and the Jth spatial mode. The meaning seems clear: the photons are in a superposition of either both being horizontally polarized or both being vertically polarized. But notice that it is not possible to write down a ket which describes the polarization state of just one of the photons: the ket must describe both photons jointly, hence the term *entangled*.

Now suppose the photons are moved far apart. If this is done carefully enough Equation (12.3.1) will continue to describe their state. But now we have a system described by a single ket whose individual parts may be space-like separated. If a measurement is made on one of the photons and it is found to be, say, horizontally polarized then the state of the other photon immediately also becomes horizontal. It appears that actions on one photon immediately effect its entangled partner seemingly in contradiction with special relativity which require any communications to be limited by the speed of light. Because of this behaviour of entangled systems quantum mechanics is often referred to as a non-local theory. In the following we will discuss quantum optical experiments which have tested these non-local effects.

12.3.1   Einstein-Podolsky-Rosen Paradox

The first major discussion of non-local effects in quantum mechanics is due to A. Einstein and his colleagues B. Podolsky and N. Rosen (EPR) [Ein35]. EPR's aim was in fact to show that quantum mechanics must be flawed. They considered two particles travelling away from each other in such a way that their transverse position was always correlated. The state of the particles can be written in the position basis as

$$|EPR\rangle = \int dx |x\rangle_A |x\rangle_B \qquad (12.3.2)$$

where $|x\rangle_K$ is a position eigenstate of the Kth particle. If a measurement is made of the position of particle A giving the result x' then we immediately know that the position of particle B is also x'. On the other hand we can also express Equation (12.3.2) in the momentum basis as

$$
\begin{aligned}
|EPR\rangle &= \int dp |p\rangle_A \langle p|_A \int dp' |p'\rangle_B \langle p'|_B \int dx |x\rangle_A |x\rangle_B \\
&= \int dp |p\rangle_A \int dp' |p'\rangle_B \int dx e^{i(p+p')x/\hbar} \\
&= \int dp |p\rangle_A |-p\rangle_B \qquad (12.3.3)
\end{aligned}
$$

where $|p\rangle_K$ is a momentum eigenstate of the Kth particle. We see that the momentum is exactly anti-correlated for this state and if a measurement of the momentum of particle A gave the result p'' then we would immediately know that the momentum of particle B was $-p''$.

EPR argued that this situation violated the spirit of the Heisenberg uncertainty relation. They held that nothing an observer did to particle A (if it was space-like separated from B) could effect the values of the position and momentum of particle B. And yet an observer in possession of particle A could choose to determine either the x or p value of B to arbitrary precision whilst only touching particle A. Because the choice of measurement was arbitrary and apparently did not affect the "real" situation of the other particle it seemed that particle B must have had well-defined values for these observables in the first place - contrary to the Heisenberg uncertainty relation [1]. The quantum description implies a connection between the particles even when far apart, what Einstein characterized as "spooky action at a distance". EPR's conclusion was that quantum mechanics was incomplete and that there must be additional "hidden variables" determining the "real" situation of each particle.

The state $|EPR\rangle$ has infinite momentum uncertainty and so is not normalizable and hence is unphysical, however any state for which

$$\Delta x_{\text{inf}}^2 \Delta p_{\text{inf}}^2 < \hbar/4 \qquad (12.3.4)$$

is upheld satisfies the basic EPR argument. Here Δx_{inf}^2 is the variance in the value of the position which can be inferred from measurement of the other particle and similarly Δp_{inf}^2 is the variance in the value of the momentum that can be inferred from measurement of the other particle.

M. Reid and P. Drummond [Rei88] pointed out that such a state can be produced in optics. Consider the Heisenberg evolution of two vacuum modes $\tilde{a}$ and $\tilde{b}$ such that

$$
\begin{aligned}
\tilde{a} &\rightarrow \sqrt{G}\,\tilde{a} + \sqrt{G-1}\,\tilde{b}^\dagger \\
\tilde{b} &\rightarrow \sqrt{G}\,\tilde{b} + \sqrt{G-1}\,\tilde{a}^\dagger \qquad (12.3.5)
\end{aligned}
$$

---

[1] Note that of course both the momentum and position of particle B can not actually be ascertained precisely because only one measurement can actually be made on particle A.

This interaction is produced by parametric amplification (see Section 6.3.2) either directly by a non-degenerate system or alternatively via degenerate parametric amplification squeezing, (see Section 6.3.2) followed by the out of phase mixing of the two modes on a 50:50 beamsplitter. The inferred variance of the amplitude quadrature of beam $\tilde{a}$ conditional on the measurement of the amplitude quadrature of beam $\tilde{b}$ is given by the conditional variance, $V1_{a|b}$ (see Section 4.6.2). From Equations (4.6.9) and (12.3.5) we obtain

$$V1_{a|b} = \frac{1}{2G-1} \tag{12.3.6}$$

Notice that as soon as we introduce some gain, that is $G > 1$, the amplitude conditional variance falls below the QNL. Similarly the inferred value of the phase quadrature of mode $\tilde{a}$ conditional on a measurement of the phase quadrature of mode $\tilde{b}$ is given by the conditional variance, $V2_{a|b}$, where

$$V2_{a|b} = \frac{1}{2G-1} \tag{12.3.7}$$

which also falls below the quantum limit for non-unity gain. The EPR argument suggests that a paradox exists when two non-commuting observables can be inferred non-locally to better precision than the uncertainty principle would predict. This is the case here as

$$V1_{a|b}V2_{a|b} = \left(\frac{1}{2G-1}\right)^2 \tag{12.3.8}$$

which will be less than the normal uncertainty bound of 1 for all non-unit gain. That this is indeed an entangled state can be seen from the state of the beams in the Schrödinger picture

$$|EPR\rangle = \frac{1}{\sqrt{G}}\left(|0\rangle_a|0\rangle_b + \frac{\sqrt{G-1}}{\sqrt{G}}|1\rangle_a|1\rangle_b + \left(\frac{\sqrt{G-1}}{\sqrt{G}}\right)^2|2\rangle_a|2\rangle_b + \cdots\right) \tag{12.3.9}$$

a result, which can be derived using the techniques of Section 6.3.4. The first experimental demonstration of an EPR state was made by the group of H.J. Kimble at Caltech in 1992 [Ou92]. The experimental layout is shown in Fig. 12.6. A crystal of potassium titanyl phosphate (KTP) was placed in a folded ring cavity arrangement and pumped at 540 [nm] by a frequency doubled Nd:YAP laser to form an optical parametric oscillator. This was run sub-threshold to give parametric amplification. The system gave a frequency and spatially degenerate output at 1080 [nm]. However the output was polarization non-degenerate. Such operation is known as type II parametric amplification. In practice it was achieved via milli-Kelvin temperature tuning of the specially cut crystal. The cavity was actively locked to the fundamental laser line at 1080 [nm] via a weak counter propagating field.

A polarizing beamsplitter was used to separate the two polarization modes. The resulting pair of beams was approximately in the EPR state described by Equation (12.3.9) or the transfer functions of Equation (12.3.5). The beams were sent to two homodyne detection systems using tap-offs from the fundamental laser line as local oscillators which could then simultaneously measure the amplitude fluctuations or the phase fluctuations of the two beams. The amplitude conditional variance was measured by feeding the weighted difference of the

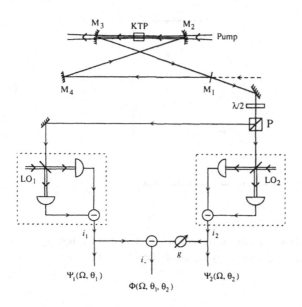

Figure 12.6: Layout of the experiment by Ou et al. [Ou92].

photo-currents from the amplitude locked homodyne systems into the spectrum analyser and finding the variance

$$\langle \, (\tilde{\mathbf{X}}1_a - g1 \, \tilde{\mathbf{X}}1_b)^2 \rangle \tag{12.3.10}$$

where $g1$ is the electronic gain factor in the sum. The conditional variance is then the minimum variance obtained as a function of $g1$. Similarly the phase conditional variance is obtained from the minimum of

$$\langle \, (\tilde{\mathbf{X}}2_a + g2 \, \tilde{\mathbf{X}}2_b)^2 \rangle \tag{12.3.11}$$

A key consideration in the experiment was to limit the total losses in the experiment - in production, propagation and detection of the beams. If the total losses exceed 50% then it is not possible to observe correlations strong enough to satisfy the EPR condition $V1_{a|b}V2_{a|b} < 1$. This can easily be confirmed by introducing beamsplitter losses to the transfer functions of Equation (12.3.5) and calculating the subsequent conditional variances. The overall efficiency of the experiment was measured to be about 63%. The experimental results are shown in Fig. 12.7. A minimum product of $V1_{a|b}V2_{a|b} = 0.7$ is observed, clearly demonstrating the EPR effect.

12.3.2 Generation of entangled CW beams

The effect of entanglement can be observed with continuous beams. Here the detection of quantum correlation is carried out via the suitable combination of the photo currents from

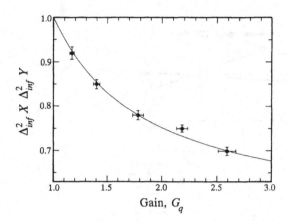

Figure 12.7: Results from the experiment by Z.Y. Ou et al. [Ou92] showing product of conditional variances $V1_{a|b}V2_{a|b}$. Here unity represents the QNL of beam a alone.

both beams, as discussed on page 93. This measurements shows that the information about the quadratures of one of the beams can be used to predict the corresponding quadratures of the other beam better then the QNL.

The technique for generating such entangled beams is to combine two squeezed beams, one with noise suppression in the X1 quadrature, the other in the X2 quadrature. The squeezed beams are generated in independent nonlinear systems pumped by one laser, such as the apparatus described on page 298. The two beams are mode matched, locked in phase and recombined on a beamsplitter and provide a pair of entangled beams.

While we state on page 258 the two individual output beams are not squeezed but rather very noisy, see Fig. 9.18, we find that information about one beam is contained in the other and vice versa. These beams can propagate and be detected separately - and one can be used to predict all properties of the other beam. This is the essence of entanglement. Experiments with pairs of photons and experiments with these pairs of beams are fully equivalent. Such CW EPR correlations were demonstrated in fibre optic systems [Sil01], where two squeezed beams in orthogonal polarizations where produced in the fibre, and from the mixing of the squeezed outputs of two sub-threshold degenerate parametric oscillators [Bow03a] and from nondegenerate above threshold parametric oscillators [Xia02].

In the experiment of W. Bowen et al. [Bow03a] the connection between entanglement and EPR correlations was explored. For pure states entanglement is a necessary and sufficient condition for the observation of EPR correlations. However, for mixed states, entanglement is only necessary, not sufficient. Bowen et al. demonstrated this experimentally by introducing attenuation, and therefore mixing, progressively into their EPR state and showing that the state remained entangled even though the EPR correlations were rapidly lost. The experimental results are shown in Fig. 12.8. The entanglement of the beams is evaluated using the criteria of L. Duan et al. [Dua00] which states that a pair of symmetric Gaussian beams (as in the experiment) will be entangled if and only if [2]

$$I = \frac{\sqrt{\langle(\tilde{X}1_a - \tilde{X}1_b)^2\rangle\langle(\tilde{X}2_a + \tilde{X}2_b)^2\rangle}}{2} < 1 \qquad (12.3.12)$$

The experimental data shows that whilst the product of the conditional variances is above 1 for loss greater than 50%, and thus no longer violates the EPR condition, entanglement, as characterized by $I < 1$, is present for all levels of loss. The maximum observed value for the EPR correlation was $V1_{a|b}V2_{a|b} = 0.58$.

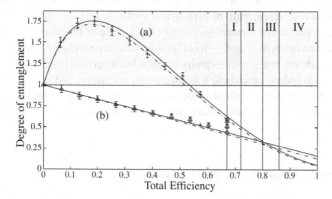

Figure 12.8: Results from the experiment by Bowen et al. [Bow03a] showing (a) the product of conditional variances $V1_{a|b}V2_{a|b}$ and (b) the entanglement criteria I, as a function of loss (symmetric between the beams). The solid and dashed lines are theory fits. Unavoidable losses in the experiment are indicated by vertical bars and were: (I) Detection efficiency, (II) Homodyne efficiency, (III) optical loss and (IV) the OPA escape efficiency.

12.3.3 Bell inequalities

Although of considerable philosophical interest, the EPR paradox did not lead to an experimentally testable difference between quantum mechanics and hidden variable theories. However, in 1965 J.S. Bell derived such a difference [Bel65] based on the properties of entangled 2-level systems [Boh51]. The beauty of Bell's argument was that it produced an inequality that must be obeyed by all local realistic theories of the type envisaged by Einstein, regardless of the actual details of the theories. But, as Bell also showed, entangled quantum systems could violate this inequality.

Consider the generic correlation experiment shown in Fig. 12.9. Correlated beams of particles are emitted from a source (S) in opposite directions, A and B. Two distinct "paths" (p and m) are available to the particles in each beam. These could be different spatial or

---

[2]Note that the form of the Duan et al criteria given here only applies when all the correlations are symmetric and we have further specifically assumed that $\tilde{X}1$ exhibits the maximal correlation and $\tilde{X}2$ exhibits the maximal anti-correlation. See [Dua00] for the more general form.

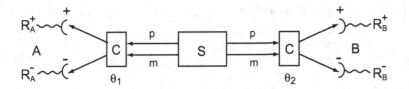

Figure 12.9: Schematic layout of a correlation experiment.

temporal paths, or orthogonal polarizations (or spins). The two paths are combined and then spatially separated to form a different pair of orthogonal paths + and −. The combiners, $C(\theta)$, are black boxes such that it is not possible, for a general value of the mixing parameter θ, to determine from measurements of + and − whether a particular particle took path p or m. Measurements are then made on the + and − paths of each beam giving results $R^+(\theta)$ and $R^-(\theta)$ respectively. In the standard case these measurements are simply the presence (1) or absence (0) of a particle in a particular path in some time interval. More generally they may represent the count rate of particles in a particular path. We can form correlation functions of the following form

$$R^{ij}(\theta_A, \theta_B) \;=\; R^i_A(\theta_A) R^j_B(\theta_B) \tag{12.3.13}$$

where $i, j = +, -$. We then construct the normalized averages

$$P^{ij}(\theta_A, \theta_B) \;=\; \frac{\langle R^{ij}(\theta_A, \theta_B)\rangle}{\sum_{k,l=\pm}\langle R^{kl}(\theta_A, \theta_B)\rangle} \tag{12.3.14}$$

It can then be shown [Bel71] that provided the P's have the form of probabilities (bounded between 0 and 1) then in any local realistic description the correlations will be bounded by the following Bell inequality

$$B = |E(\theta_A, \theta_B) + E(\theta'_A, \theta'_B) + E(\theta'_A, \theta_B) - E(\theta_A, \theta'_B)| \le 2 \tag{12.3.15}$$

where

$$E(\theta_A, \theta_B) = P^{++}(\theta_A, \theta_B) + P^{--}(\theta_A, \theta_B) - P^{+-}(\theta_A, \theta_B) - P^{-+}(\theta_A, \theta_B) \tag{12.3.16}$$

Optics offered a unique opportunity for testing the inequality of Equation (12.3.15) and hence this fundamental property of quantum mechanic. Entangled photons can be produced, widely spatially separated, and measured in the required way with relative ease compared to other quantum systems.

Consider the standard optical example of a state which violates the Bell inequality. Such a state is the number-polarization entangled state

$$|0\rangle + \frac{\chi}{\sqrt{2}}(|H\rangle_A|H\rangle_B - |V\rangle_A|V\rangle_B) \tag{12.3.17}$$

which is approximately produced by a parametric down converter operating at low conversion efficiency ($\chi \ll 1$) as will be described shortly. The requirement of low conversion efficiency is so that higher photon number terms (which appear as products of higher powers of χ) can be neglected. In the following it will be more convenient to work in the Heisenberg picture. In this picture the action of the down converter is to evolve vacuum state, input annihilation operators $C_{h,v}$ and $D_{h,v}$ according to

$$\hat{A}_{h,v} = \hat{C}_{h,v} + \chi \hat{D}^{\dagger}_{h,v}, \quad \hat{B}_{h,v} = \hat{D}_{h,v} + \chi \hat{C}^{\dagger}_{h,v} \qquad (12.3.18)$$

where as before $\chi \ll 1$ has been assumed. Our two paths, p and m, in this example are the horizontal ($\hat{A}_h$ and $\hat{B}_h$) and vertical ($\hat{A}_v$ and $\hat{B}_v$) polarization modes. The mixer C is then some combination of polarizing optics which decomposes our beams into a different, orthogonal polarization basis set ($+$ and $-$). This corresponds to the transformation

$$\hat{A}_+(\theta_A) = \cos\theta_A \hat{A}_h + \sin\theta_A \hat{A}_v, \qquad \hat{A}_-(\theta_A) = \cos\theta_A \hat{A}_v - \sin\theta_A \hat{A}_h$$
$$\hat{B}_+(\theta_B) = \cos\theta_B \hat{B}_h + \sin\theta_B \hat{B}_v, \qquad \hat{B}_-(\theta_B) = \cos\theta_B \hat{B}_v - \sin\theta_B \hat{B}_h$$

$$(12.3.19)$$

Photon counting is then performed on the beams and we define

$$\begin{aligned} R^i_A(\theta_A) &= \hat{A}^{\dagger}_i(\theta_A)\hat{A}_i(\theta_A) \\ R^i_B(\theta_B) &= \hat{B}^{\dagger}_i(\theta_B)\hat{B}_i(\theta_B) \end{aligned} \qquad (12.3.20)$$

with $i = +, -$. The definitions then follow as per Eqs. (12.3.13), and (12.3.14). An explicit calculation then gives the result

$$E(\theta_A, \theta_B) = \cos 2(\theta_A - \theta_B) \qquad (12.3.21)$$

where terms of order higher than χ^2 have been neglected. Choosing the angles $\theta_A = 3\pi/8$, $\theta'_A = \pi/8$, $\theta_B = \pi/4$, $\theta'_B = 0$ we find $B = 2\sqrt{2}$, a clear violation of Eq. (12.3.15).

The first experimental demonstrations of Bell inequality violations were actually achieved using the polarization correlated photons produced by atomic cascades. Using the $4p^2\,{}^1S_0 - 4s4p\,{}^1P_1 - 4s^2\,{}^1S_0$ cascade in calcium to produce polarization entangled visible photons, A. Aspect, P. Grangier & G. Roger [Asp81] performed the first complete test in 1981. Their experimental set-up is shown in Fig. 12.10. Two photon excitation via a Krypton ion laser at 406.7 [nm] and a Dye laser at 581 [nm] was used to selectively populate the upper state. Large aperture aspheric lenses were used to isolate and collimate beams which were sent to the analysers spatially separated by up to 6.5 [m]. The analysers consisted of plate polarizers followed by photomultipliers connected to coincidence counting electronics with a 19 [ns] coincidence window around zero time delay (see Section 7.3). The coincidence counting selects those events for which one of the photons from the entangled pair travelled to and was successfully counted by one detector, whilst the other photon of the pair similarly travelled to and was successfully detected by the other detector. The post-selected state is approximately the maximally entangled state $|H\rangle_1|H\rangle_2 - |V\rangle_1|V\rangle_2$.

The results of the experiment by Aspect et al. [Asp81] are shown in Fig. 12.11. Here the normalized coincidence rates are plotted as a function of the angle difference, ϕ, between the polarizer settings at the two analysers. The counts were found to only depend on ϕ and not on the absolute value of the settings. Under such conditions a fringe visibility in excess of $1/\sqrt{2}$ is sufficient to ensure violation of the Bell inequality (Equation (12.3.15)). The observed visibility is about 0.9 resulting in a value of $B = 2.465 \pm 0.035$ violating the inequality by more than 13 standard deviations.

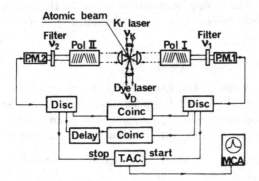

Figure 12.10: Schematic diagram of the apparatus and electronics of first Bell inequality test. The laser beams are focussed onto the atomic beam perpendicular to the figure. Feedback loops from the fluorescence signal control the krypton laser power and the dye-laser wavelength. The output of discriminators feed counters (not shown) and coincidence circuits. The multichannel analyser (MCA) displays the time-delay spectrum. After A. Aspect et al. [Asp81].

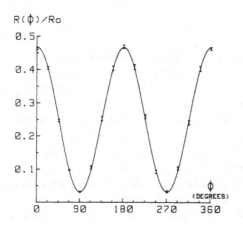

Figure 12.11: Normalized coincidence rate as a function of the relative polarizer orientation. Indicated errors are $\pm$ 1 standard deviation. The solid curve is not a fit to the data but the prediction of quantum mechanics. After A. Aspect et al. [Asp81].

The use of parametric down-conversion to produce entangled states, starting with L. Mandel's group in the late eighties [Ou88] significantly reduced the complexity of the source and increased its brightness. For example in the P. Kwiat et al experiment of 1995 [Kwi95b] 1500 coincidences per second were observed, an order of magnitude brighter than observed by Aspect et al.. A Bell violation of over a 100 standard deviations was observed in this experiment.

Novel optical arrangements are required to directly produce the polarization entangled state of Equation (12.3.17) from parametric down conversion. In the last section we discussed a type II parametric oscillator. When operated at low conversion efficiency, without a cavity, a type II crystal will produce orthogonally polarized photon pairs which radiate into conical spatial modes, which appear as rings in cross-section. Momentum conservation requires that if a photon is detected in a particular position on one ring, its paired photon will appear in the mirror image position on the other ring. By placing a pair of apertures at the appropriate positions on the rings the spatial modes thus isolated will occasionally contain a pair of photons in the state $|H\rangle_A|V\rangle_B$. They are not entangled in the polarization degree of freedom. However, if the angle between the pump beam and the optical axis of the crystal is increased, the rings will intersect. If apertures are placed at the intersection points, it is possible to obtain the pair $|H\rangle_A|V\rangle_B$ *or* the pair $|V\rangle_A|H\rangle_B$. If the walk-off due to birefringence in the crystal is compensated then these two possible pairs become only distinguishable by polarization and the effective state of the spatial modes is

$$|0\rangle + \frac{\chi}{\sqrt{2}}(|H\rangle_A|V\rangle_B + e^{i\phi}|V\rangle_A|H\rangle_B) \tag{12.3.22}$$

which is a polarization entangled state like that of Equation (12.3.17) [Kwi95b].

Seeing Bell violations over significant distances has been achieved by sending the entangled photons through optical fibre links. Preserving polarization under such conditions is difficult and so a temporal form of the Bell inequality, first derived by J. Franson [Fra89] can be used. Using this approach W. Tittel et al. were able to demonstrate a Bell violation of 16 standard deviations over a distance of almost 11 [km] [Tit98]. The experiments discussed here, plus many others, have convinced most scientists that hidden variable theories of the general type suggested by EPR are not consistent with observation.

12.4 Summary

In this chapter we have looked at quantum optical techniques and experiments which test fundamental principles of quantum mechanics. We have seen that quantum optics has been able to quantitatively test many of the gedanken experiments proposed in the early days of the formulation of quantum theory. In all cases the results have strongly supported the quantum mechanical predictions.

Bibliography

[Asp81]   Experimental tests of realistic local theories via Bell's theorem, A. Aspect, P. Grangier, G. Roger, Phys. Rev. Lett. 47, 460 (1981)

[Bel65]   On the Einstein-Podolsky-Rosen paradox, J.S. Bell, Physics 1, 195 (1965)

[Bel71]   *Foundations of Quantum Mechanics*, J.S. Bell, Ed. B. d'Espagnat, (New York: Academic) 171 (1971).

[Boh51]   *Quantum Theory*, D. Bohm, Prentice-Hall (1951).

[Boh28]   Complementarity: the quantum postulate and the recent development of atomic theory, N. Bohr, Nature 121, 580 (1928)

[Bow03a] Experimental investigation of criteria for continuous variable entanglement, W.P. Bowen, R. Schnabel, P.K. Lam, T.C. Ralph, Phys. Rev. Lett. 90, 043601 (2003)

[DeB23]   Waves and quanta, L.de Broglie, Nature (London) 112, 540 (1923)

[Dua00]   Inseparability Criterion for Continuous Variable Systems, Lu-Ming Duan, G. Giedke, J. I. Cirac, and P. Zoller, Phys. Rev. Lett. 84, 2722 (2000)

[Dur98]   Origin of quantum-mechanical complementarity probed by a 'which-way' experiment in an atom interferometer, S. Durr, T. Nonn, G. Rempe, Nature 395, 33 (1998)

[Ein05]   Über einen die Erzeugung und Verwandlung des Lichtes betreffenden heuristischen Gesichtspunkt, A. Einstein, Annalen der Physik 17,132 (1905)

[Ein35]   Can quantum-mechanical description of physical reality be considered complete?, A. Einstein, B. Podolsky, N. Rosen, Phys. Rev. 47, 777 (1935)

[Eli93]   Quantum mechanical interaction-free measurements, A. Elitzur, L. Vaidman, Found. Phys. 23, 987 (1993)

[Eng96]   Fringe visibility and which-way information: an Inequality, B-G. Englert, Phys. Rev. Lett. 77, 2154 (1996)

[Fey63]   *The Feynman Lectures on Physics*, vol. 3, R. Feynman, R.B. Leighton, M. Sands, Addison and Wesley (1963)

[Fra89]   Bell inequality for position and time, J.D. Franson, Phys. Rev. Lett. 62, 2205 (1989)

[Hon87]   Measurement of subpicosecond time intervals between two photons by interference, C.K. Hong, Z.Y. Ou, L. Mandel, Phys. Rev. Lett. 59, 2044 (1987)

[Kni01]   A scheme for efficient quantum computation with linear optics, E. Knill, L. Laflamme, G.J. Milburn, Nature, 409, 46 (2001)

[Kwi95a] Interaction-Free Measurement, P.G. Kwiat, H. Weinfurter, T. Herzog, A. Zeilinger, M.A. Kasevich, Phys. Rev. Lett. 74, 4763 (1995)

[Kwi95b] New high-intensity source of polarization-entangled photon pairs, P.G. Kwiat, K. Mattle, H. Weinfurter, A. Zeilinger, A.V. Sergienko, Y. Shih, Phys. Rev. Lett. 75, 4337 (1995)

[Man99]   Quantum effects in photon interference, L. Mandel, Review of Modern Physics, 71, S274 (1999).

[Mar92]   Stochastic interferometer, F. De Martini, L. De Dominicis, V. Cioccolanti, G. Milani, Phys. Rev. A 45, 5144 (1992)

[Ou88]   Violation of Bell's inequality and classical probability in a two photon correlation experiment, Z.Y. Ou and L. Mandel, Phys. Rev. Lett. 61, 50 (1988)

[Ou92]   Realization of the Einstein-Podolsky-Rosen paradox for continuous variables, Z.Y. Ou, S.F. Pereira, H.J. Kimble, K.C. Peng, Phys. Rev. Lett. 68, 3663 (1992)

[Pit96]   Can two-photon interference be considered the interference of two photons?, T.B. Pittman, D.V. Stekalov, A. Migdall, M.H. Rubin, A.V. Sergienko, Y.H. Shih, Phys. Rev. Lett. 77, 1917 (1996)

[Pit03]   Violation of Bell's inequality with photons from independent sources, T.B. Pittman and J.D. Franson, Phys. Rev. Lett. 90, 240401 1 (2003)

[Rar97]   Quantum interference: experiments and applications, J.G. Rarity, P.R. Tapster, Philos. Trans. R. Soc. London A 355, 2267 (1997)

[Rei88]   Quantum correlations of phase in nondegenerate parametric oscillator, M.D. Reid, P.D. Drummond, Phys. Rev. Lett. 60, 2731 (1988)

[Sch99]   Quantitative wave-particle duality and nonerasing quantum erasure, P.D.D. Schwindt, P.G. Kwiat, B-G. Englert, Phys. Rev. A 60, 4285 (1999)

[Scu91]   Quantum optical tests of complementarity, M.O. Scully, B-G. Englert, H. Walther, Nature 351, 111 (1991)

[Sil01]   Generation of continuous variable Einstein-Podolsky-Rosen Entanglement via the Kerr nonlinearity in an optical fiber, C. Silberhorn, P.K. Lam, O. Weisl, F. König, N. Korolkova, G. Leuchs, Phys. Rev. Lett. 86, 4267 (2001)

[Sum87]   Stochastic and deterministic absorption in neutron-interference experiments, J. Summhammer, H. Rauch, D. Tuppinger, Phys. Rev. A 36, 4447 (1987)

[Tit98]   Violation of Bell inequalities by photons more than 10 km apart, W. Tittel, J. Brendel, H. Zbinden, N. Gisin, Phys. Rev. Lett. 81, 3563 (1998)

[Wis95]   Uncertainty over complementarity, H. Wiseman, F. Harrison, Nature 377, 584 (1995)

[Woo79]   Complementarity in the double slit experiment: Quantum nonseparability and the quantitative statement of Bohr's principle, W.K. Wooters, W.H. Zurek, Phys. Rev. D 19, 473 (1979)

[Xia02]   Quantum dense coding exploiting a bright Einstein-Podolsky-Rosen beam, X. Li, Q. Pan, J. Jing, J. Zhang, C. Xie, K. Peng, Phys. Rev. Lett. 88, 047904 (2002)

13 Quantum Information

We discussed in Section 10.1 communicating information using light. We saw that the quantum nature of light leads to intrinsic limits on the amount of information that can be sent optically. We looked at experiments using squeezed light which could reduce those limits as much as possible. In those discussions we imagined encoding the information on the light in much the same way as one would if sending information via a classical medium.

We now consider a quite different idea: that of encoding information in ways only possible for quantum systems. Information encoded in this way is referred to as *Quantum Information*. From this point of view we will find communication and computation tasks that, rather than being limited by quantum mechanics, are actually enhanced by it.

The technical challenges in working with quantum information are formidable. Never-the-less significant experimental demonstrations have been made, a number of which we will discuss here. As we shall see quantum optics has some considerable advantages (and disadvantages) over alternative quantum systems from a quantum information point of view.

13.1 Photons as qubits

In Chapter 3 we introduced the idea of a single, polarized photon. For example we described a light beam as being in the state $|H\rangle$ if in some time interval there is unit probability that one and only one photon will be detected at the horizontal output of a horizontal/vertical polarizing beamsplitter placed in the beam path. There is zero probability a photon will be found at the vertical output in the same time interval. A beam in the state $|V\rangle$ will conversely only be found at the vertical output. It is clear that such light states could be used to carry *bits* of information. For example, we could assign the value "zero" to the $|H\rangle$ state and "one" to the $|V\rangle$ state. A string of horizontal and vertically polarized photons could then faithfully represent an arbitrary bit string.

However, being quantum objects, photons offer more possible manipulations than classical carriers of bits. In particular not only can we have zero's and one's, but we can also have superpositions of zeros and ones such as the diagonal state $|D\rangle = 1/\sqrt{2}(|H\rangle + |V\rangle)$. Indeed bits can just as effectively be encoded in such superposition states, for example using $|D\rangle$ as a zero and $|A\rangle = 1/\sqrt{2}(|H\rangle - |V\rangle)$ as a one. Because of these extra degrees of freedom we refer to information digitally encoded on quantum systems (such as photons) as quantum bits or *qubits*.

One non-classical feature of encoding in this way is the fact that different bases do not in general commute. Thus simultaneous, ideal measurements in both bases cannot be made. Furthermore any measurements which obtain any information about the bit values of one

A Guide to Experiments in Quantum Optics, 2nd Edition. Hans-A. Bachor and Timothy C. Ralph
Copyright © 2004 Wiley-VCH Verlag GmbH & Co. KGaA
ISBN: 3-527-40393-0

basis inevitably disturbs the bit values of the other basis. These features can be used to create a secure communication channel via the technique of *Quantum Key Distribution* (also referred to as quantum cryptography). A number of demonstrations of quantum key distribution have been made in optics.

Another feature of qubits is their ability to span all different bit values simultaneously. This is obviously true of a single qubit where the $|D\rangle$ state, when viewed in the horizontal/vertical basis, equally spans the two different bit values (i.e. $H = 0$ and $V = 1$). This continues to be true for multi-qubit states. For example if we start with two qubits in the state

$$|H\rangle|H\rangle \tag{13.1.1}$$

and we rotate both their polarizations by 45 degrees we end up with the state

$$|H\rangle|H\rangle + |H\rangle|V\rangle + |V\rangle|H\rangle + |V\rangle|V\rangle \tag{13.1.2}$$

which is an equal superposition of all four possible two bit values. This generalizes to n qubits where the same operation of rotating every individual qubit leads to an equal superposition of all 2^n bit values. The single qubit rotations required here are just polarization rotations and can thus be performed easily using wave plates as described in Section 5.4.2.

Although this ability to span all possible inputs simultaneously hints at the possibility of increased communication or computation power using qubits, it is not the whole story. Note in particular that analogues of the sort of superpositions represented by Equation (13.1.2) can also be created in classical optical systems as superpositions of classical waves. In order to unlock the full power of quantum information we need to create entangled states. We have seen examples of entangled states before, see Section 12.3, and observed that they have no classical analogue. In the following we will see that entanglement can be used as a resource to perform information tasks more efficiently than possible classically and discuss experimental demonstrations of these tasks using entangled optical states.

Finally we could consider performing computations using qubits instead of classical bits. To do this we need to introduce *quantum gates*. Some of these will have classical counterparts, for example the NOT gate takes $|H\rangle$ to $|V\rangle$ and vice versa. On the other hand some gates will have no classical analogue, such as the Hadamard gate which takes $|H\rangle$ to $1/\sqrt{2}(|H\rangle + |V\rangle)$ and $|V\rangle$ to $1/\sqrt{2}(|H\rangle - |V\rangle)$. As we have noted these types of operations can easily be achieved in optics using wave-plates. However, we also required two qubit gates such as the control-NOT (CNOT) which preforms the NOT operation on one qubit (the target) only if the other qubit (the control) has "one" as its logical value. Such two qubit gates can produce entanglement. Implementing two qubit gates in optics is much more difficult. We will discuss experimental efforts in this direction. Eventually, if large arrays of gate operations can be implemented efficiently, one could consider performing *Quantum Computation*. Algorithms such as Shor's algorithm [Sho94] show that for certain problems an exponential speed-up over the best known classical programs is in principle possible with a quantum computer leading to considerable interest. However, the realization of quantum computation experimentally (based on quantum optics or any other quantum platform) still remains a long way off.

13.2 Postselection and coincidence counting

Producing and detecting single photon states efficiently presents a major technological challenge. In Section 7.2 we saw that current commercial photon counting detectors are "off/on" type detectors which only "click" when there are one or more photons incident in a particular time window. Typical maximum detection efficiencies are around 70%.

Light sources which produce very low photon numbers tend to be *spontaneous sources*, that is they produce photons randomly. In the next section we will examine experimental progress towards creating light sources which produce single photons on demand. Here we will discuss how in principle experimental demonstrations can be performed using on/off detectors and spontaneous sources.

As discussed in Sections 3.4 and 12.1 single photon experiments can be performed by strongly attenuating a laser source. We can represent the state of such a laser source by the state (see Section 4.2.2)

$$|\psi\rangle = |0\rangle + \alpha|1\rangle + \frac{\alpha^2}{2!}|2\rangle + \cdots \tag{13.2.1}$$

where we are ignoring frequency dependence and normalization factors for the present. As we attenuate the source more and more, α becomes much less than one and we can write to a good approximation

$$|\psi\rangle = |0\rangle + \alpha|1\rangle \tag{13.2.2}$$

We now have a source which in any particular time interval (length determined by the frequency dependence) has a high probability of producing vacuum; some small probability of producing a single photon state and; a negligible probability of producing a multi-photon state. If a photon counter is placed at the end of the experiment and we only worry about those times when the detector "clicks" then we will *postselect* just the single photon part of the state. If the source is polarized then it can be manipulated as a qubit. Notice, however, that it is a rather inconvenient qubit source as it rarely works and you only know it worked after the fact, by evaluating at the detection record. Never-the-less this type of source has successfully been used to demonstrate single qubit type experiments such as quantum key distribution (see Section 13.1).

A major problem arises with an attenuated coherent source if we try to move to experiments requiring two qubits. One might assume we could use two highly attenuated coherent sources and then postselect only those events where two photons appear at the end of the experiment. However, the joint state of two equal power, attenuated lasers is

$$\begin{aligned}
|\psi\rangle_{ab} &= \left(|0\rangle + \alpha|1\rangle + \frac{\alpha^2}{2!}|2\rangle + \cdots\right)_a \left(|0\rangle + \alpha|1\rangle + \frac{\alpha^2}{2!}|2\rangle + \cdots\right)_b \\
&= |0\rangle_a|0\rangle_b + \alpha\left(|1\rangle_a|0\rangle_b + |0\rangle_a|1\rangle_b\right) + \\
&\quad \frac{\alpha^2}{2!}\left(2!|1\rangle_a|1\rangle_b + |2\rangle_a|0\rangle_b + |0\rangle_a|2\rangle_b\right) + \cdots
\end{aligned} \tag{13.2.3}$$

where the first ket refers to one source whilst the second one to the other source. Notice that if we go to order α^2 then there is indeed a term with a single photon state in each beam.

However, terms involving pairs of photons in one beam with vacuum in the other occur with the same probability. Postselecting on two photon events will not in general remove these terms. Hence it is not possible in general to perform two qubit experiments using highly attenuated laser sources. A more sophisticated solution is required.

As we saw in Section 12.2, since the late eighties, the solution of choice has been parametric down conversion. As was discussed in Section 6.3.4 weak parametric amplification results in the spontaneous creation of pairs of photons. If the down conversion is spatially non-degenerate then, in the Schrödinger picture, initial vacuum inputs are transformed according to

$$|0\rangle_a|0\rangle_b \rightarrow \frac{1}{\sqrt{G}} \left(|0\rangle_a|0\rangle_b + \chi'|1\rangle_a|1\rangle_b + \chi'^2|2\rangle_a|2\rangle_b + \cdots \right) \tag{13.2.4}$$

where χ' is an effective interaction strength, related to the parameters in the discussion of Section 6.3.4 by $\chi' = \frac{\chi}{\kappa}$. If we now allow χ' to be very small, which is not hard to arrange experimentally, then the state produced is given to an excellent approximation by

$$|\psi\rangle_{ab} = |0\rangle_a|0\rangle_b + \chi'|1\rangle_a|1\rangle_b \tag{13.2.5}$$

In contrast to Eq. (13.2.3) the state in Eq. (13.2.5) has only the desired two photon term to first order in χ'. If we postselect only those events from the detection record in which 2 photons are detected "simultaneously" or in coincidence (within some preset time window) then we will only record the part of the state which is due to the pairs of photons. Thus by using the combination of parametric down-conversion, the polarization degree of freedom and postselection, we can perform, at least in principle, 2 qubit experiments. Experiments carried out this way are sometimes referred to as coincidence basis experiments. In Sections 13.6 and 13.7 we will see several examples of such experiments. Note though that this source is still spontaneous, i.e. successful events are rare, random and we do not know if they have occurred until after the fact. Progress in producing sources without these drawbacks are discussed in the next section.

13.3   True single photon sources

We now discuss two distinct approaches to producing better approximations to single photon states. The first is to create a *heralded* single photon source. That is a source which, though not always producing a single photon state, produces a clear signal when successful. The second approach is to produce an *on-demand* source, which deterministically produces a single photon state when requested.

13.3.1   Heralded single photons

A heralded single photon source can be created using spatially non-degenerate down-conversion by detecting one of the output modes and only accepting the other output if a photon is detected. From an idealistic point of view the conditional state when a single photon is detected in mode a can be obtained from Equation (13.2.5) as

$$\langle 1|_a|\psi\rangle_{ab} = \chi'|1\rangle_b \tag{13.3.1}$$

indicating that a single photon state is created in mode b with probability $|\chi'|^2$. In reality things are not so simple.

The output state of the down-converter is more realistically described by

$$|\psi\rangle_{ab} = |0\rangle_a|0\rangle_b + \chi' \int dk_a dk_b F(k_a, k_b)|1\rangle_a|1\rangle_b \tag{13.3.2}$$

where k_i is the wave vector of the ith beam and the function $F(k_a, k_b)$ describes the spatio-temporal structure of the modes. The photon counter selects an ensemble of distinguishable single photon modes which can be described by the mixed state

$$\tilde{\rho}_a = \int dk_a T(k_a)|1\rangle_a\langle 1|_a \tag{13.3.3}$$

where $T(k_a)$ is the spatio-temporal distribution of the detected ensemble. The output state, conditional on a photon count, is then

$$\tilde{\rho}_b = \text{Tr}_a\left[|\psi\rangle_{ab}\langle\psi|_{ab}\rho_a\right] \tag{13.3.4}$$

In general $\tilde{\rho}_b$ is a mixed state, however if $T(k_a)$ is centred on but much "narrower" (both spatially and temporally) than $F(k_a, k_b)$, then to a good approximation the pure single photon number state

$$|\psi\rangle_b = \chi' \int dk_b T(k_b)|1\rangle_b \tag{13.3.5}$$

is produced. The above technique was demonstrated conclusively by A. Lvovsky et al. [Lvo01] at Universität Konstanz, by performing homodyne tomography on the conditionally produced photon states. A beta-barium borate crystal was used in a type I arrangement to produce frequency degenerate but spatially non-degenerate photon pairs. Transform limited pulses at 790 [nm] from a Ti:Sapphire laser were doubled and used to pump the crystal. Great care was taken to minimize any distortions of the spatial and temporal modes of the outputs due to walk-off. The trigger photons were passed through a spatial filter and a 0.3 [nm] frequency filter before being counted on a single photon detector.

A key issue in analysing the conditional photon state is that the signal detector only "looks" at the single photon mode, otherwise spurious vacuum modes will also couple to the analyser. This means that the local oscillator pulse (LO) used in the homodyne detection of the signal must be accurately mode-matched to the single photon state. Mode matching with a visibility of about 80% was achieved.

The homodyne data collected was then used to produce the Wigner distribution of the single photon state (see Sections 4.3.2 and 9.13). Normalization of the distribution was achieved by simultaneously collecting vacuum state data. The resulting Wigner function is plotted in Figure 13.1. The plot is consistent with a mixed state comprised of 55% single photon state and 45% vacuum. The source of the vacuum contribution can be identified as the non-unit LO mode-matching (an effective 35% loss) and loss due to non-unit homodyne efficiency and signal transmission of about 10%. The most salient feature of the plot is the negativity close to the origin. This demonstrates the strongly quantum mechanical nature of the detected single photon state.

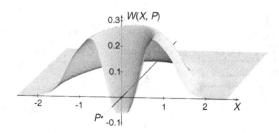

Figure 13.1: Reconstructed Wigner function from experiment of Lvovsky et al. [Lvo01] showing negative values close to the origin.

The experiment by Lvovsky et al. clearly shows that a single photon state can in principle be produced in this manner but it also highlights problems with this approach. For example, in order to obtain a conditional state which is as pure as possible, strong attenuation was applied to the trigger photon resulting in low single photon state production rates of about one photon every 4 seconds. Also, mode matching of the single photon state is seen to be a difficult problem. A more promising and practical solution would be to mode-match the single photon state into an optical fibre for subsequent use down-stream. The best results to date for this difficult problem were achieved by C. Kurtsiefer, M. Oberparleiter, and H. Weinfurter [Kur01] who obtained about 40% single photon contribution to the conditional state,

Finally, it is clearly inconvenient that the photons in these experiments are produced at random times. One solution to this problem, proposed and demonstrated in principle by T. Pittman, B. Jacobs, and J. Franson [Pit02], is to inject the single photon state into an optical fibre storage loop when the trigger photon is detected. The captured photon is then held there till required, when it is switched out of the loop. If the round trip time of the loop is matched to the repetition rate of the pulsed pump laser then a number of loops can be loaded and then released simultaneously to produce several single photons states at the same time. Currently (as well as the mode matching problem discussed above) the losses associated with the Pockell cell used to switch the photons in and out of the loop are prohibitively large.

13.3.2   Single photons on demand

Approximate single photon states can be created on demand by generating light from a single isolated emitter such as a single ion or atom. The trick here is that a single emitter can only produce a single photon "at a time" with some dead time between emissions while the source is re-excited. The effect is that the output state can be written (in an idealized fashion) as

$$|\psi\rangle = |0\rangle + \alpha|1\rangle + \tau \left(\frac{\alpha^2}{2!}|2\rangle + \cdots \right) \qquad (13.3.6)$$

where τ is a number between 0 and 1 representing the suppression of higher photon number terms. If τ is very small then α can be made large, such that there is a high probability that a single photon will be emitted, whilst the probability of multiple photon emission remains very low.

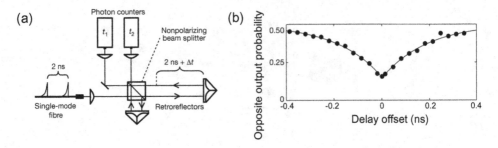

Figure 13.2: Set-up and experimental results of HOM dip due to interference of consecutive photons from the quantum dot emitter of Santori et al. [San02].

The first experiments of this kind were performed in the late seventies (as discussed in Section 3.6). However, although they clearly displayed the photon anti-bunching expected of a single photon source, they were very inefficient because they radiated into 4π steradians and being based on atomic beams there was little that could be done to improve matters. More recently various attempts have been made to create more efficient single emitters. These included: placing single neutral atoms or ions into high finesse optical cavities [Kuh02, McK03] such that the photon emission should be into a single Gaussian mode; close coupling to single solid-state emitters such as neutral vacancy (NV) centre in diamond [Bev02] and; the construction of single quantum-dot emitters integrated into distributed Bragg reflector (DBR) cavities [San02].

Perhaps the most promising of these are the quantum dot emitters. In the initial experiment of Santorini et al. [San02] at Stanford University self-assembled InAs quantum dots embedded in GaAs were sandwiched between DBR mirrors to form tiny, high finesse, monolithic cavities in the form of 5 micron high pillars. These were then cooled to 3–7 [K] and pumped by a pulsed Ti:Sapphire laser. The quantum dot emission, at around 935 [nm], was spectrally filtered (0.1 [nm]) and a single polarization was selected before being coupled into single-mode optical fibre. The efficiency of single photon production was estimated to be about 30%.

The single-photon states thus produced were characterized by their $g^{(2)}$ factor (see Section 3.6) which was typically of the order of 0.06 ($\tau^2 \approx g^{(2)}$) showing good suppression of two photon emission. To test the indistinguishability of the photon states the Hong-Ou-Mandel dip, see Section 12.2, between consecutive emissions was measured. This was achieved by forming a Michelson type interferometer with a path length difference of $\tau + \Delta t$ where $\tau = 2$ [ns] is the time interval between consecutive pulses. The set-up is shown in Fig. 13.2(a) and the resulting dip is shown in Fig. 13.2(b). The inferred visibility of the dip when measurement imperfections are taken into account is about 70%.

The idea of a push button source of single photons is alluring but many challenges remain. In particular: increasing the efficiency of photon emission (a mode matching problem both of the quantum dot emission to the DBR cavity and the emission of the cavity to an optical fibre) and; producing strong photon indistinguishability over many pulses and perhaps even between independent emitters.

13.4 Characterizing photonic qubits

We now return to the post selected domain and ask the question, given that a photon is detected, what is its state? Of course this question cannot be answered for a single event because of the probabilistic interpretation of quantum states. But given a large ensemble of detection events, corresponding to identically prepared photons, a recipe can be given for determining the state of the ensemble.

We consider polarization encoded qubits. The polarization state of the photons is most generally described by the density operator $\hat{\rho}$ (see Section 4.2.3). Our observables are the Stokes operators (corresponding to the classical Stokes parameters [Sto52])

$$
\begin{aligned}
\hat{S}_1 &= \hat{n}_H - \hat{n}_V = |H\rangle\langle H| - |V\rangle\langle V| \\
\hat{S}_2 &= \hat{n}_D - \hat{n}_A = |H\rangle\langle V| + |V\rangle\langle H| \\
\hat{S}_3 &= \hat{n}_R - \hat{n}_L = i(|V\rangle\langle H| - |H\rangle\langle V|)
\end{aligned}
\tag{13.4.1}
$$

where $\hat{n}_J$ is the number operator for the Jth polarization mode. The eigenstates of $\hat{S}_1$ are $|H\rangle$ and $|V\rangle$ with eigenvalues $+1$ and -1 respectively. Similarly the eigenstates of $\hat{S}_2$ are $|D\rangle$ and $|A\rangle$ and of $\hat{S}_3$ are $|R\rangle$ and $|L\rangle$. The expectation values of the Stokes operators are related to measurement by

$$
\begin{aligned}
\langle \hat{S}_1 \rangle &= \frac{2R_H}{R_H + R_V} - 1 \\
\langle \hat{S}_2 \rangle &= \frac{2R_D}{R_H + R_V} - 1 \\
\langle \hat{S}_3 \rangle &= \frac{2R_R}{R_H + R_V} - 1
\end{aligned}
\tag{13.4.2}
$$

where R_H and R_V are the count rates recorded at the H and V output ports respectively of a horizontal/vertical polarizing beamsplitter and similarly for the diagonal/anti-diagonal and right/left bases (see Section 3.5.1).

On the other hand we can also express the expectation values in terms of the density operator as

$$
\begin{aligned}
\langle \hat{S}_1 \rangle &= \mathrm{Tr}[\hat{\rho}\hat{S}_1] = \rho_{h,h} - \rho_{v,v} \\
\langle \hat{S}_2 \rangle &= \mathrm{Tr}[\hat{\rho}\hat{S}_2] = \rho_{h,v} + \rho_{v,h} \\
\langle \hat{S}_3 \rangle &= \mathrm{Tr}[\hat{\rho}\hat{S}_3] = i(\rho_{h,v} - \rho_{v,h})
\end{aligned}
\tag{13.4.3}
$$

where $\rho_{i,j} = \langle I|\hat{\rho}|J\rangle$ are the elements of the density matrix ρ representing the density operator in the H/V basis. Equations (13.4.3) are obtained using the ket representation of the Stokes operators given in Equation (13.4.1). Combining Equations (13.4.2) and (13.4.3) we can obtain all the elements of the density matrix in terms of the Stokes operators expectation values and hence in terms of count rates via

$$\rho_{h,h} = \frac{1 + \langle \hat{\mathbf{S}}_1 \rangle}{2} = \frac{R_H}{R_H + R_V}$$

$$\rho_{v,v} = \frac{1 - \langle \hat{\mathbf{S}}_1 \rangle}{2} = \frac{R_V}{R_H + R_V}$$

$$\rho_{v,h} = \frac{\langle \hat{\mathbf{S}}_2 \rangle + i\langle \hat{\mathbf{S}}_3 \rangle}{2} = \frac{R_D + iR_R}{R_H + R_V} - \frac{1 + i}{2}$$

$$\rho_{h,v} = \frac{\langle \hat{\mathbf{S}}_2 \rangle - i\langle \hat{\mathbf{S}}_3 \rangle}{2} = \frac{R_D - iR_R}{R_H + R_V} - \frac{1 - i}{2} \qquad (13.4.4)$$

where we have used the normalization $\rho_{h,h} + \rho_{v,v} = 1$. The density matrix contains all the information about the polarization state of the photons and properties such as the purity of the state are readily extracted. A question often asked is how similar the experimentally produced state, $\hat{\rho}$, is to some pure target state $|\phi\rangle$. A common measure of this is the fidelity, F, given by

$$F = \langle \phi | \hat{\rho} | \phi \rangle \qquad (13.4.5)$$

which is easily found in terms of the matrix elements of ρ. This technique can be extended to two, or more, photons by considering the expectation values of products of Stokes operators of each photon. For example

$$\langle \hat{\mathbf{S}}_{1,a} \hat{\mathbf{S}}_{1,b} \rangle = \rho_{hh,hh} - \rho_{hv,hv} - \rho_{vh,vh} + \rho_{vv,vv} \qquad (13.4.6)$$

where a, b label the two photons and $\rho_{ij,kl} = \langle i | \langle j | \hat{\rho} | k \rangle | l \rangle$. By considering the expectation values of all the different combinations of Stokes operator products the two photon density matrix can be characterized. Whilst 4 measurements are needed to completely characterize a single photon, 16 measurements are needed in general for two photons. Of course if the two photons are known to be in a separable state then 4 measurements on each individual photon will suffice to characterize the state. The greater number of measurements needed to characterize entangled states points to their increased complexity. The exponential increase in measurements required as a function of photon number continues with three photons requiring 64 measurements and so on.

The above techniques were developed and applied by A. White and D. James et al [Whi99], [Jam01]. One problem that arises is that, due to experimental errors, the density matrix produced from the data may be unphysical. To deal with this maximum likelihood techniques were applied such that the "closest" physical density matrix to the data can be identified [Jam01].

A large range of single and two photon states have now been analysed in this way with very high fidelity ($\geq 99\%$). In Section 13.7 we will see how these techniques can be applied to the characterization of quantum gates.

13.5   Quantum key distribution

Perhaps the simplest application of quantum information is in the field of secure communications. It is referred to variously as quantum key distribution (QKD), quantum cryptography or sometimes quantum key expansion and was initially proposed by Bennett and Brassard [Ben84]. The idea is to set up a communication channel which is secure in the sense that any attempt to eavesdrop on the communication can be detected after the fact. The channel is used to send a random number encryption key between two parties, usually referred to as Alice (the sender) and Bob (the receiver). The parties then check if an eavesdropper, called Eve, intercepted any information about the key. If no Eve was present they can proceed to use the random number key to encrypt secret messages. If they find an Eve is present they scrap that key and try again.

13.5.1 QKD using single photons

QKD's ability to detect eavesdroppers is based on the fact that any process which acquires information about an observable of a quantum mechanical system inevitably disturbs the values of any non-commuting observables. For example, first suppose Alice sends out a "zero" encoded as a horizontally polarized photon, $|H\rangle$. Eve measures in the horizontal/vertical basis, see Section 3.5.1, obtains the result "zero" and so sends a horizontally polarized photon on to Bob who will definitely get a zero if he measures in the horizontal/vertical basis. However, now suppose Alice and Bob switch to encoding in the diagonal/anti-diagonal basis without Eve knowing. Alice sends a zero as a diagonally polarized photon, $|D\rangle = 1/\sqrt{2}(|H\rangle + |V\rangle)$. Eve still measures in the horizontal/vertical basis and so has a 50/50 possibility of getting either zero or one as the result, regardless of what Alice sent. Further more, what Eve sends on to Bob is basically the mixed state $\rho = 1/2(|H\rangle\langle H| + |V\rangle\langle V|)$. So when Bob measures in the diagonal/anti-diagonal basis he also gets a random result. Thus, by measuring in the wrong basis, not only does Eve potentially get the wrong result, but she also completely erases the qubit value which is sent on to Bob who then may also get the wrong result.

The trick then is to arrange a situation in which Eve does not know in which basis the information on any particular photonic qubit has been encoded because then she is bound to make mistakes which Bob will be able to detect. A typical protocol would go as follows:

1. Alice sends a random number sequence to Bob, encoded on the polarization of single photons. She randomly swaps between encoding on the horizontal/vertical basis and encoding on the diagonal/anti-diagonal basis.

2. Bob measures the polarization of the incoming photons and records the results, but he also swaps randomly between measuring in the horizontal/vertical basis and measuring in the diagonal/anti-diagonal basis.

3. After the transmission is complete Alice and Bob communicate on a public channel. First Bob announces which basis he measured in for each transmission event. Alice tells him whether or not this corresponded to the basis in which she prepared the photon. They discard all transmission events for which their bases did not correspond.

4. Bob then reveals the bit values he measured for a randomly selected subset of the remaining data. Alice compares the values revealed by Bob with those she sent. If there are no errors then an eavesdropper could not have been present so Alice and Bob are free to use the remaining undisclosed data as a secret key. If there are many errors then an eavesdropper may have been present and so the data is discarded and they try again.

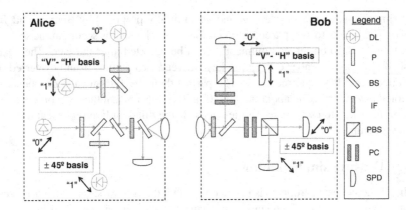

Figure 13.3: Laser and detector arrangement used in free space QKD experiments of R. Hughes et al. [Hug02].

Actually in any practical situation there will always be some errors in the transmission due to imperfections in the system. Thus what Alice and Bob do is set limits on the amount of information that Eve can have obtained based on the error rate they observe. Provided this error rate is sufficiently small, post processing of the data using techniques called error reconciliation and privacy amplification [Ben95] can be used to produce a shorter secret key. Eve's information about this shorter key can be made vanishingly small. Another important caveat is that Alice and Bob must initially share some secret information which they can use to identify each other. Otherwise Eve can fool them by pretending to be Bob to Alice and vice versa. Given these conditions QKD is provably secure [Got03]. No comparable result exists for classical communications.

The first experimental demonstration of QKD was carried out by Bennett and co-workers in 1992 over a distance of centimeters [Ben92a]. Demonstrations over distances of tens of kilometers in free space or fibre are now the state of the art. Typically highly attenuated lasers are used as the qubit source. Switching between the four input states may be achieved through electro-optic control or via the passive combination of four separate laser sources. In all cases it is crucial that the spatial and temporal modes of the four input states are identical so that no additional information is leaked to Eve. The receiver station can be a passive arrangement. A 50/50 beamsplitter is used to randomly send the incoming photons either to a horizontal/vertical analyser or a diagonal/anti-diagonal analyser. A schematics of the QKD transmitter and receiver used in the free space demonstration by R. Hughes et al. [Hug02] at Los Alamos Nation Laboratory is shown in Figure 13.3.

To increase the signal to noise of the detection system the detectors are gated, only opening for the one nanosecond or so window in which the single photon pulse is expected. Synchronization may be arranged via bright timing pulses preceding the single photon pulses or via more standard public communication links. Sophisticated reconciliation and privacy amplification algorithms then need to be implemented over the public channel. In the Hughes et al. experiment secure key rates of hundreds per second were achieved over a ten kilometer range in broad daylight, a significant achievement. Somewhat higher rates were achieved at night.

The main motivation for free space systems is to transmit secret keys to satellites securely. For terrestrial systems transmission through fibre optic networks is more desirable. Although this has the advantage of less stray light, it has the problem that optic fibre is birefringent and hence polarization encoded qubits can become scrambled. One solution is to go to a temporal mode qubit encoding. For example one could use the non-commuting encodings

$$|0\rangle \equiv |T1\rangle + |T2\rangle$$
$$|1\rangle \equiv |T1\rangle - |T2\rangle \tag{13.5.1}$$

and

$$|+\rangle \equiv |T1\rangle + i|T2\rangle$$
$$|-\rangle \equiv |T1\rangle - i|T2\rangle \tag{13.5.2}$$

where $|Ti\rangle$ represents a single photon occupying a temporal wave packet centred at time Ti. Alice could produce the state $|T1\rangle + |T2\rangle$ by allowing a single photon pulse to pass through a Mach Zehnder interferometer with unequal arm lengths, in particular where the arm length difference is $T1 - T2$. The other states are created in a similar way but where an additional phase of π (for the state $|T1\rangle - |T2\rangle$) or $\pi/2$ or $3\pi/2$ (for the other basis states) is added to one arm of the interferometer. Unfortunately, a readout by Bob would require him to have an interferometer which is phase-locked to Alice's, something that is difficult to arrange. D. Stucki et al., at the University of Geneva, came up with an elegant solution to this problem by having Bob first send a bright pulse to Alice [Stu02]. This pulse acts as a phase reference, thus avoiding the locking issue. The details are shown in Fig. 13.4. Using this arrangement they were able to transmit secure keys over distances of nearly 70 [km] at rates of about 50 [per s].

A limit on the secure key rates occurs with the use of attenuated laser sources. The initial intensity cannot be too great otherwise the probability of two-photon events will be too high. Eve can use two photon events to extract information about the key without penalty. One solution to this problem is to use a true single photon source. Beveratos et al. [Bev02] were the first to demonstrate such a scheme. They used the fluorescence from a single NV colour centre inside a diamond nano-crystal at room temperature as their single photon source. The nano-crystals were attached to a dielectric mirror and their emission was collected by a high numerical aperture microscope objective. The effective value of τ (see Equation (13.3.6)) for this source was 0.07, thus the number of two photon events is reduced by a factor of 14 over an attenuated laser source of the same intensity. A test system employing this source was able to achieve secure bit rate of $7,700$ [per s] over 50 [m] in free space.

The QKD protocol we have discussed here is called BB84. Many other protocols have been proposed and demonstrated and new protocols and demonstrations appear regularly [Gis02]. Initial steps to commercialization have already been taken.

13.5.2 QKD using continuous variables

So far we have only considered quantum information carried on the polarization of single photons; a discrete, 2-state variable. It is also possible to carry quantum information on continuous variables such as the amplitude and phase of a light mode. We now discuss how QKD can be implemented with continuous variables.

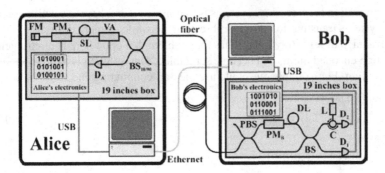

Figure 13.4: Schematics of the experiment by D. Stucki et al. [Stu02]. A strong laser pulse, emitted by Bob is first split on a beamsplitter (BS). The two pulses are recombined on a polarizing beamsplitter (PBS) after passing through a short arm and a long arm including a phase modulator (PM_B) and a 50[ns] delay line (DL). The polarisation of the short arm is rotated 90 degrees such that both pulses exit by the same port of the PBS. The pulses travel to Alice where they are reflected on a Faraday mirror, are attenuated and are sent back to Bob orthogonally polarized. As a result the pulses return along their paths in reversed order and interfere at BS. The entire system forms an autocompensated interferometer. The "0", "1" basis is encoded by Alice applying either 0 or π phase shift with PM_A and read out by Bob with no phase shift applied to the return pulse. The "+" "-" basis is encoded by Alice applying either $\pi/2$ or $3\pi/2$ phase shift with PM_A and read out by Bob with $\pi/2$ phase shift applied via PM_B to the return pulse.

As we have seen the basic mechanism used in QKD schemes is the fact that the act of measurement (by Eve) inevitably disturbs the system. This measurement back-action of course also exists for continuous quantum mechanical variables. In particular let us consider the situation in which Alice sends a series of weak coherent states to Bob whose amplitudes are picked from a two-dimensional Gaussian distribution centred on zero. Bob chooses to measure either the in-phase (amplitude) or out-of-phase (phase) projections of the states onto a shared local oscillator using homodyne detection, see Section 8.1.4. Bob will effectively see a Gaussian distribution of real amplitude coherent states when he looks at amplitude and a Gaussian distribution of imaginary amplitude coherent states when he looks at the phase. Alice can encode two different random number sequences on the amplitude and phase distributions respectively (see Section 10.1). Because the two quadrature measurements do not commute (see Section 4.3) Eve now has a similar problem as in the discrete case: any attempt to extract information about one quadrature from the beam will inevitably erase information carried on the other quadrature. If Alice and Bob compare some of the data at the end of the protocol they will thus notice increased error rates as a result of any intervention. This protocol was developed by F. Grosshans and P. Grangier [Gro02] based on earlier work [Ral00, Cer01]. A proof of security for a somewhat different continuous variable protocol has been made [Got02].

Coherent state QKD can be implemented either by sending very weak coherent pulses of light or by sending bright, quantum limited light with amplitude/phase modulation of the sidebands playing the role of the coherent states (see Section 4.5.5). The first experimental demonstration of this technique, performed by F. Grosshans et al. [Gro03] used the former technique. They were able to transmit secret bits at a rate of 75,000 [per s]in the presence of 50% line loss, albeit under laboratory conditions.

13.5.3 No cloning

An alternative way of understanding the basic mechanism of QKD is the fact that quantum information cannot be copied or *cloned*. That is, if given an unknown quantum state, it is not possible to produce an identical copy or clone of it. This was first pointed out by W.K. Wootters, W.H. Zurek [Woo82]. The best that can be achieved is to turn the original input state into a pair of output states which are approximate copies of the original. For qubits the best achievable fidelity of the clones with the original is $5/6$. For coherent states the best achievable fidelity is $2/3$.

13.6 Teleportation

When Alice and Bob share an entangled state their ability to communicate is enhanced. This is quite remarkable given that entanglement is undirected and carries no information itself. For example, suppose Alice and Bob share a pair of qubits in the entangled state $|0\rangle_a|0\rangle_b + |1\rangle_a|1\rangle_b$. Alice chooses to do one of four operations on her qubit:

1. Leave the qubit unchanged.

2. Perform a bit-flip or X operation, which takes $|0\rangle \to |1\rangle$ and $|1\rangle \to |0\rangle$.

3. Perform a phase-flip or Z operation, which leaves $|0\rangle$ unchanged but takes $|1\rangle \to -|1\rangle$.

4. Perform a bit-flip and a phase-flip.

She then sends her qubit to Bob who then has a pair of qubits in one of the four states:

1. $|\phi+\rangle = |0\rangle_a|0\rangle_b + |1\rangle_a|1\rangle_b$,

2. $|\psi+\rangle = |1\rangle_a|0\rangle_b + |0\rangle_a|1\rangle_b$,

3. $|\phi-\rangle = |0\rangle_a|0\rangle_b - |1\rangle_a|1\rangle_b$,

4. $|\psi-\rangle = |1\rangle_a|0\rangle_b - |0\rangle_a|1\rangle_b$,

respectively. These states are known as the Bell-states. The important point is that they are mutually orthogonal and so Bob can now, in principle, perform a measurement which unambiguously picks which of the four states he has (this is known as a Bell measurement). The outcome is that Alice has communicated two bits of information to Bob whilst only sending a single qubit! This effect is called *quantum dense coding*, or super dense coding, [Ben92b] and has no classical analogue.

Perhaps even more surprising is quantum teleportation [Ben93]. Here the presence of entanglement enables Alice to send Bob an unknown qubit by simply sending a classical message. Consider first what Alice can do in the absence of entanglement. Her best strategy is to measure the qubit in some basis and send a message that tells Bob to make a qubit corresponding to the result she gets. Sometimes she will be lucky and will measure in a good basis, then Bob will make a close approximation to the qubit. Other times she will measure in a bad basis and the result she gets will be completely random and Bob will make a poor approximation to the qubit. It can be shown that on average the fidelity of Bob's qubit with Alice's original is $2/3$.

If they share entanglement they can perform teleportation. This works in the following way: Alice and Bob again share an entangled pair of qubits, say $|0\rangle_a|0\rangle_b + |1\rangle_a|1\rangle_b$. Alice also has a qubit in the arbitrary state $\mu|0\rangle_{a'} + \nu|1\rangle_{a'}$ which she wishes to send to Bob. Alice does not know the state of her qubit. If we write down the state of the three qubits and then rearrange it we notice a remarkable feature: Bob's qubit can be represented as an equal superposition of four states, each differing from Alice's unknown qubit by at most a bit-flip and a phase-flip. What is more, Alice can tell which of these four states Bob actually has by making a Bell measurement on her two qubits. Explicitly:

$$
\begin{aligned}
\left(\mu|0\rangle_{a'} + \nu|1\rangle_{a'}\right)&\left(|0\rangle_a|0\rangle_b + |1\rangle_a|1\rangle_b\right) \\
&= \left(|0\rangle_{a'}|0\rangle_a + |1\rangle_{a'}|1\rangle_a\right)\ \left(\mu|0\rangle_b + \nu|1\rangle_b\right) \\
&\quad + \left(|0\rangle_{a'}|0\rangle_a - |1\rangle_{a'}|1\rangle_a\right)\ \left(\mu|0\rangle_b - \nu|1\rangle_b\right) \qquad (13.6.1) \\
&\quad + \left(|0\rangle_{a'}|1\rangle_a + |1\rangle_{a'}|0\rangle_a\right)\ \left(\mu|1\rangle_b + \nu|0\rangle_b\right) \\
&\quad + \left(|0\rangle_{a'}|1\rangle_a - |1\rangle_{a'}|0\rangle_a\right)\ \left(\mu|1\rangle_b - \nu|0\rangle_b\right)
\end{aligned}
$$

If the result of Alice's Bell measurement is $|\phi+\rangle$ she tells Bob not to do anything, if it is $|\phi-\rangle$ she tells him to do a phase-flip, if it is $|\psi+\rangle$ a bit-flip is required and finally if she measures $|\psi-\rangle$ she tells him to perform both a bit and a phase-flip. In the end Bob has turned his qubit into an exact copy of Alice's original but all Alice has sent is a two bit classical message.

13.6.1   Teleportation of photon qubits

Optics looks a good candidate for a demonstration of teleportation. We have already seen in Section 12.3.3 that spatially separated, polarization entangled states of the required form (at least after post-selection) can be generated by down-conversion and the local operations can be implemented using wave-plates. Notice though that in addition to the entangled photons we need another photon to be produced simultaneously to act as the teleportee. Also the Bell measurement represents a problem as differentiation of all four Bell states requires a non-linear Kerr interaction much stronger than is presently available.

These problems were solved in the demonstration by D. Bouwmeester et al. at the University of Innsbruck [Bou97]. Their experimental layout is shown in Fig. 13.5. Pulsed UV pumping (200 [ps] pulse length, 76 [MHz] repetition rate) of a non-linear crystal in a type II arrangement was used to produce pairs of polarization entangled photons at 788 [nm] (2 and 3). The pump pulse was then retro-reflected through the crystal such that a second pair of counter-propagating photons (1 and 4) might be produced. Teleportation can then proceed by giving Alice entangled photon 2 and photon 1 as the teleportee (after it has been prepared in some arbitrary state) and giving Bob entangled photon 3.

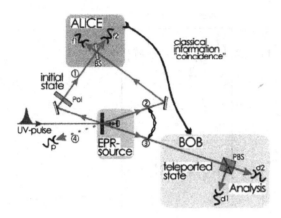

Figure 13.5: Layout of the experiment by D. Bouwmeester et al. [Bou97].

Although a complete Bell measurement is not practical, a partial Bell measurement can be achieved quite straightforwardly by passing photons 1 and 2 through a 50/50 beamsplitter and then photon counting at the outputs. The action of a beamsplitter on the Bell states is to make the photons bunch (i.e. both exit through the same port) in much the same way as for the HOM dip. For all the Bell states that is except $|\psi-\rangle$, for which case the photons always exit by different ports. Thus if Alice records a coincidence count at the output of the beamsplitter then she has unambiguously identified the $|\psi-\rangle$ Bell state and teleportation has succeeded. The experiment is arranged such that the $|\psi-\rangle$ state is the "do nothing" result. If she does not obtain a coincidence the protocol has failed.

Results from the experiment are shown in Fig. 13.6. Teleportation should have succeeded when: photon 4 is detected at p (indicating that photon 1 is on its way); a coincidence is detected at f1 and f2 (indicating the correct Bell state has been detected) and; a photon is detected at either d1 or d2 (indicating the teleported photon has been successfully detected). When these 4-fold coincidences occur we should expect the polarisation state of Bob's photon to match that of the photon prepared for Alice. This is what is shown in Fig. 13.6 for input photons in the non-orthogonal states $|A\rangle$ and $|H\rangle$ (-45 and 0 degrees respectively). At zero time delay between when photons 1 and 2 meet at their beamsplitter the original states are reproduced with a visibility of $70\% \pm 3\%$. When a time delay is introduced, photons 1 and 2 become distinguishable at the beamsplitter, the Bell measurement fails and so too does the teleportation.

This experiment was very technically challenging. The probability of four photon events was very low. To prevent any temporal distinguishability of photons 1 and 2, a frequency filtering producing a 4 [nm] bandwidth was applied, further reducing the counts. Finally the protocol itself only succeeded one quarter of the time. This resulted in roughly one successful event per minute.

A subtlety of the original experiment was that there was a significant probability for the down-converter to produce two photons each in modes 1 and 4. Then, even under perfect conditions of zero loss, a three-fold coincidence on Alice's side of the experiment does not

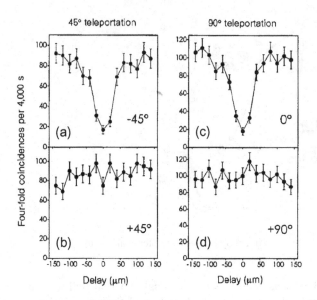

Figure 13.6: Experimental results from the Bouwmeester et al experiment [Bou97].

guarantee a photon is sent to Bob. In a later manifestation of the experiment the possibility of such errors was made negligible and fidelities of $> 80\%$ were observed, well in excess of the $2/3$ limit [Pan03].

13.6.2   Continuous variable teleportation

So far we have considered teleportation of qubits, as carried by the polarization degree of freedom of single photons. This technique will only work for single photon states. What if we wish to teleport a general field state with contributions from vacuum and higher photon number terms? The answer is to implement a teleportation protocol based on the measurement of the quadrature amplitudes of the field. Because the quadrature amplitudes are continuous, rather than discrete, variables, this is known as continuous variable (CV) teleportation. It was developed by S. Braunstein and J. Kimble [Bra98] based on earlier work by L. Vaidman [Vai94].

Consider the following situation. Alice wishes to teleport to Bob an unknown coherent state, $|\alpha\rangle$, drawn from a broad Gaussian distribution. In the absence of entanglement Alice's best approach is to divide the field into two equal parts at a beamsplitter and then measure the amplitude quadrature of one half and the phase quadrature of the other. The amplitude measurement gives an estimate of the real part of α, whilst the phase measurement gives an estimate of the imaginary part of α, however both estimates are imperfect due to noise from the vacuum field which inevitably enters through the open port of the beamsplitter. Alice sends these estimates to Bob who uses them to produce a coherent state by displacing his local vacuum state by the relevant quantities.

This situation is most easily described in the Heisenberg picture. Let the initial field mode be represented by the annihilation operator $\tilde{\mathbf{A}}$ and the vacuum entering at the 50:50 beamsplitter by $\hat{\mathbf{V}}_a$. The measurement results obtained by Alice are then represented by the quadrature operators, see Section 4.4.2.

$$
\begin{aligned}
\tilde{\mathbf{X}}1_a &= \frac{1}{\sqrt{2}}(\tilde{\mathbf{X}}1_a - \tilde{\mathbf{X}}1_{Va}) \\
\tilde{\mathbf{X}}2_a &= \frac{1}{\sqrt{2}}(\tilde{\mathbf{X}}2_a + \tilde{\mathbf{X}}2_{Va})
\end{aligned} \tag{13.6.2}
$$

These are sent to Bob who uses them to displace his vacuum field, $\tilde{\mathbf{V}}_b$ giving the output field

$$
\tilde{\mathbf{B}} = \tilde{\mathbf{V}}_b + g\frac{1}{2}(\tilde{\mathbf{X}}1_a + i\tilde{\mathbf{X}}2_a) - g\frac{1}{2}(\tilde{\mathbf{X}}1_{Va} - i\tilde{\mathbf{X}}2_{Va}) \tag{13.6.3}
$$

where g is a gain factor for the displacement. Choosing $g = 1$, unity gain, Bob's output field is

$$
\tilde{\mathbf{B}} = \tilde{\mathbf{A}} + \tilde{\mathbf{V}}_b - \tilde{\mathbf{V}}^\dagger_a \tag{13.6.4}
$$

Notice that two vacuum fields have been added to the output, one entering through Alice's measurement, the other through Bob's reconstruction. As a result Bob's state is mixed and is 3 times noisier than the QNL level of the input coherent state.

Now suppose Alice and Bob share an entangled state, in particular an EPR entangled state of the type discussed in Section 12.3.1. Alice again divides and measures her beam but this time instead of allowing vacuum to enter the empty port of her beamsplitter she sends in her half of the EPR pair. As a result her quadrature measurement results are now given by

$$
\begin{aligned}
\tilde{\mathbf{X}}1_a &= \frac{1}{\sqrt{2}}(\tilde{\mathbf{X}}1_a - \sqrt{G}\,\tilde{\mathbf{X}}1_{Va} - \sqrt{G-1}\,\tilde{\mathbf{X}}1_{Vb}) \\
\tilde{\mathbf{X}}2_a &= \frac{1}{\sqrt{2}}(\tilde{\mathbf{X}}2_a + \sqrt{G}\,\tilde{\mathbf{X}}2_{Va} - \sqrt{G-1}\,\tilde{\mathbf{X}}2_{Vb})
\end{aligned} \tag{13.6.5}
$$

where we have used Equation (12.3.5) to describe the entanglement. These results are sent to Bob who now uses them to displace his half of the EPR pair, obtaining (at unity gain)

$$
\tilde{\mathbf{B}} = \tilde{\mathbf{A}} + (\sqrt{G} - \sqrt{G-1})\tilde{\mathbf{V}}_b + (\sqrt{G} - \sqrt{G-1})\tilde{\mathbf{V}}^\dagger_a \tag{13.6.6}
$$

Now in the limit $G \to \infty$, $(\sqrt{G} - \sqrt{G-1}) \to 0$, hence in this limit Equation (13.6.6) reduces to

$$
\tilde{\mathbf{B}} = \tilde{\mathbf{A}} \tag{13.6.7}
$$

Evolution through the teleporter is the identity and so the output state is identical to the input (this is obviously true not only for the coherent input states we have been considering but for any input state).

Once again optics is an obvious candidate for demonstrating this type of teleportation. The required EPR states can be produced via a squeezing interaction (see Section 12.3.1) and the

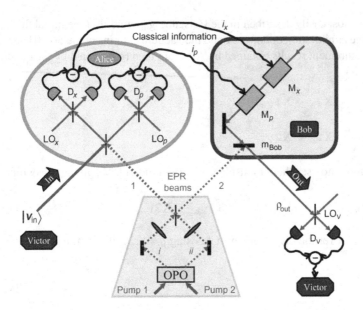

Figure 13.7: Schematical layout of the Caltech teleportation experiment. After A. Furusawa et al. [Fur98].

coherent input states can be created through side-band modulation of a quantum limited laser (see Section 4.5.5). Bob's displacement can also be achieved via modulation.

The first demonstration of this type was by A. Furusawa et al. at the California Institute of Technology [Fur98]. The experimental set-up is shown in Fig. 13.7. EPR entanglement is produced by the mixing of two out-of phase squeezed beams on a 50/50 beamsplitter. Both squeezed beams (at 860 [nm]) are generated in a single ring-cavity OPO by simultaneously pumping counter-propagating cavity modes. The cavity is actively locked and the parametric medium is temperature-tuned Potassium Niobate. One of the EPR beams is sent to Alice who mixes it with her signal beam and performs dual balanced homodyne measurements, actively locked to be 90 degrees out of phase, such that conjugate quadrature measurements are made. The photo-currents thus generated are sent to Bob who uses them to impose phase and amplitude modulations on a bright laser beam. By mixing this bright beam with his EPR beam on a highly reflective beamsplitter Bob can efficiently impose on it a displacement proportional to the modulations. All the beams in the experiment originate from a single Ti:sapph master laser, including the signal beam which has a known modulation (coherent amplitude) imposed on it before being sent to Alice. The modulation on the signal beam was imposed at 2.9 [MHz] and had an amplitude of about 25 [dB] above QNL. When, based on the signal size observed on Bob's side, unity gain was achieved, the quadrature noise floors of the teleported beam were measured by an independent balanced homodyne detector, both with and without entanglement.

The quality of Bob's reconstruction can be evaluated via the fidelity of it compared with the initial coherent state Alice sent. Provided the output is Gaussian (which it is) this fidelity

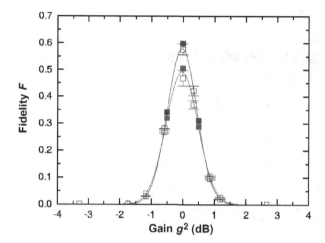

Figure 13.8: Measured fidelities from the experiment by A. Furusawa et al. [Fur98]. Results with and without entanglement are shown along with theoretical curves.

is given by

$$F = \frac{2}{\sqrt{(V1_b + 1)(V2_b + 1)}} \exp\left[-\frac{2}{\sqrt{(V1_b + 1)(V2_b + 1)}}|\alpha|^2(1-g)^2\right] \quad (13.6.8)$$

Recalling from our earlier discussion that without entanglement the quadrature variances of the outputs are $V1_b = V2_b = 3$, then we find from Equation (13.6.8) that for large α the best fidelity with no entanglement is achieved at unity gain and is $F = 0.5$. This is confirmed by Furusawa et al. who find a best fidelity without entanglement of $F_c = 0.48 \pm 0.03$. On the other hand, with entanglement, a fidelity of $F_q = 0.58 \pm 0.02$ is measured, clearly exceeding the classical bound. Experimental results as a function of gain are shown in Fig. 13.8.

A subsequent experiment by Bowen et al. at the Australian National University made significant improvements [Bow03]. These included: higher fidelities ($F_q = 0.64 \pm 0.02$); stable operation over long periods; and; a greater range of input state amplitudes. Their experiment used two independent, monolithic, sub-threshold OPO's to produce twin squeezed beams at 1064 [nm] which were then mixed on a beamsplitter to produce the required EPR entanglement. In Figure 13.9 spectra of the amplitude and phase quadratures of the input and teleported signals are shown along with time traces showing the stability of the system over several minutes.

In addition the performance of the Bowen et al, teleporter was characterized in terms of the teleportation T-V diagram, introduced by one of us (TCR) and P.K. Lam [Ral98]. The teleportation T-V diagram is in a similar vein to the QND T-V diagram, see Section 11.2. As we have seen, in the absence of entanglement strict bounds are placed on both the accuracy of measurement and reconstruction of an unknown state. These are represented by the vacuum modes that appear in Equation (13.6.4). These bounds can be expressed in such a way that, in contrast to fidelity, they are input state independent.

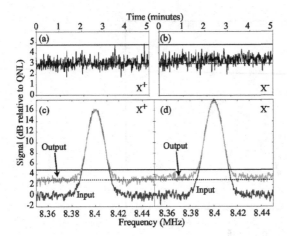

Figure 13.9: Input and output spectra for amplitude and phase quadratures from the ANU teleportation. Time traces show noise on amplitude and phase quadratures of the output over 5 [min] interval. From W. Bowen et al. [Bow03].

Alice's measurement accuracy is limited by the generalized uncertainty principle $V1_M V2_M \geq 1$ [Art88], where $V1_M, V2_M$ are the quadrature measurement penalties, which holds for *any* simultaneous measurements of conjugate quadrature amplitudes of an unknown quantum optical system. Note that in most current publications the notation has evolved into $X1 = X^+, X2 = X^-, V1 = V^+, V2 = V^-$, etc. This relationship can be re-written in terms of quadrature signal transfer coefficients, $T1 = \mathrm{SNR1_{out}}/\mathrm{SNR1_{in}}$ and $T2 = \mathrm{SNR2_{out}}/\mathrm{SNR2_{in}}$ as

$$T_q = T1 + T2 - T1\,T2\left(1 - \frac{1}{V1_{in}V2_{in}}\right) \leq 1 \qquad (13.6.9)$$

where $\mathrm{SNR1} = \alpha_1/V1$ and $\mathrm{SNR2} = \alpha_2/V2$ are the relevant signal-to-noise ratios. This expression reduces to $T_q = T1 + T2$ for minimum uncertainty input states ($V1_{in}\,V2_{in} = 1$). Without entanglement it is not possible to break the inequality given in Equation (13.6.9).

Bob's reconstruction must be carried out on a mode of the E/M field the fluctuations of which must already obey the uncertainty principle. In the absence of entanglement these intrinsic fluctuations remain present on any reconstructed field, thus the amplitude and phase conditional variances, $V1_{in|out} = V1_{out} - |\langle \delta\hat{X}1_{in}\delta\hat{X}1_{out}\rangle|^2/V1_{in}$ and $V2_{in|out} = V2_{out} - |\langle \delta\hat{X}2_{in}\delta\hat{X}2_{out}\rangle|^2/V2_{in}$ respectively, which measure the noise added during the teleportation process, will satisfy $V1_{in|out}V2_{in|out} \geq 1$. This can be written in terms of the signal transfer and quadrature variances of the output state as

$$V_q^{\mathrm{prod}} = (1 - T2)(1 - T2)\,V1_{out}V2_{out} \geq 1 \qquad (13.6.10)$$

The criteria of Equations (13.6.9) and (13.6.10) are then used to represent the quantum teleportation on the T-V graph.

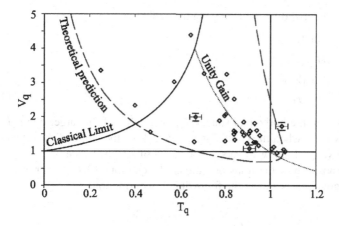

Figure 13.10: T-V graph of results from the Bowen et al experiment [Bow03].

The T_q and V_q bounds have independent physical significance. If Bob's state passes the T_q bound (Equation (13.6.9)) then he can be sure, regardless of how it was transmitted to him, that no other party can possess a copy of the state which also passes this bound (i.e. carries as much information about the original). Surpassing the V_q bound is a necessary prerequisite for reconstruction of non-classical features of the input state such as squeezing. Clearly it is desirable that the T_q and V_q bounds are simultaneously exceeded. It was first pointed out by F. Grosshans and P. Grangier [Gro01] that the cross-over point $(1, 1)$, corresponds to a fidelity of 2/3, the no-cloning limit for coherent states (see Section 13.5.3). Perfect reconstruction of the input state would result in $T_q = 2$ and $V_q = 0$.

In Figure 13.10 data from the Bowen et al experiment is plotted on a T-V diagram. Notice in particular that a number of points pass the T_q bound and one point (marginally) exceeds both T_q and V_q simultaneously.

13.7 Quantum computation

We have seen that optics has proven an excellent testing ground for many of the principles and basic tasks in quantum information. But what about the more challenging task of quantum computation? It turns out that to be able to implement arbitrary processing of information encoded on a set of qubits it is sufficient to be able to implement arbitrary operations on single qubits as well as having at least one non-trivial two qubit operation. For polarization encoded single photon qubits arbitrary single qubit operations can be implemented using a succession of half and quarter wave-plates.

An example of a non-trivial two-qubit gate is the CNOT gate. In terms of polarisation qubits its operation is summarized by the following truth table

$$
\begin{aligned}
|H\rangle_c|H\rangle_t &\rightarrow |H\rangle_c|H\rangle_t \\
|H\rangle_c|V\rangle_t &\rightarrow |H\rangle_c|V\rangle_t \\
|V\rangle_c|H\rangle_t &\rightarrow |V\rangle_c|V\rangle_t \\
|V\rangle_c|V\rangle_t &\rightarrow |V\rangle_c|H\rangle_t
\end{aligned}
\tag{13.7.1}
$$

When the control qubit is in the horizontal state, $|H\rangle_c$, the value of the target qubit $|H\rangle_t$ or $|V\rangle_t$ is unchanged. However, when the control is vertical, $|V\rangle_c$, the value of the target qubit is flipped, horizontal to vertical and vice versa. The effect of a CNOT gate on superposition states is simply a superposition of the transformations of Equation (13.7.1). For example if the control is in the diagonal basis we get the following transformations

$$
\begin{aligned}
\frac{1}{\sqrt{2}}(|H\rangle_c + |V\rangle_c)|H\rangle_t &\rightarrow \frac{1}{\sqrt{2}}(|H\rangle_c|H\rangle_t + |V\rangle_c|V\rangle_t) \\
\frac{1}{\sqrt{2}}(|H\rangle_c + |V\rangle_c)|V\rangle_t &\rightarrow \frac{1}{\sqrt{2}}(|H\rangle_c|V\rangle_t + |V\rangle_c|H\rangle_t) \\
\frac{1}{\sqrt{2}}(|H\rangle_c - |V\rangle_c)|H\rangle_t &\rightarrow \frac{1}{\sqrt{2}}(|H\rangle_c|H\rangle_t - |V\rangle_c|V\rangle_t) \\
\frac{1}{\sqrt{2}}(|H\rangle_c - |V\rangle_c)|V\rangle_t &\rightarrow \frac{1}{\sqrt{2}}(|H\rangle_c|V\rangle_t - |V\rangle_c|H\rangle_t)
\end{aligned}
\tag{13.7.2}
$$

Notice that the resulting output states are the four Bell-states, see Section 13.6. How might such an interaction between two photons be implemented? Until recently it was thought that the only solution was the use of a Kerr medium with a non-linearity so strong that a single photon could induce a π phase shift, as first suggested by G. Milburn [Mil89]. Unfortunately, massive, reversible non-linearities of such a strength are well beyond those presently available.

M. Knill, R. Laflamme and G. Milburn (KLM) found a way to circumvent this problem and implement efficient quantum computation using only passive linear optics, photodetectors, and single photon sources [Kni01]. In particular KLM showed how to make a CNOT gate that was non-deterministic, but heralded. That is, the gate does not always work, but an independent signal heralds successful operation. A somewhat simplified version of this gate is shown in Fig. 13.11. In addition to the single photon, polarisation qubits incident at ports c (control) and t (target), the gate also has ancilla inputs comprising two vacuum input ports, $v1$ and $v2$, and two single photon input ports, $p1$ and $p2$. The beamsplitter reflectivities are given by $\eta_1 = 5 - 3\sqrt{2}$ and $\eta_2 = (3 - \sqrt{2})/7$. It can be shown that when no photons are detected at outputs $vo1$ and $vo2$, and one and only one photon is detected at each of $po1$ and $po2$, then the gate has succeeded and the photon qubits exiting through co and to have had the CNOT transformation applied to them. The probability of successful operation is $\eta_2^2 \approx 0.05$.

A cascaded sequence of such non-deterministic gates would be useless for quantum computation because the probability of many gates working in sequence decreases exponentially. This problem may be avoided by using teleportation. The idea that teleportation can be used for universal quantum computation was first proposed by Gottesman and Chuang [Got99]. Consider the quantum circuit shown in Fig. 13.12(a). Two unknown qubits are individually teleported and then a CNOT gate is implemented. Obviously, but not very usefully, the result is CNOT operation between the input and output qubits. However, the commutation relations

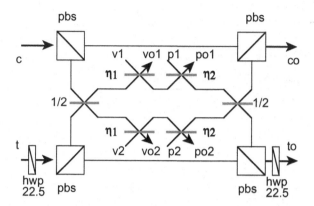

Figure 13.11: Schematic representation of non-deterministic CNOT gate. Successful operation is heralded by the detection of no photons at outputs $vo1$ and $vo2$ and the detection of one and only one photon at each of outputs $po1$ and $po2$.

between CNOT and the X and Z operations used in the teleportation are quite simple, such that in the circuits of Fig. 13.12 the alternatives (a) and (b) are in fact equivalent. But in the circuit of Fig. 13.12b the problem of implementing a CNOT gate has been reduced to that of producing the required entanglement resource. The main point is that this need not be done deterministically. Non-deterministic CNOT gates could be used in a trial and error manner to build up the necessary resource off-line. It could then be used when required to implement the gate.

A remaining issue is the performance of the Bell measurements required in the teleportation protocol. As we have seen in the previous section these cannot be performed deterministically with linear optics. KLM showed that by using the appropriate entangled resource the teleportation step can be made near deterministic: the success probability approaching one as the number of ancilla photons used in the entanglement is increased. This near deterministic teleportation protocol requires only linear optics, photon counting and fast feedforward, albeit with a significant resource overhead. The important thing to note though is that this resource overhead does not grow exponentially with the size of the overall circuit. Hence the claim of scalability.

Let us return to the basic non-deterministic CNOT gate of Fig. 13.11. Even at this level the technical requirements are demanding. Four photons need to simultaneously enter the circuit. The detections at $po1$ and $po2$ have to distinguish between zero, one or two photons. Any inefficiency in the production or detection of photons will lead to mistakes and rapidly erase the operation of the gate. High visibility single photon and two photon (HOM type) interference are required simultaneously: as a result excellent mode-matching and photon indistinguishability are essential.

A significantly simpler CNOT design can be realized by working in coincidence [Ral02]. In particular we can allow the photon qubits to be their own ancilla, such that only two photons are required. The gate is shown schematically in Fig. 13.13. Consider the input $|H\rangle_c|H\rangle_t$.

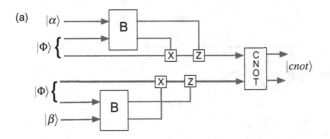

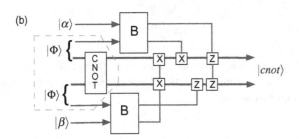

Figure 13.12: Schematic representation of gate operation via teleportation. Figures (a) and (b) are equivalent, yet in (b) a non-deterministic CNOT gate is sufficient as failure only destroys the entanglement: the operation can be repeated till successful without losing the qubit.

The target waveplate produces the transformation

$$|H\rangle_c|H\rangle_t \rightarrow \frac{1}{\sqrt{2}}|H\rangle_c(|H\rangle_t + |V\rangle_t)$$

The polarizing beamsplitters then spatially separate the polarization modes of the two beams. An array of different possibilities are present after the middle beamsplitters, however, we select (by postselection) only those where a photon arrives at both the target and control outputs. There are two ways for this to happen: the control photon must take the top path and reflect off beamsplitter 1; the target photon may take its upper path and reflect off beamsplitter 2 or take the bottom path and reflect off beamsplitter 3. In both cases the effect is just to reduce the amplitude of the successful components by a factor of $1/3$. The output state is then transformed by the second target waveplate such that

$$\frac{1}{3}\frac{1}{\sqrt{2}}|H\rangle_c(|H\rangle_t + |V\rangle_t) \rightarrow \frac{1}{3}|H\rangle_c|H\rangle_t$$

Similarly the input $|H\rangle_c|V\rangle_t$ is unchanged by passage through the circuit, other than a $1/3$ reduction in amplitude.

Things are different when the control is in the vertical state. Consider the input state $|V\rangle_c|H\rangle_t$. The target waveplate produces the transformation

$$|V\rangle_c|H\rangle_t \rightarrow \frac{1}{\sqrt{2}}|V\rangle_c(|H\rangle_t + |V\rangle_t)$$

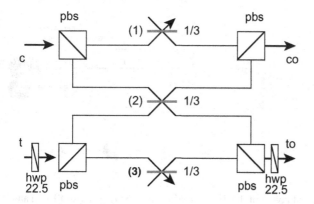

Figure 13.13: Schematic representation of a coincidence basis CNOT.

Now there are three ways for a successful detection to occur. The control takes its lower path. If the target photon takes the bottom path then both must reflect off their respective beamsplitters as before, simply reducing the amplitude by $1/3$. However, if the target photon follows its upper path then there are two possibilities at beamsplitter 2: either both photons may be reflected, giving an amplitude of $1/3$, or; both may be transmitted, giving an amplitude of $-2/3$. If the photons are indistinguishable then these amplitudes are added giving a total amplitude for that component of $-1/3$! Thus when the polarization modes are recombined the state carries a minus sign on one target component and the second target waveplate makes the transformation

$$\frac{1}{3}\frac{1}{\sqrt{2}}|V\rangle_c(|H\rangle_t - |V\rangle_t) \rightarrow \frac{1}{3}|V\rangle_c|V\rangle_t$$

and the value of the target qubit is flipped as required. Similarly the circuit does the transformation $|V\rangle_c|V\rangle_t \rightarrow 1/3|V\rangle_c|H\rangle_t$. Hence CNOT operation is realized whenever a coincidence is recorded. The probability of success is $(1/3)^2 = 1/9$.

J. O'Brien and G. Pryde et al at the University of Queensland used this technique to demonstrate CNOT operation for single photon qubits [OBr03]. Their experimental set up is shown in Figure 13.14. A pair of polarisation beam displacers was used to spatially separate and then recombine the polarisation modes of the qubits in an interferometrically stable configuration. A single waveplate oriented at 62.5 degrees and inserted between the displacers then played the role of all three one-third beamsplitters in the conceptual diagram (Figure 13.13). Photons pairs at 702 [nm] were produced via type II down conversion in beam-like modes. The pairs were not polarisation entangled. They were spatially filtered through optical fibres before being prepared in particular qubit states via a sequence of half and quarter-wave plates and injected into the circuit. An automated state tomography system could then be used to obtain the density operator of the output state for comparison with the expected outcome, see Section 13.4. After passing through 0.36 [nm] frequency filters, coincidences detected within a 5 [ns] time window were excepted.

The truth table performance ranged from about 95% correct for "control-off" operation, to 73% for "control-on" operation. The difference is attributed to the need for both single

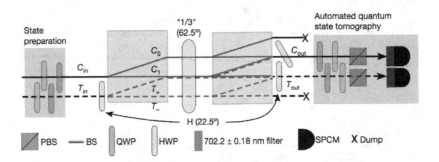

Figure 13.14: Experimental layout of the demonstration of a CNOT gate by O'Brien and Pryde et al. [OBr03].

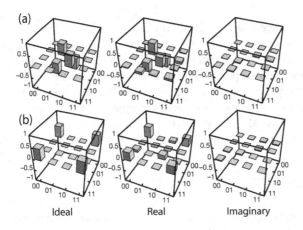

Figure 13.15: Density matrices for two of the Bell states produced by the CNOT gate of O'Brien and Pryde et al. [OBr03]: (a) real part of density matrix for $|\psi-\rangle$ Bell state (imaginary part zero) and real and imaginary parts of corresponding measured density matrix; (b) real part of density matrix for $|\phi+\rangle$ Bell state (imaginary part zero) and real and imaginary parts of corresponding measured density matrix.

and two-photon interference when the control is on. All four Bell-states were produced as described in Eq. (13.7.2) with fidelities ranging from 77% to 87%, thus demonstrating the ability of the gate to produce entanglement. Experimental density matrices for two of the Bell-states are shown in Fig. 13.15.

13.8 Summary

Remarkable progress has been made in implementing quantum information protocols in quantum optics. The progress in QKD in particular is sufficiently advanced that commercial applications are seriously considered. Teleportation, of a quality clearly exceeding the limits set in the absence of entanglement, has been demonstrated in both the discrete and continuous domains. In both cases the best results hover around the more demanding no-cloning limit. The demonstration of basic two-qubit quantum gates is promising but is a long way short of full-scale quantum computation. Continued advances along this path require technical solutions to the problem of efficient single photon production and detection. They also require theoretical solutions to the presently impractical complexity of scale-up. One promising direction may be in unifying discrete and continuous variable quantum optical techniques through the use of coherent superposition states [Ral03].

Further reading

Quantum computation and quantum information, M. Nielsen and I. Chuang, Cambridge University Press, Cambridge, UK (2000).

Quantum Information with continuous variables, S.L. Braunstein and A.K. Pati, (Eds, Kluwer, Netherlands 2003).

Bibliography

[Art88]   Quantum Correlations: A generalized Heisenberg uncertainty relation, E. Arthurs, M.S. Goodman, Phys. Rev. Lett. 60, 2447 (1988)

[Ben84]   C.H. Bennett, G. Brassard, Proceedings of IEEE International Conference on Computers, Systems and Signal Processing, Bangalore, India, p. 175 (1984)

[Ben92a] Experimental quantum cryptography, C.H. Bennett, F. Bessette, G. Brassard, L. Salvail, J. Smolin, J. Cryptol 5, 3 (1992)

[Ben92b] Communication via one- and two-particle operators on Einstein-Podolsky-Rosen states, C.H. Bennett, S.J. Wiesner, Phys. Rev. Lett. 69, 2881 (1992)

[Ben93]   Teleporting an unknown quantum state via dual classical and Einstein-Podolsky-Rosen channels, C.H. Bennett, G. Brassard, C. Crżpeau, R. Jozsa, A. Peres, W.K. Wootters, Phys. Rev. Lett. 70, 1895 (1993)

[Ben95]   Generalized privacy amplification, C.H. Bennett, G. Brassard, C. Crepeau, U.M. Maurer, IEEE Trans. Inf. Theory 41, 1915 (1995)

[Bev02]   Single Photon Quantum Cryptography, A. Beveratos, R. Brouri, T. Gacoin, A. Villing, J.-P. Poizat, P. Grangier, Phys. Rev. Lett. 89, 187901 (2002)

[Bou97]   Experimental quantum teleportation, D. Bouwmeester, J.-W. Pan, K. Mattle, M. Eibl, H. Weinfurter, A. Zeilinger, Nature 309, 575 (1997)

[Bow03]   Experimental investigation of continuous-variable quantum teleportation, W.P. Bowen, N. Treps, B.C. Buchler, R. Schnabel, T.C. Ralph, H-A. Bachor, T. Symul, P.K. Lam, Phys. Rev. A 67, 032302 (2003)

[Bra98]   Teleportation of Continuous Quantum Variables, S.L. Braunstein, H.J. Kimble, Phys. Rev. Lett. 80, 869 (1998)

[Cer01]   Quantum distribution of Gaussian keys using squeezed states, N.J. Cerf, M.Lżvy, G.V. Assche, Phys. Rev. A 63, 052311 (2001)

[Fur98]   Unconditional quantum teleportation, A. Furusawa, J.L. Sörensen, S.L. Braunstein, C.A. Fuchs, H.J. Kimble, E.S. Polzik, Science, 282, 706 (1998)

[Gis02]   Quantum cryptography, N. Gisin, G. Ribordy, W. Tittel, and H. Zbinden, Rev. Mod. Phys. 74, 145 (2002)

[Got99]   Demonstrating the viability of universal quantum computation using teleportation and single-qubit operations, D. Gottesman, I.L. Chuang, Nature 402, 390 (1999)

[Got02]   Secure quantum key distribution using squeezed states, D. Gottesman, J. Preskill, Phys. Rev. A 63, 022309 (2001)

[Got03]   Proof of security of quantum key distribution with two-way classical communications, D. Gottesman, H.-K. Lo, IEEE Transactions on Information Theory 49, 457 (2003)

[Gro01]   Quantum cloning and teleportation criteria for continuous quantum variables, F. Grosshans, P. Grangier, Phys. Rev. A 64, 010301 (2001)

[Gro02]   Continuous Variable Quantum Cryptography Using Coherent States, F. Grosshans, P. Grangier, Phys. Rev. Lett. 88, 057902 (2002)

[Gro03]   Quantum key distribution using gaussian-modulated coherent states, F. Grosshans, G.V. Assche, J. Wenger, R. Brouri, N.J. Cerf, P. Grangier, Nature, 421, 238 (2003)

[Hug02]   Practical free-space quantum key distribution over 10 km in daylight and at night, R.J. Hughes, J.E. Nordholt, D. Derkacs, C.G. Peterson, New J. Phys. 4, 43 (2002)

[Jam01]   Measurement of qubits, D.F.V. James, P.G. Kwiat, W.J. Munro, A.G. White, Phys. Rev. A 64, 052312 (2001)

[Kni01]   A scheme for efficient quantum computation with linear optics, E. Knill, R. Laflamme, G.J. Milburn, Nature 409, 46 (2001)

[Kuh02]   Deterministic single-photon source for distributed quantum networking, A. Kuhn, G. Rempe, Phys. Rev. Lett. 89, 067901 (2002)

[Kur01]   High-efficiency entangled photon pair collection in type-II parametric fluorescence, C. Kurtsiefer, M. Oberparleiter, H. Weinfurter Phys. Rev. A 64, 023802 (2001)

[Lvo01]   Quantum state reconstruction of the single-photon Fock state, A.I. Lvovsky, H. Hansen, T. Aichele, O. Benson, J. Mlynek, S. Schiller Phys. Rev. Lett. 87, 050402 (2001)

[McK03]   Experimental realization of a one-atom laser in the regime of strong coupling, J. McKeever, A. Boca, A.D. Boozer, J.R. Buck, H.J. Kimble, Nature, 425, 268 (2003)

[Mil89]   Quantum optical Fredkin gate, G.J. Milburn, Phys. Rev. Lett. 62, 2124 (1989)

[OBr03]   Demonstration of an all-optical controlled-NOT gate, J.L. O'Brien, G.J. Pryde, A.G. White, T.C. Ralph, D. Branning, Nature (2003)

[Pan03]   Experimental realization of freely propagating teleported qubits, J.-W. Pan, S. Gasparoni, M. Aspelmeyer, T. Jennewein, A. Zeilinger, Nature 421, 721 (2003)

[Pit02]   Single photons on pseudodemand from stored parametric down-conversion, T.B. Pittman, B.C. Jacobs, J.D. Franson, Phys. Rev. A 66, 042303 (2002)

[Ral98]   Teleportation with Bright Squeezed Light, T.C. Ralph and P.K. Lam, Phys. Rev. Lett. 81, 5668 (1998)

[Ral00]   Security of continuous-variable quantum cryptography, T.C. Ralph, Phys. Rev. A 62, 062306 (2000)

[Ral02]   Linear optical controlled-NOT gate in the coincidence basis, T.C. Ralph, N.K. Langford, T.B. Bell, A.G. White, Phys. Rev. A 65, 062324 (2002)

[Ral03]   Quantum computation with optical coherent states, T.C. Ralph, A. Gilchrist, G.J. Milburn, W.J. Munro, S. Glancy, Phys. Rev. A 68, 042319 (2003)

[San02]   Indistinguishable photons from a single-photon device, C. Santori, D. Fattal, J. Vuckovic, G.S. Solomon, Y. Yamamoto, Nature, 419, 594 (2002)

[Sho94]   P.W. Shor in Proceedings of the 35th Annual Symposium on the Foundations of Computer Science, 124-133 (IEEE Computer Society Press, Los Alamitos, California 1994), Shor's quantum algorithm for factorising numbers, Rev. Mod. Phys. 68, 733 (1996)

[Sto52]   G.G. Stokes, Trans. Cambridge Philos. Soc. 9, 399 (1852)

[Stu02]   Quantum key distribution over 67 km with a plug and play system, D. Stucki, N. Gisin, O. Guinnard, G. Ribordy, H. Zbinden, New J. Phys. 4, 41 (2002)

[Vai94]   Teleportation of quantum states, L. Vaidman, Phys. Rev. A 49, 1473 (1994)

[Whi99]   Nonmaximally entangled states: production, characterization, and utilization, A.G. White, D.F.V. James, P.H. Eberhard, P.G. Kwiat, Phys. Rev. Lett. 83, 3103 (1999)

[Woo82]   A single quantum cannot be cloned, W.K. Wootters, W.H. Zurek, Nature 299, 802 (1982)

14 Summary and outlook

Demonstration of quantum properties of light

The idea of optical quanta, conceived by Planck at the turn of the century, has opened the way to a remarkable range of effects that defy classical interpretation.

The invention of the laser has led to experiments that can clearly distinguish between quantum and classical description, and has shown that the concept of quantum optics is necessary to explain the properties of light. The laser has led to many applications in our daily life that we take for granted. A remarkable demonstration of the unpredictability and usefulness of fundamental research. Optics has developed into an area where ideas in quantum theory can be rigorously tested. Technology has advanced to the point where fundamental quantum effects such as the fluctuations of the vacuum field and the discreteness of light can be observed in undergraduate labs.

Quantum limits in electro-optical instruments

The main consequence of the quantum nature of light in optical instruments is the presence of quantum noise. This noise limits the range of signals that can be used. In addition, the performance of the system is different for very large signals, or modulations, and for small signals close to the quantum noise level. For example feedback controllers, optical amplifiers and even passive cavities show quite different properties near the quantum noise level. However, all these effects can be described in one simple, consistent approach: quantum transfer functions can be used which describe the complete system and include all quantum effects. This engineer's approach is as easy to use as conventional classical transfer functions. With this tool optical systems can be determined and optimized – the quantum effects can be fully integrated into the optical and electronic design.

Squeezed light

Some of the limitations introduced by quantum noise can be avoided by using non-classical, or squeezed, states of light. The experimental techniques developed over the past decade have strongly improved and reliable, strong noise suppression is now a reality. One can expect in future a suppression of 10 [dB], a factor of ten, once all the components and materials have been optimized.

What are the likely applications? The farsighted ideas of C. Caves are certainly turning into reality. But the early euphoria has calmed down and the emphasis has shifted. It is clear that applications of squeezed light will only occur in those areas where the instrument is

A Guide to Experiments in Quantum Optics, 2nd Edition. Hans-A. Bachor and Timothy C. Ralph
Copyright © 2004 Wiley-VCH Verlag GmbH & Co. KGaA
ISBN: 3-527-40393-0

already quantum noise limited and losses are extremely small. For example, in optical communications this is not yet the case – and applications are not likely in the next few years. But the possibility of quantum cryptography has opened completely new opportunities for non-classical light and created significant scientific and even commercial interest. The situation with optical sensors is also promising – in laboratory instruments such as polarimeters or interferometers the improvement of the signal to noise ratio with squeezed light has already been demonstrated. And special devices with improved sensitivity can certainly be expected. The Gravitational wave detector is a clear example: The use of squeezed light is possible, it compares to the increase of the laser power by a factor of 2 to 3. Whether squeezing will be used in the operation of these instruments will depend on the actual configuration used, but it is a firm option for any future GW detector.

One should not judge the future too early, a rapid development has taken place and it is reasonable to join into the optimism of R. Slusher and B. Yurke, some of the pioneers of experimental quantum optics who said in 1990:

> *The field is still young and full of surprises. It is therefore highly likely that the most useful applications of squeezed light have not yet been imagined.* [SLU90].

Certainly one would have to count the recent demonstrations of teleportation using entanglement generated from squeezed light as one such surprise, who knows what else is to come.

Photon counting experiments

We have seen that at the level of single photons even more exotic effects appear: non-local effects; negative "probability" distributions; wave-particle duality. We have also seen how these effects may be harnessed to perform useful tasks such as the distribution of fundamentally secure secret keys.

The work of Knill, Laflamme and Milburn showing that massive non-linearities can be conditionally induced by photon counting and then boosted to near deterministic operation by teleportation was a recent surprise. This has opened the door to many new possibilities including that of an all-optical quantum computer. The speed with which these new ideas were turned into experimental demonstrations bodes well for the future.

The future

What lies in the future? This is a risky thing to speculate about. Still two areas seem promising. Firstly, the ever more precise control of the matter-light interaction, whether between light and; atoms or; ions or even; "artificial atoms" such as quantum dots, looks likely to produce exciting new possibilities in the future. The ability to strongly and cleanly interact single emitters with single optical modes will make realistic many of the gedanken experiments of early quantum optics theory.

The second possibility is suggested by the dichotomy between the "bright" light and single photon experiments presented here. In the bright experiments, wave effects dominate, coherence between fields dominates, quantum effects are "noise". In single photon experiments particle effects dominate, fields are incoherent and only interfere through quantum indistinguishability. What if we could produce fields which displayed both strong coherence and strong quantization? An example of such a field would be a superposition of coherent states, e.g. $|\alpha\rangle + |-\alpha\rangle$. To manufacture and manipulate such fields would require an ambitious marrying of bright and single photon technologies, but the pay-off could be a new and more powerful type of quantum optics.

15 Appendices

Appendix A: Gaussian functions

Gaussian functions are used in many places in this book, they have properties which simplify the mathematical formalism tremendously. A normalized 1-dimensional Gaussian is defined as:

$$G(x) = \frac{1}{\pi} \exp\left(-\frac{(x - x_m)^2}{\Delta x}\right) \tag{15.0.1}$$

where Δx is the standard deviation of the distribution and x_m defines the centre of the distribution. The full width half maximum of this curve is equal to the standard deviation. A Gaussian distribution of a variable has a variance of Δx^2. It should be noted that a Poissonian and a Gaussian distribution are identical if the mean values x_m are large. In this case $\Delta x = \sqrt{x_m}$

A 2-dimensional Gaussian function is described by the product of two orthogonal 1-dimensional Gaussian functions

$$G(x, y) = G(x)G(y) = \frac{1}{\pi^2} \exp\left(-\frac{x^2}{\Delta x}\right) \exp\left(-\frac{y^2}{\Delta y}\right) \tag{15.0.2}$$

Any cross-section through this 2-dimensional function, in arbitrary direction, is again a 1-dimensional function. The contour at half maximum is an ellipse, and for the special case $\Delta x = \Delta y$ it is a circle. For numerical simulations the following approximation is useful [Pre80]: Random deviates with a normal, Gaussian, distribution can be generated by

$$
\begin{aligned}
y1 &= \sqrt{-2\ln(x1)}\,\cos(2\pi x2) \\
y2 &= \sqrt{-2\ln(x1)}\,\sin(2\pi x2)
\end{aligned}
\tag{15.0.3}
$$

where x1, x2 are two quantities with uniform random distribution in the interval $(0, 1)$. This equation was used for the simulations in Chapter 9.

A Guide to Experiments in Quantum Optics, 2nd Edition. Hans-A. Bachor and Timothy C. Ralph
Copyright © 2004 Wiley-VCH Verlag GmbH & Co. KGaA
ISBN: 3-527-40393-0

Appendix B: List of quantum operators, states and functions

For completeness most quantum operators, states and functions used in quantum optics are listed here in one place.

The operators, individually or as a combination, describe the effect of optical systems. The fundamental operator are the pairs $\tilde{a}$, $\tilde{a}^\dagger$ and $\tilde{A}$, $\tilde{A}^\dagger$ for the annihilation and creation of photons. Here lower case indicates a cavity field whilst upper case indicates a travelling field. Note that these operators are not Hermitian and thus do not represent observables. A second group of operators are combinations of these, such as the quadrature operators. These are Hermitian and thus do describe observable properties of a mode. The effect of optical processes such as modulation or squeezing can be described by unitary operators. Note that the phase operator is listed but cannot be defined in a closed form. It can be defined only in the limit of a series of operators.

All quantum optical states can be expanded in terms of the number states $|n\rangle$. The coherent state is generated by the displacement operator, the squeezed state by a combination of the squeezing and displacement operator. The description of the states contain exactly the minimum number of parameters which are required to describe the state uniquely: two parameters (one complex number) for a coherent state, four parameters for a minimum uncertainty squeezed state. The states are eigenfunctions of certain operator, for example the coherent state is eigenstate of the annihilation operator.

The functions are used to evaluate the properties of quantum states. There are two groups. The first group contains functions which describe the statistical properties of the light. These are; the variance of the quadratures and the variance of the photon number. Similarly, the correlation functions $g^{(n)}$ are used to describe the statistical and coherence properties of the light. The second group of functions provides two dimensional representations of the states. There is a whole continuum of such representations. In practice, however, only a limited number of functions are employed. In this text we only use the Q and Wigner functions.

OP	Name	Expression	Type
$\tilde{a}$	annihilation	$\tilde{a}$	non-Hermitian
$\tilde{a}^\dagger$	creation	$\tilde{a}^\dagger$	non-Hermitian
$\tilde{A}$	annihilation	$\tilde{A}$	non-Hermitian
$\tilde{A}^\dagger$	creation	$\tilde{A}^\dagger$	non-Hermitian
$\hat{A}$	annihil. in time	$\tilde{A}^\dagger$	non-Hermitian
$\tilde{D}$	displacement	$\tilde{D}(\alpha) = \exp(\alpha\tilde{a}^\dagger - \alpha^*\tilde{a}^\dagger)$	unitary
$\tilde{D}^\dagger$	displacement	$\tilde{D}^\dagger(\alpha) = \tilde{D}^{-1}(\alpha) = \tilde{D}^\dagger(\alpha)$	unitary
$\tilde{S}$	squeezing	$\tilde{S}(\xi) = \exp(\frac{1}{2}\xi^*\tilde{a}^2 - \frac{1}{2}\xi(\tilde{a}^\dagger)^2)$	unitary
$\tilde{S}^\dagger$	squeezing	$\tilde{S}^\dagger(\xi) = \tilde{S}^{-1}(\xi) = \tilde{S}(-\xi)$	unitary
$\tilde{n}$	number	$\tilde{a}^\dagger\tilde{a}$	Hermitian
$\tilde{N}$	number	$\tilde{A}^\dagger\tilde{A}$	Hermitian
$\tilde{\Phi}$	phase	only as limit of series	Hermitian
$\tilde{X}1(\theta)$	quadrature	$\tilde{a}e^{-i\theta} + \tilde{a}^\dagger e^{i\theta}$	Hermitian
$\tilde{X}1$	amplitude qu.	$\tilde{a} + \tilde{a}^\dagger$	Hermitian
$\tilde{X}2$	phase qu.	$i(\tilde{a}^\dagger - \tilde{a})$	Hermitian

STATE	Name	Expression	Significance
$|0\rangle$	vacuum	$\tilde{a}|0\rangle = 0|0\rangle$	Ground state
$|n\rangle$	Fock	$\tilde{a}^\dagger|n\rangle\sqrt{n+1}|n+1\rangle \quad \tilde{n}|n\rangle = |n\rangle$	complete, orthogonal
$|\alpha\rangle$	coherent	$|\alpha\rangle = \tilde{D}(\alpha)|0\rangle \quad \tilde{a}|\alpha\rangle = \alpha|\alpha\rangle$	overcomplete, orthog.
$|\alpha,\xi\rangle$	squeezed	$|\alpha,\xi\rangle = \tilde{D}(\alpha)\tilde{S}^\dagger(\xi)|0\rangle$	minimum uncertainty

FUNCTION	Name	Expression
Vn	variance of n	$\langle(\hat{a}^\dagger\hat{a})^2\rangle - \langle\hat{a}^\dagger\hat{a}\rangle^2$
$V1(\Omega)$	noise spectrum	$\langle|\delta\tilde{X}1|^2\rangle$
$g^{(1)}(\tau)$	1.order correlation	$\langle\tilde{X}1(t)^\dagger\tilde{X}1(t+\tau)\rangle/\langle\tilde{X}1(t) + \tilde{X}1(t)\rangle$
$g^{(2)}(\tau)$	2.order correlation	$\langle\tilde{X}1^\dagger(t)^2\tilde{X}1(t+\tau)^2\rangle/\langle\tilde{X}1^\dagger(t)^2\tilde{X}1(t)^2\rangle$
$S(\Omega)$	squeezing spectrum	$\mathrm{Var}(\tilde{X}1(\theta)$ with Θ optimized for each Ω))
Q	Q function	$Q(\alpha) = 1/\pi\langle\alpha|\tilde{\rho}|\alpha\rangle$
P	P function	$\rho = \int P(\alpha)|\alpha\rangle\langle\alpha|d^2\alpha$
W	Wigner function	$1/\pi^2 \int e^{\mu^*\alpha - \mu\alpha^*} X(\mu)d^2\mu$

Appendix C: The full quantum derivation of quantum states

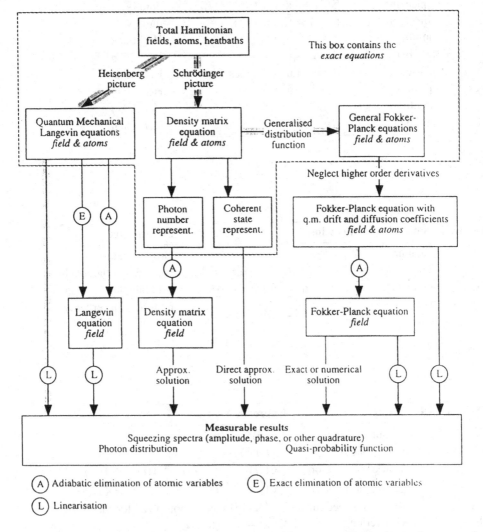

Figure 15.1: Overview of the different techniques for deriving measurable results, such as squeezing spectra, photon distributions and Wigner-function. Adapted from Haken [Hak66].

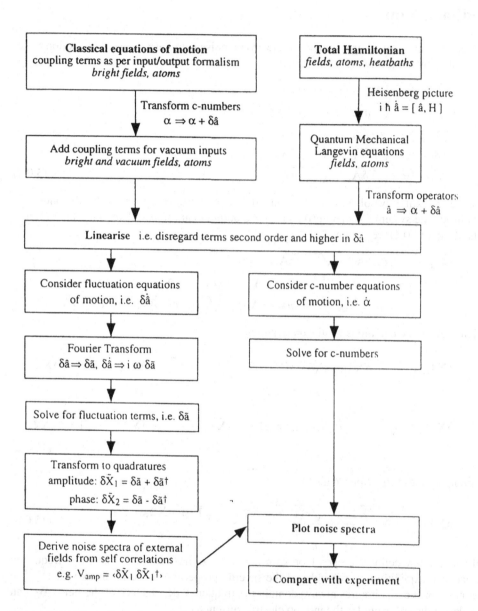

Figure 15.2: The individual steps taken in the derivation of the linearized quantum operators model. This diagram shows the analogy between using the classical equations of motion and the Hamiltonian in the Heisenberg picture. Summarized by A.G. White [Whi97].

Appendix D: Calculation of of the quantum properties of a feedback loop

A electronic intensity control system, or electronic noise eater, as discussed in Section 8.3.2, is described by the following set of equations.

$$
\begin{aligned}
\delta \tilde{A}_{il} &= \delta \tilde{A}_{las} + g\sqrt{\epsilon \eta}(\Omega)\, \delta \tilde{A}_{el} \\
\delta \tilde{A}_f &= \sqrt{\epsilon}\, \delta \tilde{A}_{il} + \sqrt{1-\epsilon}\, \delta \tilde{A}_{vac} \\
\delta \tilde{A}_{el} &= \sqrt{\eta}\, \delta \tilde{A}_f - \sqrt{1-\eta}\, \delta \tilde{A}_{vac} \\
\delta \tilde{A}_o &= \sqrt{1-\epsilon}\, \delta \tilde{A}_{il} + \sqrt{\epsilon}\, \delta \tilde{A}_{vac} \\
\delta \tilde{A}_{out} &= \sqrt{\eta}\, \delta \tilde{A}_o - \sqrt{1-\eta}\, \delta \tilde{A}_{vac}
\end{aligned}
\tag{15.0.4}
$$

Here we are explicitly solve these equations to derive the transfer functions for the photocurrents inside the system (il, f and el)) and for the output beam (out). Combining the different parts of Eq. (15.0.4) we obtain

$$
\begin{aligned}
\delta \tilde{A}_{el} &= \sqrt{\eta}\, \delta \tilde{A}_f + \sqrt{1-\epsilon}\, \delta \tilde{A}_{vac} \\
&= \sqrt{\eta \epsilon}\, \delta \tilde{A}_{il} - \sqrt{\eta(1-\epsilon)}\, \delta \tilde{A}_{vac} - \sqrt{1-\eta}\, \delta \tilde{A}_{vac} \\
&= \sqrt{\eta \epsilon}(\delta \tilde{A}_{las} + \sqrt{\epsilon \eta}\, g(\Omega)\, \delta \tilde{A}_{el}) - \sqrt{\eta(1-\epsilon)}\, \delta \tilde{A}_{vac} - \sqrt{1-\eta}\, \delta \tilde{A}_{vac}
\end{aligned}
$$

which leads to the equations for the quadratures

$$
\delta \tilde{X}1_{el} = \sqrt{\eta \epsilon}\left(\delta \tilde{X}1_{las} + \sqrt{\epsilon \eta}\, g(\Omega)\, \delta \tilde{X}1_{el}\right) - \sqrt{\eta(1-\epsilon)}\, \delta \tilde{X}2_{vac} - \sqrt{1-\eta}\, \delta \tilde{X}2_{vac}
$$

$$
\delta \tilde{X}1_{out} = \sqrt{\eta(1-\epsilon)}\left(\sqrt{\epsilon \eta}\, g(\Omega)\, \delta \tilde{X}1_{el} + \delta \tilde{X}1_{las}\right) + \sqrt{\eta \epsilon}\, \delta \tilde{X}2_{vac} - \sqrt{1-\epsilon}\, \delta \tilde{X}2_{vac}
\tag{15.0.5}
$$

Rearranging these equations leads to

$$
\delta \tilde{X}1_{el} = \frac{\sqrt{\eta \epsilon}\, \delta \tilde{X}1_{las} - \sqrt{\eta(1-\epsilon)}\, \delta \tilde{X}2_{vac} - \sqrt{1-\eta}\, \delta \tilde{X}2_{vac}}{1 - \eta \epsilon g(\Omega)}
\tag{15.0.6}
$$

and we can now define the open loop gain $h(\Omega) = \eta \epsilon\, g(\Omega)$ where we keep the frequency dependence explicitly as a reminder of the specific properties of the feedback system. The operators for the quadratures always contain the frequency dependence of the field. We can now derive the spectrum for the in-loop electric current as:

$$
\begin{aligned}
V1_{el}(\Omega) &= \frac{\eta \epsilon V1_{las}(\Omega) + \eta(1-\epsilon) + (1-\eta)}{|1 - \eta \epsilon g(\Omega)|^2} \\
&= \frac{\eta \epsilon (V1_{las}(\Omega) - 1) + 1}{|1 - h(\Omega)|^2}
\end{aligned}
\tag{15.0.7}
$$

The spectrum of the outcoming laser beam can be derived by substituting Equation (15.0.6) into Equation (15.0.5) resulting in

$$
\delta\tilde{X}1_{\text{out}} = \frac{\sqrt{\eta(1-\epsilon)}\,\delta\tilde{X}1_{\text{las}} + (1-\eta g(\Omega))\sqrt{\eta\epsilon}\,\delta\tilde{X}2_{\text{vac}}}{1 - h(\Omega)}
$$
$$
- \frac{g(\Omega)\sqrt{\eta^2\epsilon(1-\epsilon)(1-\eta)}\,\delta\tilde{X}2_{\text{vac}} - (1-h(\Omega))\sqrt{1-\eta}\,\delta\tilde{X}2_{\text{vac}}}{1 - h(\Omega)}
$$

which leads to

$$
V1_{\text{out}}(\Omega) = \frac{\eta(1-\epsilon)V1_{\text{las}}(\Omega) + |1-\eta^2\epsilon g(\Omega)|^2}{|1-h(\Omega)|^2}
$$
$$
+ \frac{|g(\Omega)|^2\eta^2\epsilon(1-\epsilon)(1-\eta) + |1-h(\Omega)|^2(1-\eta)}{|1-h(\Omega)|^2}
$$
$$
= 1 + \frac{1-\epsilon}{\epsilon}\frac{\eta\epsilon(V1_{\text{las}}(\Omega)-1) + |h(\Omega)|^2}{|1-h(\Omega)|^2} \tag{15.0.8}
$$

This is obviously a spectrum which is, in most situations, larger than the QNL. This result shows the effect of the feedbackloop adding vacuum noise which is entering the system through the unused port of the beamsplitter and at the detectors.

Appendix E: Symbols and abbreviations

Nomenclature	Symbol
Electric Field	E
Optical amplitude	α
Optical phase	ϕ
Intensity	I
Intensity fluctuation	$\delta I(t)$
Variance	$\Delta I(t)^2 = \text{VAR}_i = [\delta I(t)^2] - [\delta I(t)]^2$
Optical power	$P(t)$
Current	$i(t)$
RF noise power	$p_i(\Omega)$
Gaussian beam diameter	$2W$
Gaussian beam parameter	z_0
Optical frequency [Hz]	ν
Detection frequency [Hz]	Ω
Mirror reflectivity	ϵ
Quantum efficiency	η
Current junction division	ζ
Correlation function	$C\,(E_1, E_2)$
Visibility	$\text{VIS}\,(E_1, E_2)$
Normalized visibility	$\mathcal{V}$
Signal to noise ratio	$M(\Omega)^2/\Delta I(\Omega)^2$
General operator	$\tilde{\chi}$
Variance of operator	$\text{Var}\,(\tilde{\chi}) = \langle \tilde{\chi}^\dagger \tilde{\chi} \rangle - \langle \tilde{\chi} \rangle^2$
Creation and annihilation operator	$\hat{\mathbf{a}},\ \hat{\mathbf{a}}^\dagger$
Travelling wave, time domain	$\hat{\mathbf{A}},\ \hat{\mathbf{A}}^\dagger$
Travelling wave, frequency domain	$\tilde{\mathbf{A}},\ \tilde{\mathbf{A}}^\dagger$
Number state	$|n\rangle$
Coherent state (internal)	$|\alpha\rangle$
Propagating coherent state	$|A_{\text{out}}\rangle$
Expectation value	$\langle \hat{\mathbf{a}} \rangle = \langle \alpha | \hat{\mathbf{a}} | \alpha \rangle = \alpha$
Operator for fluctuations, fr. domain	$\tilde{\delta}\mathbf{a}, \tilde{\delta}\mathbf{a}^\dagger$
Linearized operator	$\tilde{\mathbf{a}}_1 = \alpha + \tilde{\delta}\mathbf{a}$
Photon number operator	$\tilde{\mathbf{n}} = \tilde{\mathbf{a}}^\dagger \tilde{\mathbf{a}}$
Photon flux operator	$\tilde{\mathbf{N}} = \tilde{\mathbf{A}}^\dagger \tilde{\mathbf{A}}$
Photon number	$n_{\alpha_1} = \langle \alpha_1 | \tilde{\mathbf{n}} | \alpha_1 \rangle = |\alpha_1|^2$
Variance of photon number	$V n_1 = \text{Var}\,(\tilde{\mathbf{n}})$ evaluated for $|\alpha_1\rangle$
Generalized quadrature	$\tilde{\mathbf{X}}(\theta) = \tilde{\mathbf{a}}^\dagger \exp(i\theta) + \tilde{\mathbf{a}} \exp(-i\theta)$
Amplitude quadrature operator	$\tilde{\mathbf{X}}\mathbf{1} = \tilde{\mathbf{X}}(0) = \tilde{\mathbf{a}} + \tilde{\mathbf{a}}^\dagger$
Phase quadrature operator	$\tilde{\mathbf{X}}\mathbf{2} = \tilde{\mathbf{X}}(\frac{\pi}{2}) = -i(\tilde{\mathbf{a}} - \tilde{\mathbf{a}}^\dagger)$

Measured amplitude noise	$V1_{out}(\Omega) = \langle|\delta\tilde{\mathbf{X}}\mathbf{1}|^2\rangle = $ Fourier Transform $[V_{out}(0)]|_\Omega$
Squeezing parameter	$\xi = r_s \exp(i2\theta_s)$
Squeezing angle	θ_s
Squeezing degree	r_s
Squeezed quadrature amplitude	$\tilde{\mathbf{Y}}\mathbf{1} = \tilde{\mathbf{X}}(\theta_s) = \tilde{\mathbf{a}}^\dagger \exp(i\theta_s) + \tilde{\mathbf{a}} \exp(-i\theta_s)$
Squeezing spectrum	$S(\Omega) = \mathrm{Var}\,(\tilde{\mathbf{X}}(\theta))$ optimized for each Ω
Photon number probability	$\mathcal{P}_\alpha(n)$
Fano number	$f_{A_{out}} = VN_{A_{out}}/N_{A_{out}} = V1_{out}$

Abbreviations

APD	Avalanche photodiode
AOM	Acousto optical modulator
CCD	Charge coupled device
CW	Continuous wave
CV	Continuous variable
EOM	Electro-optic modulator
EPR	Einstein-Podolski-Rosen
ESA	Electronic spectrum analyser
FSR	Free spectral range
HOM	Hong-Ou-Mandel
OPA	Optical parametric amplifier
OPO	Optical parametric oscillator
QKD	Quantum key distribution
QND	Quantum non-demolition
QNL	Quantum noise limit
PBS	Polarising beam splitter
PDH	Pound-Drever-Hall
PID	Proportional, intergral, differential
PMT	Photomultiplier tube
RBW	Resolution bandwidth
RF	Radio frequency
RMS	Root mean square
RRO	Resonant relaxation oscillation
SHG	Second harmonic generation
SNR	Signal to noise ratio
SQL	Standard quantum limit
TTL	Transistor-transistor logic
VBW	Video bandwidth
4WM	Four wave mixing

Bibliography

[Hak66] *Handbuch der Physik Vol. XXII*, H. Haken, Springer (1966)

[Pre80] *Numerical Recipes*, W.H. Press, 202 Cambridge University Press (1980)

[Whi97] PhD Andrew White, Australian National University (1997)

Index